# Developing Brain and Behaviour:
## The Role of Lipids in Infant Formula

# Developing Brain and Behaviour: The Role of Lipids in Infant Formula

Edited by

## John Dobbing
*Emeritus Professor of Child Growth and Development,
Department of Child Health,
University of Manchester,
Manchester,
UK*

ACADEMIC PRESS
San Diego   London   Boston
New York   Sydney   Tokyo   Toronto

This book is printed on acid-free paper

Copyright © 1997 by ACADEMIC PRESS

*All Rights Reserved*
No part of this publication may be reproduced or transmitted in any form or by any means, electronic or mechanical, including photocopy, recording, or any information storage and retrieval system, without permission in writing from the publisher.

Academic Press
525 B Street, Suite 1900, San Diego, California 92101-4495, USA
http://www.apnet.com

Academic Press Limited
24–28 Oval Road, London NW1 7DX, UK
http://www.hbuk.co.uk/ap/

ISBN 0-12-218870-5

A catalogue record for this book is available from the British Library

Typeset by J&L Composition Ltd, Filey, North Yorkshire
Printed in Great Britain by WBC Book Manufacturers, Bridgend, Mid Glamorgan

97 98 99 00 01 02 EB 9 8 7 6 5 4 3 2 1

# Contents

| | |
|---|---|
| List of Contributors | vii |
| Dobbing Workshops | ix |
| Preface | xi |

*Clinical Studies*

| | |
|---|---|
| Long-chain polyunsaturated fatty acids, infant feeding and cognitive development<br>Alan Lucas | 3 |
| Long-chain polyunsaturated fatty acid supplementation of preterm infants<br>Susan E. Carlson | 41 |
| Polyunsaturated fatty acid nutrition in infants born at term<br>Sheila M. Innis | 103 |
| Statistically significant versus biologically significant effects of long-chain polyunsaturated fatty acids on growth<br>William C. Heird | 169 |

*Methodology: Clinical Studies*

Methodological considerations in longitudinal studies of infant risk
Lynn T. Singer — 209

Grating acuity cards: validity and reliability in studies of human visual development
D. Luisa Mayer and Velma Dobson — 253

*Behavioural Science Considerations*

Long-chain polyunsaturated fatty acids and the measurement and prediction of intelligence (IQ)
Robert B. McCall and Clay W. Mash — 295

Individual differences in infant cognition: methods, measures, and models
John Colombo — 339

Early nutrition and behavior: a conceptual framework for critical analysis of research
Patricia E. Wainwright and Glenn R. Ward — 387

The effects of early diet on synaptic function and behavior: pitfalls and potentials
Christopher A. Shaw and Jill C. McEachern — 427

*Appendices: General Commentary on Behavioural Science Implications and Methodology*

Appendix A: Nutrition and development – observations and implications
Marc H. Bornstein — 475

Appendix B: Design, measurement, and statistical approaches
Mark Appelbaum — 511

Appendix C: General commentary
Martha Neuringer and Sydney Reisbick — 517

Index — 529

# Contributors

Mark Appelbaum, PhD, *Department of Psychology-0109, University of California, San Diego, La Jolla, CA 92093-0109, USA*

Marc H. Bornstein, PhD, *Child and Family Research, NICHD, Building 31 – Room B2B15, 9000 Rockville Pike, Bethesda, MD 20892-2030, USA*

Susan E. Carlson, PhD, *Departments of Pediatrics, Obstetrics, Gynecology & Biochemistry, University of Tennessee — Memphis, 853 Jefferson Avenue, Room 201, Memphis, TN 38163, USA*

John Colombo, PhD, *Department of Human Development, University of Kansas, 4001 Dole Center, Lawrence, KS 66045-2133, USA*

John Dobbing DSc, FRCP, FRCPath, Hon FRCPCH, *Higher Cliff Farm, Birch Vale, High Peak SK22 1DL, UK*

Velma Dobson,* PhD, *College of Medicine, Department of Ophthalmology, The University of Arizona, Health Sciences Center, 1801 N. Campbell Avenue, Tucson, AZ 85719, USA*

William C. Heird, MD, *Children's Nutrition Research Center, Department of Pediatrics, Baylor College of Medicine, Texas Medical Center, 1100 Bates Street, Houston, TX 77030, USA*

Sheila M. Innis, PhD, *Department of Pediatrics, BC Research Institute for Child & Family Health, The University of British Columbia, 950 West 28th Avenue, Vancouver, BC, Canada V5Z 4H4*

Alan Lucas, MD, FRCP, *MRC Childhood Nutrition Research Centre, Institute of Child Health, 30 Guilford Street, London WC1N 1EH, UK*

Clay W. Mash,* MS, *Office of Child Development, UCSUR/121 University Place, University of Pittsburgh, Pittsburgh, PA 15260, USA*

D. Luisa Mayer, MEd, PhD, *Department of Ophthalmology, Boston Children's Hospital, Harvard Medical School, 300 Longwood Avenue, Boston, MA 02115, USA*

Robert B. McCall, PhD, *Office of Child Development, UCSUR/121 University Place, University of Pittsburgh, Pittsburgh, PA 15260, USA*

Jill C. McEachern,* BSc, *Department of Ophthalmology, c/o Department of Anatomy, 2177 Wesbrook Mall, University of British Columbia, Vancouver, BC, Canada V6T 1Z3*

Martha Neuringer, PhD, *Section of Clinical Nutrition & Lipid Metabolism, Department of Medicine, Oregon Health Sciences University, and Oregon Regional Primate Research Center, 505 NW 185th Avenue, Beaverton, OR 97006, USA*

Sydney Reisbick,* PhD, *Section of Clinical Nutrition & Lipid Metabolism, Department of Medicine, Oregon Health Sciences University, and Oregon Regional Primate Research Center, 505 NW 185th Avenue, Beaverton, OR 97006, USA*

Christopher A. Shaw, PhD, *Department of Ophthalmology, c/o Department of Anatomy, 2177 Wesbrook Mall, University of British Columbia, Vancouver, BC, Canada V6T 1Z3*

Lynn T. Singer, PhD, *General Pediatrics, Stop 1038, Rainbow Babies & Children's Hospital, 2101 Adelbert Road, Cleveland, OH 44106, USA*

Patricia E. Wainwright, PhD, *Department of Health Studies & Gerontology, University of Waterloo, Waterloo, Ontario, Canada, N21 3G1*

Glenn Ward,* PhD, *Department of Health Studies & Gerontology, University of Waterloo, Waterloo, Ontario, Canada, N21 3G1*

\* Authors who did not attend the Workshop.

# Dobbing Workshops

Dobbing, J. (ed.) (1981) Maternal Nutrition in Pregnancy: Eating for Two? Academic Press, London.

Dobbing, J. (ed.) (1983) *Prevention of Spina Bifida and other Neural Tube Defects.* Academic Press, London.

Dobbing, J. (ed.) (1984) *Scientific Studies in Mental Retardation.* Macmillan Press, London.

Dobbing, J. (ed.) (1985) *Maternal Nutrition and Lactational Infertility.* Raven Press, New York.

Dobbing, J. (ed.) (1987) *Sweetness.* Springer-Verlag, London.

Dobbing, J. (ed.) (1987) *Early Nutrition and Later Achievement.* Academic Press, London.

Dobbing, J. (ed.) (1987) *Food Intolerance.* Baillière Tindall, Bristol.

Dobbing, J. (ed.) (1988) *A Balanced Diet?* Springer-Verlag, London.

Dobbing, J. (ed.) (1989) *Dietary Starches and Sugars in Man – A Comparison.* Springer-Verlag, London.

Dobbing, J. (ed.) (1990) *Brain, Behaviour, and Iron in the Infant Diet.* Springer-Verlag, London.

Dobbing J. (ed.) (1992) *Lipids, Learning and the Brain: Fats in Infant Formula.* Ross Laboratories, Columbus, OH.

Dobbing, J. (ed.) (1997) *Developing Brain and Behaviour: the Role of Lipids in Infant Formula.* This volume.

# Preface

The origin of the hypothesis which forms the main topic of this monograph was the statement that certain long-chain polyunsaturated fatty acids (LCPUFAs), in particular arachidonic acid (AA) and docosahexaenoic acid (DHA), were essential components for the nutrition of infants, especially of preterm infants, in the literal sense that they could not be synthesized by the infant organism, and therefore had to be supplied in the diet. Breast milk was known to contain these 'essential' substances, albeit in widely varying quantities. The idea might have been thought to be challenged by the fact that manufactured infant formulae, consumed by the majority of infants in industrialized countries, contain neither LCPUFAs, and yet the infants did not noticeably suffer as a consequence. When it was pointed out that these fatty acids were constituents of structural membranes, notably in the brain, alarm bells rang, and it was predicted, mostly by biochemists, that infants thereby deprived would suffer deficits of 'brain function', by which the non-neurobiologist, or the non-behavioural scientist, like the lay person, usually takes to mean cognitive function. Since the retina is one part of the brain which contains large amounts of DHA, it was also predicted that the infant deprived of a dietary supply, by being brought up on formula, would suffer visual disturbances, particularly of visual acuity.

The absence of LCPUFAs, especially DHA in formula was, therefore, considered to be of some importance. It is for this reason that very considerable research time and effort has been devoted to a search, in human infants, non-human primates and other animals, for evidence of deficiencies in measured behaviour that might relate to intelligence (IQ) and in visual function, following the administration of diets free from DHA and its metabolic precursors, compared with diets supplemented with supposedly adequate quantities; or with breast-feeding, which normally supplies LCPUFAs.

It is interesting that the hypothesis was 'powered from behind', in the sense that it derived from the then biochemical knowledge and the predictions derived therefrom, rather than the safer and more usual 'powering from in front': i.e. the noticing of the manifestations, disease states, or the abnormalities, *followed* by a search for their aetiologies or for antecedent events. At no stage had any deficit in IQ been known or suspected in the formula-fed compared with the breast-fed; and no disturbance of visual acuity had been associated with a similar dietary history. However, as a result of diligent searching, using modern, very sensitive techniques, deficiencies such as these have now been claimed, as this monograph will relate: but it is still disputed whether they *matter*.

Throughout the book the important practical difference between two meanings of the word 'significance' will be referred to. One is a statistical concept. The other answers the question '*Does it matter?*' For example, I find it almost impossible to lift a sack of corn. If I remove a scoop-full and re-weigh the sack on modern scales, it will be found to have a different weight with an infinitesimally small $p$ value: but, in spite of this highly significant difference in the statistical sense, I will not notice the difference: *it does not matter*.

The 'powering from behind' of a hypothesis is feasible, but often extremely hazardous. In the example of DHA and cognition or visual acuity there are far too many important potential *non sequiturs*, which it would take too long to describe here in detail. Early in the chain of reasoning important metabolic truths are often ignored, which mainly have to do with quantitative considerations. How much of these fatty acids needs to be transferred to the brain each day, week or month? And what about the well-known 'sparing mechanisms' with which the developing brain is notably endowed, so that what little of a substance or its precursor is circulating is reserved preferentially for the developing brain? Certainly the post mortem brains of a handful of formula-fed infants have been analysed and found to be somewhat LCPUFA deficient: but the deficiency was small, and the infants were manifestly very ill or they would not have died, and they had not been tested in life for the presumed functional effects; nor could they have been. Nor, as will be pointed out, were many surveys of formula-fed children's IQs or disturbances of visual acuity always conducted with a proper regard for statistical power, and for the elimination of confounders and so forth. The very nature of intelligence, its physical basis in the brain, and the extent to which it is represented by IQ is barely understood: and when it comes to *'Does it matter?'* the prodigious capacity of developing brain and behaviour to compensate for minor restriction by means of one of the manifold other contributors to human achievement is quite unreasonably ignored. Further, it is comparatively recently that attention has been paid to the distinct possibility that there might even be harmful effects of adding new substances indiscriminately to formula.

So, as we shall see, the apparently simple question must be enormously expanded from the simplistic formulation of the original idea outlined above, important as the question itself is.

A further obstacle in the way of progress is the almost unparalleled multidisciplinary nature of the enquiry. This is also recognized in our monograph, and we have tried to begin to tackle it. No individual scientist can be expected to summon such an array of different expertise as is required at this stage to illuminate the problem further. We have striven to bring biochemists and neurochemists together with several sorts of behavioural scientist to discover what remains of their work when it has been dropped in the acid of the other major disciplines on which their momentous conclusions depend. I choose to single out behavioural science, since it is too often the poor relation of other sciences. Many non-behavioural scientists seem happy to do behavioural research as well as biochemical or clinical research. Would a biochemist, for example, attach any importance whatever to biochemical research performed by a behavioural scientist? The proposition is ludicrous, and yet behavioural science is *at least* as complex and specialized as biochemistry! My singling out of biochemistry and behavioural science may seem tendentious, for the same strictures apply to almost any other pair of the major disciplines involved in our question. These include clinical neonatology and care of the low-birth-weight baby, neurochemistry, developmental neurobiology, nutrition, auxology (the study of growth), ophthalmology and its possible bearing on cognition, epidemiology, developmental behavioural science, biochemistry, research design and statistics, cognitive science, sociology, physiological psychology, an understanding of cross-species extrapolation, and so forth. Add to this a facility to handle a further order of complexity: the multiple interactions between most of the above-listed disciplines. Faced with such an array of required wisdom and understanding it will surely be a brave and humble person, or an exceedingly foolish one, who will travel the auditoria of the world declaiming that the infant formula companies are in dereliction of their duties not to fortify their product with DHA!

The task of this book is to convey, in very modest outline, the present state-of-the-art in this area. To do this we operated a fairly sophisticated design of workship, which we believe is more productive in beginning to unravel such a complex, multidisciplinary and contentious subject as this. It is a design which has already been quite productive on 12 previous occasions. It is important for the reader to appreciate the salient features of the design in order to be persuaded that this is not 'just another book'. Everything you read here, except this Preface and the three special appendices at the back, have been subjected to a peer-review process like no other: far more severe, even, than a research paper published in the most responsible scientific journal. In a bid to persuade you of this we will reproduce below a description of the so-called 'Dobbing Workshop' design as described in the previous volume and as it has evolved over many years.

All those who contributed Chapters wrote them some weeks before we met. They were precirculated to all the other authors, whose business it was to write well-considered, referenced Commentaries on each. These, too, were precirculated to everybody, so that when we met we had all read and fully considered all the draft Chapters and Commentaries. The final process, taking three days, was to subject each to an exhaustive discussion, during which extensive modifications were made to all the Chapters and their Commentaries. During this time it frequently happened that new points arose, and these were incorporated into the final text. The result is a report whose Chapters have been subjected to considerably more thorough peer review than is the case with most original scientific papers, and certainly much more than occurs with the usual 'books of meetings'. The surviving Commentaries have also been reproduced after each Chapter and the reader is encouraged to pay these at least as much attention as the Chapters themselves. They were just as extensively refereed.

At the same time it must be made clear that whatever appears under an author's name is what survived, in that person's judgement, at the end of our discussions. No attempt was made to arrive at or impose a consensus. This would have been a travesty on the frontiers of such an open field. Neither is the reader told about the considerable discussion that led to modifications, additions or deletions, any more than he usually is about such changes in original scientific papers that are modified by referees' reports and are published in scientific Journals. Suffice it to say that everything herein has been very thoroughly peer reviewed by the whole authorship.

The Workshop itself was held in late February, 1997.

The reader will perhaps feel entitled to ask for a conclusion, or a summary statement. What has been decided? Is it desirable, prudent or a solemn duty to fortify infant formula with DHA/LCPUFAs? Or doesn't it much matter? Unfortunately it is the nature of a Dobbing Workshop, or rather of the type of question most productively examined by this procedure, that no such declaration is forthcoming. Instead, we hope that you will find, by reading the chapters, especially including the surviving appended commentaries, the means of coming to some sort of conclusion of your own.

A few further explanations: (i) the three appendices at the end were not discussed by the participants. Like this Preface, therefore, they are not peer-reviewed. (ii) The original intention was to disallow non-peer-reviewed references such as abstracts, 'personal communications', discussions of current research, chapters in non-peer-reviewed books, etc. In such a dynamic, on-going field this proved impossible to achieve. For the non-professional scientist, therefore, not used to reserving an opinion until the definitive work is properly published, and even then exercising a well-honed critical faculty, we issue a serious warning: be very suspicious and careful of this type of reference until you see it quoted in full in a reputable journal. The same, seemingly uncharitable warning must emphatically be proclaimed in the case of this Preface. It is entirely the Editor's own, and has not even been seen by the other participants.

It remains for me to thank a number of people without whom the Workshop project could not have been brought to fruition. First among these are the participants. Few will ever understand, apart from myself and they, what an enormous extra amount of work has been demanded of them compared with attendance at 'yet another meeting'. Secondly, the costs of the project were borne by Ross Products Division of Abbott Laboratories. It is a mark of their tolerance and integrity that they have at no time sought to influence my management of the process. Clarence Washington acted as postman, both electronic and by air mail, organizing the reception and distribution of the huge volume of written material. He was, indeed, the hub around which the practical wheel turned. Our creature comforts when we met were ably ministered to by Clint Johnson, Director of Professional Services, and his team at Ross. The sponsorship was ably administered by John Benson, Nancy Auestad and Robin Halter. And, finally, I particularly thank my wife, Dr Jean Sands, for acting as 'scientific secretary' and as a calming influence on me at times of occasional tension, in the run-up period and at the Workshop.

John Dobbing

* This monograph is a sequel to a 'Dobbing Workshop' held in Adelaide in 1992, entitled *Lipids, Learning, and the Brain: Fats in Infant Formulas*, and published in 1993 by Ross Laboratories, Columbus, Ohio 43216 as the Report of the 103rd Ross Conference on Pediatric Research

# Clinical Studies

# Long-Chain Polyunsaturated Fatty Acids, Infant Feeding and Cognitive Development

ALAN LUCAS

*MRC Childhood Nutrition Research Centre, Institute of Child Health, 30 Guilford Street, London WC1N 1EH, UK*

## Introduction

Whether or not infant formulas should be supplemented with long-chain polyenoic fatty acids (LCPUFAs) including docosahexaenoic acid (DHA; 22:6n-3) and arachidonic acid (AA; 20:4n-6) has become a major issue in infant nutrition. Views on the subject are polarized, international conferences have been convened to discuss the matter, official advisory bodies have devoted an unusual degree of attention to the problem and there has been substantial media and public interest. A major impetus for research in this area comes from the presence of LCPUFAs in breast milk (1) and the observation that a number of studies suggest children who were breast-fed may have higher cognitive performance than those fed formula (2-6). This evidence now forms part of a broader range of data that now bear on the issue of whether LCPUFAs should be added to infant formula. In this article, a balanced appraisal of current evidence is attempted, culminating in an assessment of whether current knowledge justifies official recommendations.

## Biological Perspective

The principal scientific focus in this field has been on the possible essentiality of dietary LCPUFAs for central nervous system development in early infancy. The main outcomes of interest have been retinal and cognitive function. The latter will be given most attention in this paper.

The issue of whether deficiency of a nutrient or group of nutrients could have an impact on neurodevelopment is a matter of broader biological importance. Considerable research effort has been directed to the concept of 'programming', a term suggested by Lucas (7) to denote the idea that a stimulus or insult, when applied at a 'critical' or 'sensitive' stage of development, may have a long-lasting or permanent effect on the structure or function of the organism. Whilst there are many examples in developmental biology of a short-lived early event having life-time significance, one question has been whether early *nutrition* could act in this programming way. The evidence for this in animals is now well established; brief periods of early dietary manipulation during critical windows have been shown to have life-time effects on a wide range of health outcomes (for instance blood pressure, diabetes, obesity and atherosclerosis) and on behaviour and learning (8–13). Substantial research now addresses whether such sensitive or 'vulnerable' periods for early nutrition exist in *humans* with respect to their later health and, of interest in the context of this article, cognitive function (7,14–16).

The questions of the essentiality of LCPUFAs for cognitive development must be framed, then, against the broader issue of whether it is biologically *plausible* that early nutrition *in general* could influence or 'program' later cognitive function. Most of the evidence in this area relates to the outcome of children from developing countries who survived early malnutrition (14). Such evidence has been flawed by experimental design problems, which has made it difficult to extricate the effects of malnutrition from the concomitant adverse effects on the infant of poverty, poor social circumstances and lack of stimulation. More recently, experimental nutritional interventions with strict randomization of groups and adequate follow-up, have provided more compelling data that early nutrition could affect neurodevelopment (16,17–21). Perhaps the most surprising studies are those in premature babies indicating that as little as 4 weeks of randomized dietary intervention in the newborn period could have a long-term and perhaps permanent impact on cognitive function (17,22).

Such studies have involved deficiencies in a range of macro- and micronutrients. Whether more specific nutrient deficiencies could have an impact on neurodevelopment, has also received attention. Maternal iodine deficiency has been shown convincingly to affect cognitive function in offspring (23), though substantial efforts to establish that early iron deficiency could affect later cognitive function has provided more inconclusive results (24–26).

Thus, whilst it is not implausible that early nutritional deficiencies during a 'sensitive' period of early brain development could result in delayed development or more long-lasting effects on cognitive function, such effects are hard to prove and require rigorous experimental approaches. Whether or not LCPUFAs have been convincingly shown to be critical nutrients in this respect is considered in this article.

## Background

Biochemical pathways for long-chain fatty acid synthesis from essential fatty acid precursors of the n-3 and n-6 series are complex (1) and are linked to pathways for the eicosanoids, including prostaglandins. Of numerous LCPUFAs, AA and DHA have received particular attention.

AA is formed by $\Delta^6$ desaturation, elongation and $\Delta^5$ desaturation of the parent n-6 fatty acid linoleic acid (LA; 18:2n-6). The pathway for DHA synthesis from its parent n-3 fatty acid, $\alpha$-linolenic acid (LNA; 18:3n-3) involves $\Delta^6$ desaturation, elongation and $\Delta^5$ desaturation to eicosapentanoic acid (EPA; 20:5n-3), and then elongation to 24:5n-3, $\Delta^6$ desaturation to 24:6n-3 and chain shortening to DHA (27–28). The enzymes required for the biosynthesis of AA and DHA from their precursors (LA and LNA) are present in the central nervous system and liver (29–31).

AA and DHA are believed to be functionally critical components of membrane phospholipids (32) and major constituents of the nervous system. DHA is particularly abundant in the retina, comprising 30–40% of fatty acids in some phospholipids of the rod outer segment membranes (33).

The key issue here is whether LCPUFAs, notably AA and DHA, are essential nutrients for optimal development in human infants; or, alternatively, whether adequate supply of their precursors, LA and LNA, would provide sufficient substrate for endogenous LCPUFA synthesis. This question is of clear importance in infant formula design, since the precursor 'essential fatty acids' are generally supplied in modern formulas, whereas LCPUFAs are usually not.

## The Argument for LCPUFA Supplementation

Areas of research that have a bearing on whether or not dietary LCPUFAs are required by human infants include:

(i) studies on animals rendered n-3 deficient;
(ii) LCPUFA content of breast milk;
(iii) comparison of cognitive scores and visual function in breast-fed versus formula-fed infants;
(iv) plasma and red cell levels of AA and DHA in breast-fed and formula-fed infants;
(v) autopsy data on the LCPUFA content of the brains of formula-fed versus breast-fed infants;
(vi) randomized trials of LCPUFA supplementation in formula-fed infants.

In preterm infants it has been suggested that there is a special need for LCPUFAs. Further areas that have been considered in relation to this include:

(i) the rapid brain growth of premature infants;
(ii) the maturity of LCPUFA biosynthetic pathways;
(iii) the frequent state of negative energy balance in this group that might result in LCPUFA precursors being used as energy sources, limiting LCPUFA synthesis.

These areas are evaluated in the following sections, but with particular emphasis given to examination of outcome data from clinical trials of LCPUFA supplementation.

### Animal Studies

The principal focus of this review is the evidence in humans. However, the extensive animal data (34–38) require brief comment. In rats and monkeys diets extremely low in n-3 fatty acids (trace amounts of LNA and no DHA) result in reduced DHA content in the brain and retina, associated with impairment in retinal function and development of visual acuity and alterations in behaviour, though the latter needs careful interpretation (see Wainwright (39) and the chapter by Wainwright and Ward in this volume).

However, the studies in rats, piglets and monkeys report normal accumulation of DHA in the brain and retina, provided there is adequate intake of LNA of at least 0.7% of dietary energy.

There are two important considerations here. Firstly, the animal data generally do not provide evidence for the essentiality of dietary DHA for neurodevelopment, but rather point to the importance of providing adequate precursor (LNA). Secondly, the complexity of human cognitive function and the interspecies differences in the timing of the brain growth spurt, make it speculative to extrapolate animal findings on the neurobiological effects of LCPUFAs to humans.

## LCPUFAs in Breast Milk

Human milk is a highly complex biological source of dietary lipid, containing cell membrane covered fat globules, at least 160 fatty acids, over 100 triglycerides and a range of phospholipids (40). Human milk fats also have unique stereoisomeric structure, providing for instance a high proportion of palmitic acid in the Sn-2 (central) position on the glycerol backbone. This factor alone may have major influence on lipid handling in the gut (41). Moreover, lipids in breast milk are uniquely packaged within the milk fat globule membrane, which is composed of a proteinaceous inner coat, a biomembrane and a complex outer glycocalyx of unknown function (42).

In addition to LA and LNA, longer chain n-6 and n-3 derivatives are present in breast milk, including AA and DHA (32). However, LCPUFAs are not present in fixed amounts or ratios. AA content varies from around 0.3% to 0.8% milk fats. DHA content varies to an even greater extent, up to 15-fold, from as little as 0.1% to 1.5% of milk lipids (43), and is influenced by diet, being highest in those traditionally eating foods rich in n-3 fatty acids. These extremes in concentration would be too large to be substantially offset by the infant's regulation of breast milk volume intake, which has a coefficient of variation of about 30%. Breast milk lipid content has changed during this century along with secular changes in dietary lipid consumption (43,44). Recently it has been suggested that the DHA content of breast milk in North America, which is lower than in some other countries, is suboptimal (45).

From these observations it is clear that it is not possible for manufacturers to design an infant formula with the same lipid structure as breast milk. Moreover, the oils available to supplement infant formulas with LCPUFAs have not yet been fully characterized. In algal or fungal oils two or three of the fatty acids on the glycerol backbone may be AA or DHA, a situation not found in breast milk (46).

In view of the variability of breast milk LCPUFA content, it is not clear what the appropriate composition model should be. Does this variability reflect redundancy, or is the bottom end of the breast milk LCPUFA range undesirably low? These issues are not yet resolved.

Breast milk contains a wide variety of constituents not present in formula. Many of these have unknown roles. The presence of a factor in breast milk does not necessarily imply essentiality. Indeed the most prevalent constituent, lactose (around 70 g/l), which is unique to mammalian milk, has not been shown to be an essential nutrient.

Since replication of breast milk composition is neither a realistic nor necessarily an essential goal for formula manufacturers, a more practical target in formula design is optimization of *functional* outcome in the infant.

## Cognitive and Visual Function in Individuals Fed Breast Milk or Formula

A critical argument used in favour of the essentiality of breast milk LCPUFA content is the higher cognitive and visual function observed in those who have received breast milk. A number of studies have examined cognitive or visual outcomes in breast- and formula-fed infants during infancy or at longer term follow-up (2–6).

Of the studies performed in *infancy*, some have failed to show any difference between groups (47,48), even without adjusting for any potential confounding factors. A clear case for the superiority of breast milk in promoting short-term development and visual function cannot be made. The inconsistency in outcome between different studies could relate to the population selected. However, it has also been noted that it was in those studies where the formula used was relatively low in LNA (the essential fatty acid precursor for DHA) that a disadvantage in visual development for the formula-fed group was observed (47). This might be coincidence or could reflect the essentiality of LNA rather than of dietary LCPUFAs (DHA).

The studies which document cognitive function in *childhood* or later, however, do include a majority that show small or moderate benefits for those fed breast milk in early life. However, the major problem of interpretation of breast versus formulas studies, at least in more recent decades, is that mothers who choose to breast-feed have higher education, higher socio-economic status and show a greater degree of positive health behaviour (15). Since these factors may independently influence neurodevelopment, comparative studies involving a breast-fed cohort are generally confounded.

Some have found advantages for the breast-fed group even after attempts to adjust for socio-biological confounding. In other cases such adjustments have eliminated the apparent advantage. Where advantages for the breast-fed group have existed, they have generally been of the order of 0.25–0.5 SD of cognitive scores. The advantage has perhaps been less in term infants than in preterm infants.

The problem has been how to get at causation. The pursuit of this has been revitalized recently by those who cite the cognitive advantage of the breast-fed infants as evidence in favour of an essential role for dietary LCPUFAs in neurodevelopment.

Repeated verification of findings is not an optimal approach to causation; and certainly there are enough studies showing a cognitive advantage for breast-fed babies, even after adjustment for confounding factors, for there to be limited value in repeating that exercise. A more useful approach is to identify novel circumstances in which the causation issue can be examined afresh. One approach is to study populations in which it was the mothers with *lower* education and socio-economic status who breast-fed. Recently, Gale and Martyn (49) published such a study; the raw data showed that the previously breast-fed group, now in adult life, still performed more highly on cognitive tests, despite the social disadvantage of the breast-fed group in that era (50). In Hoeffer and Hardy's study (50) on subjects born in the USA between 1915 and 1921, the breast-fed group was socially disadvantaged, yet still performed better than the group fed formula; though if babies were exclusively breast-fed for more than 9 months, they did less well. However, bottle feeding earlier this century was quite different to that practised now; modern formulas have only been developed over the past 25 years. Thus these earlier findings need replicating in modern cohorts.

Our own approach has been to explore this issue in premature babies where the problems of confounding can be tackled in a new way. The two major proposed sources of confounding in this area are: (i) the breast-feeding effect might be due to the act of breast-feeding itself on mother–infant interaction and hence development; and (ii), as discussed above, the socio-biological advantage of modern breast-feeders. We have devised two studies, each of which *avoids* one of the above types of confounding. The first study (15) involved the follow-up of 300 children from a five-centre study who were divided according to whether or not they had received their own mother's milk. Using the WISC-R intelligence quotient (IQ) test at 7.5 years, those whose mothers provided breast milk had a 10-point higher IQ. Adjustment for mother's education (as a proxy for mother's IQ), social class and sex reduced the difference to 8.3 points.

This residual 8.3-point difference in IQ at 7.5 years was large, although, at least in part, it might still have been explained by residual social and educational confounding. However, the novel circumstance that assists the causation issue here is that we cannot argue that any advantage was conferred by *breast-feeding itself*, since these infants were too immature to suck and were fed by nasogastric tube. Indeed an analysis on those babies who received breast milk (for an average of a month in hospital) but did not go home breast-feeding, still had a near 8-point advantage in subsequent IQ.

The second study (51) was perhaps the most compelling. Here only preterm babies whose mothers had chosen not to provide their own milk were studied. In one randomized trial, babies fed on a standard term formula during, on average, their first 4 weeks, had a 15-point deficit in psychomotor development index (Bayley Scales) at 18 months compared with those fed a nutrient-enriched preterm formula, demonstrating the importance of meeting early nutrient needs. In another parallel trial, however, babies fed donated breast milk from unrelated donors had a similar psychomotor development score at 18 months to those fed the same preterm formula. Since donor breast milk, like standard formula, does not meet the nutrient needs of preterm infants, it was surprising that infants fed on it were not apparently disadvantaged, especially since those fed donor milk had substantially poorer somatic and brain growth performance in the neonatal period than those fed preterm formula (51). We hypothesized therefore that a factor or factors in human milk might be offsetting the potentially adverse effects on later development of its suboptimal nutrient content. To test this hypothesis we compared infants fed on the standard term formula from the first trial with infants fed on donor milk in the second trial. The purpose here was to compare two diets of similar *nutrient* content, one of which was human milk and the other a formula. In fact the protein and energy content of donor breast milk was somewhat lower than those in the term formula (17,51) and neonatal growth rate was marginally (though not significantly) lower (12.6 $\pm$ 0.4 SE g/kg per day vs 13.0 $\pm$ 0.6 SE g/kg per day). This was not a randomized comparison, but was nevertheless a legitimate one given there was the 'internal standard' of the same preterm formula used in both trials. Since both groups had mothers who had chosen not to breast-feed, the usual sources of potential sociobiological confounding found in other studies were eliminated. Indeed babies fed donor milk or formula had similar social class (17,51), the major predictor of cognitive function in our studies. Yet, the babies fed banked milk had a significant near 9-point advantage in psychomotor development index over those fed term formula. These data indicate, therefore, that even without the confounding effect of maternal choice to provide breast milk, preterm infants fed diets of similar nutrient content were neurodevelopmentally advantaged at 18 months after consuming human milk during, on average, their first 4 postnatal weeks.

These two studies were done on preterm infants, and although they provide more compelling evidence than previously for an effect of breast milk itself on later cognitive function, they do not provide confirmation, since in neither case was the study a randomized one. Also, the findings cannot necessarily be extrapolated to the term infant born during a less rapid stage of brain growth or indeed to infants fed modern formulas which contain higher LNA concentrations.

It should be emphasized, however, that even if there was a breast milk advantage, it could not be assumed to be due to the presence of LCPUFAs: there are indeed many candidate factors in human milk (thyroid hormone, growth factors, etc.) that might exert a biological effect.

In summary, because of the problem of potential confounding, whether or not breast milk itself influences cognitive and visual development is unproved in full-term infants. If there is an advantage, it could be explained by factors other than LCPUFAs in breast milk, including low LNA intake, hormones or growth factors (52). In preterm infants, born at a more rapid stage of brain development, the evidence that human milk affects cognitive function is stronger, though still not proven, and whether or not such an advantage could relate to breast milk LCPUFAs is unknown. Interestingly, in the most robust studies on preterm infants by Lucas *et al.*, the formula-fed groups received trivial amounts of LNA (since the formulas were designed in the early 1980s), so that these studies cannot in any case address whether LCPUFAs (DHA) are independently essential nutrients.

## Declining Levels of AA and DHA in the Plasma and Red Cells of Formula-fed Infants

Numerous studies have shown that plasma and red cell membrane levels of AA and DHA decline in infants fed formulas without added LCPUFAs (53–61). The question is whether this decline is simply a biochemical marker for the consumption of a diet with a different pattern of lipids to that in breast milk, or alternatively reflects a deficiency of functional significance. Current literature is divided on this point. Some studies indicate that circulating LCPUFA status correlates with functional outcomes such as retinal function or growth (59–61). However, other studies fail to show any such relationship (47,48,56,57). For instance, a large prospective randomized LCPUFA supplementation study of 200 infants (56) showed that, despite large differences between formula and breast milk groups in plasma or erythrocyte AA and DHA, there were no group differences in retinal function, visual function or neurodevelopment. The LCPUFA concentration in the circulation is at least an uncertain and unreliable index of visual or cognitive function. As a corollary, it cannot be assumed from current evidence that manipulation of blood levels of AA and DHA by dietary intervention will favourably affect vision or neurodevelopment.

### Autopsy Data on Brain LCPUFA Content

Farquharson *et al.* (62) in the UK and Makrides *et al.* (63) in Australia showed that formula-fed babies had lower contents of DHA in the cerebral cortex than those fed on breast milk. In the Makrides study, however, there was no influence of diet on DHA content in the retina. In the Farquharson study the formula-fed group had higher AA content in the cerebral cortex than the breast-fed group; and in the Makrides study diet had no impact on cerebral cortex AA. Some of the formulas used by infants in the Makrides and Farquharson studies had less than the currently recommended LNA content of >0.7% of dietary energy. More importantly, however, it must be emphasized that in these two post mortem studies, there are no functional data, so that the significance of these biochemical findings in humans is unknown. Indeed, currently, the *range* of human cerebral DHA contents that is associated with *normal* function is unknown. Indeed, in mice, over a wide range of brain DHA contents (75–120% of control values) there appears to be little relationship with spatial learning ability (Wainwright, personal communication).

## LCPUFA Supplementation Trials in Formula-fed Infants

It is apparent that the five lines of evidence considered above cannot independently establish a case for the essentiality of LCPUFAs and, therefore, a clear need for LCPUFA supplementation of infant formula. The most powerful tool available to explore the importance of dietary LCPUFAs is the randomized trial of infants fed formulas with and without LCPUFAs, but replete in LNA and LA. Randomization is of critical importance to counter the major confounding influences on neurodevelopment that may occur in nutritional studies, particularly those involving non-randomized comparison of breast- and formula-fed infants (see above).

The studies considered in this section are randomized trials in which neurodevelopment or visual development, with or without growth performance, were primary outcomes. These studies represent the principal ones used as a basis for current recommendations and debate, although several further studies, including major ones, are currently under way or in the literature only in abstract form.

*Preterm Studies*

There have been four published, strictly randomized studies in preterm infants (though several others are in progress or near completion, including a large trial involving over 300 subjects conducted by the author).

Between 1987 and 1996, Uauy, Birch and co-workers conducted a three-group randomized study on 71 preterm infants (64–66). Formula A contained medium-chain triglyceride (MCT), coconut oil and corn oil, providing LA (18:2n-6) and low LNA (18:3n-3) as in older formulas. Formula B contained MCT, coconut oil and soy oil, providing both LA and LNA. Formula C was similar to B, but contained fish oils providing EPA (0.6%) and DHA (0.4%). The subjects were followed at 36 and 57 weeks postconception, the latter corresponding to 4 months post-term. By this 4-month follow-up, 26 subjects had been lost to follow-up, leaving 45 (average of 15 per group). A reference group of 12 breast milk fed preterm infants were included, only 8 of whom remained to the 4-month follow-up (a significant proportion of babies in the breast milk group were in fact supplemented with formula C). The study showed that visual acuity (forced-choice preferential looking and visual evoked potentials) and retinal function (electroretinogram) was improved in the group fed formula C with fish oil, providing DHA, compared with visual acuity in group A. Generally, infants fed the fish oil formula performed similarly to those fed breast milk. The group fed soybean oil containing LNA (group B) did not have lower retinal function. No group difference in growth was found, though sample size was small. The fish oil formula group (C) had longer bleeding time (mean 2.2 minutes) compared with group A (1.8 minutes) and group B (1.9 minutes), though all values were said to be within the acceptable clinical norms. No differences in red cell lipid peroxidation were found.

Carlson performed two studies on preterm infants. The first (54,55,67,68) involved use of a preterm formula (Similac Special Care) and then a commercially available standard formula after discharge up to 9 months of age, with or without fish oil, providing DHA (0.2%), but also a high content of EPA (0.3%).

Both the preterm and standard formulas contained >3% LNA. The study started with 67 subjects; by 12 months 28 subjects in each group were tested for cognitive function. Visual acuity measured by Teller Acuity cards was higher in the fish oil supplemented group at 2 and 4 months, but by 6.5 months visual acuity of the unsupplemented group was similar to that in the LCPUFA-supplemented one; and at 9 and 12 months follow-up there was also no difference between groups. Weight, length and head circumference measures, however, were all significantly *reduced* in the fish oil supplemented groups at all time periods to the 12-month follow-up.

At 12 months the infants were tested using the Bayley Scales. Mental development index (28 subjects in each group) was approximately half a standard deviation (7.3 points) lower in the fish oil supplemented group compared with controls, though this did not reach statistical significance. However, Bayley psychomotor development (23 subjects only per group) was about 0.5 SD lower (8 points) in the fish oil group (95.2 ± 15.6 vs 87.2 ± 15.2; $p < 0.09$). However, it should be emphasized that the sample size in this study would only allow 0.83 SD difference (about 13 points) in Bayley scores to be detected at the 5% significance level with adequate power.

The Fagan Infantest was selected as an additional measure of cognition (69). This test is based on the tendency of the infant to prefer looking at a novel rather than familiar stimulus. In Carlson's study, novelty preference was the same for the two groups at 6.5 and 9 months, but significantly lower in the fish oil group at 12 months. There were apparently no differences between groups on repeated measures ANOVA, and in any case the significance of Fagan test scores in terms of later functioning requires further research. In addition to differences in novelty preference, shorter look duration was reported at 12 months in the fish oil fed group, which was interpreted as more rapid information processing in the light of recent studies on monkeys made n-3 deficient. However, the predictive value of these findings in humans requires further investigation.

In a further Carlson study (68,70), preterm formula followed by term formula (with >3% LNA) was given with or without a fish oil with lower EPA content (0.06%) than the previous study. This time the intervention stopped at 2 rather than 9 months post-term. The rationale for this was that in the first study high EPA intakes were associated with lower AA, which in turn correlated with lower growth performance. The shorter intervention period in the second trial was intended to reduce any depressant effect of n-3 fatty acids on AA status and growth. Infants were enrolled as in the previous study, except that now infants who developed bronchopulmonary dysplasia were not rejected. Also, the postdischarge formula was more nutrient-enriched (to meet the needs of preterm infants) than in the previous study. The attrition rate in the study was substantial. Initially there were 94 subjects randomized in two equal groups. Of these, 35 were dropped from the study before 2 months, leaving only 59 subjects (33 controls and 26 supplemented). Of the 35 subjects that left the study, 19 had sepsis or necrotizing enterocolitis (NEC); of concern, this comprised only 5 subjects (3 NEC) in the controls versus 14 (9 NEC) in the supplemented group (see later discussion).

In order to make up the relatively small numbers of the original cohort followed and the subsequent loss to follow-up, subjects were added to replace lost infants (next available infant of the same sex and gestation was recruited to replace lost subjects, a procedure that disturbs the randomized, blind design). In all, 26 controls and 25 supplemented were seen at both 2 and 12 months. It is unclear how the additional infants were assigned to diet (randomized or not), bearing in mind the differential dropout rate.

Visual acuity was greater in the LCPUFA-supplemented group at 2 months but no different to that in the supplemented group at 4, 6, 9 and 12 months. Bayley mental development index was increased in the LCPUFA-supplemented group by 11.7 points (controls $98.0 \pm 19$ SD, $n = 21$; supplemented $109.7 \pm 15.3$ SD, $n = 22$). Supplemented infants had more looks to familiar and shorter looks to novel stimuli, which was interpreted by the investigators as indicative of faster information processing. However, the LCPUFA-supplemented group had *lower* weight at 6 and 9 months, *smaller* head circumference at 9 months and *lower* weight for length at 2, 6, 9 and 12 months.

In 1996, Montalto *et al.* (71) reported a study of similar size to the Carlson trial using preterm and then term formula postdischarge to 6 months with or without high DHA, low EPA fish oil supplementation. In a preplanned analysis by gender, slower growth rate was found in males.

*Term Studies*

In 1995, Makrides *et al.* (61) published a small trial on 26 infants randomly assigned to a standard formula ($n = 14$) or formula ($n = 12$) supplemented with fish oil, providing EPA and DHA, and primrose oil, to provide γ-linolenic acid (GLA; 18:2ω-6) as a precursor of AA. The subjects were studied from birth to 7 months. The placebo group had lower visual acuity (using visual evoked potentials) at 4 and 7 months compared with the treatment group (and also a breast-fed reference group). No influence of supplementation on growth was seen.

One of the largest studies (56,57,60,72) was a three-centre randomized trial of 135 subjects. One group was fed a control formula, one a formula with additional DHA from low EPA fish oil, and a third group a formula containing egg phospholipid as a source of both DHA and AA. The three formulas all contained 1% of energy as LNA. The formulas were given up to 12 months of age and subjects were followed to 14 months. Neither addition of DHA nor DHA plus AA conferred any benefit in visual acuity (measured by visual evoked potential and acuity cards at 2, 4, 6, 9 and 12 months), in retinal function (assessed by electroretinogram at 4 months) or in development (Bayley Scales at 12 months; MacArthur's Communicative Developmental Inventories at 14 months). The only significant outcome finding was that in the group supplemented only with fish oil (DHA) vocabulary scores were actually *reduced* compared with the control group. Infants supplemented with DHA had higher red cell DHA levels than did those breast-fed from 4 months onwards; those supplemented with both DHA (0.12%) and AA (0.4%) had red cell DHA and AA levels similar to those of the breast-fed group.

Carlson *et al.* (73) randomized 39 term infants to a control formula or a similar formula to the one supplemented with DHA and AA in the three-centre study above. Visual acuity using acuity cards was not influenced overall, but there was a transient increase in the unsupplemented group at 2 months that was not seen at 4, 6, 9 or 12 months.

Agostoni *et al.* (74) randomly assigned 56 infants to a standard formula or one providing DHA and AA from egg lipid. Developmental quotient was lower at 4 months in the unsupplemented group, but developmental differences were transient and not seen at either 1 or 2 years (75). In any case, LNA content in the unsupplemented formula was only 0.35% of dietary energy.

In a further trial on 56 subjects, Willatts *et al.* (76), also comparing a control formula with one containing DHA and AA, examined total, mean and peak visual fixation time, finding no significant difference between groups. One subgroup with late peak fixation (a measure of poor attention control) apparently benefited from supplementation; this post-hoc observation generates a new hypothesis for prospective testing.

An unpublished study by Birch and co-workers on term infants shows improvements in visual development up to 4 months in LCPUFA-supplemented infants. In one of our own (unpublished) trials of over 300 infants (the first of four current LCPUFA trials to complete randomization of subjects, and to our knowledge the largest trial undertaken), we failed to find any impact on 9 month developmental scores or growth using a formula supplemented with 0.3% AA and 0.3% DHA.

## Discussion

It is difficult to make firm inferences on the essentiality of LCPUFAs from breast milk composition, or from the lipid content of blood or post mortem brain, since in none of these instances are the biochemical findings clearly and reliably linked to function. Comparative studies of visual or developmental performance in breast- and formula-fed infants are non-randomized, often prone to sociobiological confounding and any advantages apparent in some studies for breast-fed babies cannot necessarily to attributed to LCPUFAs rather than the many other factors found in breast milk and not formula. Experimental findings on the effects of LCPUFA status on behaviour in animals cannot necessarily be extrapolated to human cognitive function.

For these reasons, therefore, the data from randomized trials of LCPUFA supplementation in formula-fed infants provide the most important evidence. These trials provide provocative data, but also pose a number of concerns, and some have significant shortcomings, as discussed below.

### Inconsistent Findings

The preterm trials in particular do not provide uniform evidence of benefit from LCPUFA supplementation. There is evidence of early benefit for visual function and retinal function. However, in studies of more than 4 months' duration, the visual effects have proved transient (68,70) and of unknown significance for later function. Of more concern is the evidence from three studies using fish oil that growth was suppressed (54,68,70,71); in two studies beyond the period of dietary assignment. Developmental tests have also produced inconsistent results, with possibly deleterious effects and possibly beneficial effects depending on the study.

In term infants, the largest published study (the three-centre study of Auestad *et al.* (56)) showed no benefit for LCPUFA supplementation in visual, retinal or cognitive development, and the only significant finding was a *deficit* in vocabulary development at age 14 months (60) – one of the longest periods of follow-up of the studies currently published. A number of smaller studies have shown short-term or transient benefits for visual development. No studies have reported either a benefit or disadvantage in growth rate for LCPUFA-supplemented full-term infants. One study showed a transient beneficial effect on development at 4 months (74), when test procedures are crude and scores are often poorly predictive of later performance; but later development up to 2 years of age was unaffected by LCPUFA supplementation (75).

## Duration of Follow-up

As yet no study in either term or preterm infants has been of sufficient duration to establish whether the early benefits or deficits following LCPUFA supplementation have any lasting significance for visual or cognitive performance. Nor have the cohorts with 'neutral' results been followed long enough to exclude 'later emergence' of beneficial or detrimental effects that might become apparent at an age when more sophisticated neurodevelopmental testing is possible.

## Sample Size

The differences in visual and cognitive development identified between breast- and formula-fed babies in some studies provide a basis for calculated sample size needed to detect plausible effects in LCPUFA formula supplementation studies. From the studies reported, an average 'effect size' of about one-third of a standard deviation of cognitive developmental score (around 5 points) is found in those studies that show an advantage for the breast-fed group. Even if *all* of this advantage was due to LCPUFAs, sample size in a formula supplementation study targeted to detect this difference would be 144 subjects per group for 5% significance and 80% power. If it was hypothesized that only half, say, of the breast-feeding 'effect' was due to LCPUFAs, then the required sample size would rise to 576 per group. No study in this field even approaches this size; even the largest study having 45 subjects per group (the three-centre study that showed no benefit for supplementation). In our own series of formula supplementation trials, each involving a different strategy for LCPUFA supplementation in preterm or term infants, we have taken account of the breast versus formula studies to calculate realistic sample sizes: four large studies are under way and two more are at the planning stage. All these studies have planned long term follow-up. The first of these studies is currently being analysed.

The possibility of type I error in undersized studies is a real one. In those studies that have shown an apparent benefit for LCPUFA supplementation, the mean sample size per group has been only 25 ($\pm$ 6 SD) subjects. In preterm infants, in particular, the groups are likely to be heterogeneous. In assessing the extent to which some published findings in small studies could be due to chance, it would be important to know to what extent negative publication bias has operated, a well-recognized problem in the early stages of a new field. As more studies reach completion, it is likely that this issue will be resolved with appropriate meta-analyses performed on smaller studies and with the emergence of larger, 'free-standing' studies.

## Other Outcomes

So far published studies have focused largely on a limited range of outcomes: neurodevelopment, visual functional and growth. LCPUFAs, however, are recognized to have a wide variety of potential roles. They may, for instance, influence gene expression, gut function, insulin resistance, blood pressure, bone metabolism and immune function (77–82). The possibility that LCPUFA supplementation at a critical early stage of development could result either in beneficial or deleterious effects on systems other than those traditionally explored has received little attention.

## Safety

The major thrust in LCPUFA studies has been to test for biochemical and functional efficacy. Less attention has been paid to the formal investigation of safety. Some aspects of safety require larger sample sizes than are required to test for efficacy, since large numbers of subjects may be needed to detect a differential incidence of a relatively rare adverse event, such as proven sepsis. No currently published LCPUFA supplementation study has been of adequate size for realistic appraisal of safety. Yet there are some safety indicators within the body of published work that now require formal appraisal. In Carlson's second study (70), 94 subjects were randomized to regular or LCPUFA-containing formula. In all, 35 were taken out of the study before 2 months. Of these, 19 had infection or necrotizing enterocolitis (NEC). The aspect of concern was the differential incidence of infection and NEC between randomized groups: the control formula group had three cases of NEC versus nine cases in the fish oil supplemented group ($p = 0.06$); and two of infection amongst the cases versus five in the supplemented group. It would be speculative, though not implausible, to hypothesize that a common mechanism, impaired immunity, explained both the NEC and the infection. If this was a valid assumption, combining these two conditions and assuming an equal or near-equal split of the original 94 subjects (as confirmed by the author; personal communication), 14/47 (30%) of subjects in the supplemented groups versus 5/47 (11%) in the control group had NEC or infection ($\chi^2$ 5.3, $p = 0.02$). Whether the data on NEC and infection are combined or considered separately, the small sample precludes a firm conclusion on causation. As the author indicates, a larger study would be needed to confirm or allay the concerns raised; this would seem urgent.

In Uauy's preterm studies (83), some effort was made to examine safety aspects albeit on small numbers. He found that subjects supplemented with fish oil had a significantly increased bleeding time compared with those on the other two formulas. Whist all subjects were thought to be in the normal range for bleeding time, a larger study would be needed to exclude the possibility that fish-oil-fed preterm infants were not shifted into a higher risk category for haemorrhagic events. Indeed in Uauy's study, subjects were well, relatively large preterm infants and safety studies need to be focused also on the sicker, smaller infants often seen in neonatal units.

The impairment of growth, notably of head and therefore brain growth, in LCPUFA-supplemented preterm infants, is also a significant area for concern, particularly since the apparently disadvantageous effects on growth in Carlson's trials extended beyond the period of LCPUFA supplementation.

Exploration of these safety aspects is of critical importance to our appraisal of the value of LCPUFA supplementation. For instance, if, hypothetically, improvement in look duration and transient effects on acuity were to be weighed against a major increase in NEC (were that to be confirmed) it is likely that paediatricians would be more swayed by the latter, a serious gut disease with a 20–40% mortality. However, whether the safety concerns raised by previous studies reflect chance findings on small samples or unbalanced addition of n-3 and n-6 LCPUFAs, or alternatively reflect unavoidable risks of LCPUFA supplementation, can only be learned from new, larger randomized interventions targeted to explore safety as well as efficacy.

## Are Preterm Infants a Special Case?

It has been argued (see above) that preterm infants are a special group that have a greater need for LCPUFA supplementation. Indeed, the European Society for Paediatric Gastroenterology and Nutrition (ESPGAN) in 1991 recommended that preterm formulas contain LCPUFAs. It is certainly the case that their rate of brain growth is rapid compared to the term infant, though Martinez points out that AA and DHA deposition in the brain after 40 weeks is only minimally less than before (84). Preterm infants are frequently in poor energy balance, which might theoretically result in oxidation of 18-carbon LCPUFA precursors (LA and LNA) as energy sources. The argument that preterm infants are unable to generate LCPUFAs from precursors, however, is not supported by two recent studies that examine the conversion of stable isotope labelled precursor to LCPUFAs in vivo (85,86).

Nevertheless, the best guide to whether preterm infants would benefit from LCPUFA supplementation in formula is not the theoretical arguments about vulnerability, but the clinical trial data. The findings on LCPUFA supplement trials reviewed here provide, if anything, data of more concern than do the corresponding studies on full-term infants. The outcome findings are inconsistent and include, theoretically, beneficial effects (but where long-term benefit has not been confirmed), possible disadvantageous medium term effects on growth and development, transient effects and no effects. The short-term safety issues, particularly NEC and sepsis, raised by the Carlson study remain to be resolved conclusively.

## Should Firm Recommendations be Made for Addition of LCPUFAs to Formula?

In any area of clinical intervention or therapeutics there are two fundamental issues, efficacy and safety. The modern regulatory climate for infant formula research and development is increasingly centred around those issues. A new Committee of Medical Aspects document, shortly to be released in the UK, to guide researchers on appropriate conduct of clinical trials on infant formulas, again is modelled around the need for researchers to establish efficacy and safety for novel products and new formulations.

So, has efficacy or safety been established for the various strategies used to enrich formulas with LCPUFAs? Current published data do not comfortably support that position. Certainly the efficacy data on visual and retinal development and on attention that have emerged from several small studies in term and preterm infants are provoking, promising and warrant further investment. These exciting findings, however, are tempered by the fact that the clinical significance of some of these observations is not yet clear; outcome findings in some studies have been neutral or of concern and studies have generally been too small. Safety has not been satisfactorily explored, and some observations to date raise concerns. The inconsistency of findings, in the face of some positive findings, if nothing else indicates that researchers have not yet devised a predictable strategy for achieving outcome benefit. Indeed, some potential strategies for LCPUFA supplementation have not been adequately researched at all.

The wave of enthusiasm for LCPUFA research has swayed some official bodies into making provisional or more firm recommendations for

LCPUFA addition to formula, notably for preterm infants (87). Some of the key data now available and shortly to be released were not available to these bodies when they came to their conclusions. Moreover, the quality of evidence now required by regulatory bodies to approve a novel addition to infant formula is much higher than it was. A few years ago theoretical arguments (for instance based on the composition of breast milk) or endpoints of unknown functional significance (for instance plasma biochemistry) could be used effectively to underpin a new scientific advance in infant feeding. This is no longer the case.

Thus, whilst there is no doubt that individual formula companies will follow their own research leads and act individually, the current uncertainty in the field makes it hard for *official* recommendations for addition of LCPUFAs in formula to be defended fully. We are entering the next phase of LCPUFA research during which the results of much more robust outcome studies will be announced; these will be based on more realistic samples, targeted outcomes of clearer functional significance and demonstration of safety as a key objective. At this time it would seem justified to adopt a watchful and conservative approach and for those official bodies that have already made or considered recommendations to review their position periodically.

## References

1. Innis SM. Human milk and formula fatty acids. *J Pediatr* 1992; **120**: S56–61.
2. Florey C du V, Leech AM, Blackhall A. Infant feeding and mental and motor development at 18 months of age in first born singletons. *Int J Epidemiol* 1995; **24**: S21–6.
3. Rogan WJ, Gladen BC. Breast feeding and cognitive development. *Early Human Dev* 1993; **31**: 181–93.
4. Fergusson DM, Beautrais AL, Silva PA. Breast feeding and cognitive development in the first seven years of life. *Soc Sci Med* 1982; **16**: 1705–8.
5. Rodgers B. Feeding in infancy and later ability and attainment: a longitudinal study. *Dev Med Child Neurol* 1978; **20**: 421–6.
6. Hoeffer C, Hardy MC. Later development of breast-fed and artificially fed infants. *JAMA* 1929; **92**: 615–9.
7. Lucas A. Role of nutritional programming in determining adult morbidity. *Arch Dis Child* 1994; **71**: 288–90.
8. Smart J. Undernutrition, learning and memory: review of experimental studies. In: Taylor TG, Jenkins NK, eds. *Proceedings of XII International Congress of Nutrition*. London: John Libbey, 1986: 74–8.
9. Snoek A, Remacle C, Ruesens B, Moet J. J. Effect of a low protein diet during pregnancy of the fetal rat endocrine pancreas. *Biol Neonate* 1990; **57**: 107–18.
10. Barker DJP, Gluckman PD, Godfrey KM, Harding JE, Owens JA, Robinson JS.

Fetal nutrition and cardiovascular disease in adult life. *Arch Dis Child* 1993; **341**: 938–41.
11. Hahn P. Effect of litter size on plasma cholesterol and insulin and some liver and adipose tissue enzymes in adult rodents. *J Nutr* 1984; **114**: 1231–4.
12. Lewis DS, Bartrand HA, McMahan CA, McGill HC Jr, Carye KD, Masoro EJ. Pre-weaning food intake influences the adiposity of young adult baboons. *J Clin Invest* 1986; **78**: 899–905.
13. Mott GE, Lewis DS, McGill HC. Programming of cholesterol metabolism by breast or formula feeding. In: Bock G. R., Whelan J., eds. *The Childhood Environment and Adult Disease* (CIBA Foundation Symposium No. 156). Chichester: Wiley, 1991: 56–76.
14. Grantham-McGregor S. Field studies in early nutrition and later achievement. In: Dobbing J, ed. *Early Nutrition and Later Achievement*. London: Academic Press, 1987: 128–74.
15. Lucas A, Morley R. Breast milk and subsequent intelligence quotient in children born preterm. *Lancet* 1991; **339**: 261–4.
16. Lucas A, Morley RM, Cole TJ, Gore SM. A randomized multicentre study of human milk versus formula and later development in preterm infants. *Arch Dis Child* 994; **70**: F141–6.
17. Lucas A, Morley R, Cole TJ *et al*. Early diet in preterm babies and developmental status at 18 months. *Lancet* 1990; **335**: 1477–81.
18. Waber DP, Vuori-Christiansen L, Oritz N *et al*. Nutritional supplementation, maternal education and cognitive development of infants at risk of malnutrition. *J Clin Nutr* 1981; **34**: 807–13.
19. Super CM, Herrera MG. The cognitive outcomes of early nutritional intervention in the Bogota study. *Soc Res Child Dev* 1991.
20. Grantham-McGregor SM, Powell CA, Walker SP, Himes JH. Nutritional supplementation, psychosocial stimulation and mental development of stunted children: the Jamaican study. *Lancet* 1991; **338**: 1–5.
21. Husaini MA, Karyadi L, Husaini YK, Karyadi D, Pollitt E. Developmental effects of short-term supplementary feeding in nutritionally at risk Indonesian infants. *Am J Clin Nutr* 1991; **54**: 799–804.
22. Lucas A. Is future health 'programmed' by infant nutrition? *MRC News* 1996; **71**: 34–7.
23. Fierro-Benieter R, Casar R, Stanbury J *et al*. Long-term effects of correction of iodine deficiency on psychomotor development and intellectual development. In: Dunn J., Pretell E., Dara C., Viteri F., eds. *Towards the Eradication of Endemic Goiter, Cretinism and Iodine Deficiency*. Washington, DC: Pan American Health Organization, 1986.
24. Lozoff B, Jimeniz E, Wolf AW. Long-term developmental outcome of infants with iron deficiency. *N Engl J Med* 1991; **325**: 687–94.
25. Moffatt ME, Longstaffe S, Besant J, Dureski C. Prevention of iron deficiency and psychomotor decline in high risk infants through use of iron fortified formula: a randomized trial. *J Pediatr* 1994; **125**: 527–34.
26. Walter T. Impact of iron deficiency on cognition in infancy and childhood. *Eur J Clin Nutr* 1993; **47**: 307–16.
27. Sprecher H. Interconversions between 10- and 22-carbon ω-3 and ω-6 fatty acids via 4-desaturase independent pathways. In: Sinclair A, Gibson R, eds.

*Essential Fatty Acids and Eicosanoids.* Champaign, IL: American Oil Chemists Society, 1992: 18–20.
28. Voss A, Reinhart M, Sankaraappa S, Sprecher H. The metabolism of 7,10,13,16,19-docosapentanoic acid to 4,7,10,13,16,19-docoshexaenoic acid in rat liver is independent of a 4-desaturase. *J Biol Chem* 1991; **266**: 19 995–20 000.
29. Cook HW. In vitro formation of polyunsaturated fatty acids by desaturation in rat brain: some properties of the enzymes in developing brain and comparisons with liver. *J Neurochem* 1978; **30**: 1327–34.
30. Moore SA, Yoder E, Murphy S *et al*. Astrocytes, not neurons, produce docosahexaenoic acid (22:6 ω-3) and arachidonic acid (20:4 ω-6). *J Neurochem* 1991; **56**: 518–24.
31. Sprecher H. Interconversions between 20- and 22-carbon n-3 and n-6 fatty acids via 4-desaturase independent pathways. In: Sinclair A, Gibson R, eds. *Essential Fatty Acids and Eicosanoids.* Champaign, IL: American Oil Chemists Society, 1993: 18.
32. Innis SM. Essential fatty acids in growth and development. *Prog Lipid Res* 1991; **30**: 39–103.
33. Pleisler SJ, Anderson RE. Chemistry and metabolism of lipids in vertebrate retina. *Prog Lipid Res* 1983; **22**: 79–131.
34. Bourre JM, Durand G, Pascal G, Youyou A. Brain cell and tissue recovery in rats made deficient in n-3 fatty acids by alterations of dietary fat. *J Nutr* 1989; **119**: 15–22.
35. Bourre JM, François M, Youyou A *et al*. The effect of dietary alpha linolenic acid on the composition of nerve membranes, enzymatic activity, amplitude of electrophysiological parameters, resistance to poisons and performance of learning tasks in rats. *J Nutr* 1989; **119**: 1880–2.
36. Neuringer M, Connor WE, Vanm Petten C *et al*. Dietary omega-3 fatty acid deficiency and visual loss in infant rhesus monkeys. *J Clin Invest* 1984; **73**: 272–6.
37. Neuringer M, Conner WE, Lin DS *et al*. Biochemical and functional effects of prenatal and postnatal s-3 fatty acid deficiency on retina and brain in rhesus monkeys. *Proc Natl Acad Sci USA* 1986; **83**: 4021–5.
38. Arbuckle LD, MacKinnon MJ, Innis SM. Formula 18:2(n-6) and 18:3(n-3) content and ratio influence long-chain polyunsaturated fatty acids in the developing igliet liver and central nervous system. *J Nutr* 1994; **124**: 289–98.
39. Wainwright PE. Do essential fatty acids play a role in brain and behavioral development? *Neurosci Biobehav Rev* 1992; **16**: 193–205.
40. Jensen RG, Hagerty MM, McMahon KE. Lipids of human milk and infant formulas: a review. *Am J Clin Nutr* 1978; **31**: 990–1016.
41. Quinlan PT, Lockton S, Irwin J, Lucas AL. The relationship between stool hardness and stool composition in breast- and formula-fed infants. *J Pediatr Gastroenterol Nutr* 1995; **20**: 81–90.
42. Mather IH, Jack LJ. A review of the molecular and cellular biology of butyrophilin, the major protein of bovine milk fat globule membrane. *J Dairy Sci* 1993; **76**: 3832–50.
43. Innis S. Human milk and formula fatty acids. *J Pediatr* 1992; **120**(2): S56–61.
44. Benson JD, MacLean WC Jr, Pinder DL, Sauls HR. Modifications of lipids in infant formulas: concerns of industry. In: Sinclair A, Gibson R, eds. *Essential*

*Fatty Acids and Eicosanoids.* Champaign, IL: American Oil Chemists Society, 1992: 218–21.
45. Jorgensen MH, Hernell O, Lund P, Holmer G, Michaelsen KF. Visual acuity and erythrocyte docosahexaenoic acid status in breast-fed and formula-fed term infants during the first four months of life. *Lipids* 1996; **31**: 99–105.
46. Myher J, Kuksis A, Geher K, Park PW, Diersen-Schade DA. Stereospecific analysis of triacylglycerols rich in long chain polyunsaturated fatty acids. *Lipids* 1996; **31**: 207–15.
47. Innis S, Nelson CM, Lwanga D, Rioux FM, Waslen P. Feeding formula without arachidonic acid and docosahexaenoic acid has no effect on preferential looking acuity or recognition memory in healthy full-term infants at 9 months of age. *Am J Clin Nutr* 1996; **64**: 40–6.
48. Innis SM, Nelson CM, Rioux FM, King DJ. Development of visual acuity in relation to plasma and erythrocyte ω-6 and ω-3 fatty acids in healthy term gestation infants. *Am J Clin Nutr* 1994; **60**: 347–52.
49. Gale CR, Martyn CN. Breast feeding, dummy use and adult intelligence. *Lancet* 1996; **347**: 1072–5.
50. Hoefer C, Hardy MC. Later development of breast fed and artificially fed infants. *JAMA* 1929; **92**: 615–19.
51. Lucas A, Morley R, Cole TJ, Gore SM. A randomized multicentre study of human milk versus formula and later development in preterm infants. *Arch Dis Child* 1994; **70**: F141–6.
52. Sack J. Hormones in milk. In: Filer S, Fidelman AI, eds. *Human Milk, its Biological and Social Value.* Amsterdam: Excerpta Medica, 1980: 56–61.
53. Carlson SE, Rhodes PG, Rao VS, Goldgar DE. Effect of fish oil supplementation on the n-3 fatty acid content of red blood cell membranes in preterm infants. *Pediatr Res* 1987; **21**: 507–10.
54. Carlson SE, Cooke RJ, Werkman SH, Tolley EA. First year growth of preterm infants fed standard compared to marine oil (fish oil) n-3 supplemented formula. *Lipids* 1992; **27**: 901–7.
55. Carlson SE, Werkman SH, Peeples JM *et al.* Arachidonic acid status correlates with first year growth in preterm infants. *Proc Natl Acad Sci USA* 1993; **90**: 1073–7.
56. Auestad N, Montalto MB, Hall RT *et al.* Visual acuity. Erythrocyte fatty acid composition and growth in term infants fed formula with long chain polyunsaturated fatty acids for one year. *Pediatr Res* 1997; **41**: 1–10.
57. Hartmann EE, Neuringer M. Longitudinal behavioral measures of visual acuity in full-term human infants fed different dietary fatty acids. *Invest Ophthalmol Vis Sci* 1995; **36**: S869.
58. Innis S, Lupton BA, Nelson CM. Biochemical and functional approaches to study of fatty acid requirements for very premature infants. *Nutrition* 1994; **10**: 72–6.
59. Makrides M, Simmer K, Goggin M, Bigson RA. Erythrocyte docosahexaenoic acid correlates with the visual response of healthy, term infants. *Pediatr Res* 1993; **33**: 425–7.
60. Jorgensen MH, Hernell O, Lund P *et al.* Visual acuity and erythrocyte docosahexaenoic acid status in breast-fed and formula-fed term infants during the first four months of life. *Lipids* 1996; **31**: 99–105.
61. Makrides M, Neumann M, Simmer K, Pater J, Gibson R. Are long-chain

polyunsaturated fatty acids essential nutrients in infancy? *Lancet* 1995; **345**: 1463–8.
62. Farquharson J, Cockburn F, Patrick WA *et al.* Infant cerebral cortex phospholipid fatty acid composition and diet. *Lancet* 1992; **340**: 810–13.
63. Makrides M, Neumann MA, Byard RW *et al.* Fatty acid composition of brain, retina, and erythrocytes in breast- and formula-fed infants. *Am J Clin Nutr* 1994; **60**: 180–94.
64. Birch E, Birch D, Hoffman DR, Uauy R. Dietary essential fatty acid supply and visual acuity development. *Invest Ophthalmol Vis Sci* 1992; **33**: 3242–53.
65. Birch E, Birch D, Hoffman DR *et al.* Breast feeding and optimal visual development. *J Pediatr Ophthalmol Strabismus* 1993; **30**: 33–8.
66. Uauy RD, Birch DG, Birch EE *et al.* Effect of dietary omega-3 fatty acids on retinal function of very low birthweight infants. *Pediatr Res* 1990; **28**: 485–92.
67. Carlson SE, Werkman SH, Rhodes PG, Tolley EA. Visual acuity development in healthy, preterm infants: effect of marine oil (fish oil) supplementation. *Am J Clin Nutr* 1993; **58**: 35–42.
68. Carlson SE, Werkman SH, Peeples JM, Wilson WM. Long chain fatty acids and early visual and cognitive development of preterm infants. *Eur J Clin Nutr* 1994: **48**(Suppl 2): S27–30.
69. Werkman SH, Carlson SE. A randomized trial of visual attention of preterm infants fed docosahexaenoic acid until 9 months. *Lipids* 1996; **31**: 91–7.
70. Carlson SE, Werkman SH, Tolley EA. Effect of long chain n-3 fatty acid supplementation on visual acuity and growth of preterm infants with and without bronchopulmonary dysplasia. *Am J Clin Nutr* 1996; **63**: 687–97.
71. Montalto MB, Mimouni FB, Sentipal-Walerius J *et al.* Reduced growth in hospital-discharged low birthweight infants fed formulas with added marine oil (fish oil). *Pediatr Res* 1996; **39**: 316A.
72. Neuringer M, Fitzgerald KM, Weleber RG *et al.* Electroretinograms in four-month-old full term human infants fed diets differing in long chain n-3 and n-3 fatty acids. *Invest Ophthalmol Vis Sci* 1995; **36**: S48.
73. Carlson SE, Ford AJ, Werkman SH. Visual acuity and fatty acid status of term infants fed human milk and formulas with and without docosahexaenoate and arachidonate from egg yolk lecithin. *Pediatr Res* 1996; **39**: 882–8.
74. Agostoni C, Trojan S, Bellu R. Neurodevelopmental quotient of healthy term infants at 4 months and feeding practice: the role of long-chain polyunsaturated fatty acids. *Pediatr Res* 1995; **38**: 262–6.
75. Agostoni C, Trojan S, Riva E, Bellu R, Bruzzese MG, Giovannini M. Early type of diet and neurodevelopment outcome of term infants: results of a 2-year follow up. In: *Proceedings of the 17th International Symposium on Neonatal Intensive Care*, San Remo. 1996: 25A.
76. Willatts P, Forsyth JS, DiMondugno MK, Varma S, Colvin M. The effects of long chain polyunsaturated fatty acids on infant habituation at three months and problem solving at nine months. *PUFA in Infant Nutrition: Consensus and Controversies. Proceedings of American Oil Chemists Society*, Barcelona. Champaign, IL: American Oil Chemists Society, 1996: 42A.
77. Danesch U, Weber PC, Sellmayer A. Arachidonic acid increases *c-fos* and *Egr*-1 mRNA in 3T3 fibroblasts by formation of prostaglandin $E_2$ and activation of protein kinase C. *J Biol Chem* 1994; **269**: 27 258–63.

78. Clarke SD, Jump DB. Dietary polyunsaturated fatty acid regulation of gene transcription. *Annu Rev Nutr* 1994; **14**: 83–98.
79. Toft I, Bonaa KH, Ingebretsen OC, Nordoy A, Jenssen T. Effects of n-3 polyunsaturated fatty acids on glucose homeostasis and blood pressure in essential hypertension: a randomized controlled trial. *Ann Intern Med* 1995; **123**: 911–18.
80. Storlien LH, Jenkins AB, Chishold DJ *et al*. Influence of dietary fat composition on development of insulin resistance in rats: relationship to muscle triglyceride and omega-3 fatty acids in muscle phospholipid. *Diabetes* 1991; **40**: 280–9.
81. Claassen N, Cowtzer H, Steinmann CM, Kruger MC. The effect of different n-6/n-3 essential fatty acid ratios on calcium balance and bone in rats. *Prostaglandins Leukotrienes Essential Fatty Acids* 1995; **53**: 13–19.
82. Sperling RI, Benincaso AI, Knoell CT *et al*. Dietary omega-3 polyunsaturated fatty acids inhibit phosphoinositide formation and chemotaxis in neutrophils. *J Clin Invest* 1993; **91**: 651–60.
83. Uauy R, Birch DG, Birch EE, Hoffman D, Tyson J. Visual and brain development in infants as a function of essential fatty acid supply provided by the early diet. In: *Lipids, Learning and the Brain: Fats in Infant Formulas. Report of the 103rd Ross Conference on Pediatric Research*. 1993: 215–30.
84. Martinez M, Pineda M, Vidal R, Martin. Docosahexaenoic acid: a new therapeutic approach to peroxisomal patients. Experience with two cases. *Neurology* 1993; **43**: 1389–97.
85. Salen N Jr, Wegher B, Mena P, Uauy R. Arachidonic and docosahexaenoic acids are biosynthesized from their 18-carbon precursors in human infants. *Proc Natl Acad Sci USA* 1996; **93**: 49–54.
86. Carnielli VP, Wattimena JL, Lluijendijk IHT *et al*. The very low birthweight premature infant is capable of synthesizing arachidonic and docosahexaenoic acids from linoleic and linolenic acids. *Pediatr Res* 1996; **40**: 169–74.
87. Aggett PJ, Haschke F, Heine W *et al*. Comment on the content and composition of lipids in infant formula. ESPGAN Committee on Nutrition. *Acta Paediatrican Scand* 1991; **80**(8–9): 887–96.

# Commentary

**Carlson**: You contend that a major impetus for research on LCPUFAs in infant nutrition has been the presence of LCPUFAs in breast milk coupled with the observation that breast-fed babies have higher cognitive performance than those fed formula. This view is not supported by the animal literature from the 1970s and early 1980s or the human work that immediately followed. Also it is my understanding that 'programming' has been used for many years to describe the impact of compounds at a critical stage of early development, which have physiological effects later in life. The concept that early nutrition can influence later development and the recognition that nutrition interacts with environmental stimulation to influence development has been around at least since Cravioto in the 1960s.

As far as I know, no one has looked at animals to determine if they need DHA for neurodevelopment.

None of the references cited (your references (2) to (6)) appear to deal with visual outcomes in breast- and formula-fed infants. Also, one cannot argue for the essentiality of breast milk LCPUFA based on higher cognitive and visual function in those who have received breast milk.

You refer to the non-randomized comparison of babies in your study who were fed either term formula in one trial or banked human milk in another trial. Although you say that their mothers did not choose to provide human milk, thus eliminating concerns about socio-economic status (SES) and maternal IQ, it would be helpful to have actual data about SES or IQ confirming your point. Also, it would be nice to know if the studies were done in the same or different hospitals, how far apart they were in time, if there were exclusion and retention criteria, the weight range of the infants studied, what diseases occurred in the study groups and with what frequency, etc. These details would allow the reader to decide if the comparisons are 'legitimate'.

If I recall correctly, the composition of your 'standard' term formula is quite low in protein (and many other nutrients) compared with commercially available term formulas in the USA. Also, aren't the preterm formula in your studies more equivalent to term formulas in the USA and much lower in nutrients than US preterm formulas? Protein, zinc and many other nutrients could have limited growth in the infants fed your term formula and accounted for the low psychomotor development index.

Your references 59–61 do not show a correlation between LCPUFA status and growth. It might be worth pointing out that infants apparently received much higher amounts of DHA in the diet in references 59–61 compared with reference 56 (57 is the same study as 56). In relation to this and your later comments about the preterm studies, it is important to look at the details of studies that find no effect, just as for studies that reject the null hypothesis (some would argue more important). Casual acceptance of 'no effect' in studies that are not validly designed to look for an effect can have significant adverse consequences for infants by delaying changes supported by more carefully done studies.

With regard to the Fagan data, it would be better to cite the actual report, which shows novelty preference, look number and look duration at all three ages (1). There was no overall effect of diet on novelty preference but there were highly significant effects of diet on look number and look duration across all ages (ANOVA).

The comment from Innis implies that the Fagan test uses different cards at different ages, which is not true. Also, look duration and look number are not derived. They are measured and averaged with our software. The decrease in time allowed for familiarization and test is the mechanism used by Fagan to normalize across age; therefore, look duration and novelty preference automatically standardize across various familiarization and test times. My comments to Singer indicate the high correlation between ages on the number of looks, evidence that these are not independent tests any more than are anthropometric measures at term, 2 and 4 months. The only measures indicated in Tables 3 and 4 (85) that cannot be interpreted are clearly identified.

I do not understand the meaning from Innis that one is not allowed to look at data when there is no interaction between diet and age (diet $\times$ age). Generally, one would prefer not to have an interaction because interpretation becomes somewhat less straightforward. Perhaps she meant to say diet rather than diet $\times$ age? There is indeed a rule that if there is no overall diet effect, one should not compare ages in a repeated measures design (unless this was part of the planned design and the ANOVA is used to generate the correct error term as we have done in our studies of visual acuity). The analysis shows there is no effect on novelty preference, only one of my objections to the citation of lower novelty preference at 12 months as a negative effect (Innis, this volume). On the other hand, there is a highly significant effect of diet on look duration and look number, allowing comparisons of diet groups at individual ages (although toward what end I am not sure).

Innis has commented to you about loss of infants from our second trial. The study was designed to be a comparison between healthy infants who did or did not receive n-3 LCPUFA. There was a single exception to this in that infants with bronchopulmonary dysplasia (BPD) who could maintain enteral feeding were retained and the data were analysed by diet–BPD group. There is no reason to suggest, as the statement implies, that healthy babies remaining in the n-3 LCPUFA group were healthier than those remaining in the control group. What we know of the effects of disease on the development of preterm infants is most reasonably attributed to the disease itself. It is not necessary to invoke a hypothesis that infants destined to score lower on developmental tests are also more likely to develop necrotizing enterocolitis (NEC) or sepsis. While this may be true, it seems unlikely and is certainly not supported by any evidence. On the other hand, you may legitimately be suspicious that the infants who were lost could have scored differently on developmental tests. This has been discussed as an issue of internal versus external validity that affects all of the randomized trials in preterm infants (Carlson, this volume).

We have said that there was no overall effect on weight, length or head circumference, but that there is a consistent decrease in weight-for-length. I believe this is the only fair way to present the data. Because the study had adequate power to measure growth, it is not appropriate to pull out random measurements at different ages that appear to be spurious

You mention the possibility of a type I error in undersized studies. Of course, one trial (whether large or small) cannot be regarded as proof of anything. As reported in my chapter, however, there has been replication of higher visual acuity and shorter look duration in the preterm studies that have been published. The higher Bayley mental development index score in one of our trials may have been corroborated in a trial by Damli et al. (2), but I am withholding judgment until I see a manuscript. Yes, type I errors are possible, but they are not usually considered probable in studies that reach the expected conclusion. And, when they are replicated in separate trials, the results must logically be considered to have validity.

It is important to point out that the sample sizes you mention for studies of long-term development are required only for determining differences given the influence of many variables that increase variables over time. Many investigators do not feel comfortable with, nor are they trained to do, such long-term studies, and may prefer instead to do carefully controlled trials that minimize variability.

It should also be pointed out that studies with very large numbers of infants often do not control for many variables or collect information many feel is important finally to claim that differences in nutrition are involved, e.g. biochemical measures of LCPUFA or other nutritional indices. Besides limiting the ability to interpret the findings in this way, the paucity of biochemical information severely restricts the number and quality of new hypotheses that are the goal of any good scientific study.

While I think it is important to resolve issues of safety, in studies of preterm infants this is more complicated than one might readily assume. I have discussed this in my chapter and in at least one other commentary and so will not repeat myself here. Suffice it to say that it is important to resolve concerns about safety, in particular related to NEC and sepsis. Toward this end, it is important for us to focus on what those concerns are. There are too many people raising spectres about sources of LCPUFAs without intent to refute or substantiate the charges. I find this a very disturbing intrusion of politics into science.

1. Werkman SH, Carlson SE. A randomized trial of visual attention of preterm infants fed docosahaenoic acid until nine months. *Lipids* 1996; **31**: 91–7.
2. Damli A, von Schenck U, Clausen U, Koletzko B. Effects of long-chain polyunsaturated fatty acids (LCPUFA) on early visual acuity and mental development of preterm infants. In: *American Oil Chemists Conference on PUFA in Infant Nutrition: Consensus and Controversies Program and Abstracts*, 7–9 November 1996. Champaign, Il: American Oil Chemists Society.

**Innis** You suggest it is 'speculative to extrapolate animals to human'. Why should the human be different from other animals? Is there an example of another nutrient which is essential in the diet of the human infant, but not essential in the diet of other species or adults. This might help illustrate your suggested need for caution. It would also be useful here to note how difficult it is to create an n-3 fatty acid deficiency ($\alpha$-linolenic acid) sufficient to cause measurable functional change in other species. Thus, even if DHA (or AA) is uniquely essential in the human infant, is it likely that we can create a biochemical deficiency of sufficient magnitude that there will be a measurable functional deficit?

The Fagan test is a separate test, with different cards and times for the test for infants of different ages. Thus, you cannot do a repeated measures study to consider a change in measured or derived parameters age. In the Werkman and Carlson paper (1) (page 94, Table 3), the diet $\times$ age term in the ANOVA is not significant. You state that look duration is shorter at 12 months. Is it valid to do a test for effects of diet when the ANOVA tells you there is no diet $\times$ age effect? Note, as the infants get older, the test allows less time (less time available for looking). Therefore the apparent 'significant improvement' with age cannot be interpreted as 'improvement' with age.

1. Werkman SH, Carlson SE. A randomized trial of visual attention of preterm infants fed docosahaenoic acid until nine months. *Lipids* 1996; **31**: 91–7.

**Wainwright**: This paper provides an informative and critical appraisal of the evidence currently available from studies in preterm and term infants on the putative relationship between LCPUFAs and cognitive function in humans. In so doing, Lucas emphasizes the importance in these trials of comparing formulae with adequate levels of LNA and LA with LCPUFA supplementation. He also provides cogent arguments for considering other possible outcomes, such as effects on immune functioning. Although I do agree with Lucas that ultimately decisions on whether LCPUFAs influence cognitive development in human infants must be based on studies conducted in humans, I nevertheless do think that there may be more to be gained from animal studies than he allows, particularly when it comes to issues of mechanisms, as well as those of toxicology and safety.

It is true that in my review, cited by Lucas, I question some of the behavioral effects reported in animals on methodological grounds (small sample sizes and inappropriate statistical treatment of the data). But I also have concerns about the interpretations applied to behavioural outcomes, particularly those relating to learning, on which I elaborate further in this volume. Providing that the study design takes into account species differences in the relevant parameters, I do believe that there remains a contribution to be made by animal studies in understanding the neurobiology of DHA. I think that this is well-illustrated in the chapter by Shaw and McEachern in this volume. The rat, for example, is a respected model in behavioral neuroscience, and one on which there is already a great deal of information highly relevant to illuminating the mechanisms of brain–behaviour relationships.

While the range of human cerebral DHA contents that is associated with normal function is currently unknown, we do have evidence from our recent work in mice that over a wide range of brain DHA levels (75–120% of control values), there appears to be little relationship with spatial learning ability as assessed in the place version of the Morris water maze (1).

1. Wainwright PE, Xing H-C, Mutsaers L, McCutcheon D, Kyle D. Arachidonic acid offsets the effects on mouse brain and behavior of a diet with a low (n-6):(n-3) ratio and very high levels of docosahexaenoic acid. *J Nutr* 1997; **127**: 184–193.

**Singer**: The difficulties inherent in proving that LCPUFAs are critical nutrients which have durable effects on cognitive function are summarized in this chapter. As in Innis' chapter, the possibility of deleterious effects, as well as the lack of conclusive studies of the benefits of LCPUFAs in developmental outcome, are comprehensively considered. Of note is the contrast in interpretation of the growth data with Heird, who noted that the lower growth rates found in preterm, LCPUFA-supplemented infants were not of greater concern, while this chapter and that of Innis indicate that the reduction in growth may be of clinical importance.

**Shaw and McEachern**: Lucas makes a strong case that the addition of particular molecules, e.g. LCPUFAs, could be more detrimental than beneficial, and that functional tests have focused on visual tasks and growth, while LCPUFAs may affect gene expression, gut function, insulin resistance, blood pressure, bone metabolism, immune function, etc. Innis (this volume) cites evidence that LCPUFAs may increase lipid peroxidation and evidence for DHA-induced oxidative stress on retinal cells has been cited (1). These latter points are potentially of great interest since free radical action may be particularly damaging to neurons (2) versus other cell types, and especially so in infants since antioxidant defences are not fully developed (3) in early life.

1. Neuringer M. The relationship of fatty acid composition to function in the retina and visual system. In: Dobbing J, ed. *Lipids, Learning, and the Brain: Fats in Infant Formulas*. Columbus, OH: Ross, 1996: 134–63.
2. Evans PH. Free radicals in brain metabolism and pathology. *Br Med Bull* 1993; **49**: 577–87.
3. Kudo H, Kokunai T, Kondoh, T, Tamaki N, Matsumoto S. Quantitative analysis of glutathione in rat central nervous system: comparison of GSH in infant brain with that in adult brain. *Brain Res* 1990; **511**: 326–8.

CARLSON: In the absence of evidence to the contrary, where potential negative effects of LCPUFAs are invoked one must also consider equally the potential for positive effects (e.g. gene expression, gut function, insulin resistance) that you mention here. The same may be said for the potential of LCPUFAs to induce oxidative stress on retinal cells. The reference you cite discussed the theoretical advantages as well as the disadvantages of DHA in this regard. For example, Reme *et al.* (1) showed that isolated retinas from marine-oil-fed rats had enhanced susceptibility to lipid peroxidation in vitro, but at the same time the retinas of intact animals fed marine oil were *less* rather then *more* susceptible to light-induced damage compared with controls. Just as one could hypothesize negative effects of higher DHA, one could hypothesize potential protective effects against light- or oxygen-induced damage, e.g. lower production of proinflammatory eicosanoids and platelet activating factor.

1. Reme CE, Malnoe A, Jung HH, Wei P, Munz K. Effect of dietary fish oil on acute light-induced photoreceptor damage in the rat retina. *Invest Ophthalmol Vis Sci* 1994; **35**: 78–90.

**Heird**: This appraisal of current knowledge concerning the essentiality of LCPUFAs in infancy includes a few misleading, or perhaps inadequately refuted, assumptions. In addition, it raises some philosophical issues that might be worthy of further discussion.

The first misconception concerns the adequacy during early development of the enzymes required for biosynthesis of AA and DHA in the central nervous system and liver. This is probably true for the human infant, particularly with respect to the liver, but all references cited are based on studies in either rats or single-cell systems and these results may or may not be applicable to the human infant. We (1) and others (2–4) have shown that both preterm and term infants can convert ALA and LNA to AA and DHA, respectively; however, these data are based entirely on measurements in circulating plasma lipids and, presumably, reflect hepatic synthesis. Whether conversion of LA and ALA to the longer-chain polyunsaturated fatty acids occurs in the central nervous system of the human species is not known.

Another possible misconception is the assumption by most, not necessarily you, that the greater rate of brain growth of preterm versus term infants necessitates a greater demand for LCPUFA. Based on the data of Martinez *et al.* (5), deposition of AA and DHA in the brain of term infants, expressed as mg/day, is only minimally less after 40 weeks' gestation than before.

Your discussion of the types of studies needed to establish the efficacy and safety of LCPUFA in infant formulas raises the philosophical issue of responsibility for resolving the many unresolved issues. Suppose, for example, that your large randomized trial demonstrates an advantage of whatever combination of LCPUFAs the trial is evaluating and, further, that the supplement evaluated is safe. This does not mean that all combinations and/or all sources of LCPUFAs are equally efficacious or safe. Are similar studies required for every source of supplementation and/or every combination of LCPUFA supplementation? Or, does whatever combination and/or source your study demonstrates to be efficacious and safe become the 'standard' for supplementation? Assume, on the other hand, that your large study demonstrates no advantages of LCPUFA supplementation but also no safety concerns. This type of result, regardless of your sample size, is unlikely to resolve the issue of whether some infants might benefit from LCPUFAs. In this situation, are further studies required? If so, what is the responsibility for companies versus government agencies to support the required studies?

Finally, you imply that formula companies are responsible for designing and supporting the required studies. I would argue that it is the responsibility of agencies such as the National Institutes of Health in the USA and the Medical Research Council in the UK to support studies to demonstrate essentiality of a specific nutrient. Once essentiality (or, even, desirability) is determined, it is the manufacturer's responsibility to demonstrate safety of whatever component it chooses as the source of the essential (or desirable) nutrient in its formula.

Your discussion of safety issues is particularly relevant. As you point out, none of the published studies of LCPUFA supplementation, except that by Uauy *et al.* (6), has even remotely addressed this issue. Considering the difficulty of using the methodologies currently available for assessing visual and neurodevelopmental function to establish with certainty that LCPUFA supplementation is efficacious, many might (and do) argue that the circumstantial evidence for efficacy is sufficient, provided addition of LCPUFA to formulas is 'safe'. In this regard, studies focusing on the incidence of diseases such as bronchopulmonary dysplasia, necrotizing enterocolitis, sepsis, etc., are important, but so are studies addressing more specific outcomes such as immune function, eicosanoid metabolism, insulin resistance, etc. For example, it is conceivable that LCPUFAs could enhance visual and/or central nervous system development, as well as increase the incidence of necrotizing enterocolitis, as you suggest. However, unless the study also included some mechanistic outcomes, it provides no insight into possible strategies to optimize the beneficial effects by overcoming the undesirable ones. Hopefully, your large study that is now in progress will help resolve some of these issues.

1. Sauerwald TU, Hachey DL, Jensen CL, Chen H, Anderson RE, Heird WC. Effect of dietary α-linolenic acid intake on incorporation of docosahexaenoic and arachidonic acids into plasma phospholipids of term infants. *Lipids* 1996; **31**: S131–5.
2. Demelmair H, Rinke U, Behrendt E, Sauerwald T, Koletzko B. Estimation of arachidonic acid synthesis in full term neonates using natural variation of $^{13}$C-abundance. *J Pediatr Gastroenterol Nutr* 1995; **21**: 31–6.
3. Carnielli VP, Wattimena DJL, Luijendijk IHT, Boerlage A, Degenhart HJ, Sauer PJJ. The very low birth weight premature infant is capable of synthesizing arachidonic and docosahexaenoic acid from linoleic and linolenic acid. *Pediatr Res* 1996; **40**: 169–74.
4. Salem N Jr, Wegher B, Mena P, Uauy R. Arachidonic and docosahexaenoic acids are biosynthesized from their 18-carbon precursors in human infants. *Proc Natl Acad Sci USA* 1996; **93**: 49–54.
5. Martinez M, Pineda M, Vidal R, Martin B. Docosahexaenoic acid: a new therapeutic approach to peroxisomal patients. Experience with two cases. *Neurology* 1993; **43**: 1389–97.
6. Uauy, RD, Birch DG, Birch EE, Tyson JE, Hoffman DR. Effect of dietary

omega-3 fatty acids on retinal function of very-low-birth-weight neonates. *Pediatr Res* 1990, **28**: 485–92.

**McCall and Mash**: We have only two thoughts regarding this paper, one pertaining to the concept of 'essentiality' and the other with respect to the interpretation of human- versus formula-fed infants and mental development.

Lucas attempts to deal with whether LCPUFAs are essential for human development, coming to the conclusion that the data do not warrant a firm conclusion. But I wonder whether a conclusion of essentiality would lead to a recommendation to supplement infant formula? Even if LCPUFAs were essential for human development, how much is essential and when is it essential to have such amounts? Is more of an essential element better, or could it be worse?

Lucas describes the best argument available from his own data for the possible role of LCPUFAs in improving intelligence at age 18 months. He reports, for example, that infants whose mothers had chosen not to provide human milk were randomly assigned to receive banked human milk versus term formula, and those who received the human milk had a 9 point IQ advantage at 18 months of age. However, if I read these papers correctly, two other facts bear on the implications of this finding. First, those infants who were fed banked human milk did not show any advantage over those infants fed the enriched preterm formula. This comparison suggests that the advantage conferred by human milk could also be conferred by enriched nutrition without LCPUFAs; but the second comparison shows that the LCPUFAs are not necessary to produce the same result. Second, in all the other comparisons at all of the ages at which infants were later assessed mentally, infants whose mothers chose not to provide human milk but were fed banked human milk did not show any advantage over those infants who were fed with appropriately enriched formulas. It seems to me that LCPUFAs are not essential to providing what benefits were observed, and in most of the strongest comparisons human milk did not provide advantage over appropriately enriched formulas with respect to mental performance.

**Neuringer and Reisbick**: Full term infants have rapid brain growth postnatally also, just less of it. The more critical point is that preterms are at an earlier, and therefore probably more vulnerable, stage of brain development when they become dependent on dietary lipid supply.

With regard to the absence of effects on retinal DHA in the Makrides study, it is important to appreciate that about 90% of the DHA in the retina is in photoreceptor outer segment membranes, which are extremely fragile and deteriorate within hours post mortem. Therefore post-mortem analysis of retinal fatty acid composition must be viewed with skepticism. Fatty acid levels in post mortem cortical tissue are more stable. The fact that cerebral cortex AA was higher in the formula-fed infants should not be treated as surprising or failing to confirm the decrease in DHA; it is what one would expect, as n-3 and n-6 LCPUFAs generally change reciprocally. Replacement of DHA with n-6 LCPUFA including AA, 22:4n-6 and 22:5n-6 (all of which increased in the Farquharson study) is the hallmark of n-3 fatty acid deficiency, resulting in conservation of the degree of unsaturation.

Although there are no functional data on the same infants, the compositional data from these infants are the best available estimate (and probably a fairly accurate one) of brain composition in living infants fed similar formulas and for whom functional data are available. It is unlikely that it would ever be possible to get both kinds of information on the same human infants. Also, comparable information for preterm infants is not available: data have been published for only four formula-fed preterm infants, two in the Farquharson paper and two in Martinez's paper (1). Their DHA levels were substantially lower than for full-term infants, but poor food intake or absorption could have been factors. Nevertheless, it would be expected that DHA levels would be more affected in preterm infants. The bottom line is that formula-fed infants do appear to have significantly lower cortex DHA than breast-fed infants. Of course, as you state, the more critical question is whether this compositional change is functionally significant.

1. Martinez M. Tissue levels of polyunsaturated fatty acids during early human development. *J Pediatr* 1992; **120**: S129–38.

**Appelbaum**: At what stage do you feel we are in research on LCPUFAs?

AUTHOR'S REPLY: In assessing evidence concerning any dietary need for LCPUFAs in infancy, it is helpful to examine the stage of evolution of this field. As we see it, the development of a new area of clinical therapeutic science is generally a three-stage process. In stage I, early observations raise the question 'Is this worth pursuing?' If this is thought to be the case, the field enters stage II, in which pertinent observational data are collected, epidemiological and mechanistic studies undertaken, animal experiments, including toxicological evaluations, may take place, and small, pilot therapeutic trials may be performed. If such studies generate clinically important hypotheses, these may be taken forward into stage III, in which more definitive, robust intervention experiments are undertaken. The key objective in stage III is to address the generic question: 'Does the intervention really *matter* in terms of human health or development?' The critical issues to be resolved in stage III studies (generally randomized, controlled and blinded trials) are the establishment of *clinical* efficacy and also safety of the proposed intervention. Generally a change in clinical or public health practice can only be defended by stage III studies. Indeed, in a complex field, numerous stage III studies are required to examine a range of efficacy and safety issues, to compare a range of specific therapeutic strategies, and often to identify those subgroups that serve to benefit most from the proposed intervention. Continued work in stage II is usually required to underpin new stage III interventions.

Research into the treatment of high blood pressure is a good example of a field that has progressed over several decades through these stages. As a result of stage III research, we now know that treatment of high blood pressure does indeed matter in terms of reduction in morbidity and increase in survival. Moreover, the efficacy and safety of a range of therapeutic agents are now well researched and there are considerable data on which population subgroups would benefit from treatment.

Given this preamble, the question now is: 'What stage are we at in the field of LCPUFA supplementation in infancy?' In fact none of the published data are in stage III. The information that currently exists is typical of stage II. It concerns demographic and descriptive data; mechanistic studies in animals; physiological and biochemical investigations; and small, pilot therapeutic trials in humans. Based on these studies, clinically relevant hypotheses have been and are being formulated which are suitable for stage III research. The decision to move into stage III is, however, a complex one. A high level of evidence is required to take this plunge because of the major investment in resources required, and because many other fields entering stage III will legitimately compete for such resources. Moreover, a premature move into stage III poses potentially serious issues. The most plausible outcomes may not have been worked out and, more importantly, essential background work, including toxicological studies in animals, may not yet have fully paved the way for large-scale trials in humans.

Around 6 years ago, one major European committee (ESPGAN) recommended that LCPUFAs should be put into preterm formulas. Nowadays clinical nutrition policy tends more to be based on stage III studies. However, back in 1991 nutritional policy and discussion on what should be regarded as semi-essential nutrients tended to be based on stage II data, including biochemical findings. The ESPGAN recommendations were acted on, and now all preterm formulas in continental Europe contain LCPUFAs. Given this move, stage III studies immediately became important and urgent to demonstrate efficacy and more importantly safety in a field that had 'taken off' somewhat faster than it would have done today. Indeed now, in 1997, North American and British committees are still actively debating what recommendations would be justified by current data.

# Long-chain Polyunsaturated Fatty Acid Supplementation of Preterm Infants

SUSAN E. CARLSON

*Departments of Pediatrics, Obstetrics, Gynecology & Biochemistry, University of Tennessee – Memphis, 853 Jefferson Avenue, Room 201, Memphis, TN 38163, USA*

Introduction

Studies published 15 years ago established that term infants fed commercially available infant formulas had significantly lower erythrocyte long-chain polyunsaturated fatty acids (LCPUFAs) compared with infants fed human milk (1,2). Initially, it was not clear why this difference occurred. Although samples of human milk, but not infant formula, contained LCPUFAs, their amounts were very small compared with the amounts of their essential fatty acid precursors (1,2,3). Later, feeding studies established that even small amounts of dietary LCPUFAs when fed continuously had quite large effects on erythrocyte LCPUFAs. Other variables were also found to influence erythrocyte LCPUFAs, but to a smaller degree.

Preterm infants also had declines in erythrocyte LCPUFAs after birth, and, like term infants, they had higher LCPUFAs when fed human milk instead of commercially available formulas (4). It was already known that brain and retinal phospholipids were highly enriched in the n-3 fatty acids, in particular docosahexaenoic acid (5,6), and that the last intrauterine trimester was very important for brain docosahexaenoic acid accumulation (7). Other reports showed that fatty acids of the n-3 family, in particular, were necessary for normal retinal physiology (8,9), visual acuity (10), and behavior (11–13). These observations led to speculation that lower erythrocyte docosahexaenoic acid in formula-fed infants could be a sign of lower docosahexaenoic acid accumulation in tissues such as the brain and retina and, therefore, lead to suboptimal sensory and cognitive function. Based on the results of studies in animals (8–13), we proposed that randomized trials of docosahexaenoic acid supplementation and functional outcomes such as visual acuity in supplemented infants were needed to determine if declining postnatal erythrocyte phospholipid docosahexaenoic acid was related to suboptimal docosahexaenoic acid status (4). There have now been three published trials of LCPUFAs supplementation in preterm infants, and these will be discussed in this chapter.

Biochemists are taught that 'structure is function', but, not uncommonly, they take the view that the presence of an enzyme pathway for synthesis eliminates any further concern for normal structure. Nutritionists conceived the idea that under certain physiological conditions a compound may become a conditionally essential nutrient; i.e. despite the presence of an enzymic pathway, the compound may be synthesized in amounts less than optimal for normal function, and, therefore, may be needed in the diet. Several compounds have been considered conditionally essential strictly on the basis of blood levels. However, in the strictest sense some lower function is considered necessary. Conditional essentiality is more likely to apply to immature than mature groups of humans. The question of conditional essentiality of LCPUFAs seemed to apply in the case of the formula-fed preterm infants because the biochemical evidence of poor docosahexaenoic acid status (decreasing erythrocyte docosahexaenoic acid) occurred even though formulas contained linolenic acid, the nutritionally essential precursor of docosahexaenoic acid (Fig. 1). In this regard, formula-fed infants differed significantly from animals that had poor docosahexaenoic acid status because they consumed diets deficient in α-linolenic acid.

## Essential Fatty Acids and Formation of LCPUFAs

There are three main families of unsaturated fatty acids (Fig. 1): the n-9 (ω-9), the n-6 (ω-6), and the n-3 (ω-3) fatty acids. The numerical designation of each indicates the position of the first double bond from the methyl

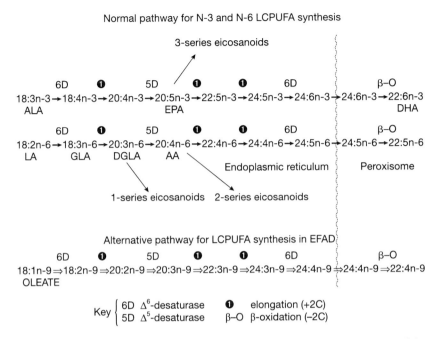

Fig. 1. Pathway for elongation and desaturation of the nutritionally essential fatty acids revised according to Sprecher and co-workers (28,29) and indicating the sources for the series of eicosanoids. Beyond 20:3n-9, the alternative pathway is speculative.

end of the fatty acid. The parent fatty acids of these respective families are oleic acid, linoleic acid and α-linolenic acid. Oleic acid, linoleic acid, and α-linolenic acid are frequently described with the numerical designations 18:1n-9, 18:2n-6, and 18:3n-3, respectively, indicating that they have 18 carbon atoms with, respectively, 1, 2, and 3 double bonds positioned at the ninth (n-9), sixth (n-6), and ninth (n-6), and third, sixth and ninth (n-3) carbon atoms from the methyl end of the fatty acid molecule (Fig. 1). Plants can synthesize oleic acid, linoleic acid, and α-linolenic acid because they have enzymes that catalyze the insertion of double bonds (desaturases) in these positions. Of these three fatty acids, however, only oleic acid can be synthesized by higher animals.

Higher animals and humans must consume linoleic acid and α-linolenic acid, but they have the enzymes required to synthesize their longer, more unsaturated products (n-6 and n-3 LCPUFAs) through a sequence of alternating chain desaturation and elongation reactions occurring at the carboxy end of the molecule (Fig. 1). Because the methyl end of the molecule is not altered during mammalian elongation and desaturation, the LCPUFA products of desaturation and elongation retain the same double bond configuration (n-3, n-6, and n-9) as their parent fatty acids. Little conversion of oleic acid to n-9 LCPUFAs, mainly 20:3n9, occurs by the pathway shown in Fig. 1 unless the diet is deficient in linoleic acid and α-linolenic acid. If linoleic acid and α-linolenic acid are deficient, the ratio of this triene, 20: 3n-9, to the n-6 tetraene, 20:4n-6, increases. In plasma, a triene/tetraene ratio of $\geq$ 0.4 has been used clinically as evidence of essential fatty acid deficiency.

Human milk generally contains between 6% and 14% of total fatty acids as linoleic acid and 0.6–2.0% of total fatty acids as α-linolenic acid (the approximate range for per cent of energy can be determined by decreasing these numbers by half). In addition, human milk also contains 0.3–1.0% arachidonic acid and 0.1–0.9% docosahexaenoic acid as well as other 20 and 22 carbon atom n-6 and n-3 LCPUFAs (14). Infants maintained on a skimmed milk diet developed large and frequent stools and progressively more severe skin changes from dryness to thickening and later desquamation and oozing (15). These changes were reversed by a diet with 2.0% of energy intake from linoleic acid. In 1985, American Academy of Pediatrics Committee on Nutrition (AAP/CON) recommended linoleic acid and α-linolenic acid intakes of at least 2.7% and 0.3% of energy (16). These recommendations were not changed in the 1993 revision (17). Based on these numbers, a typical infant formula with half of its energy from fat would need to have at least 5.4% and 0.6% of total fatty acids as linoleic acid and α-linolenic acid, respectively. Functional studies published since these recommendations were made suggest that 0.3% energy from α-linolenic acid is probably too low even for term infants (18–20). In 1992, α-linolenic acid recommendations of 1.0% of energy were made for preterm infants (21). Since then, intakes in this range (1.35–2.5% energy) have been found to result in lower visual acuity, longer look duration, and lower scores on the Bayley mental developmental index compared with formulas containing α-linolenic acid and as little as 0.1% of energy from docosahexaenoic acid (22–27), evidence that docosahexaenoic acid is a conditionally essential nutrient for preterm infants.

Until recently, mammals were believed to have three microsomal desaturases, the $\Delta^6$-, $\Delta^5$-, and $\Delta^4$-desaturase, that catalyzed the insertion of double bonds in linoleic acid and α-linolenic acid to form LCPUFAs. The putative $\Delta^4$-desaturase has not been confirmed; however, synthesis of docosahexaenoic acid (22:6n-3 or $\Delta$4,7,10,13,16,19–22:6) and docosapentaenoic acid (22:5n-6 or $\Delta$4,7,10,13,16–22:5) have been shown to occur via a pathway that adds two sequential two carbon atom units (elongations) to the respective n-3 and n-6 family products of $\Delta^5$-desaturase, eicosapentaenoic acid (20:5n-3) and arachidonic acid (20:4n-6) (28). The 24 carbon atom intermediates thus formed (24:5n-3 and 24:4n-6) receive one more double bond at the $\Delta^6$-position to form 24:6n-3 and 24:5n-6 before they are transferred to the peroxisomes to undergo partial β-oxidation, which removes two carbon atoms from each molecule to form 22:6n-3 (docosahexaenoic acid) and 22:5n-6, respectively (Fig. 1). Because peroxisomal enzymes are not available to continue β-oxidation of 22:6n-3 and 22:5n-6, their transfer out of the peroxisome to the endoplasmic reticulum is favored and they become available for construction of membrane lipids (Fig. 1) (29).

## Functions of LCPUFAs

The essential n-6 and n-3 fatty acids are found in the membranes of all organs and tissues. Arachidonic acid (20:4n-6) and docosahexaenoic acid (22:6n-3) are the predominant membrane LCPUFAs from these fatty acid families. All organs and tissues have fatty acid profiles with some unique characteristics due to tissue enzyme specificity, but each tissue is also subject to effects of other variables, such as differences in n-3 and n-6 fatty acid intake. The high proportion of total fatty acids found as docosahexaenoic acid and arachidonic acid in brain and as docosahexaenoic acid in the retina encouraged investigators to identify their functional roles. For the most part, these studies were done by studying the effects of decreasing their presence in brain and retina through the use of essential fatty acid deficient or n-3 fatty acid deficient diets.

### Essential Fatty Acid Deficiency

It is possible to decrease brain docosahexaenoic acid and arachidonic acid by feeding fat-free diets or fats deficient in both α-linolenic acid and linoleic acid to growing animals. Such diets are referred to as essential fatty acid deficient diets and the condition they cause is referred to as essential fatty acid deficiency. The earliest studies of essential fatty acid deficiency used diets deficient in both linoleic acid and α-linolenic acid; however, we know now that the symptoms reported for essential fatty acid deficiency were due mainly to n-6 fatty acid deficiency. These n-6 fatty acid deficiency symptoms included growth retardation, loss of dermal integrity, poor wound healing, reproductive failure, liver and kidney degeneration (30), and increased susceptibility to infection (31).

## n-3 Fatty Acid Deficiency

The variability in linoleic acid and α-linolenic acid content among plant oils also makes it possible to create n-3 fatty acid deficiency without n-6 fatty acid deficiency. In specific n-3 fatty acid deficiency, the retina and brain docosahexaenoic acid decreases and 22:5n-6, a structurally similar fatty acid with one less double bond originating from linoleic acid, increases. Safflower oil, sunflower oil, peanut oil and corn oil are examples of plant oils that are deficient in α-linolenic acid. The biochemical and physiological effects of essential fatty acids have been determined by feeding essential fatty acid deficient and n-3 fatty acid deficient diets to animals. The results of these studies have directed us to outcomes that have been meaningful for studies to determine if docosahexaenoic and arachidonic acid are conditionally essential in human infants (9,10,32), because failure to accumulate these acids in tissues is the basis for the effects of these fatty acid deficiencies.

Unlike essential fatty acid deficiency, n-3 fatty acid deficiency does not result in retardation of growth, reproductive failure, or visible signs of ill health, so the functional effects of n-3 fatty acid deficiency went unrecognized for nearly 50 years after n-6 fatty acids were known to be important physiologic mediators.

Lamptey and Walker (11) were the first to report behavioral effects of feeding an n-3 fatty acid deficient diet. In their studies, the n-3 deficient rats had fewer correct responses in a Y-maze requiring black and white discrimination to locate food. More than 10 years later, analogous studies by Yamamoto *et al.* (12,13) also found proportionately fewer correct responses in rats fed n-3 fatty acid deficient diets using a brightness discrimination learning task for food reinforcement. These and other early animal studies of n-3 fatty acid deficiency established the importance of n-3 fatty acids in behavior; however, the interpretation of these visually mediated tasks is equivocal because other studies have found sensory impairment with n-3 fatty acid deficiency. Furthermore, many of the results of long-term n-3 fatty acid deficiency can be explained by other performance variables such as motivation or reactivity to stimuli (33).

Many of the animal studies of n-3 fatty acid deficiency have required detection or discrimination of sensory cues. n-3 Fatty acid deficient animals with lower performance are then interpreted as having poorer learning (cognition) when, in fact, their poor performance could be due to sensory deficits from n-3 fatty acid deficiency. Benolken, Wheeler, and colleagues

(8,34) first linked n-3 fatty acid deficiency to changes in the retinal response to light, a sensory deficit, and showed that α-linolenic acid was better able to reverse this effect of a fat free diet than was linoleic acid. Neuringer and colleagues (9,10) subsequently studied both retinal function and early visual acuity in rhesus monkeys. Neuringer (35) has pointed out that their choice of the rhesus monkey for studies of n-3 fatty acid deficiency was based in part on the many similarities between the monkey and human visual systems, including the relatively high proportion of cone photoreceptors (necessary for daylight vision, perception of detail, and color vision) and the high degree of retinal centralization, allowing sharp visual acuity, features that also characterize humans. In their studies, n-3 fatty acid deficient monkeys had lower a-wave amplitudes of both cone and rod responses but no effect on b-wave amplitude early in infancy. These differences in a-wave amplitude were not found at 2 years of age. However, other differences in the retinal responses in n-3 fatty acid deficient monkeys, notably an increased latency to the peak of the b-wave and slower recovery of the electroretinograms to repeated stimulation, became larger with age and did not recover even after the animals were fed docosahexaenoic acid to produce a normal retinal fatty acid profile (37).

Neuringer and colleagues (10) also measured visual acuity in n-3 fatty acid deficient and n-3 fatty acid fed control monkeys using forced-choice preferential looking at 1, 2, and 3 months of age. The method used was similar to that originally developed by Teller *et al.* (38) for research testing of human infants. Compared with the control group, visual acuity in the n-3 fatty acid deficient group of monkeys was about one octave lower at 2 and 3 months of age, the oldest ages tested, and the lower acuity was not due to refractive errors. Besides reduced retinal and visual function, n-3 fatty acid deficient monkeys have polydipsia/polyuria (39). Recently, Reisbick *et al.* (32,40) have also found that n-3 fatty acid deficient monkeys compared to controls have a longer look duration during visual paired-comparison tests that occurs in the absence of effects on visual recognition memory. These effects of lower docosahexaenoic acid status in the rhesus monkey, notably impaired retinal function, lower visual acuity, and increased look duration during visual paired comparison tests, have also been related to poorer docosahexaenoic acid status in preterm infants, as will be discussed later in this chapter.

### Relationship between Disease and LCPUFAs

The peroxisomal diseases, including Zellweger's syndrome, neonatal adrenoleukodystrophy and infantile Refsum's disease, result in lower docosahexaenoic acid in the human central nervous system as well as in other tissues (41). In the peroxisomal disorders, very long chain 24 and 26 carbon atom saturated and monounsaturated fatty acids accumulate in plasma and organs because liver peroxisomes are the main site for their β-oxidation. At the same time, docosahexaenoic acid and other 22 and 24 carbon atom fatty acids decrease in the brain or fail to accumulate in brain myelin. The discovery by Sprecher and co-workers (28,29) that peroxisomes are needed for docosahexaenoic acid synthesis probably explains why docosahexaenoic acid accumulates so poorly in these disorders (41) and may explain, at least in part, the severe neurological and motor impairments that are characteristic of these disorders.

Recently, Martinez (42) has reported increased erythrocyte docosahexaenoic acid and dramatic improvements in vision, liver function, muscle tone, and social interaction among patients with less severe variants of Zellweger syndrome who were fed docosahexaenoic acid. The changes in vision were not assessed using standardized procedures for acuity. Rather, infants presumed blind began to follow light and objects. One of the most interesting and surprising effects of feeding docosahexaenoic acid has been the change toward normal of several signs of abnormal lipid metabolism that characterize these diseases. With docosahexaenoic acid intake, very long chain fatty acids in plasma decreased, the brain showed increases in myelination on magnetic resonance imaging, and erythrocyte plasmalogen increased toward normal (42). Other chronic diseases have also been associated with lower plasma and erythrocyte docosahexaenoic acid and/or arachidonic acid, evidence of poorer LCPUFA status, with diets that are not essential fatty acid deficient. For example, arachidonic acid and/or docosahexaenoic acid have been reported to be lower in diabetes (43), prematurity (44), and attention deficit hyperactivity disorder (45). When low membrane LCPUFAs are found associated with a disease, it is not known if lower LCPUFAs are a cause or the result of the disease, and only randomized trials of LCPUFA supplementation can answer this question.

*LCPUFAs and Growth*

It has been nearly 70 years since total essential fatty acid deficiency was found to retard growth in animals (46) and 40 years since a similar observation was made in human infants (15), but the molecular mechanisms by which essential fatty acid deficiency decreases somatic growth are still unclear. The most likely mechanism seems to be an increase in resting metabolism. Mitochondrial respiratory studies have found impaired norepinephrine-stimulated oxidative capacity in essential fatty acid deficiency, with some evidence pointing to changes in mitochondrial membrane function as a cause (47). Normally the cellular mitochondria couple the β-oxidation of fatty acids with production of energy in the form of adenosine triphosphate through a process known as oxidative phosphorylation. When oxidation is uncoupled from adenosine triphosphate production, energy is wasted as heat and so is unavailable for growth. Sellmayer *et al.* (48) have also reported that arachidonic acid, but not n-3 LCPUFAs, can directly affect growth by affecting the expression of the growth-related early genes *c-fos* and *Egr*-1 via prostaglandin $E_2$ (a lipid-derived mediator from arachidonic acid) and subsequent activation of protein kinase C. Diets with arachidonic acid more rapidly reverse the growth effects of essential fatty acid deficiency than do diets with linoleic acid, thus providing evidence that the effects of essential fatty acid deficiency on growth are related to arachidonic acid status.

*LCPUFAs as a Source of Lipid-derived Mediators*

Not only do LCPUFAs play a structural role in biomembranes, as related earlier, but they also serve as substrates for some lipid-derived mediators that have critical roles in both cell–cell communication and signal transduction. Lipid-derived mediators include the eicosanoids (prostaglandins, prostacyclin, leukotrienes, and lipoxins), platelet activating factor, diacylglycerides, and ceramide. These compounds act like hormones in that they are present in very low concentrations, bind to specific receptors, and produce large biological effects, but, unlike hormones, their action is local, and they are cleared from the circulation very rapidly. The release of a molecule of fatty acid from phospholipid or sphingomyelin by a phospholipase is the first step in the formation of lipid-derived mediators. Cytokines help direct the production of specific lipid-derived mediators, which can play a role in both normal and abnormal physiological responses.

The specific LCPUFA released by the action of a phospholipase further determines the lipid-derived mediator that is formed and, therefore, the physiological response of the organism. Arachidonic acid is the predominant substrate, but dihomo-γ-linolenic acid (20:3n-6) and eicosapentaenoic acid are also sources of independent series of mediators that can have quite different physiological effects compared with those derived from arachidonic acid. There have been a number of attempts to modify physiological responses in humans by modifying the balance of membrane arachidonic acid and dihomo-γ-linolenic acid or arachidonic acid and eicosapentaenoic acid with dietary dihomo-γ-linolenic acid or eicosapentaenoic acid. Many of these studies have involved inflammatory disease because mediators from arachidonic acid are more likely to be pro-inflammatory or proliferative and those from dihomo-γ-linolenic acid and n-3 LCPUFAs to be anti-inflammatory and anti-proliferative (49).

The physiological effects of eicosanoids depend upon their balance with other eicosanoids (e.g. prostaglandins and leukotrienes), the balance among the eicosanoid series (those derived from arachidonic acid, eicosapentaenoic acid, and dihomo-γ-linolenic acid), and the organ or tissue studied. Among the physiological effects mediated by the arachidonic acid derived eicosanoids are modulation of vascular tone, cytoprotection of mucosal cells, the inflammatory response, the production of pain and fever, the regulation of blood pressure, the induction of blood clotting, the control of reproductive function, and regulation of the sleep/wake cycle.

## Sources of LCPUFAs in Preterm Infants

Three primary factors determine the final accumulation of docosahexaenoic acid and arachidonic acid in any given tissue or organ of an infant: (i) the supply of α-linolenic acid and linoleic acid and the synthesis of LCPUFAs from them by the fetus/infant; (ii) the intrauterine transfer of LCPUFAs from mother to fetus; and (iii) the postnatal transfer of LCPUFAs from mother to infant in human milk. A fourth option, the inclusion of LCPUFAs in artificial feedings, has been considered by several randomized trials of LCPUFA supplementation of preterm infants.

### Synthesis

Mammalian brain and liver can synthesize n-6 and n-3 LCPUFAs from their precursors (50,51). Recently, several groups of investigators have proven conclusively that the conversion of α-linolenic acid to docosahexaenoic acid and linoleic acid to arachidonic acid occurs in preterm infants (52–54). Although these stable isotope studies have contributed to our understanding of the process of LCPUFA biosynthesis, they have not been able to quantify the amount of LCPUFA that could come from synthesis. However, studies have shown marked variability in LCPUFA biosynthesis among preterm infants fed the same diet (53). Earlier, plasma phospholipid concentrations of arachidonic acid and docosahexaenoic acid were observed to vary several fold among preterm infants fed the same diet at all ages up to 12 months (55). While these data were consistent with highly variable synthesis among preterm infants, possible differences in lipid transport, lipid uptake into tissues, and intrauterine LCPUFAs accumulation could not be discounted. Moreover, other factors such as birth weight and length of parenteral nutrition might have had some influence at early ages. The evidence of variability in LCPUFA biosynthesis suggests that some infants might benefit from the inclusion of LCPUFAs in the diet more than others.

## Maternal-to-fetal Transfer of LCPUFAs

In human infants the most rapid rate of neuronal n-3 and n-6 LCPUFAs accumulation occurs during the last intrauterine trimester, between 26 and 40 weeks' postmenstrual age (7,41), and in the early months after birth (41). LCPUFAs accumulate by increases in both brain size and brain LCPUFA concentration. The fetal circulation has significantly higher levels of DHA and AA than the maternal circulation, with lower levels of linoleic acid and α-linolenic acid (40,56), suggesting that the fetus selectively obtains LCPUFAs from the mother. This idea is further supported by a study demonstrating cord blood arterial–venous differences in erythrocyte arachidonic acid and docosahexaenoic acid, but not in linoleic acid and α-linolenic acid (57). In general, both arachidonic acid and docosahexaenoic acid increase in cord blood plasma phospholipids as infants approach term (58,59), but, after birth, LCPUFAs decline in plasma and red blood cell phospholipids, regardless of gestational age at birth (4). These observations suggest that LCPUFAs are transferred to the infant during the last trimester and that the transfer is interrupted by birth. Most infants born at 26 weeks' postmenstrual age survive but miss the opportunity for maternal-to-fetal transfer of LCPUFAs.

Regardless of their gestational age, infants dying in the last intrauterine trimester have considerable individual variability in brain n-3 and n-6 LCPUFA accumulation (7,41). The causes of this variability are unknown, but it could be related to the variability in cord blood n-3 and n-6 LCPUFAs. This variability might also be a determinant of the inconsistent functional responses of term infants to LCPUFA supplementation among randomized trials (60–62). There is no evidence that α-linolenic acid and linoleic acid are inadequate in the diets of Western women whose infants have been studied, and some have considered that differences in apparent LCPUFA status at birth might be the result of differences in maternal LCPUFA intake. Given the multiplicity of patterns of food intake worldwide, it is reasonable to suggest that LCPUFA intake is quite variable; however, no one has attempted to compare brains of infants whose mothers can be expected to have consumed very different amounts of n-3 LCPUFAs during pregnancy.

Now that dietary sources of arachidonic acid and docosahexaenoic acid are becoming available, prospective studies of the biochemical and functional effects of varying maternal LCPUFA intake during pregnancy are feasible. Only two studies have reported the effects of fish intake on maternal and cord blood erythrocyte LCPUFAs, and they have found apparently opposite effects of n-3 LCPUFA intake on maternal and infant n-3 LCPUFAs. The population that routinely consumed fish (63) did not respond to gradations in fish intake with similarly graded maternal docosahexaenoic acid or infant arachidonic acid and docosahexaenoic acid, suggesting a possible ceiling effect. However, plasma and erythrocyte docosahexaenoic acid were increased in mothers eating a normal American diet and their infants when one gram of docosahexaenoic acid/day was consumed from sardines and additional fish oil for 9 weeks of pregnancy (64). Reisbick and Neuringer (33) have supported the idea of using natural and dietary-based variations in LCPUFAs of mother's milk and infant (and child) tissue levels to test hypotheses about the effects of fatty acids on behavior.

There is some evidence that variables other than differences in LCPUFA intake influence maternal and infant LCPUFA status. These include: (i) pregnancy and lactation history (65); (ii) smoking and alcohol use (59); and (iii) a number of variables with the potential to influence infant LCPUFA status (e.g. zinc status (66), *trans* fatty acid intake (67–69), pre-eclampsia and other causes of hypertension (70), diabetes (43), and differences in biosynthetic ability (53)). We noted a highly significant correlation between maternal height and infant plasma arachidonic acid concentration long after birth (24), which suggests that differences in the ability to synthesize n-6 LCPUFAs may also be genetically determined.

*Maternal-to-infant Transfer in Human Milk*

Transfer of LCPUFAs from mother to infant in human milk can be regarded as a continuation of the maternal LCPUFA contribution that begins in utero. The same variables that influence the availability of LCPUFAs in utero may also influence the amount of LCPUFAs in human milk. When formula-fed infants were first noted to have lower erythrocyte arachidonic acid and docosahexaenoic acid compared with infants fed human milk (1,2), there was little known about n-3 and n-6 LCPUFA variability in human milk. In the past 15 years, there have been many reports of both milk fatty acid composition and of plasma and erythrocyte arachidonic acid and docosahexaenoic acid of infants after milk and formula feeding. The highest levels of milk docosahexaenoic acid have been found in populations consuming high n-3 LCPUFAs from marine mammals (71) and in women supplemented with fish oil containing docosahexaenoic acid (72), while the lowest have been reported in vegan vegetarians (73). Jensen *et al.* (14) have compiled data from the literature on the LCPUFA content of milk samples. No equivalent compilations seem to exist for the published data on infant erythrocyte arachidonic acid and docosahexaenoic acid after feeding human milk and formula.

Do infants who consume milk with 0.4% docosahexaenoic acid accumulate four times as much docosahexaenoic acid from milk as infants who consume milk with 0.1% docosahexaenoic acid? Do high intakes of LCPUFAs shut down new synthesis while lack of it enhances synthesis with little difference in the final concentration? If docosahexaenoic acid and arachidonic acid are retained in higher amounts with higher intakes of LCPUFA intake, where do they go? The increase in brain docosahexaenoic acid in the two reports of mostly term infants fed human milk compared with formula suggest that docosahexaenoic acid from human milk could produce an increase of about 20% in brain docosahexaenoic acid compared with formulas containing higher α-linolenic acid (74,75), although bigger differences were found with low α-linolenic acid formula (74) and might be expected also in preterm infants. Furthermore, storage of samples prior to analysis could have decreased the absolute amount of LCPUFAs and reduced true differences between groups. Clandinin *et al.* (76) calculated that term infants accumulated quite a large proportion of their total body stores of LCPUFAs in adipose tissue. Storage of LCPUFAs above what is needed in adipose tissue could create a reservoir of LCPUFAs available for growing infants and children. While highly speculative, such a possibility may be worth studying now that low LCPUFA status has been identified as a characteristic of some children with attention deficit hyperactivity disorder (45).

*Dietary Sources of Arachidonic Acid and Docosahexaenoic Acid for Studying LCPUFA Supplementation*

When it was first suggested that preterm infants might benefit from dietary docosahexaenoic acid (4), fish oils were the only readily available sources of docosahexaenoic acid, and sources of arachidonic acid were not available for study. The chemical and physical stability of fish oils in fat emulsions and their biological availability to preterm infants were also unknown, and these required evaluation before functional trials began (77,78). The number of dietary sources of docosahexaenoic acid and arachidonic acid now available for feeding trials are but one example of the increase in sophistication that has occurred in a single area of LCPUFA research during the last 10 years. Egg phospholipids (with and without egg cholesterol) have been fed as a source of arachidonic acid and docosahexaenoic acid, and algal and fungal oils have also been produced as sources of arachidonic acid and docosahexaenoic acid. These sources are currently available in infant formulas in some countries. Chicken eggs with virtually any desired ratio of docosahexaenoic acid and arachidonic acid have been produced by varying the diet of chickens, most recently by feeding different amounts of the biomass remaining after the extraction of single-cell oils containing arachidonic acid and/or docosahexaenoic acid. Results of feeding studies with docosahexaenoic acid enriched eggs should be available soon. Some companies are now attempting to produce genetically engineered plant sources of LCPUFAs by introducing the genes from single-cell organisms responsible for LCPUFA synthesis. These novel and traditional sources of LCPUFAs, whether fed alone or in combination, have the potential for improving LCPUFA status and function, should further research indicate this is desirable.

## LCPUFA Supplementation Trials of Preterm Infants

There have been three published randomized trials designed to look at functional outcomes in LCPUFA-supplemented preterm infants. The question asked in each was: Can diets with n-3 LCPUFAs and α-linolenic acid, compared with diets containing 1.35–2.5% of energy from α-linolenic acid (i) maintain plasma and erythrocyte docosahexaenoic acid at approximately the same level found in cord blood and following human milk feeding, and (ii) produce more mature sensory, cognitive, and motor development in preterm infants as assessed by retinal function, visual acuity development, and tests of early cognitive and motor function?

The questions related to sensory and cognitive function can be considered plausible because:

(i) Animals deficient in n-3 fatty acids compared with animals fed n-3 fatty acids have lower concentrations of docosahexaenoic acid in the retina and brain that are accompanied by longer latencies and lower amplitudes of responses to light (8,35), slower visual acuity development (10), and changes in behavior (11–13).

(ii) Very low birth weight infants (<1500 g) have little docosahexaenoic acid in brain and retina and little docosahexaenoic acid available in body stores at birth compared to term infants (76).

(iii) After birth, plasma and erythrocyte docosahexaenoic acid decline in preterm infants who are fed commercially available infant formulas with α-linolenic acid, evidence of worsening status.
(iv) Very low birth weight infants fed commercially available formulas have to depend on new synthesis for all docosahexaenoic acid accumulated after birth, even though synthesis of docosahexaenoic acid from α-linolenic acid is considerably less efficient than dietary docosahexaenoic acid for brain docosahexaenoic acid accumulation (50).

*Trial Design*

Each of the three published trials was a randomized study with subjects assigned to diet and assessed by investigators unaware of their diet (double-blind). In each trial a group of supplemented infants was compared with a group of infants fed formula with at least 1.35% energy as α-linolenic acid (range 1.35–2.5%). Uauy and co-workers (22) also studied preterm infants fed human milk and a corn-oil-based (α-linolenic acid deficient) formula (0.25% energy as α-linolenic acid). In all three trials, fish oil was fed as a source of n-3 LCPUFAs, and the formulas were manufactured as a ready-to-feed liquid. In the first two trials the fish oil provided eicosapentaenoic acid at concentrations approximately 50% higher than the concentration of docosahexaenoic acid (high eicosapentaenoic acid fish oil). The third trial performed later at one of the original sites provided the same amount of docosahexaenoic acid as the first trial (0.2%) but only 20% as much eicosapentaenoic acid (low eicosapentaenoic acid fish oil). In all three trials, the randomized groups were enrolled from hospitals serving populations from the lowest socio-economic groups. A brief description of the trials, including the levels of docosahexaenoic acid and eicosapentaenoic acid, the α-linolenic acid and linoleic acid in the experimental formulas, the study duration, and the planned outcomes and assessment ages are shown in Tables 1 and 2.

Two of the three studies were designed as multiple outcome trials to evaluate first-year LCPUFA status (LCPUFAs in several phospholipids of plasma and erythrocytes), growth, visual grating acuity, early cognitive function (visual recognition memory, number and length of discrete looks at 6.5, 9, and 12 months, 12-month Bayley mental developmental index), and gross motor function (12-month Bayley psychomotor developmental index). The power of the study was determined from the number of infants needed to detect a mean difference in growth of about 0.8 S.D. Differences in growth were not expected, but growth is a standard assessment in

Table 1. Design of the three randomized trials to study LCPUFA supplementation of preterm infants.

CARLSON STUDY I
- Informed consent obtained when enteral intake of a preterm formula >100 kcal/kg/day, if eligible (mean 23.5 days)
- Randomization to diet immediately after consent obtained
- Diet from enrollment to 1800 g birth weight:
    Group I:    Similac Special Care (3.0% ALA)
    Group II:   Similac Special Care with added fish oil (3.1% ALA, 0.2% DHA, 0.3% EPA, 0% AA)
- Diet from 1800 g to 79 weeks PMA (9 months corrected):
    Group I:    Similac with Iron (4.8% ALA)
    Group II:   Similac with Iron with added fish oil (4.9% ALA, 0.2% DHA, 0.3% EPA, 0% AA)
                Both groups began to receive other foods at 57 weeks
- Diet from 79 weeks:
    Groups I and II: Cows' milk and a mixed diet
- Ages when follow-up visits were scheduled: 38 weeks (term), 48 weeks (2 months), 57 weeks (4 months),* 68 weeks (6.5 months),* 79 weeks (9 months),* 92 weeks (12 months)*

CARLSON STUDY II
- Informed consent when enteral intake of a preterm formula started (2–5 days of age)
- Randomization to diet immediately after consent obtained
- Diet from enrollment to 48 weeks PMA:
    Group I:    Similac Special Care (3.0% ALA)
    Group II:   Similac Special Care with added fish oil (3.1% ALA, 0.2% DHA, 0.06% EPA, 0% AA)
- Diet from 48–92 weeks PMA (2–12 months corrected):
    Groups I and II: Similac with Iron (4.8% ALA)
                Both groups began to receive other foods at 57 weeks
- Ages when assessments were scheduled: 39 weeks (term), 48 weeks (2 months), 57 weeks (4 months),* 68 weeks (6.5 months),* 79 weeks (9 months),* 92 weeks (12 months)*

UAUY/BIRCH
- Informed consent while mother in hospital
- Randomization to diet by day 10 of life, if eligible
- Formula groups matched by birth weight (1000–1250 and 1250–1500 g) and gender
- Diets from enrollment to 57 weeks PMA
    Group I:    Human milk (not randomized)
    Group II:   Corn-oil-based formula (0.5% ALA)
    Group III:  Soybean-oil-based formula (2.7% ALA)
    Group IV:   Soybean-oil-based plus fish oil (1.4% ALA, 0.35% DHA, 0.65% EPA, 0% AA)
- Ages when assessments were scheduled: 36 wk and 57 wk PMA

AA, arachidonic acid; ALA, α-linolenic acid; DHA, docosahexaenoic acid; EPA, eicosapentaenoic acid; PMA, postmenstrual age.
* Infants were not required to return at these ages to be considered as study completers. Fatty acid percentages are total fatty acids. The fatty acid contribution to energy is approximately one-half the number provided.

Table 2. Study planned outcomes of the three randomized trials to study LCPUFA supplementation of preterm infants.

CARLSON STUDY I
- Plasma and red blood cell individual* phospholipid fatty acids: 38, 48, 57, 68, 79, and 92 weeks PMA
- Anthropometric assessments – weight, length, head circumference: 38, 48, 57, 68, 79, and 92 weeks PMA
- Visual acuity – Teller acuity cards: 38, 48, 57, 68, 79 and 92 weeks PMA
- Fagan Infantest – visual recognition memory, look number (and duration): 68, 79, and 92 weeks PMA
- Bayley MDI and PDI: 92 weeks PMA

CARLSON STUDY II
- Plasma and red blood cell individual* phospholipid fatty acids: 39, 48, 57, 68, 79, and 92 weeks PMA
- Anthropometric assessments – weight, length, head circumference: 39, 48, 57, 68, 79, and 92 weeks PMA
- Visual acuity – Teller acuity cards: 39, 48, 57, 68, 79, and 92 weeks PMA
- Fagan Infantest – visual recognition memory, look number (and duration): 92 weeks PMA
- Bayley MDI and PDI: 92 weeks PMA

UAUY/BIRCH
- Plasma and red blood cell total phospholipid fatty acids: 36 and 57 weeks PMA
- Retinal function – electroretinogram studies: 36 and 57 weeks PMA
- Visual acuity – visual evoked potentials, forced-choice preferential looking: 36 and 57 weeks PMA

MDI, mental development index; PDI, psychomotor development index; PMA, postmenstrual age.
* Phosphatidylethanolamine; phosphatidylcholine; sphingomyelin.

feeding studies of growing animals and infants who are sensitive to numerous toxic and nutritional influences. (Repeated measurements on the same infants were made throughout the year, which allowed correction of the data for baby effects. Thus, the studies actually detected a much smaller mean difference in growth of about 0.4–0.6 SD as significant.) Assessments at sequential ages during the first year also allowed correction for baby effects on visual acuity, visual recognition memory, and look number (look duration) in the ANOVA for effects of diet and time.

Because all three published studies were designed primarily to determine if there were effects of LCPUFAs on growth and development, there were objective criteria for inclusion and retention in the study. The intent of these criteria was to eliminate any condition with a known or theoretical effect on infant growth and development, except for prematurity itself. As can be appreciated from Table 3, the exclusion criteria were designed to eliminate factors expected to impact infant growth and development, while the criteria for retention were designed to assure that infants consumed their assigned diets. Enrollment and random assignment to diet occurred after eligibility was confirmed, and the decision to remove an infant from the study (other than by death) was always made by an investigator unaware of the infant assignment.

For the most part, these studies were designed with adequate power to answer the questions posed, i.e.: (i) to determine the influence of dietary n-3 LCPUFAs on the outcomes; (ii) to determine the ages at which specific sensory and cognitive outcomes were sensitive to differences in n-3 LCPUFA status; and (iii) to determine relationships between sensory and cognitive outcomes. Effects seen in the trial could be expected to occur in other equivalent groups of healthy preterm infants (internal validity). However, none of the three trials had external validity for very low birth weight infants who were very small or sick, although one of these trials included infants with chronic lung disease or bronchopulmonary dysplasia, extending the validity to one illness that allowed normal enteral feeding (24). In the other two trials, exclusion of infants who needed mechanical ventilation at enrollment effectively eliminated bronchopulmonary dysplasia as a variable.

*Biochemistry*

The biochemical effects of feeding fish oil to preterm infants were established in two short-term (4–6 weeks' duration) feeding trials before functional studies were undertaken (77,78). One concern was that preterm infants might absorb little fish oil because of their known lower pancreatic lipase, bile salts, and gastrointestinal motility compared with term infants. While bolus administration did result in apparently poor digestion/absorption (77), this problem was overcome by the use of sonication to disperse the fish oil in formula as microdroplets of oil (78). When formulas containing fish oil were first manufactured enabling infants to consume n-3 LCPUFAs after hospital discharge, another short-term feeding study confirmed the effectiveness of 0.2% docosahexaenoic acid to maintain docosahexaenoic acid in erythrocyte phospholipids at a level somewhere between those of cord blood and the blood of infants fed human milk. This 'physiological range' of docosahexaenoic acid then could be defended for longer term feeding studies capable of addressing questions about function.

When diets with 0.2% of total fatty acids were fed to preterm infants for

Table 3. Exclusion and retention criteria for the three randomized, double-blind trials of n-3 LCPUFA supplementation of preterm infants.

| Exclusion after randomization | Removal from Study |
|---|---|
| **CARLSON STUDY I** | |
| Birth weight <725 g and >1400 g | |
| First intake ≥100 kcal/kg/day after 6 weeks | |
| Mechanical ventilation at enrollment | Illness precluding enteral feeding for >1 week |
| IVH > grade 2 | Death |
| ROP > stage 2 | Parent's failure to comply with appointments or assigned feeding, or parent's request to leave the study before completion of 4-month visit |
| Gastrointestinal surgery for NEC or other gastrointestinal problems | |
| Birth weight <5th percentile for age | |
| Maternal alcohol or drug abuse | |
| Gross congenital malformations | |
| **CARLSON STUDY II** | |
| Birth weight <725 g and >1300 g | |
| Birth weight <5th percentile | All above, except follow-up through 2-month visit (when the experimental formula was discontinued) instead of 4-month visit |
| Congenital infections | |
| IVH > grade 2 | |
| ROP > stage 2 | |
| Gross congenital malformations | |
| **UAUY** | |
| Birth weight <1000 g and >1500 g | Necrotizing enterocolitis |
| Birth weight <10th percentile for age | Surgery for patent ductus arteriosus |
| Enteral intake <70 kcal/kg/day by day 10 | Parent's failure to comply with appointments |
| IVH > grade 2 | |
| ROP any stage | |
| Mechanical ventilation >7 days | |
| Gross congenital malformations | |

IVH, intraventricular hemorrhage; ROP, retinopathy of prematurity.

many months after birth, both plasma and erythrocyte docosahexaenoic acid remained in this physiological range (79). However, docosahexaenoic acid declined in these same tissues of infants fed commercially available infant formulas with 1.5–2.5% of energy from α-linolenic acid. The nadir in erythrocyte phospholipid docosahexaenoic acid occurred at approximately 4 months corrected age, after which docosahexaenoic acid remained low and unchanged for the next 8 months. Arachidonic acid decreased after birth in both plasma and erythrocyte phospholipids, regardless of diet, but the decrease was greater when the formulas with n-3 LCPUFAs were given.

Because of concerns related to poorer growth among preterm infants fed n-3 LCPUFAs (80) and the demonstration that these were related to lower plasma phospholipid arachidonic acid (81), a second randomized trial in the same site provided the same amount of docosahexaenoic acid (0.2%), but with much less eicosapentaenoic acid (0.06%), so that the total n-3 LCPUFA intake in the second study was only half that of the first study. The experimental formula was also fed for a shorter time (until 2 rather than 9 months corrected age, i.e. for roughly 5 months rather than 11 months) compared with the first trial. Further, the experimental formula provided only half as much linoleic acid as in the first trial, which also appeared to result in higher circulating phospholipid arachidonic acid (44). Finally, a nutrient-enriched (preterm) formula was fed from discharge until 2 months corrected age (about 12 weeks) rather than until discharge, as in the first trial. The result of these collective changes was that plasma phospholipid arachidonic acid concentration did not decrease with n-3 LCPUFA supplementation in the second trial done in our center, although the ratio of arachidonic acid to docosahexaenoic acid was still lower than among infants fed commercial formula (24).

*Visual Acuity*

In all three randomized trials of n-3 LCPUFA supplementation, visual acuity was higher in infants fed formula with n-3 LCPUFAs even though the control formulas contained recommended amounts of α-linolenic acid (22–24). The use of different measures for visual acuity assessment strengthened this observation. Birch *et al.* (22) measured visual evoked potential acuity at $-1$ month and 4 months corrected age and reported significantly higher grating acuity in infants fed the n-3 LCPUFA-supplemented formula compared with the formula prepared with soybean oil (an excellent source of α-linolenic acid). Grating acuity in their studies also appeared higher with another assessment measure, forced-choice preferential looking, but did not reach statistical significance, perhaps due to inadequate power because fewer than 15 infants were studied in three of the four groups (82). However, the group fed n-3 LCPUFA did have higher preferential looking acuity than the group fed an α-linolenic deficient formula prepared with corn oil.

We measured preferential looking grating acuity with the Teller acuity cards (83) at 0, 2, 4, 6, 9, and 12 months corrected age. More than 30 infants were tested in each group at each age, except at term (27 infants/group) and 12 months (29 infants/group). The ability to correct for individual baby effects in the context of repeated measurements of acuity probably further enhanced the power of the study to detect differences in acuity between groups. Compared with infants fed formula without LCPUFA, mean grating acuity was approximately one-half octave higher in the n-3 LCPUFA-fed infants at 2 and 4 months (a significant effect) but not at any of the other ages (23). A similar half-octave higher mean visual acuity occurred at 2 months in healthy infants fed a low eicosapentaenoic acid source of docosahexaenoic acid. However, at 4 months, 2 months after the experimental diet was discontinued, there was no effect of diet on visual acuity. At both 2 and 4 months, infants with bronchopulmonary dysplasia had lower visual acuity than infants who did not have bronchopulmonary dysplasia (24). There are at least two explanations possible for the lack of effect of dietary n-3 LCPUFAs on visual acuity at 6, 9, and 12 months. (i) Visual function becomes completely normal perhaps because the portions of the retina and brain necessary for visual function ultimately accumulate enough docosahexaenoic acid by synthesis from α-linolenic acid. (ii) High contrast grating acuity becomes normal but other aspects of visual function do not.

*Visual Attention*

Infant visual attention has been used to infer cognitive abilities such as processing and visual recognition memory. Processing has been inferred from infant visual behavior during tests of habituation, which involve two components: (i) presentation of a stimulus for long enough that the infant reaches a defined criterion of lower response, and (ii) presentation of a new stimulus to ensure that the decrease in interest is due to habituation rather than fatigue or state change. Visual recognition memory is inferred from the proportion of attention directed to a novel stimulus viewed simultaneously with a familiar stimulus.

The Fagan Test of Infant Intelligence (Infantest Corporation, Cleveland, OH) (84) is a paired comparison test designed to test visual recognition memory. It has been standardized for administration at 67, 69, 79, and 92 weeks postconceptional age (6, 6.5, 9, and 12 months). The Fagan Test consists of a series of familiarizations to pictures of faces (often two copies of the same face are shown simultaneously) followed by a paired comparison of one new and one familiar picture. Although the same pictures are used for familiarization and recognition memory at all ages, the Fagan Test has been standardized to eliminate age effects on novelty preference by manipulating the time required for familiarization and permitted for test. The time allowed for both decreases with age so that normal infants should have a preference for novelty above chance (>57%) and similar mean scores at all ages.

Although the Fagan Test was designed to test visual recognition memory, it also affords an opportunity to study the visual attention of infants during both familiarization and test (paired-comparison phases). Dr Susan Rose of Albert Einstein College of Medicine felt information from attentional shifts during the Fagan Test could be a worthwhile measure in our LCPUFA-supplementation trials. At her suggestion, we obtained software from the Infantest Corporation designed to determine the looking behavior of infants (number and duration of looks) during familiarizations and paired comparisons. These aspects of visual attention rather than visual recognition memory (novelty preference) were influenced by docosahexaenoic acid supplementation during infancy (25,26), and we appreciate her suggestion.

The effects of n-3 LCPUFA-supplemented and control formula on look number and novelty preference were presented in preliminary form at an earlier Dobbing Workshop. Simple ANOVA indicated higher look number during the test phase at 6.5, 9 and 12 months, and lower novelty preference at one age (12 months) in the n-3 LCPUFA group (85). It could be added that the significant difference between diet groups at 12 months was not because of a decrease in novelty preference in the n-3 LCPUFA group but rather to a higher novelty preference in the control group at 12 months, i.e. it was the control group that failed to conform to the expected 12-month novelty preference. Neuringer (36) commented that infants normally increase their number of looks between 4 and 8 months, and thus increases in number of looks in the n-3 LCPUFA group were consistent with more advanced development. However, none of the experts present at that Workshop knew how to interpret the lower novelty preference in this group of infants at 12 months. In the end there was no basis for interpreting the data on look number in the context of any known effect of n-3 fatty acid deficiency on visual attention. At the same time, the lower novelty preference at one age was impossible to ignore given that most of the Workshop participants were inclined to interpret higher scores as better scores.

More recently we have attempted to point out that there is little support for this view, especially at 12 months (see also Singer, this volume). Furthermore, this expectation goes counter to increasing shifts of attention between novel and familiar stimuli as infants mature (25,26,36).

Subsequently, Reisbick *et al.* (32,40) reported that n-3 fatty acid deficient monkeys compared with n-3 fatty acid fed monkeys had longer average look duration but similar novelty preference during paired-comparison tests similar to the Fagan Test. We reasoned that because the duration of the test phase was fixed in the Fagan Test, our earlier observation of increased numbers of looks (85) likely meant average look duration was shorter in the n-3 LCPUFA-supplemented group, which proved to be the case (25). During the test phase, infants fed n-3 LCPUFAs had significantly more total looks ($p < 0.03$) and shorter looks ($p < 0.03$) (25). The increase in total looks represented more looks to both the novel ($p < 0.03$) and to the familiar ($p < 0.03$) faces. Infants fed n-3 LCPUFAs also tended to have a shorter look duration during familiarizations ($p < 0.12$) (25). The look duration during familiarization and test were highly correlated (see my response to Singer, this volume). We think it is possible that the effects are more likely to be seen during test than familiarization in our infants and in Reisbick's monkey study because the familiarization is much longer than would be required for all but the most impaired infants to habituate to a defined criterion. We are presently testing this theory in other trials. Like dietary n-3 LCPUFAs, higher age also led to shorter look duration during the familiarization and paired-comparison phases of the Fagan Test, evidence that this outcome represented more advanced development, as suggested by Neuringer (36).

Approximately half of the infants in the second randomized trial received the same version of the Fagan Test at 12 months as infants in the first trial. The other half received a newer version that resulted from a change to different software that defined the shortest look as 0.05 s rather than 0.1 s, resulting in a marked increase in look number and a decrease in look duration relative to the original. This resulted in a loss in power in the second trial. Infants who received the same test as those in the first trial had virtually identical effects of n-3 LCPUFA supplementation on mean number of looks and look duration, even though only two outcomes reached statistical significance (26), but infants who received the newer version had significantly more looks, shorter look duration, and demonstrated no effect of diet (see also my responses to Singer for possible reasons). This second trial confirmed results of the first, which suggested an increase in processing speed (86), ability to disengage (87), or decreased reactivity (40) with higher docosahexaenoic acid status in preterm infants. This study also provided the first evidence that there are persistent effects of higher docosahexaenoic acid status early in infancy on later cognition. When these effects of diet on look duration were observed at 12 months corrected age, erythrocyte phospholipid docosahexaenoic acid was the same in both groups, reflecting their recent diet.

The second study was designed to look for changes in incidence of bronchopulmonary dysplasia and the effects of bronchopulmonary dysplasia on visual and cognitive outcomes. 12-Month visual attention was analyzed for effects of diet and bronchopulmonary dysplasia. Despite the smaller than planned numbers of infants who received the original version of the test, the effects of bronchopulmonary dysplasia were highly significant (evidence that the effect is quite robust). Infants with PPD had fewer looks during familiarization and significantly longer duration looks during both familiarization and paired-comparison tests (26). The Fagan Test has also been used by Jacobson and co-workers (88) to study infants whose mothers consumed alcohol during pregnancy. High intrauterine alcohol exposure increased look durations during familiarization (85), without affecting visual recognition memory. Previously, exposure to polychlorinated biphenyl (PCB) reduced visual recognition memory with no effect on look duration (89).

*Are Visual Acuity and Visual Attention Related?*

Because there are well-known sensory effects of n-3 fatty acids, both Neuringer (36) and Wainwright (90) have commented on the difficulty of assigning differences in behavior to cognition or learning with certainty when n-3 LCPUFA status differs. This is also the case for n-3 LCPUFA-fed preterm infants observed to have higher visual acuity (sensory function) at 2 and 4 months and shorter look duration at 12 months (apparent higher cognitive function).

In an attempt to gain some insight into this question, we compared visual acuity at 2 months and look duration at 12 months for all 84 infants who received the earlier version of the Fagan Test. Surprisingly, there was no correlation between higher visual acuity at 2 months and shorter look duration at 12 months, even though these were both outcomes of n-3 LCPUFA supplementation (91). We found only two variables that significantly and consistently were related to shorter look duration at 12 months. The first was erythrocyte phosphatidylethanolamine docosahexaenoic acid at 2 months (a marker for early docosahexaenoic acid status), and the second was the combined number of years of formal parent education. These two unrelated variables ($r = 0.01$) accounted for between 12% and

21% of the variance in look duration at 12 months of age (91). While these data suggest that differences in 12-month look duration are not due to obvious sensory differences early in infancy, subtle heretofore undetected differences in visual function might still exist at 12 months due to earlier differences in n-3 LCPUFA intake.

*Retinal Physiology*

Uauy and co-workers (92) investigated the effects of n-3 LCPUFA supplementation on retinal function of preterm infants. The choice of this functional outcome was also based on evidence of low retinal docosahexaenoic acid and functional changes in the retina of n-3 fatty acid deficient rats and monkeys (9). Approximately one month before infants should have been born, the group fed corn oil (an n-3 deficient diet) had lower maximum b-wave and a-wave amplitudes and an increase in the rod b-wave threshold compared with any of the other diet groups: human milk, soybean-oil-based formula, or formula with soybean oil and fish oil. These effects were transient, disappearing by 57 weeks postconception (4 months corrected age). At 4 months the only persistent effect of corn oil (the n-3 deficient diet) was a delay in the light-adapted oscillatory potentials, evidence that function of retinal neurons receiving input from cones was altered (93). There was no suggestion that the electroretinograms were abnormal in the group fed 1.35% energy from $\alpha$-linolenic acid at either age (92,93). Neuringer and co-workers have noted that the effects of n-3 fatty acid deficiency in monkeys on a-wave amplitude disappear with time, but that other measures of retinal function became progressively worse (9). Differences in dietary docosahexaenoic acid that are reflected in the brain are not found in the retina of term infants (75). This may also be the case for preterm infants after some months, but the persistent effects with an n-3 deficient diet suggest that function is still not normal.

*Global Scores of Mental Development*

The Bayley mental developmental index is a standardized test (94) that samples perception, memory, learning, problem solving, and vocalization, as well as early verbal communication and abstract thinking (95). The test was originally normalized with healthy term infants to a mean score of 100 with a standard deviation of 26. The original version of the test was administered at 12 months in two of the original preterm trials and in another prospective study of term infants from the same hospital. Mean test scores have steadily increased with the original version. A renormed version of the test is now available, and lower mean scores should be expected as a result (96).

In our first randomized trial, infants fed commercial formulas and formulas supplemented with n-3 LCPUFAs had similar mean scores on the Bayley mental developmental index. More importantly, because the numbers of infants studied were too small to reject the null hypothesis, the mean values for each group were similar to our historical mean for infants in the same birth weight range, i.e. they did not suggest any real effect of diet. In the second trial, several changes in design were made and the n-3 LCPUFA-supplemented group scored significantly higher (n-3 LCPUFAs, 109.2; control, 96.2; center norms – all preterm infants, 97.2; term breast-fed, 114.9; term formula-fed, 107.5). Although Lucas has pointed out that this could represent a type I error, a group mean of 109.2 is an unprecedented score for preterm infants in our center (see also my response to Bornstein, indicating an even larger difference when infants with bronchopulmonary dysplasia are removed from consideration). Recently, Damli *et al.* (97) have also reported significantly higher scores on the 15-month Bayley mental developmental index scores for preterm infants fed formulas with arachidonic acid and docosahexaenoic acid (111.4) compared to those fed formula without LCPUFAs (104.2). Because this final report is not yet available, the other details of the study such as the nutrient content of the formulas, the biochemical indices of nutritional status, and the social or contextual factors for this study are not yet known.

*Nutrient Interaction*

One of the changes in design in our second compared with our first trial (Table 1) was use of a nutrient-enriched base formula with or without n-3 LCPUFAs for several months after discharge. If, as we have speculated, higher n-3 LCPUFA status interacted with higher nutritional status to improve this global score (27), a single nutrient cannot be targeted because virtually all nutrients except iron are higher in preterm than in term formulas. We do have direct evidence of higher status of vitamin A and zinc with the use of preterm formula. Biochemical indices of vitamin A status were determined in both studies, and they show that plasma retinol concentrations were normal months earlier in infants from the second compared with the first study (98,99). Plasma zinc was also normal in the majority of infants by term (100). Although plasma zinc was not measured in our first study, the published literature suggests that zinc status also remains suboptimal for months after preterm infants are discharged on a term formula (101,102). Vitamin A and zinc, along with many other nutrients, play important roles in central nervous system development.

## Growth and Motor Development

As noted previously, one of the effects of n-6 LCPUFA deficiency is slower growth. Compared with preterm infants fed commercial formulas, those fed a high eicosapentaenoic acid fish oil source of docosahexaenoic acid for 11 months (from 2 months before term until 9 months corrected age) had poorer arachidonic acid status (lower plasma phosphatidylcholine arachidonic acid concentration) and lower weight, length, and head circumference beginning at around 40 weeks postmenstrual age (80). Scores on the Bayley psychomotor developmental index tended to be lower in the n-3 LCPUFA-supplemented group ($p < 0.09$). The normalized linear growth of the entire group of infants (control and n-3 LCPUFAs) was highly correlated with performance on the Bayley psychomotor developmental index, but not with the Bayley developmental index (85). The Bayley psychomotor developmental index measures gross motor abilities such as sitting, walking, standing, and stair climbing, and hand and finger manipulatory skills (95).

Preterm infants had generally poor arachidonic acid status, which was related to poorer normalized first-year growth. Feeding n-3 LCPUFAs only exacerbated the low arachidonic acid status of preterm infants and the effects of low arachidonic acid status on growth (81). Therefore, these findings signaled caution for feeding n-3 LCPUFAs without n-6 LCPUFAs. Ideally, an experimental formula with both arachidonic acid and docosahexaenoic acid should have been fed in the next randomized trial in our site, but sources of arachidonic acid were still not available for feeding when that trial began. Instead, efforts were made to minimize the effect of n-3 LCPUFAs on arachidonic acid status by feeding a low eicosapentaenoic acid fish oil that provided the same amount of docosahexaenoic acid, but with half as much n-3 LCPUFA, and by decreasing by half the interval at which the experimental formula was fed (Table 1). As noted previously, these efforts successfully prevented reductions in plasma and erythrocyte phospholipid arachidonic acid related to n-3 LCPUFA intake, even though the ratio of n-6 LCPUFAs to n-3 LCPUFAs in plasma and erythrocyte phospholipid was still lower in the experimental group because their n-3 LCPUFAs increased (24). The smaller effect on LCPUFAs status relative to our first study (79) was related to a smaller effect on growth (24,80). In fact, the only consistent effect of n-3 LCPUFAs feeding on normalized growth was a lower weight-for-length at several ages. Overall, weight, length, and head circumference were unaffected by diet. If anything, infants in the experimental group in this study were slightly taller than controls by one year of age, and they had higher scores on the Bayley psychomotor developmental index, but again not significantly so. Jensen et al. (19) have recently reported a relationship between lower weight and weight-for-length among term infants fed 1.6% compared with 0.2% of energy from α-linolenic acid, and related differences in weight to arachidonic acid status among four groups fed a range of linolenic acid from 0.2% to 1.6% of energy.

A number of centers are now feeding experimental diets that include variable amounts and ratios of arachidonic acid and docosahexaenoic acid in preterm infants. Until the results of these studies are available, it will not be known if our speculation was correct that preterm infants might grow better if fed diets with arachidonic acid (81).

## Diseases of Premature Infants

All of the published randomized trials in preterm infants had exclusion and retention criteria for infants who were eligible for enrollment, making them a select group among the population of hospitalized preterm infants. In two of the studies, infants were not randomized to treatment until their risk of developing many of the diseases of the preterm infant was low (23,82). The exclusion criteria for the studies (Table 3) were appropriate given that the intent was to determine if preterm infants could benefit from n-3 LCPUFAs by demonstrating higher visual acuity and cognition.

However, if formulas with n-3 LCPUFAs were made available for preterm infants, they would likely be fed to all, including many smaller and sicker infants. Formulas with n-3 LCPUFAs or n-3 and n-6 LCPUFAs have the potential both to increase and to decrease the incidence of disease in this vulnerable group. A single study to resolve all these possibilities would have to be very large. In our second trial, infants were randomly assigned to diet from their first enteral feeding (2–5 days of age). As a result, subsequent disease could be viewed as an outcome of the study. However, because only 47 infants in each group were enrolled and randomized, the study was able to provide only suggestive evidence of effects that could be studied in larger trials. No significant effects of diet on disease were noted, but there were nonsignificant trends ($p > 0.05$ and $p < 0.10$) for both a lower incidence of bronchopulmonary dysplasia and a higher incidence of necrotizing enterocolitis in the group fed the n-3 LCPUFA-supplemented formula. Power estimates indicated that at least 100 infants per group would have had to be studied to determine a significant effect of diet on either disease.

*Are LCPUFAs Conditionally Essential for Preterm Infants?*

Proof of the conditional essentiality of a compound is convincing only when its intake improves function(s) shown previously to depend upon its nutritionally essential precursor, and higher function(s) is/are found in comparison with a group fed the nutritionally essential precursor. By these criteria, the three published n-3 LCPUFA supplementation studies offer clear evidence that n-3 LCPUFAs are conditionally essential nutrients for preterm infants. First, all of the functional effects due to higher docosahexaenoic acid status after feeding n-3 LCPUFAs are analogous to functional effects of higher tissue docosahexaenoic acid in monkeys fed a diet with low rather than high α-linolenic acid (Table 4). Second, the effect of n-3 LCPUFAs on visual acuity has been confirmed in two centers with two different measures of acuity. Third, shorter average look duration with dietary n-3 LCPUFAs has been confirmed in two trials in one center.

## The Next Generation Trials

The three initial trials discussed in this chapter involved relatively small numbers of infants, but, for the most part, they were large enough to answer the questions they were designed to ask. However, they have raised additional questions that can only be answered by larger trials. The first of these questions is whether or not all healthy preterm infants would benefit from dietary LCPUFAs. It seems likely that they would, because the population and management variables that influence infant LCPUFA status are unlikely to be as influential as the loss of the normal maternal-to-fetal transfer of LCPUFAs in utero. For example, although Heird and co-workers found quite variable LCPUFA synthesis among infants (53), infants fed only precursors of LCPUFA nevertheless had much lower LCPUFA than those fed human milk (19) the second unanswered question is whether all preterm infants, healthy or not, can be expected to benefit from dietary LCPUFAs. If our study of bronchopulmonary dysplasia infants is any example, sicker preterm infants will not necessarily respond in the same way as healthy preterm infants. More likely, they will have variable responses to LCPUFA supplementation, with some outcomes being improved and others unaffected. A third question raised by these studies that can only be answered in much larger trials is whether and how the

Table 4. Effects of higher docosahexaenoic acid status on sensory and cognitive behaviors in monkey and preterm infants

| Outcome | Monkey infants | Preterm infants |
| --- | --- | --- |
| Retinal function | Higher a-wave amplitudes | Higher a-wave amplitude at −1 month, not at 4 months |
| | Lower implicit times (4 months and 2 years) | Higher amplitude and lower threshold of the b-wave at −1 month; not at 4 months |
| | Faster recovery following repeated stimulation (4 months and 2 years) | Lower implicit times and cone oscillatory potentials (human milk vs corn oil) (4 months) |
| Visual grating acuity | Higher at 1, 2, and 3 months | Higher at 2 and 4 months in healthy infants (Carlson Study I) |
| | | Higher at 2 months in infants without BPD (Carlson Study II) |
| | | Higher at −1 month and 4 months in healthy infants (Uauy) |
| Stereoacuity | Not measured | Higher at 3 years with human milk vs corn oil (Uauy/Birch) |
| Look duration | Shorter look duration in infancy (4, 8, 12, and 16 weeks) | Shorter look duration during paired comparisons ($p < 0.03$) of the Fagan Test at 6, 9, and/or 12 months (Carlson Study I and II) |
| Recognition memory | Unaffected | Unaffected |
| Bayley MDI | Not applicable | Higher in conjunction with preterm formula fed until 48 weeks PMA (Carlson Study II) but not 36 weeks PMA (Carlson Study I) |

BPD, bronchopulmonary dysplasia; PMA, postmenstrual age.

early effects of n-3 LCPUFAs on sensory and cognitive function are related to analogous functions in childhood. Finally, concerns about safety will require larger studies. These concerns must be resolved before LCPUFAs can be fed to all preterm infants.

Two kinds of safety issue have been raised for LCPUFAs in the diet of infants. First, sources of arachidonic acid and docosahexaenoic acid need to be carefully evaluated to ensure that they have no contaminants with adverse effects. Second, concerns need to be resolved that risk for diseases of the preterm is not increased by feeding docosahexaenoic acid and arachidonic acid. Because lipid-derived mediators have been invoked as part of the etiology of most of the 'alphabet soup' of preterm diseases (intraventricular hemorrhage/periventricular hemorrhage, bronchopulmonary dysplasia, necrotizing enterocolitis, retinopathy of prematurity, as well as for infection), it is clearly plausible that n-3 LCPUFAs could have an influence on these diseases, favorable or unfavorable, because lipid mediators from the n-3 and n-6 fatty acid families often have antithetical effects, however, it is good to keep in mind that the balance of n-3 and n-6 LCPUFAs could be critical to the final outcome. At this time, it is highly unlikely that infants will be fed diets with only n-3 LCPUFAs, and so some of the issues raised about their roles in protection and risk of disease in this chapter and by Lucas (this volume) may be resolved by feeding a balance of n-6 and n-3 LCPUFAs.

## Summary

In addition to providing background on the role of essential fatty acids and the factors that influence the ultimate accumulation of their LCPUFAs in biomembranes, this chapter has reviewed the evidence that contributed to the first randomized trials of n-3 LCPUFAs in infants and their results. These first trials have provided leads for later studies designed to determine risk and benefit for disease and for long-term behavioral outcomes. To a greater or lesser degree, they have also been a guide for randomized trials designed to look at outcomes in term infants (reviewed in this volume by Innis). Unlike the preterm trials, studies in healthy term infants do not raise the same degree of concern for increasing risk for disease. On the other hand, they require much greater control of variables that could influence LCPUFA status and functional outcomes (work in this area is also very much needed), because these variables are more likely to assume prominence when maternal-to-fetal LCPUFA transfer is permitted. Finally, the term studies would do well to look to the preterm trials for functional tests and ages when effects of LCPUFAs can be demonstrated in order to conclude validly an absence of LCPUFA effects.

New trials in preterm infants of equivalent or larger size are being conducted in a number of centers. Most have included both docosahexaenoic acid and arachidonic acid, and some of the results and suggestive evidence obtained from the studies that fed n-3 LCPUFAs without n-6 LCPUFAs may differ from those in the newer trials. Nevertheless, the newer trials are expected to investigate some of the same functional effects studied in the preterm trials discussed in this chapter and to obtain other behavioral measures that assess different domains of infant development. These trials also need to look critically at the incidence of diseases of the preterm where the relationships of LCPUFA status to disease is plausible, e.g. necrotizing enterocolitis, bronchopulmonary dysplasia, retinopathy of prematurity, and infection, which can occur after weeks of feeding. On the other hand, higher grades of intraventricular hemorrhage/periventricular hemorrhage (IVH/PVH) and congenital infections might best be exclusion criteria because they occur prenatally or, in the case of IVH/PVH, shortly after birth before enteral feeding has started (103). Not only can these conditions not be considered outcomes of diet, but they are well recognized for their potential to have devastating and unpredictable effects on neurodevelopment (104).

## References

1. Sanders TAB, Naismith DJ. A comparison of the influence of breast-feeding and bottle-feeding on the fatty acid composition of the erythrocytes. *Br J Nutr* 1979; **41**: 619–23.
2. Putnam JC, Carlson SE, DeVoe PW, Barness LA. The effect of variations in dietary fatty acids on the fatty acid composition of erythrocyte phosphatidylcholine and phosphatidylethanolamine in human infants. *Am J Clin Nutr* 1982; **36**: 106–114.
3. Bitman J, Wood DL, Hamosh M, Hamosh P, Mehta P, Mehta NR. Comparison of the lipid composition of breast milk from mothers of term and preterm infants. *Am J Clin Nutr* 1983; **38**: 300–12.
4. Carlson SE, Rhodes PG, Ferguson MG. Docosahexaenoic acid status of preterm infants at birth and following feeding with human milk or formula. *Am J Clin Nutr* 1986; **44**: 798–804.
5. Svennerholm J. Distribution and fatty acid composition of phosphoglycerides in normal human brain. *J Lipid Res* 1968; **9**: 570–9.
6. Crawford MA, Casperd NM, Sinclair AJ. The long chain metabolites of linoleic and linolenic acids in liver and brain in herbivores and carnivores. *Comp Biochem Physiol* 1979; **54B**: 395–401.
7. Clandinin MT, Chappell JE, Leong S, Heim T, Swyer PR, Chance GW. Intra-

uterine fatty acid accretion rates in human brain: implications for fatty acid requirements. *Early Human Dev* 1980; **4**: 121–9.
8. Benolken RM, Anderson RE, Wheeler TG. Membrane fatty acids associated with the electrical response in visual excitation. *Science* 1973; **182**: 1253–4.
9. Neuringer M, Connor WE, Lin DS, Barstad L, Luck S. Biochemical and functional effects of prenatal and postnatal omega-3 fatty acid deficiency on retina and brain in rhesus monkeys. *Proc Natl Acad Sci USA* 1986; **83**: 4021–5.
10. Neuringer M, Connor WE, Van Petten C, Barstad L. Dietary omega-3 fatty acid deficiency and visual loss in infant rhesus monkeys. *J Clin Invest* 1984; **73**: 272–6.
11. Lamptey MS, Walker BL. A possible essential role for dietary linoleic acid in the development of the young rat. *J Nutr* 1976; **106**: 86–93.
12. Yamamoto N, Saitoh M, Moriuchi A et al. Effect of dietary α-linolenate/linoleate balance on brain lipid compositions and learning ability of rats. *J Lipid Res* 1987; **28**: 144–51.
13. Yamamoto N, Hashimoto A, Takemoto Y et al. Effect of dietary α-linolenate/linoleate balance on lipid compositions and learning ability of rats: II. Discrimination process, extinction process, and glycolipid compositions. *J Lipid Res* 1988; **29**: 1013–21.
14. Jensen RG, Bitman J, Carlson SE, Cavel SC, Hamash M, Newburg DS. Milk lipids. A. Human milk lipids. In: Jensen RG, ed. *Handbook of Milk Composition*. San Diego, CA: Academic Press, Inc., 1995; 495–542.
15. Hansen AE, Wiese HF, Boelsche AN, Haggard ME, Adam DJD, Davis H. Role of linoleic acid in infant nutrition. *Pediatrics* 1963; **31**: 171–92.
16. Forbes GB, Woodroff CW, eds. Fats and fatty acids. In: *Pediatric Nutrition Handbook*, 2nd edn. Elk Grove Village, IL: American Academy of Pediatrics, 1985: 105–10.
17. Barness LA, ed. Fats and fatty acids. In: *Pediatric Nutrition Handbook*, 3rd edn. Elk Grove Village, IL: American Academy of Pediatrics, 1993: 107–14.
18. Birch EE, Birch D, Hoffman D et al. Breast-feeding and optimal visual development. *J Pediatr Ophthalmol Strabismus* 1993; **30**: 33–8.
19. Jensen CL, Prager TC, Fraley JK, Anderson RE, Heird WC. Effects of dietary linoleic/α-linolenic acid ratio on growth and visual function of term infants. *J Pediatrics* 1997; in press.
20. Courage M, Friel J, Andrews W, McCloy U, Adams R. Dietary fatty acids and the development of visual acuity in human infants. *Invest Ophthalmol Vis Sci* 1995; **36**: S48.
21. Innis S (with commentary by Gross SJ, Hamosh M, Koletzko B, Uauy R). In: Tsang RC, Lucas A, Uauy R, Zlotkin S, eds. *Nutritional Needs of the Preterm Infants: Scientific Basis and Practical Guidelines*. Baltimore, MD: Williams & Wilkins, 1992: 65–86.
22. Birch EE, Birch DG, Hoffman DR, Uauy R. Dietary essential fatty acid supply and visual acuity development. *Invest Ophthalmol Vis Sci* 1992; **33**: 3242–53.
23. Carlson SE, Werkman SH, Rhodes PG, Tolley EA. Visual acuity development in healthy preterm infants: effect of marine oil (fish oil) supplementation. *Am J Clin Nutr* 1993; **58**: 35–42.
24. Carlson SE, Werkman SH, Tolley EA. The effect of long chain n-3 fatty acid supplementation on visual acuity and growth of preterm infants with and without bronchopulmonary dysplasia. *Am J Clin Nutr* 1996; **63**: 687–97.

25. Werkman SH, Carlson SE. A randomized trial of visual attention of preterm infants fed docosahexaenoic acid until 9 months. *Lipids* 1996; **31**: 91–7.
26. Carlson SE, Werkman SH. A randomized trial of visual attention of preterm infants fed docosahexaenoic acid until 2 months. *Lipids* 1996; **31**: 85–90.
27. Carlson SE, Werkman SH, Peeples JM, Wilson WM III. Growth and development of premature infants in relation to n-3 and n-6 fatty acid status. In: Galli C, Simopoulos AP, Tremoli E, eds. *Fatty Acids and Lipids: Biological Aspects, World Review of Nutrition and Dietetics*, Vol. 75. Basel: Karger, 1994: 63–9.
28. Sprecher H, Luthria DL, Mohammed BS, Baykousheva SP. Re-evaluation of the pathways or the biosynthesis of polyunsaturated fatty acids. *J Lipid Res* 1995; **36**: 2471–7.
29. Luthria DL, Mohammed BS, Sprecher H. Regulation of the biosynthesis of 4,7,10,13,16,19-docosahexaenoic acid. *J Biol Chem* 1996; **271**: 16 020–5.
30. Holman RT. Essential fatty acid deficiency. *Prog Chem Fats and Other Lipids* 1968; **9**(2): 258–348.
31. Hansen AE, Haggard ME, Boelsche AAN, Adam DJD, Wiese HF. Essential fatty acids in infant nutrition. III. Clinical manifestations of linoleic acid deficiency. *J Nutr* 1958; **66**: 565–76.
32. Reisbick SW, Neuringer M, Gohl E, Wald R. Visual attention in infant monkeys: effects of dietary fatty acids and age. *Dev Psychol* (in press).
33. Reisbick S, Neuringer M. Omega-3 fatty acid deficiency and behaviour: a critical review and directions for further research. In: Yehuda S, Mostofsky DI, eds. *Handbook of Essential Fatty Acid Biology: Biochemistry, Physiology and Behavioural Neurobiology*. Totowa, NJ: Humana Press, 1997: 397–425.
34. Wheeler TJ, Benolken RM, Anderson RE. Visual membranes: specificity of fatty acid precursors for the electrical response in visual excitation. *Science* 1975; **188**: 1312.
35. Neuringer M. The relationship of fatty acid composition to function in the retina and visual system. In: Dobbing J, ed. *Lipids, Learning, and the Brain: Fats in Infant Formulas, Report of the 103rd Ross Conference on Pediatric Research*. Columbus, OH: Ross Laboratories, 1993: 134–58.
36. Neuringer M. Commentaries on Carlson and Wainright. In: Dobbing J, ed. *Lipids, Learning, and the Brain: Fats in Infant Formulas. Report of the 103rd Ross Conference on Pediatric Research*. Columbus, OH: Ross Laboratories, 1993: 212–4, 88–94.
37. Neuringer M, Connor WE, Barstad L et al. Supplementation with docosahexaenoic acid (22:6n-3) fails to correct electroretinogram abnormalities produced by omega-3 fatty acid deficiency in rhesus monkeys. *Invest Ophthalmol Vis Sci* 1990; **31**: 426.
38. Teller DY, Morse R, Borton R et al. Visual acuity for vertical and diagonal gratings in human infants. *Vision Res* 1974; **18**: 561.
39. Reisbick S, Neuringer M, Hasnain R, Connor WE. Polydipsia in rhesus monkeys deficient in omega-3 fatty acids. *Physiol Behav* 1990; **47**: 315–23.
40. Reisbick S, Neuringer M, Connor WE. Effects of n-3 fatty acid deficiency in nonhuman primates. In: Bindels JG, Goedhart AC, Visser HKA, eds. *Recent Developments in Infant Nutrition*. Dordrecht: Kluwer, 1996: 157–72.
41. Martinez M. Developmental profiles of polyunsaturated fatty acids in the brain of normal infants and patients with peroxisomal diseases: severe deficiency of

docosahexaenoic acid in Zellweger's and pseudo-Zellweger's syndromes. *World Rev Nutr Diet* 1991; **66**: 87–102.
42. Martinez M. Docosahexaenoic acid therapy in docosahexaenoic acid-deficient patients with disorders of peroxisomal biogenesis. *Lipids* 1996; **31**: S145–52.
43. Shin CS, Lee MK, Park KS, et al. Insulin restores fatty acid composition earlier in liver microsomes than erythrocyte membranes in streptozotocin-induced diabetic rats. *Diabetes Res Clin Pract* 1995; **29**: 93–8.
44. Carlson SE. Arachidonic acid status of human infants: influence of gestational age at birth and diets with very long chain n-3 and n-6 fatty acids. *J Nutr* 1996; **126**: 1092S–8S.
45. Stevens LJ, Zentall SS, Deck JL et al. Essential fatty acid metabolism in boys with attention-deficit hyperactivity disorder. *Am J Clin Nutr* 1995; **62**: 761–8.
46. Burr GO, Burr MM. A new deficiency disease produced by the rigid exclusion of fat from the diet. *J Biol Chem* 1929; **82**: 345–67.
47. Goubern M, Yazbeck J, Senault C, Portet R. Non-shivering thermogenesis and brown adipose tissue activity in essential fatty acid-deficient rats. *Arch Int Physiol Biochem* 1990; **98**: 193–9.
48. Sellmayer A, Danesch U, Weber PC. Effect of different polyunsaturated fatty acids on growth-related early gene expression and cell growth. *Lipids* 1996; **31**: S37–40.
49. Kremer JM. Effects of modulation of inflammatory and immune parameters in patients with rheumatic and inflammatory disease receiving dietary supplementation of n-3 and n-6 fatty acids. *Lipids* 1996; **31**: S243–7.
50. Sinclair AJ, Crawford MA. The incorporation of linolenic acid and docosahexaenoic acid into liver and brain lipids of developing rats. *FEBS Lett* 1972; **26**: 127–9.
51. Pawlosky RJ, Ward G, Salem N Jr. Essential fatty acid uptake and metabolism in the developing rodent brain. *Lipids* 1996; **31**: S103–7.
52. Salem N, Wegher B, Mena P, Uauy R. Arachidonic and docosahexaenoic acids are biosynthesized from their 18-carbon precursors in human infant. *Proc Natl Acad Sci USA* 1996; **93**: 49–54.
53. Sauerwald T, Jensen CL, Chen HM, Anderson RE, Heird WC, Hachey DL. Effect of dietary 18:ω-3 intake and postnatal age on the kinetics of elongation and desaturation of 18:ω-6 and 18:ω-3. *Pediatr Res* 1995; **37**: 319A.
54. Carnielli VP, Sauer PJ. Essential fatty acid metabolism in neonates. In: Bindels JG, Goedhart AC, Visser HKA, eds. *Recent Developments in Infant Nutrition*. Dordrecht: Kluwer, 1996: 173–81.
55. Carlson SE, Cooke RJ, Rhodes PG, Peeples JM, Werkman SH. Effect of vegetable and marine oils (fish oils) in preterm infant formulas on blood arachidonic and docosahexaenoic acids. *J Pediatr* 1992; **120**; S159–67.
56. Crawford MA, Hassam AG, Williams G. Essential fatty acids and fetal brain growth. *Lancet* 1976; **i**: 452–3.
57. Ruyle M, Connor WE, Anderson GJ, Lowensohn RI. Placental transfer of essential fatty acids in humans: venous-arterial difference for docosahexaenoic acid in fetal umbilical erythrocytes. *Proc Natl Acad Sci USA* 1990; **87**: 7902–6.
58. Leaf AA, Leighfield MJ, Costeloe KL, Crawford MA. Long chain polyunsaturated fatty acid and fetal growth. *Early Human Dev* 1992; **30**: 183–91.
59. Smuts CM, Dhansay MA, Tichelar HY *et al*. Cord blood essential fatty acid metabolic status in infants of different gestational ages: Implications for infant

nutrition. In: *American Oil Chemists' Society Conference on PUFA in Infant Nutrition: Consensus and Controversies.* Champaign, IL: American Oil Chemists Society, 1996, 38–9.
60. Makrides M, Neumann M, Simmer K, Pater J, Gibson R. Are long-chain polyunsaturated fatty acids essential nutrients in infancy? *Lancet* 1995; **345**: 1463–8.
61. Carlson SE, Ford AJ, Werkman SH, Peeples JM, Koo WWK. Visual acuity and fatty acid status of term infants fed human milk and formula with and without docosahexaenoate and arachidonate from egg yolk lecithin. *Pediatr Res* 1996; **39**: 1–12.
62. Auestad N, Montalto MB, Hall RT et al. Visual acuity, erythrocyte fatty acid composition, and growth in term infants fed formulas with long-chain polyunsaturated fatty acids for one year. *Pediatr Res* 1997; **41**: 1–10.
63. Sanjuro P, Matorras R, Ingunza N, Rodriquez-Alarcon J, Perteagudo L. Blue fish intake and percentual levels of polyunsaturated plasmatic fatty acids at labor in the mother and the newborn infant. *J Perinat Med* 1994; **22**: 337–44.
64. Connor WE, Lowensohn R, Hatcher L. Increased docosahexaenoic acid levels in human newborn infants by administration of sardines and fish oil during pregnancy. *Lipids* 1996; **31**: S183–7.
65. Hornstra G, Al MDM, Van Houwelingen AC, Foreman-Van Drongelen MMHP. Essential fatty acids, pregnancy and pregnancy outcome. In: Bindels JG, Goedhart AC, Visser HKA, eds. *Recent Developments in Infant Nutrition.* Dordrecht: Kluwer, 1996: 51–63.
66. Eder K, Kirchgessner M. Dietary fat influences the effect of zinc deficiency on liver lipids and fatty acid in rats force-fed equal quantities of diet. *J Nutr* 1994; **124**: 1917–26.
67. Koletzko B. *Trans* fatty acids may impair biosynthesis of long-chain polyunsaturates and growth in man. *Acta Paediatr* 1992; **81**: 302–6.
68. Desci T, Koletzko B. Do *trans* fatty acids impair linoleic acid metabolism in children? *Ann Nutr Metab* 1995; **39**: 36–41.
69. Ayyagari A, Peeples JM, Carlson SE. Relationship of isomeric fatty acids in human cord blood to n-3 and n-6 status. *Pediatr Res* 1996; **39**: 304A.
70. Al MD, Van Houwelingen AC, Badart-Smook A, Hasaart TH, Roumen FJ. The essential fatty acid status of mother and child in pregnancy-induced hypertension: a prospective longitudinal study. *Am J Obstet Gynecol* 1995; **172**: 1605–14.
71. Innis SM, Kuhnlein HV. Long-chain n-3 fatty acids in breast milk of Inuit women consuming traditional foods. *Early Human Dev* 1988; **18**: 185–9.
72. Harris WE, Connor WE, Lindsey S. Will dietary ω-3 fatty acid change the composition of human milk? *Am J Clin Nutr* 1984; **40**: 780–5.
73. Sanders TAB, Ellis FR, Dickerson JWT. Studies of vegans: the fatty acid composition of plasma choline phosphoglycerides, erythrocytes, adipose tissue, and breast milk, and some indicators of susceptibility to ischemic heart disease in vegans and omnivore controls. *Am J Clin Nutr* 1978; **31**: 805–13.
74. Farquarson J, Jamieson EC, Abbasi KA, Patrick WJ, Logan RW, Cockburn F. Effect of diet on the fatty acid composition of the major phospholipids of infant cerebral cortex. *Arch Dis Child* 1995; **72**: 198–203.
75. Makrides M, Neumann MA, Byard RW, Simmer K, Gibson RA. Fatty acid

composition of brain, retina and erythrocytes in breast- and formula-fed infants. *Am J Clin Nutr* 1994; **60**: 189–94.
76. Clandinin MT, Chappell JE, Heim T, Swyer PR, Chance GW. Fatty acid utilization in perinatal de novo synthesis of tissues. *Early Human Dev* 1981; **5**: 355–66.
77. Carlson SE, Rhodes PG, Rao V, Goldgar DE. Effect of fish oil supplementation on the omega-3 fatty acid content of red blood cell membranes in preterm infants. *Pediatr Res* 1987; **21**: 507–10.
78. Liu C-CF, Carlson SE, Rhodes PG, Rao V, Meydrech EF. Increase in plasma phospholipid docosahexaenoic and eicosapentaenoic acids as a reflection of their intake and mode of administration. *Pediatr Res* 1987; **22**: 292–6.
79. Carlson SE, Cooke RJ, Rhodes PG, Peeples JM, Werkman SH, Tolley EA. Long-term feeding of formulas high in linolenic acid and marine oil (fish oil) to very low birth weight infants: phospholipid fatty acids. *Pediatr Res* 1991; **30**: 404–12.
80. Carlson SE, Cooke RJ, Werkman SH, Tolley EA. First year growth of preterm infants fed standard compared to marine oil (fish oil) n-3 supplemented formula. *Lipids* 1992; **27**: 901–7.
81. Carlson SE, Werkman SH, Peeples JM, Cooke RJ, Tolley EA. Arachidonic acid status correlates with first year growth in preterm infants. *Proc Natl Acad Sci USA* 1993; **90**: 1073–7.
82. Hoffmann DR, Birch EE, Birch DG, Uauy RD. Effects of supplementation with 3 long-chain polyunsaturated fatty acids on retinal and cortical development in premature infants. *Am J Clin Nutr* 1993; **57**: 807S–12S.
83. Teller DY, McDonald MA, Preston K et al. Assessment of visual acuity in infants and children: The acuity card procedure. *Dev Med Child Neurol* 1986; **28**: 779–89.
84. Fagan JF, Shepherd PA. *The Fagan Test of Infant Intelligence Training Manual (Version 4.0)*. Cleveland, OH: Infant Corporation, 1987.
85. Carlson SE. Lipid requirements of very-low-birth-weight infants for optimal growth and development. In Dobbing J, ed. *Lipids, Learning, and the Brain: Fats in Infant Formulas, Report of the 103rd Ross Conference on Pediatric Research*. Columbus, OH: Ross Laboratories, 1993: 188–207.
86. Colombo J, Mitchell DW, Caldron JT, Freeseman LJ. Individual differences in infant visual attention: are short lookers faster processors or feature processors? *Child Dev* 1991; **62**: 1247–57.
87. Johnson MH, Posner MI, Rothbart MK. Components of visual orienting in early infancy: contingency learning, anticipatory looking and disengaging. *J Cognitive Neurosci* 1991; **3**: 335–44.
88. Jacobson SW, Jacobson JL, Sokol RJ, Martier SS, Ager JW. Prenatal alcohol exposure and infant information processing ability. *Child Dev* 1993; **64**: 1706–21.
89. Jacobson SW, Fein GG, Jacobson JL, Schwartz PM, Dowler JK. The effect of intrauterine PCB exposure on visual recognition memory. *Child Dev* 1985; **56**: 853–60.
90. Wainwright P. Lipids and behavior: the evidence from animal models. In: Dobbing J, ed. *Lipids, Learning, and the Brain: Fats in Infant Formulas, Report of the 103rd Ross Conference on Pediatric Research*. Columbus, OH: Ross Laboratories, 1993: 69–101.

91. Carlson SE. Early visual acuity does not correlate with later evidence of visual processing in preterm infants although each is improved by the addition of docosahexaenoic acid to infant formula. In: *Second International Congress of the International Society for the Study of Fatty Acids and Lipids Congress Program and Abstracts*. Champaign, IL: American Oil Chemists' Society, 1995: 53.
92. Uauy RD, Birch DG, Birch EE, Tyson JE, Hoffman DR. Effect of dietary omega-3 fatty acids on retinal function of very-low-birth-weight neonates. *Pediatr Res* 1990; **28**: 485–92.
93. Birch DG, Birch EE, Hoffman DR, Uauy RD. Retinal development in very low birth weight infants fed diets differing in omega-3 fatty acids. *Ophthalmol Vis Sci* 1992; **33**: 23 650–76.
94. Bayley N. *Manual for the Bayley Scales of Infant Development*. San Antonio, TX: The Psychological Corporation, 1969.
95. Anastasi A. Tests of general intellectual level. In: *Tests for Special Populations Psychological Testing*, 6th edn. New York: Macmillan, 1988: 271–309.
96. Bayley N. *Manual for the Bayley Scales of Infant Development*, 2nd edn. San Antonio, TX: The Psychological Corporation, 1993.
97. Damli A, von Schenk U, Clausen U, Koletzko B. Effects of long-chain polyunsaturated fatty acids (LCPUFA) on early visual acuity and mental development of preterm infants. *American Oil Chemists' Society Conference on PUFA in Infant Nutrition: Consensus and Controversies Abstracts*. Champaign, IL: American Oil Chemists' Society, 1996: 14.
98. Peeples JM, Carlson SE, Werkman SH, Cooke RJ. Vitamin A status of preterm infants during infancy. *Am J Clin Nutr* 1991; **53**: 1455–9.
99. Carlson SE, Peeples JM, Werkman SH, Koo WWK. Plasma retinol and retinol binding protein concentrations in premature infants fed preterm formula past hospital discharge. *Eur J Clin Nutr* 1995; **49**: 134–6.
100. Rajaram S, Carlson SE, Koo WWK, Braselton EW. Plasma mineral concentrations of preterm infants fed a nutrient-enriched formula after hospital discharge. *J Pediatr* 1995; **126**: 791–6.
101. Tyrala EE, Manser JI, Brodsky NL et al. Serum zinc concentrations in growing premature infants. *Acta Paediatr Scand* 1983; **72**: 695–8.
102. Tyrala EE. Zinc and copper balances in preterm infants. *Pediatrics* 1986; **77**: 513–7.
103. Anderson GD, Bada HS, Sibai BM et al. The relationship between labor and route of delivery in the preterm infant. *Am J Obstet Gynecol* 1988; **158**: 1382–90.
104. Leviton A, Paneth N. White matter damage in preterm newborns – an epidemiological perspective. *Early Human Dev* 1990; **24**: 1–22.

Commentary

**Lucas:** The apparently firm conclusion that LCPUFAs are conditionally essential nutrients is liable to be misinterpreted by the reader as meaning that current data support a dietary requirement for LCPUFAs in formula-fed preterm infants. If essentiality is based on functional outcomes then there are two general categories of such outcomes (which could be subdivided further): (i) neurodevelopment, including mental and motor development plus looking behavior (as an indicator of cognitive function); and (ii) visual function, including acuity and retinal physiology. To these we could add growth as a further outcome of fundamental interest in nutritional interventions and of possible relevance to function (see Heird, this volume). Of these outcomes, growth rate appears to be lower in the LCPUFA-supplemented group in at least two and possibly three of four trials in preterm infants (see Lucas, this volume). With regard to neurodevelopmental function, Carlson's first trial shows a lower mental developmental index (MDI) (not significant; but see comment below) and lower psychomotor developmental index (PDI) ($p < 0.09$) in the LCPUFA-supplemented group. In her second trial these two scores were higher in the supplemented group. The tests of looking behavior (Fagan) used as a measure of cognitive function have yielded at least controversial results. Innis (this volume) challenges the validity of the analysis of the look duration data in Carlson's first trial. But, regardless of this, Carlson's interpretation of her results on look duration as reflecting a biological advantage for the supplemented group is based, as I understand it, on a study in monkeys given a test analogous to the Fagan Test after rendering them n-3 deficient. In the absence of long-term data in humans on the effect of LCPUFA supplementation on later cognitive function, the interpretation of the findings on look duration in this context must surely remain speculative. The effect of LCPUFA supplementation on the final outcome, visual function, is more compelling, though the observed effects have been transient or short term, and any lasting significance unexplored. Taking an overview here, it would be hard to argue that functional essentiality of LCPUFAs has been proven. What we have at present is a mixed bag of negative, neutral, and positive findings based on short-term, undersized (see below) studies. I am not suggesting that further, larger studies, perhaps with more optimal level of n-6 and n-3 LCPUFAs, will not demonstrate convincing essentiality of dietary LCPUFAs; indeed they may, but this remains to be seen. The longer term effects on cognitive function are particularly important to explore. To my knowledge, of all the trials in preterm and term infants, the few data on cognitive function beyond one year of age either show no impact of early LCPUFA supplementation or an adverse effect.

In the section on 'next generation trials', Carlson states that the three trials done on preterm infants so far were, for the most part, 'large enough to answer the questions they were designed to ask'. This statement needs to be reconsidered, since it appears not to be the case. Indeed, Carlson herself puts considerable weight on possible cognitive outcomes (look duration) in her argument that dietary LCPUFAs are conditionally essential. As pointed out in my own chapter, any plausible hypothesis on the cognitive benefits of n-3 LCPUFA supplementation (based on any cognitive advantage of breast-fed versus formula-fed infants) would need to be tested on hundreds and not tens of infants. For instance, to detect a plausible one-third of an SD difference between feed groups, requires 144 subjects per group at 5% significance and 80% power. None of the studies begins to approach this size. It is not implausible that the apparent difference in general cognitive (MDI) and psychomotor function (PDI) between groups in the small studies done so far could represent 'noise' or type I error (particularly since the differences between groups are in both directions: favoring and not favoring the supplemented group).

Carlson states that in her first randomised trial infants fed commercial formulas and formulas supplemented with n-3 LCPUFAs had similar mean scores on the Bayley MDI. This statement is incomplete and statistically unsound. The Carlson trial was not of adequate size to test for a plausible difference in cognitive scores (see above) so that a statement of 'no difference' is misleading. If there had been, say, five subjects per feed group, clearly it would have been impossible to detect anything but the most implausibly massive difference if it was present: that is, a type II error would have been virtually inevitable. In fact, Carlson's study sample size is little less extreme than that, given that hundreds rather than tens of subjects would be needed (see above). The second point is that, in fact, the supplemented group had a mean MDI that was 7.3 points lower than the controls. This is around half a standard deviation. Were this to have been confirmed as significant in an appropriately sized study, it would have reflected a major cognitive deficit for the n-3 LCPUFA-supplemented group. A difference of this magnitude should be quoted in Carlson's text and the phase 'similar mean scores' deleted.

Carlson puts growth and psychomotor development in the same category. This is contentious, and has the effect of 'sidelining' the suggested adverse effect of n-3 LCPUFA supplementation on the Bayley PDI in the first Carlson trial. A better case could be made for categorizing the Bayley PDI and MDI together as 'neurodevelopmental outcomes'. They are indeed linked. In the absence of significant language development at 12 months, conventional cognitive testing is highly dependent on age-appropriate motor skills, so that it is difficult, at this age, to extricate psychomotor and mental skills. Indeed, had Carlson correlated the PDI with the MDI rather than with linear growth, as she chose to do, she might have found much better correlation. In our own study (see references 17 and 51 in my chapter, this volume) on preterm infants, albeit at 18 months rather than at 12 months, the Bayley MDI and PDI were significantly correlated ($r = 0.5$, $n = 695$; unpublished), whereas the correlation coefficients between PDI and (a) current body length or (b) length gain in the previous 9 months were, respectively, only 0.2 and 0.1.

Another general comment, leading on from the last point, is the somewhat unguarded use of 'correlations' in the text which may sometimes be taken to infer cause; for instance, correlations between the Bayley PDI and linear growth, and arachidonic acid and growth. Significant correlations are frequently present between variables in data sets and may not have any causal relationship whatsoever. When correlations are used in a sense that could imply cause, then more full discussion is required on the other factors that would support a causal relationship (e.g. biological plausibility, consistency with other findings, and temporal relationship).

Carlson points out that the balance of n-3 to n-6 LCPUFAs could be critical to final outcome in terms of safety, making the safety issues raised so far (for instance in my own chapter) irrelevant. This is speculative, given the absence of published data at this stage from trials that Carlson envisages. Moreover, if alterations in the balance of LCPUFAs in the diet can indeed make a large difference to outcome, as Carlson speculates, then there will always be potential safety issues to be addressed as each different formulation is introduced.

I would question Carlson's statement where she appears to contend that long-term studies on the impact of LCPUFA supplementation on later cognitive and sensory outcomes are being done to determine better the predictive value of infant tests for later function. If I have understood what is being said here, it is 180° from my own view (and indeed that expressed in Wainwright's chapter). The purpose of the long-term studies is to validate the efficacy of LCPUFA additions at a stage in development when the biological significance of the outcomes is in less doubt than it is in infancy. As I and Wainwright (this volume) point out, it is not valid to use previous correlations between infant and later test results as a means of predicting how in future studies (using different experimental circumstances) infant tests will indicate long-term (more definitive) functional efficacy. Long-term studies will always be required. In any case, the predictive value of all cognitive tests in infancy including those based on looking behavior, is disappointingly low (see the chapters by McCall and Colombo, this volume).

The subsequent statement made by Carlson that long-term studies are not necessary to resolve the issue of whether preterm infants would benefit from dietary LCPUFAs needs clarification, and I would find this statement, as it stands, difficult. Obviously some short-term findings may have clear contemporary significance: any adverse health effects identified with LCPUFA supplementation would be a good example. Also, adverse effects on behavior in infancy could, arguably, have contemporary biological significance. But, since the major interest in the LCPUFA field concerns outcomes such as cognitive function that are difficult to test satisfactorily in infancy, it would seem that long-term testing would always be needed to demonstrate meaningful functional efficacy.

The initial trials by Carlson and Uauy have undoubtedly been key contributions to this field, and it is valuable to see them put into broader context in this review. It is clear that the 'next generation trials', as Carlson terms them, will be of critical value in testing formally the important hypotheses raised and in underpinning practice in infant nutrition.

AUTHOR'S REPLY. My view that LCPUFAs are conditionally essential is based on functional evidence of positive effects. The negative effects on growth should be a solvable problem, but they are, of course, in the area of unresolved safety issues. I have written of the need to resolve these issues in my chapter.

As I understand the comment of Innis, it is based on her misunderstanding that the Fagan test uses different stimuli at different ages. The effects of n-3 LCPUFA on look duration were found in both trials, clearly not suggesting a power issue. As we have written, there is a completely plausible explanation for the higher (significant) Bayley MDI on our second trial compared with our first (no effect, although you suggest it may be due to studying too few infants).

I do not take issue with your statement that hundreds of babies must be studied to detect differences in the Bayley MDI and PDI between groups. Your power calculation appears to be based upon the relatively small difference you find in older infants who were formerly breast- or bottle-fed. The differences in our second trial and in a recent report from Damli et al. (reference 93 in Carlson's chapter) suggest that larger effects of LCPUFA supplementation on the Bayley MDI may occur in preterm infants.

We do not see how grouping growth and psychomotor development could be seen as contentious. Moreover, your comment seems to imply that you do not consider the Bayley PDI to be as important as the Bayley MDI. Your experience with how these measures correlate is valuable to include here, but I see it as only one further way to look at this type of data. We have been in the habit of combining linear growth with the PDI only because they are correlated at 12 months in preterm infants, whereas the MDI and linear growth are not.

**McCall and Mash**: After reading the first several pages, I wondered (as I have after reading some other chapters) whether the results from deficiency studies are relevant to the issue of adding supplements? It seems to me that this issue at least ought to be discussed, and it should be related to the concept of conditional essentiality discussed here, and in some other chapters. Please excuse my ignorance of biochemical systems, but it seems at least possible that a substance can be essential for development, as demonstrated by deficiency studies, but that the vast majority of infants receive a sufficient minimum that is adequate to keep them above the danger point and that additional amounts of the substance are somehow not used or eliminated by the system, thereby making supplementation rather ineffectual. Furthermore, it is possible that there are limits on the infant's system to regulate the amounts of the substance above the sufficient minimum, such that there is also a sufficient maximum, such that amounts of the substance that exceed this sufficient maximum cannot be regulated by the infant's system and may produce undesirable outcomes. The issue would seem to rest on the absolute amounts of the substance in deficiency versus supplement studies. If preterm infants who are to be supplemented have presupplementation levels of the substance in the range of those produced in experimental studies of deficiency, then the literature on deficiency might well apply to supplementation studies. However, if the range of values produced by deficiency studies is substantially lower than the levels naturally found in preterm infants who are to be supplemented, then the deficiency literature might not have relevance to the supplementation literature. Presumably, these levels are known (and may be in this article), but I would like to see a direct comparison and a little discussion of this issue (which may be less obvious to those of us who lack biochemical training).

You provide evidence that pertains to the issue underlying my first comment. That is, I think some discussion, albeit speculative, should be presented on the possible biological models of how the infant takes up and regulates LCPUFAs and the consequences of normal and abnormal levels of LCPUFAs and abilities of the infant to regulate them. Here you describe the very great variability of both the mother's contribution of LCPUFAs to the infant prenatally and through her milk, as well as the very great variability among infants to synthesize what is provided. The result is great variability within infants, even among those who are breast-fed, in the levels of docosahexaenoic acid and arachidonic acid, for example. Any concern about deficiencies or overdoses of LCPUFAs would seem to need to deal with the fact that great variability in the levels of these substances exist within otherwise normal or healthy infants. This is why a study that relates individual differences in the levels of these substances in mother's milk and in the infant to physical, sensory, and behavioral outcomes would seem to be of great value. If there are no correlations, at least across a substantial range of such values, then presumably the system may have a sufficient minimum, even though the substances are conditionally essential. Also, what would a result that says there is no correlation between the levels of these substances in mother's milk and the levels in the infant say about the wisdom or utility of supplementation for infants whose systems contain these substances within the range studied? You speculate a bit about these issues, but it seems to me that this is a topic that deserves concentrated attention in preparation for a decision on whether LCPUFAs should be added to formula.

You provide some evidence pertaining to this question in which a marker for docosahexaenoic acid status at 2 months is (I assume) correlated $r = 0.35$ (12% of the variance) with look duration at 12 months. You indicate that this is based upon all 84 infants in both randomized trials; are these correlations across both supplemented and control groups? If they are not calculated separately within these groups, they may only reflect the mean differences between groups produced by the supplementation and not reflect any relationship that occurs as a function of natural individual differences in the levels of these substances within these groups. In other words, I for one would like to see the correlations calculated separately within supplemented and control groups, because it is statistically possible for the within-group correlations to be 0.00 but nevertheless have the correlations you report when these two groups are combined into one.

Finally, your conclusions seem to suggest that you would advocate the addition of LCPUFAs to formulas for all healthy preterm infants. I think it would be useful, since others in our group may take different positions, for you to indicate why you are persuaded of this decision and, perhaps more importantly, how you handle the apparent concerns of others who are less positive about the need or wisdom of adding LCPUFAs to formula. Do you feel that all the concerns are eliminated if supplementation is restricted to healthy preterms (which helps to deal with potential health risks to infants with certain diseases or disabilities as well as the possible negative outcomes for full-term babies)?

AUTHOR'S REPLY: The variability of LCPUFAs in human milk and the variability of LCPUFA synthesis have not been shown to be a factor in the growth and development of normal infants. The same variability in preterm infants may be a factor in them, although the evidence to date suggests that if there are some infants who can acquire optimal docosahexaenoic acid from synthesis or by some other means, they are probably a small proportion of the total. A number of people have found correlations of visual acuity with red blood cell docosahexaenoic acid across a standard range in preterm infants. Higher red blood cell docosahexaenoic acid at 2 months of age was also related to shorter look duration at 12 months of age in our two randomized trials. These relationships are discussed somewhat later in the chapter, but do provide evidence that at least for preterm infants some infants have better docosahexaenoic acid status than others with functional consequences.

One of my main points was to suggest that above a certain intake of LCPUFA there may be no choice but to store the material in adipose tissue or to use it for energy. We have a pretty good idea that membranes can be saturated with both n-3 and n-6 LPUFAs such that further intakes will not increase membrane n-3 and n-6 LCPUFAs further. Again, however, the main issue with infants is how to guarantee that none has inadequate LCPUFA accumulation. The other issues could have to do with how to determine that we give enough but not too much. This is, of course, important to determine.

Parent years of formal education and red blood cell phosphatidylethanolamine docosahexaenoic acid at 2 months adjusted age were significantly inversely related to look duration at 12 months when both diets were combined. At your suggestion a similar stepwise regression analysis was done for each diet independently. The total years of parent formal education in the control group was inversely related to look duration ($r = -0.442$, $F$ to remove 9.524) but RBC PE DHA was not (keeping in mind that RBC PE DHA becomes quite low after months of a diet without DHA and there is little variation). On the contrary, red blood cell phospholipid DHA in the n-3 LCPUFA supplemented group was inversely related to shorter look duration ($r = -0.453$, $F$ to remove 10.318) while parent years of formal education were not. The significant relationship between this indicator of DHA status found only in the group fed DHA is analogous to our earlier report that visual acuity at 2 and 4 months was also related to RBC PE DHA at 2 months in the supplemented and not in the control group. We interpret these data as evidence that 0.2% DHA is probably not enough to provide optimal DHA for all preterm infants.

I have tried to make it clear in my chapter that, despite evidence that preterm infants benefit from inclusion of DHA in their formulas, it is important to decide first what the true safety issues are related to LCPUFAs and to resolve them.

My comments about less than healthy preterm infants related more to the fact that the most impaired infants probably will not benefit from dietary LCPUFAs. For example, it would seem unreasonable to expect LCPUFAs to improve developmental outcomes in infants damaged by grade 3 and 4 intraventricular hemorrhage.

As I have indicated in my reply to Singer's comment, there is a potential downside in trials with preterm infants that 'take all comers': i.e. in raising irrelevant safety issues. For example, our studies included a select group of infants who were healthier than the whole population of preterm infants in our nursery. Nevertheless, the infants studied were of very low birth weight. These infants have a higher incidence of infection than term infants, and may also have retinopathy of prematurity, grade I or II intraventricular hemorrhage, or chronic lung disease.

**Mayer and Dobson**: The unique role of DHA in photoreceptor membranes is reviewed by Neuringer (1) in a previous Dobbing Workshop proceedings. In a section titled 'Disadvantages of high retinal levels of DHA', Neuringer cites evidence from research with nonprimate animals that high retinal DHA may contribute to light-induced retinal damage, and she suggests these findings 'may have important implications for premature human infants, who are often exposed to high levels of constant light and oxygen . . .'. This possibility should certainly be considered in weighing the risks versus the benefits of manipulating DHA levels in preterm nutritional formulas. Moreover, one wonders whether any of the published studies of the effects of nutrition on visual functions in preterm infants have examined light exposure of individual infants in the neonatal intensive care unit as a potential confounder.

1. Neuringer M. The relationship of fatty acid composition to function in the retina and visual system. In: Dobbing J, ed. *Lipids, Learning, and the Brain: Fats in Infant Formulas, Report of the 103rd Ross Conference on Pediatric Research.* Columbus, OH: Ross Laboratories, 1993: 134–58.

AUTHOR'S REPLY: We have not seen any evidence of retinal damage with supplemental DHA in our studies based on early visual acuity (consistently higher in DHA-supplemented infants) and incidence of retinopathy of prematurity based on visual acuity or retinopathy of prematurity (see below). Are these outcomes relevant to the issue you raise here? In our first randomized trial, supplementation began at about 4 weeks of age in infants with low or no supplemental oxygen requirements, 26.5% (9/34) of control infants and 21.2% (7/33) of fish-oil-supplemented infants had some degree of retinopathy of prematurity (ROP) (all stage 1 or 2). In our second study, in which supplementation was started in the first week of life and many infants both control and supplemented had a continued need for supplemental oxygen and later developed bronchopulmonary dysplasia (BPD) (48% of the controls (16/33), 46% of fish-oil-supplemented infants (12/26) with ROP). Only two (both controls) developed stage 3 ROP and one of these required laser surgery. Clearly, many more infants who develop BPD develop ROP than infants who do not develop BPD, but this is not necessarily directly related to their increased need for supplemental oxygen. However, there is no evidence that the inclusion of fish oil n-3 LCPUFA further increased the incidence of ROP.

**Heird**: You state in the section 'The next generation trials' that, 'It would be interesting to determine the effects of LCPUFAs on later behavior ... to determine what behaviors in infancy best relate to what functional behaviors in childhood'. Although developmental psychologists appear to have done considerable work in this area, I agree that more such studies are needed. However, you continue: '... such studies are not necessary to resolve the issue of whether preterm infants would benefit from dietary LCPUFAs'. I may be naive but your basis for this statement is not obvious. If the small differences in visual acuity and behavior that have been documented during the first year of life are no longer apparent in childhood, what is the evidence that preterm infants would benefit from LCPUFAs?

I also have a problem with your conclusion that supplementation of preterm formulas with both arachidonic acid (AA) and docosahexaenoic acid (DMA) will make concerns about the balance of lipid mediators derived from the n-3 and n-6 LCPUFAs irrelevant. I am aware of no published data from preterm (or term) infants fed formulas supplemented with n-6 and n-3 LCPUFAs. Thus, in the absence of data, your statement seems a bit strong. In general, the statement probably is correct if formulas are supplemented with the appropriate mixture of n-6 and n-3 LCPUFAs. However, considering the many factors other than diet that affect plasma and tissue levels of LCPUFA, arriving at the appropriate mixture is not likely to be easy. In this regard, it is not clear that the mixture of these fatty acids in human milk is as helpful as usually assumed. In addition to the variability in fatty acid patterns of human milk, human milk contains at least small amounts of all intermediates in conversion of LA and α-linolenic acid to LCPUFA and it is likely that these are effective precursors of the two LCPUFAs that have received the most attention. Thus, I am not convinced that supplementation of formulas with the mean or median amounts of AA and DHA in human milk will necessarily result in the same plasma and tissue levels of these fatty acids observed in infants fed human milk.

AUTHOR'S REPLY: I meant only that in my view there is already evidence that preterm infants benefit from dietary LCPUFAs in infancy. This is the question that has been asked by the extant trials. Longer term trials can answer other questions, e.g. if dietary LCPUFA in infancy has later benefits, but we need to keep in mind that there is a tremendous potential for type II errors in such studies, especially in the hands of those without training in development. How convinced would anyone be by negative data generated in studies conducted by physicians or nutritionists? Would not such studies simply lead to more speculation that we have not yet found the proper outcome to measure? Long-term studies carried out by individuals with expertise in neonatology, development/behavior, and nutrition/metabolism are required.

I certainly did not mean to say that supplementation of preterm formulas with both AA and DHA will make concerns about the balance of lipid mediators derived from the n-3 and n-6 LCPUFAs irrelevant. On the contrary, I have repeatedly stated that the results of studies in preterm infants with both n-3 and n-6 LCPUFAs are not yet available and that any suggestion that the inclusion of both AA and DHA in formula will be desirable is speculation.

When we talk about the variability of LCPUFAs in human milk, it is important to remember that this is not particularly relevant or helpful to the question of whether or not infants fed formula need LCPUFAs. I think they do, and am more inclined to view the variability of LCPUFAs in human milk as evidence that wide ranges of safe and effective intakes of n-3 and n-6 LCPUFAs probably also exist. The questions of whether or not infants benefit from increasingly higher amounts of LCPUFAs is a different one that has been asked in only one study as far as I know: Makrides and coworkers increased the amount of DHA in human milk by feeding a source of DHA during lactation. Despite this, they reported during discussion at the American Oil Chemist's Society Meeting on PUFA in Infant Nutrition: Concern and Controversies that visual acuity did not increase further when milk DHA exceeded 0.2% of total fatty acids.


**Singer**: Average fixation duration is the measure reported by Colombo *et al.* (1) to have some stability and predictive validity. What evidence is there for stability and predictive validity in the derived measures used in your trials for preterm infants of n-3 supplements, i.e. number of total looks, number of looks to novel, number of looks to familiar, and time/ novel and familiar looks?

In preterm study II (2) (I believe) and even in study 1 (3), the numerous comparisons (10 derived, interrelated measures) during paired comparisons on the Fagan Test and 6 during familiarizations (Tables 3 and 4) (2) with high attrition (27 subjects from 52) at 12 months strongly capitalizes on chance, making any interpretation highly speculative. In addition, the measure of look duration used by Jacobson *et al.* (4) in their alcohol studies averaged look duration for each problem by dividing *total* duration looking time by number of looks across *all* problems at each age and then across the two ages. Perhaps your data could be re-analyzed using this measure. In study I (3), for example, look duration was significantly lower only at 9 months, raising questions about its significance if groups did not differ at the other two ages. How, also, can the findings of shorter look duration only at 9 months (3) be reconciled with findings of lower head circumference, a more well validated marker of developmental outcome (5,6), also only seen at 9 months (7)? Apart from the concerns about the large numbers of comparisons which increase the possibility of type I error, I have additional concerns about the possibility of the influence of confounding medical variables common in very low birth weight populations (see Singer, this volume) which have not been assessed. With such small sample sizes, the influence of one or two factors, such as ROP patent ductus arteriosis, is likely to be nonrandomized and may affect your outcomes in unspecified ways.

In study II (2), you indicate that results from 27 infants are reported who were tested on the same version of the Fagan Test used in study I. It would appear that about an equal number of infants were tested using a different version. Were findings similar for both sets of infants? Your paper indicates that this version (old) of the Fagan Test accrued looks at ≤0.1 s, while Colombo's chapter (this volume) indicates that his studies using infant control procedures require at least a 1 s look to begin/terminate a trial. If findings regarding look duration are robust across both procedures, should not this measure be valid across both versions of the Fagan Test?

1. Colombo J, Mitchell D, O'Brien M, Horowitz FD. The stability of visual habituation during the first year of life. *Child Dev* 1987: **58**: 474–87.
2. Carlson S, Werkman S. A randomized trial of visual attention of preterm infants fed DHA until two months. *Lipids* 1996; **31**, 85–90.
3. Werkman S, Carlson SE. A randomized trial of visual attention of preterm infants fed DHA until 9 months. *Lipids* 1996; **31**, 91–7.
4. Jacobson S, Jacobson J, Sokol R, Martier S, Ager J. Prenatal alcohol exposure and infant information processing ability. *Child Dev* 1993; **64**: 1706–21.
5. Ross G, Lipper EG, Auld DM. Growth achievement of very low birthweight premature children at school age. *J Pediatr* 1990; **117**, 307–9.
6. Hack M, Breslau N, Weissman B, Aram D, Klein N, Borowski E. Effects of VLBW and subnormal head size on cognitive abilities at school age. *N Engl J Med* 1991; **325**: 231–7.
7. Carlson S, Werkman S, Trotley E. Effects of long chain n-3 fatty acid supplementation on visual acuity and growth of preterm infants with and without BPD. *Am J Clin Nutr* 1996; **63**: 687–97.

AUTHOR'S REPLY: I am assuming you mean derived in the sense of measured during the Fagan Test? Related to stability of look number/duration, I indicated (my comment on Colombo, page 377) that we find a correlation among all three test ages on look number ($r = 0.48$ to $0.68$). However, except for a weak correlation between 9 and 12 months ($r = 0.361$), there is no correlation among the three test ages on visual recognition memory, the intended purpose of the Fagan Test. Although Fagan has implied that the test gives equivalent results (i.e. is stable) for visual recognition memory at these three ages, this appears to be generally true for groups but not for individuals.

As you know, we cannot address the issue of predictive validity, because the infants in our study were not followed after 12 months corrected age. However, Jacobson finds a correlation between the average look duration at 12 months (derived by averaging look duration during familiarization and test) and IQ in childhood (see my reference 88). We have looked at look duration during familiarization and test independently, but they are highly correlated, especially at 9 and 12 months: 9 months, $r = 0.809$ ($n = 51$, study I); 12 months, $r = 0.535$ ($n = 57$, study I), $r = 0.704$ ($r = 27$, study II), $r = 0.578$ ($n = 84$, combined studies I and II).

Colombo (this volume) has indicated that look durations during familiarization might be conceptualized similarly to look durations during paired comparisons. This seems to be supported by his reanalysis of some of his early data (Colombo, this volume) as well as by the studies of Neuringer and Reisbick (commentary, this volume), reanalysis of the Jacobson data (discussant in the session on at the 10th International Conference on Infant Studies), and the correlations from our study above. All the above find a positive correlation between look duration during habituation (or familiarization) and look duration during test (or recovery).

You raise questions here about a number of issues related to our use of the Fagan Test of infant intelligence: (i) the number of interrelated measures reported; (ii) the loss of half of the sample in study II because a newer version of Fagan software led to different results; (iii) the averaging of look duration across familiarization and test and the suggestion that we analyze our data using this averaged measure; (iv) the overall significant effect of diet on look duration versus significance at an individual time point; (v) potential confounding medical variables common in very low birth weight populations; and (vi) differences between the two versions of the Fagan Test software used in our study II. The following responses address these issues.

(i) The study was designed a priori to look at the effects of n-3 LCPUFAs on look number across time. The additional expression of the data as look duration does not involve different data, merely a different means of expressing the same data. That is because look duration is the reciprocal of look number/time when time exposure to the stimulus is fixed as in the Fagan Test familiarization, and virtually identical to the reciprocal of look number/time when the time of exposure to the stimuli is extremely short, as in the Fagan Test. The literature on behavior includes examples of both means of expressing this type of data. Since becoming aware of the data of Reisbick and Neuringer, we have provided information as look duration as well as look number. Jacobson has also expressed similar data from alcohol-exposed infants given the Fagan Test as look duration.

We give credit to Reisbick and Neuringer for demonstrating that failure to accumulate docosahexaenoic acid in the retina and central nervous system increased look duration of rhesus monkey infants ((1), subsequently reported in my reference 85). Although a preliminary report showed look number was increased in preterm infants fed docosahexaenoic acid (my reference 85), we could not interpret our data as an effect of n-3 fatty acid status, because n-3 fatty acid deficiency had not been reported to influence look number/duration.

The other, highly interrelated measures reported in these papers, which you refer to here, are experimental. They were presented to supply information that might enhance the understanding of the look duration/look number data. This information would otherwise have been lost. While the analysis of these interrelated measures is not the most conservative, readers may use the information to generate new hypotheses or to enhance their understanding of the effects of docosahexaenoic acid on look duration.

(ii) Study II does not stand alone because of the loss of the last half of the sample due to the new version of Fagan software. Study II does, however, support study I. The effects of diet on visual attention in the two studies are remarkably similar. Moreover, despite the loss of power, some effects in study II reach statistical significance, evidence that these effects of diet on look duration are fairly robust. Regardless of the limitations of study II, therefore, it confirms and even expands the findings of study I.

(iii) There is a significant overall effect of diet on look duration during test across all ages (ANOVA) and look duration during test and familiarization are highly correlated (as detailed above). There seems to be little value in calculating yet another interrelated measure.

(iv) The repeated measures ANOVA indicates that there is a highly significant overall effect of diet on look duration across all ages. Individual comparisons at different ages may be made, but cannot be assumed to have the same power as the overall test. When statistical significance is not reached at individual time points, the assumption would generally be that there is not sufficient power at those ages rather than that there is not a difference at those ages.

There was no overall effect of diet on head circumference in this study and normalized head circumference was smaller only at one age (9 months). This one-time finding may or may not be spurious. However, regardless, we know these infants do not have developmental delay. Three months later, this group scored 13 points higher on the Bayley MDI and 5 points higher on the Bayley PDI than controls. You refer to two studies in which lower head circumference in childhood was related to poorer developmental outcomes. However, the implied correlation between lower head circumference and development in both of the studies you cite appears to have been due to abnormal head growth after infancy in a small subset of the very low birth weight population: Ross *et al.* (your reference 5) classified previously low birth weight infants as neurologically abnormal or normal and then compared head circumference, while Hack *et al.* (your reference 6) classified infants as having subnormal or normal head size and then compared their cognitive abilities.

(v) As shown in Table 5, our studies I and II did not exclude infants with grade I and II intraventricular hemorrhage (IVH) or stage 1–3 ROP. However, there is no evidence that these occurred other than randomly between

Table 5.

| Study I | Control | Fish-oil-supplemented |
|---|---|---|
| No IVH | 26 | 28 |
| IVH grade 1 | 6 | 4 |
| IVH grade 2 | 2 | 1 |
| Total | 34 | 33 |
| No ROP | 25 | 27 |
| ROP stage 1 | 7 | 5 |
| ROP stage 2 | 2 | 0 |
| ROP stage 3, total (laser) | 0 (0) | 1 (1) |
| Total | 34 | 33 |

| Study II | Control | | Fish-oil-supplemented | |
|---|---|---|---|---|
| | No BPD | BPD | No BPD | BPD |
| No IVH | 13 | 9 | 16 | 7 |
| IVH grade 1 | 4 | 5 | 2 | 1 |
| IVH grade 2 | 1 | 1 | 0 | 0 |
| Total | 18 | 15 | 18 | 8 |
| No ROP | 13 | 4 | 12 | 2 |
| ROP stage 1 | 4 | 6 | 5 | 5 |
| ROP stage 2 | 1 | 3 | 1 | 1 |
| ROP stage 3, total (laser) | 0 (0) | 2 (1) | 0 | 0 |
| Total | 18 | 15 | 18 | 8 |
| *Study II (subset studies for visual attention)* | | | | |
| No IVH | 5 | 3 | 7 | 5 |
| IVH grade 1 | 1 | 1 | 2 | 1 |
| IVH grade 2 | 1 | 1 | 0 | 0 |
| Total | 7 | 5 | 9 | 6 |
| No ROP | 4 | 1 | 6 | 1 |
| ROP stage 1 | 2 | 2 | 3 | 5 |
| ROP stage 2 | 1 | 1 | 0 | 0 |
| ROP stage 3, total (laser) | 0 (0) | 1 (0) | 0 (0) | 0 (0) |
| Total | 7 | 5 | 9 | 6 |

the diet groups in study I. In study II, a somewhat higher proportion of controls than fish-oil-supplemented infants developed BPD. The control BPD infants were somewhat more likely to have had IVH grades 1 and 2, but this was not the case in the subgroup who received the same version of the Fagan Test as infants in study I. The proportion of infants with some stage of retinopathy of prematurity was virtually identical in both diet groups in studies I and II.

Even when diseases appear to be distributed equally between groups, however, they do not necessarily have predictable effects on an outcome such as look duration. This is why it is important not to reach any conclusion on a single study. The replication in our second study of the effect of dietary docosahexaenoic acid on look duration observed in our first study increased our confidence that the effects on look duration in both studies were due to docosahexaenoic acid intake. The ability to determine the independent effects of diseases of the preterm on look duration are another reason for doing larger trials. We think such information could even lead to better hypotheses about the role of docosahexaenoic acid in central nervous system development.

(vi) Approximately the same number of infants were tested with both versions of the Fagan Test at 12 months. The newer version (compared with the one that was reported in your reference 3) resulted in significantly higher novelty preference, number of looks and shorter look duration. There was no suggestion of an effect of diet on look duration with the second version of the Fagan Test, even though the effects seen with the first version in an earlier randomized trial were confirmed. I have questioned what might be considered a meaningful look in my response to Innis.

1. Reisbick S, Neuringer M, Gohl E. Increased look duration in paired comparisons by rhesus monkey infants with n-3 fatty acid deficiency (n-3 FAD). *Soc Neurosci Abstr* 1994; **20**: 1696.

**Wainwright and Ward**: In discussing the original studies conducted using learning tasks in rats fed n-3 deficient diets, Carlson emphasizes that it is important to consider effects on components of the task unrelated to learning ability per se, such as sensory abilities and/or motivation. In the brightness discrimination task used by Yamamoto *et al.*, these authors argue that differences in vision cannot account for their findings because they used light intensities beyond the minimum threshold. Nevertheless another concern remains with respect to the interpretation of these findings.

In describing this study, Carson states that there were fewer correct responses in animals fed n-3 deficient diets. However, a more accurate description of these findings is that these animals showed a lower ratio of correct to incorrect responses, i.e. they were less efficient in their overall pattern of response. In fact, as the n-3 deficient animals showed an overall higher level of responding, the absolute number of correct responses (and therefore their number of food rewards) do not appear to be very different from the controls. This suggests therefore that differences in response-inhibition may be what underlie the observed differences in learning performance on this task.

One of the problems associated with repeated-measures designs is that of subject attrition, i.e. the analysis is based only on the number of subjects who provide data at all time-points. Thus the power gained by controlling for individual differences can be threatened by smaller numbers at some ages. The alternative approach of conducting separate ANOVAs at each time point (in the absence of a diet × time interaction in a repeated measures analysis), may lead to increased probability of type I error. Is there any way around this dilemma, such as, for example, assigning the missing data the average value for their group, or does this too have problems?

AUTHOR'S REPLY: This is an excellent question about a method of analysis that seems to be frequently misunderstood. When subjects are studied at more than one time-point, a repeated-measures design is implicit. Several ways to do repeated-measures ANOVA have been developed. It is rarely correct to do an analysis as though the data were cross-sectional (different subjects at each age) by conducting separate ANOVAs at each time-point because this analysis does not exploit the full power of the design, and can lead to type I errors, as you point out. A repeated-measures ANOVA based only on subjects who provide data at all time-points could also be a problem. For example, this procedure would result in the loss of a subject if only 5 of 6 planned assessments were obtained. It can be valid to estimate missing data based on other data from that subject, but not, so far as I know, by using group averages (as one might do for handling missing values in a two-way factorial arrangement of treatments).

In our own longitudinal studies, we have used a design involving repeated measures with each infant serving as his or her own control. We have analyzed the resulting data as a split-plot design, and as a consequence do not lose cases due to missing observations. Comparing least-squares means retains the advantages of a repeated-measures ANOVA, but neither estimates missing data nor excludes available data from subjects who cannot be tested at all ages. Generally, we also provide information about how many infants were tested at each time point, because this information allows the reader to determine how many measurements were used to obtain the least-squares means and error terms at each age. I might add that individual infants are rarely 'dropouts' but rather absent at random assessment times.

Other important points about such a split-plot analyses are as follows:

(a) The null hypotheses is that there is no effect of diet on outcomes at individual ages. The design is to compare the effects of diet at different ages and such comparisons are not post hoc and governed by the 'rules' of multiple comparisons. The overall ANOVA does not have to be significant to make the preplanned comparisons specified by the design.
(b) Moreover, $F$-tests for the overall ANOVA need to be interpreted with caution because the overall ANOVA may not be significant if the infants are measured at ages when differences should not be present (e.g. growth, biochemistry, or visual function at the time subjects are randomized to treatment) or have already disappeared (e.g. visual acuity after 4 months in some of the trials reported at this meeting).

1. Sokal RR, Rohlf FJ. *Biometry*, 2nd edn. New York: WH Freeman, 1981.
2. Searle SR, Speed FM, Milliken GA. Population marginal means on the linear model: an alternative to least squares means. *Am Statist* 1980; **34**: 216–21.

**Shaw and McEachern**: We concur that in cases where sensory deficits are discovered no conclusion about cognitive function can be reached. What then do you make of the animal and human literature reporting effects of diet on ERG and VEP in relation to presumed indices of cognition (e.g. visual habituation)? Can any inference be made regarding cognition if sensory effects persist? Points 1(i) and 1(ii) raised in our commentary on Heird are relevant here.

Your references to your work on vitamin A (1) and the effects of various minerals (2) indicate that some of the infants in each study also received LCPUFA-supplemented formula. Do your LCPUFA studies reported here reflect the same infant population? If so, this raises two questions. (i) Although internal validity may not be affected since all infants received vitamin A/minerals, it may prove difficult to compare your 'controls' with the 'controls' from other studies where such additions were not included in formula. Might not such differences account for the failure of other groups to find effects of n-3 FA supplementation? Might not possible interactions between these various molecules lead to different thresholds for function/dysfunction on both visual and cognitive tasks? (ii) The broader question is as follows: since all infant formula is a 'cocktail' of ingredients, how can one ever be certain that a particular LCPUFA is alone affecting neural development, rather than the interaction of LCPUFAs with various other molecules? For example, DHA might have only a small deleterious effect on retinal development; and vitamin A might have a similarly small effect, such that neither on its own would cause significant functional impairment. The two might act synergistically when administered together in formula, however, particularly when added in forms never occurring in nature (e.g. in breast milk). The converse situation also applies, in that a particular molecule might have large effects on its own but be constrained by the presence of other molecules (e.g. the apparent limiting of DHA-related deficits by AA). This highlights once again the importance of maintaining molecules in appropriate contexts and ratios in dynamic systems.

1. Peeples JM, Carlson SE, Werkman SH, Cooke, RJ. Vitamin A status of preterm infants during infancy. *Am J Clin Nutr* 1991; **53**: 1455–9.
2. Rajaram S, Carlson SE, Koo WWK, Braselton WE. Plasma mineral concentrations in preterm infants fed a nutrient-enriched formula after hospital discharge, *J Pediatr* 1995; **126**: 791–6.

AUTHOR'S REPLY: In addition to sensory and cognitive explanations, Reisbick and co-workers (my reference 40) indicate that differences between diet groups in motivation and reactivity to stimuli could account for some of the behavioral differences found in infants with lower n-3 LCPUFA status. On the other hand, Neuringer and Reisbick (commentary, this volume) find no concurrent relationship between visual acuity and look duration, even though both are affected by n-3 fatty acid deficiency. In our studies, significant effects of DHA supplementation on visual acuity were found early (2 and 4 months) and effect on look duration were found late (6, 9, and 12 months) after the effects of diet on visual acuity could no longer be detected. Neither early or later visual acuity were related to look duration. These studies could be used to argue that differences in sensory function do not underlie the observed effects of n-3 LCPUFA status on look duration.

However, as I read Colombo (this volume), visual pathways may influence look duration. If this were the case, subtle effects of n-3 LCPUFA status on sensory function might still be responsible for differences in look duration found in the n-3 fatty acid studies.

You comment 'that any molecule added to a complex system out of context to its normal constraints . . . for neurons at least . . . would, at best, be ineffective'. We actually have some evidence from between-trial comparisons that DHA does not influence the Bayley MDI when it is provided to infants who have marginal status of other essential nutrients (Carlson, this volume). Higher nutrient intake alone also had no effect on the Bayley MDI. However, the group of infants who received both higher nutrient intake and dietary DHA had Bayley MDI scores equivalent to those of term, formula-fed infants. The mean score was higher than for any group of preterm infants studied previously in our center. Given the known importance of protein, zinc, and vitamin A as well as other nutrients for central nervous system development, it seems plausible that DHA might not be as useful to infants who have marginal status of many of these nutrients.

As I see it, the myriad molecules in breast milk not included in formula is not really the issue. The mixture of ingredients within a given formula is the real issue, and could be important to outcome. As has been pointed out in several chapters and commentaries, infant formulas are not standardized products. From this, we could speculate that a comparison between different formulas could result in quite different growth and developmental outcomes if compared with each other (in fact this speculation is quite plausible). We really have little choice but to compare outcomes between groups fed the same formula with and without a given molecule. Our conclusions about whether the effects are positive or negative must be based on the results of individual studies because we know that differences among outcomes of studies may be attributed to the base formula as well as to the added molecules.

While it is an interesting rhetorical question to ask if any formula can possibly be tailored to fit the probable wide range of variations across infants and within infants at different ages, the question ignores the reality that existing infant formulas have at best undergone only the crudest tailoring. In some places, the choice of ingredients is based almost solely on cost.

Other than the suggestion that AA might be needed to balance added DHA for growth, immune function, etc. (a theory that is under investigation but has not been proven), there is no evidence that adding n-3 LCPUFA to infant formula creates an imbalance in a complex system that might lead to pathology. I think the key to deciding issues such as you raise (when cumulative risk of DHA exceeds its possible benefits) must be based upon an awareness of physiologically normal ranges, the plausibility that harm might occur (based upon the known functions of n-3 and n-6 LCPUFAs and their interactions with their precursors and other nutrients), and data from randomized trials as discussed above.

**Bornstein**: Carlson discusses three published trials of LCPUFA supplementation in preterm infants. The questions for each were, in Carlson's words: 'Can diets with n-3 LCPUFAs and α-linolenic acid compared with diets containing 1.35–2.35% of energy from α-linolenic acid (a) maintain plasma and DHA at approximately the same level found in cord blood and following human milk feeding and (b) produce more mature sensory, cognitive, and motor development in preterm infants as assessed by retinal function, visual acuity development, and tests of early cognitive and motor function'.

The study of preterm infants is not uninteresting from this point of view since, as Carlson notes, 'most infants born at 26 weeks menstrual age survive but miss the opportunity for maternal-to-fetal transfer of LCPUFAs'. The increase in brain DHA of term infants fed human milk compared with formula suggests that DHA from human milk produces at most 10–15% increase in brain DHA compared with formula (Carlson's references 74 and 75) although bigger differences might be expected with preterm infants. Very low birth weight infants (less than 1500 g) have little DHA in brain and retina and little DHA available in body stores at birth compared to term infants (Carlson's reference 76). Carlson provides a sanguine view of the patterns of conflicting results among these three experiments in an effort to interpret the role of LCPUFAs in infant functional development. For example, the first assessment failed to find any differences between infants fed commercial formulas and infants supplemented with n-3 LCPUFAs on the Bayley MDI; the second trial found the n-3 LCPUFA-supplemented group scored higher. Is this a failure to replicate? Or a failure to control for exogenous factors that may be influential in preterm infant performance?

Design criteria, in the author's words, were included to 'eliminate any condition with a known or theoretical effect on infant growth and development, except for prematurity itself'. Unfortunately, if Table 3 is the guide for such exclusivity criteria, virtually no common social or contextual factors were in evidence or controlled for. Goldberg and DeVitto (1) show, for one example, that maternal intervention strategies are greater in preterm than in term infants. Variables alternative to biochemical ones may play a perhaps larger role than LCPUFAs in preterm cognitive recovery.

Although preterm and low birth weight infants may have lower IQs later in life, properly timed and titrated supplementation and stimulation appear to offset 'biological' factors: a contemporary study is that reported by the Infant Health and Development Program (2). This research team conducted an early intervention with preterm infants recruited from eight geographical and demographic sites around the USA. A total sample of 985 infants who were born at an average gestational age of less than 37 weeks and who weighed less than 5.5 pounds at birth participated. The sample was divided into an intervention group and a control group. Infants in the intervention group were given quality pediatric care, and their homes were visited weekly by trained caseworkers in the first year and biweekly in the second and third years. Parents were provided with instructions in how to care for and enhance the development of their babies, and also had the opportunity to attend professionally led support groups. During the children's second year, parents themselves attended child development centers whose staff were specially trained. Children in the control group received only pediatric care during the same period. Normally, infants born preterm might show a variety of neurological and sensorimotor, feeding, and sleep dysfunctions, and experience diverse problems making their day-to-day care more difficult. In this study, infants in the intervention group had significantly higher mean IQ scores than the control group at 3 years, and their mothers reported significantly fewer behavioral problems than did mothers of control-group infants. Thus, environmental supports proved effective in mitigating biological vulnerabilities and disadvantages of infancy.

In overview, subsequent remediation appears to compensate for early dietary deprivation. A real-world difficulty, of course, is that in the natural order of things malnutrition during pregnancy is often accompanied by the compounding effects of poverty, inadequate medical care, disease, postnatal malnutrition, and poor education.

Carlson enumerates a variety of physiological effects mediated by AA deprivation including: (i) modulation of vascular tone, (ii) cytoprotection of mucal cells, (iii) the inflammatory response, (iv) the production of pain and fever, (v) the regulation of blood pressure, (vi) the induction of blood clotting, (vii) the control of reproductive function, and (viii) the induction of sleep–wake cycle. Perhaps these sites are at least meritorious of exploratory investigation as outcomes alternative to acuity and IQ (Bornstein, this volume).

Carlson also articulates such safety factors and questions as: 'Do sources of AA and DHA contain other contaminants which may have adverse effects?' and 'Might AA and DHA might increase the risk of other diseases in preterms?' These admonitions bear repeating.

1. Goldberg S, DiVitto B. Parenting children born preterm. In: Bornstein MH, ed., *Handbook of Parenting*, Vol. 1. Mahwah, NJ: Lawrence Erlbaum Associates, 1995: 209–31.
2. Infant Health and Development Program. Enhancing the outcomes of low-birth-weight, premature infants: a multisite, randomized trial. *JAMA* 1990; **263**: 3035–42.

AUTHOR'S REPLY: To your comment in the second paragraph, I would add the possibility that the higher scores in the second trial occurred because marginal deficiencies of nutrients known to influence central nervous system were corrected by continued feeding of a formula with a higher nutrient content for an additional period of 12 weeks postdischarge.

This use of higher nutrient content was a major change in design between the first and second trials. The change in design was made because of concern that the results of n-3 LCPUFA supplementation might be found specifically because of marginal nutritional status and disappear once marginal nutritional status was eliminated. This did not occur at least so far as the effects on early visual acuity (1) and look duration (2), which were replicated, are concerned. Moreover, effects of n-3 LCPUFA on the Bayley MDI not found in the first trial were found in the second trial (higher Bayley MDI compared with controls). Because we confirmed that biochemical indices of two important nutrients, zinc and vitamin A (and likely many others), became normal at least several months earlier in that trial than expected from historical data, replication of the results of the earlier trial might not be considered an expected outcome.

When we compare the results of both trials to our historical data with the 12-month Bayley MDI in Memphis, the mean score of the group with higher nutrient intake and n-3 LCPUFA (109) is higher than any found previously for preterm infants in this birth weight range. Moreover, if we compare the control and fish-oil-supplemented groups in our second trial with the equivalent groups in the first trial (i.e. those infants without BPD fed control and fish-oil-supplemented formula), the difference between the control and fish-oil-supplemented infants is nearly 20 points (control, 95.6, $n = 14$; fish-oil-supplemented, 114.5, $n = 15$) despite similar disease characteristics (detailed in my response to Singer). Some of the social or contextual factors that we have not controlled for could be influential, but it is difficult to imagine how they could account for an 18-point increase in the Bayley MDI compared with our historical means for infants in the same birth weight range.

Your discussion of the social and contextual factors involved in infant development is a valuable one for inclusion in this volume. The study of the effects of maternal intervention strategies on outcome of preterm infants quantifies what has been understood about the effects of environment on outcome of malnourished children for many years. I would only add that just as stimulation may substitute for poor nutrition, nutrition may substitute for lack of stimulation. Ironically, given the real-world difficulty of providing such stimulation on a background of poverty, poor medical care, disease, and poor education, nutrition might be easier to provide (of course it is a fantasy to think that LCPUFA could compensate for deprivation).

In the extant trials, these physiological effects mediated by AA have been of interest primarily for safety reasons because the n-3 LCPUFA often have antithetical effects compared with the n-6 LCPUFA, AA. Now that there is evidence that AA may not be adequate for preterm infants either, these effects seen in animals and in humans at other ages are potential outcomes of trials in that include supplementation of preterm infants with AA. I agree with your idea that we should be looking for effects of LCPUFA other than acuity and IQ.

Also see my comment to Innis. It is important to have some balance in looking at this issue. Some of the comments made by participants of this meeting imply an assumption that infant formula has met the same tests for safety that formulas with added LCPUFAs are now being held to.

I meant that diets with DHA produce positive effects on early visual acuity and, possibly, cognitive function above those found when its precursor essential fatty acid, $\alpha$-linolenic acid, is fed; and those effects are analogous to the known functional roles of n-3 fatty acids.

1. Carlson SE, Werkman SH, Tolley EA. The effect of long chain n-3 fatty acid supplementation on visual acuity and growth of preterm infants with and without bronchopulmonary dysplasia. *Am J Clin Nutr* 1996; **63**: 687–97.
2. Carlson SE, Werkman SH. A randomized trial of visual attention of preterm infants fed docosahexaenoic acid until 2 months. *Lipids* 1996; **31**: 85–90.

# Polyunsaturated Fatty Acid Nutrition in Infants Born at Term

SHEILA M. INNIS

*Department of Pediatrics, BC Research Institute for Child & Family Health, The University of British Columbia, 950 West 28th Avenue, Vancouver, BC, Canada V5Z 4H4*

## Introduction

The evidence that differences in development occur between breast-fed infants and infants fed formula which are causally related to long-chain polyunsaturated fatty acids (LCPUFAs) found in human milk but absent from formula will be examined in three phases. Firstly, the argument will be presented that there is reasonable evidence for a difference in neurodevelopment between breast-fed infants and infants fed formula which is the specific result of the method of feeding. Next, evidence to show that dietary polyunsaturated fatty acids can alter the central nervous system fatty acid composition and function will be summarized to establish a framework from which to consider the probability that fatty acid nutrition can alter neurodevelopment in healthy term infants. This section will end with a review of studies undertaken for the purpose of demonstrating that fatty acid nutrition alters neurodevelopment in term gestation infants. A brief reference to the effects of dietary polyunsaturated fatty acid deficiency in animals is useful because it provides information as to which physiological systems may be sensitive to deficiency of particular fatty acids, and the extent, duration, and stage of development at which deficiency may reasonably be expected to perturb the developing central nervous system. Similarly, reference to results from studies with infants who are prematurely delivered to the extrauterine environment is of value because these studies provide information about a less mature (and thus perhaps more susceptible) central nervous system with greater deviations in fatty acid supply from the anticipated physiological norm. Detailed reviews of studies with premature infants and with animals can be found elsewhere in this volume. The final sections of this chapter will introduce issues relating to the supplementation of formula for infants born at term with LCPUFAs, including the choice of levels and sources of the fatty acids to be supplemented.

## The Breast-fed versus Bottle-fed Infant

Cognitive development is a complex process which is influenced individually and interactively by many genetic and environmental factors. Generally, the incidence of breast-feeding in industrialized countries is positively associated with social and economic advantages, and higher levels of education and occupation (1). Consequently, the demonstration of higher achievement on tests of intelligence for groups of children who were breast-fed than for groups of children who were bottle-fed as infants should not be unexpected. However, when the effects of confounding variables such as family and social characteristics are considered, breast-feeding still seems to confer a small, but significant developmental advantage (2–7). This advantage could be due to one or more of the many nutritional components, growth factors or other biologically active substances (e.g. LCPUFAs, cholesterol, peptide hormones, prostaglandins, and lactoferrin) in human milk but not formula (8). On the other hand, the benefits could be due to other aspects of breast-feeding such as maternal–infant bonding; or perhaps the socio-demographic and educational information on family background reflected in simple measures of socio-economic status does not accurately or completely reflect the important influences of the breast-feeding home environment. Another variable in studies which extend beyond the breast- or bottle-feeding period is whether the weaning and subsequent family diet of families who breast-feed differ from that of families who bottle-feed their infants. Setting aside these considerations, and accepting that causal mechanisms may not be identified, there is a body of literature from studies in different countries, with children at different ages of testing and from both socially disadvantaged and privileged groups with covariate control, to support the possibility of a positive influence of breast-feeding on developmental outcome (2–7).

## The Biochemical Framework

The rapid expansion of knowledge of the biochemistry, metabolism and functional roles of fatty acids provides a framework for considering certain polyunsaturated fatty acids as important nutrients which influence normal growth and development. Polyunsaturated fatty acids are divided into classes designated by the position of the first double bond from the methyl terminal of the fatty acid carbon chain (9). Fatty acids with the first double bond appearing at carbon number 3 or 6, designated n-3 or n-6, respectively, are considered essential dietary nutrients. This definition of nutritional essentiality is based on the absence of the enzymes needed to insert a double bond at the n-6 or n-3 position of a fatty acid carbon chain, and the appearance of deficiency signs in humans or animals fed diets lacking these fatty acids (9,10). Many of the physiological functions of the n-6 and n-3 fatty acids are not due to linoleic acid and $\alpha$-linolenic acid themselves. Rather, many important physiological and biochemical processes involve longer chain, more highly unsaturated n-6 and n-3 fatty acids and their oxygenated derivatives collectively termed eicosanoids. Some of the best known of the longer chain n-6 and n-3 fatty acids are arachidonic acid of the n-6 series, and eicosapentaenoic acid and docosahexaenoic acid of the n-3 series. These fatty acids are found predominantly in the structural lipids (glycerophospholipids) of cell and intracellular membranes, and it is from this position that they play their important physiological roles.

Arachidonic acid and docosahexaenoic acid can be formed from the dietary essential fatty acids, linoleic acid and $\alpha$-linolenic acid, respectively, through alternating desaturation and elongation (9,10). The levels of arachidonic acid and docosahexaenoic acid found in tissue glycerophospholipids is the net result of the amounts of the preformed arachidonic acid and docosahexaenoic acid in the diet, the rates of endogenous synthesis from linoleic acid and $\alpha$-linolenic acid, respectively, and, more specifically, the activity of the acyl transferases involved in glycerophospholipid synthesis and remodeling. In vitro studies and studies with other species have shown that all the enzymes needed for the synthesis of arachidonic acid and docosahexaenoic acid are present in liver, brain, eye, and retina (11–15).

Early human brain development includes growth and differentiation at the cellular, morphologic, and biochemical levels, allowing for age-specific patterns of neurodevelopment. The non-myelin membranes of the central nervous system contain high proportions of arachidonic acid and docosahexaenoic acid, as well as other carbon chain 20 and 22 n-6 and n-3 fatty acids, and some longer chain metabolites (16,17). An unusual feature of the brain and retina is that levels of the precursor fatty acids linoleic acid and α-linolenic acid are very low (generally less than 2% of total fatty acids). In other tissues, linoleic acid can readily exceed 20% of the total fatty acids, and levels increase with increasing dietary intake. Unfortunately, no definitive information is as yet available on the preferred pathway by which the brain and eye normally acquire arachidonic acid and docosahexaenoic acid, i.e. by uptake and further metabolism of linoleic acid and α-linolenic acid, or by uptake of arachidonic acid and docosahexaenoic acid from the circulation (9). Unlike the brain and eye, tissues such as heart and kidney appear to lack some desaturase activities (18,19). These tissues, therefore, depend on uptake of arachidonic acid and docosahexaenoic acid from the circulation in order to maintain membrane phospholipid levels of these fatty acids. Similarly, recent in vitro evidence suggests that intestinal cells may not be able to form docosahexaenoic acid, although synthesis of arachidonic acid does occur (20).

Recent studies have provided evidence that brain synthesizes most, if not all of the palmitic acid (16: 0) and cholesterol incorporated into brain lipids (21,22). This suggests brain has a high degree of selectivity and specificity for fatty acid uptake, and does not utilize pathways of receptor-mediated lipoprotein uptake, or lipoprotein lipase-mediated hydrolysis of lipoprotein triacylglycerols in a similar manner to that used by liver, adipose tissue, or muscle.

Numerous papers are available on the importance of arachidonic acid and its eicosanoid metabolites in a variety of biochemical and physiological processes in the central nervous system (23–25). These include roles as intra- and intercellular messengers for the action of a variety of hormones, neurotransmitters, and growth factors on neural and non-neural cells. Current information, however, does not provide a mechanism by which dietary n-6 fatty acid deficiency could lead to immediate or long-term deficits in neural function. Some recent information is available to suggest that docosahexaenoic acid may also be involved in neurotransmitter metabolism and receptor function (26,27). The biochemical mechanisms of these effects, however, are as yet unknown.

Research in the 1970s identified reduced a- and b-wave amplitudes and reduced brain and retina levels of docosahexaenoic acid in rats fed an α-linolenic acid deficient diet (28,29). Subsequently, studies with rhesus monkeys fed α-linolenic acid deficient diets confirmed electroretinogram abnormalities and also showed reduced visual (preferential looking) acuity, as well as polydipsia and a change in stereotyped behavior (30–33). The decrease in central nervous system docosahexaenoic acid which results from dietary α-linolenic acid deficiency is accompanied by a reciprocal increase in carbon chain 20 and 22 n-6 fatty acids (9). This lends credence to a theory that docosahexaenoic acid has specific functional roles in the visual, and perhaps other, neural processes which extend beyond contributing to the physical properties of retinal or synaptic terminal membranes. In this regard, for example, recent studies have shown that docosahexaenoic acid is closely associated with rhodopsin in rod outer-segment membranes (34).

Large amounts of arachidonic acid and docosahexaenoic acid, are incorporated into the central nervous system during development (35–37) in a membrane- and lipid-class-specific pattern (17). This, together with the expanding knowledge on the roles of arachidonic acid and docosahexaenoic acid in the central nervous system, suggests that a reduced supply of these fatty acids might alter the normal course of central nervous system development and in so doing lead to measurable differences in performance on psychometric or visual function tests. The questions are whether dietary arachidonic acid and docosahexaenoic acid are necessary, or whether dietary linoleic acid and α-linolenic acid can be converted to arachidonic acid and docosahexaenoic acid and so fulfil tissue needs for n-6 and n-3 fatty acids during growth and development.

Further important issues relate to the timing of the so-called 'sensitive (also known as critical) periods' when the brain is sensitive to long-term functional compromise (38), in this case due to inadequate or unbalanced n-6 and/or n-3 fatty acid supply. Anatomical, biochemical, and functional maturation of the central nervous system proceed at different rates and at different ages in its different regions. It is therefore reasonable to believe that both the timing and the duration of the so-called sensitive periods varies among different regions of the central nervous system. Consequently, the anatomical and biochemical processes that are developing at the time at which the suspected nutritional deficiency is imposed, the duration of the deficiency, and whether any subsequent tests measure functions associated with the potentially damaged systems are all pertinent questions.

Studies with animals have clearly shown that the brain and retina tenaciously retain arachidonic acid and docosahexaenoic acid, even during extended and prolonged dietary n-6 and n-3 fatty acid deficiency

(9,39–41). Consequently, quite severe experimental conditions of diet composition and duration have been used in developing animals to generate a docosahexaenoic acid deficiency in the central nervous system sufficient to explore the functional effects. For example, rodent studies illustrating differences in behavior, visual function, and resistance to toxins, utilized third-generation rats fed diets with 0.004% energy from α- linolenic acid (42). The often-cited studies with monkeys illustrating the importance of dietary n-3 fatty acids in visual function used monkeys fed an α-linolenic acid deficient diet from 2 months before conception, then through gestation and postnatal development (30,31,43). At 12 weeks of age, the n-3 fatty acid deficient infant monkeys had a plasma phospholipid docosahexaenoic acid of 0.14% compared to 2.35% in the control group fed soybean oil (31); at 22 months of age the occipital cortex and retina phosphatidylethanolamine docosahexaenoic acid was 6% and 7%, respectively, in the deficient monkeys and 34% and 37% in the control animals ($n = 2$ per group) (30). Analysis of tissue from two fetuses born from 1 month before term (165 days) and three infants of up to 16 days in the n-3 fatty acid deficient group, and from one fetus born 5 days before term and four infants of 1–18 days of age in the control group found mean levels of docosahexaenoic acid of 4% and 15% in the occipital cortex, and 8.6% and 17.3% in the retina, respectively (30). The plasma phospholipids of 12-week-old breast-fed infants generally contain about 4–5% docosahexaenoic acid compared to about 2% in infants bottle-fed with formula lacking docosahexaenoic acid (44–46). This suggests that the plasma phospholipid levels of docosahexaenoic acid in the rhesus monkey fed soybean oil (control) is similar to that in the infant fed formula without docosahexaenoic acid.

The most sensitive time for assessing the acquisition of new skills or of maturing functions in an infant or child may not necessarily correspond with the time when function is susceptible to long-term alteration in development. For example, the dorsal lateral geniculate nucleus 'sensitive period' in development of the visual system is the period during which the visual system is sensitive to visual image deprivation (47). The effects of deprivation can be reversed, provided reversal (allowing visual stimulation) is within the sensitive period. Deprivation of pattern imaging in the human infant between birth and 4 months of age does not adversely affect development of the visual system; the period from 4 months to 3 years, on the other hand, is extremely sensitive. Preferential acuity, however, shows rapid improvement during the first 4 months after birth, then plateaus between about 6 and 12 months of age (48). Consequently, the most sensitive time to measure changes in the rate of early maturation of preferential looking acuity is during the first 4 months after birth. The timing of different phases of maturation, and of sensitive periods in relation to birth, must also be considered when extrapolating data from animal studies to human infants. For example, brain development is more advanced in the monkey, but less advanced in the rat at birth than in the human (49–51). Important work on visual function and dietary α-linoleic acid deficiency has been done with monkeys (30,31). In this species, the sensitive period for reversal of visual image deprivation is birth to about 2 months of age (47). Thus, nutritional deprivation within the first 2 months after birth may have long-term consequences for visual function in the monkey, but in the human infant such early differences may be fully accommodated by plasticity of the developing central nervous system.

Studies with rodents and monkeys have identified altered electroretinogram responses and visual acuity and some behavioral changes in animals fed diets containing safflower oil, or other oils with extremely low amounts of α-linolenic acid (9,30–33,42,52). These studies may be offered as evidence for the possibility that dietary n-3 fatty acid intake could alter brain and visual function in humans. Extrapolation of this postulate to the human necessitates critique of the α-linolenic acid intake and the control diets with regard to the likelihood that such intakes are relevant to the human situation. Safflower or sunflower oil, as used in many studies on experimental n-3 fatty acid deficiency, contains about 77% linoleic acid and up to 0.3% α-linolenic acid. There is no specific information on whether dietary linoleic and α-linolenic acid intakes are best expressed as a percentage of dietary fat, or as a percentage of total dietary energy. Fat generally represents about 50% of the energy in human milk and infant formula. Consequently, an infant formula containing safflower oil blended with a source of saturated fatty acids would provide about 23% of dietary energy as linoleic acid and about 0.1% as α-linolenic acid. Animals fed diets containing the same oils but, for example, with 13% or 30% energy from fat as in maternal and infant rhesus monkey diets, respectively (31), or 4% from fat as used in rodent diets (42) would contain proportionally lower amounts of n-6 and n-3 fatty acids, i.e. about 0.01%, 0.03%, and 0.004% energy as α-linolenic acid, respectively. The control diets used in studies of experimental n-3 fatty acid deficiency have contained soybean or other vegetable oils high in α-linolenic acid. There are no studies to show that adding arachidonic acid or docosahexaenoic acid to diets containing adequate linoleic and α-linolenic acid confers any benefit to visual or other neurological functions of any mammal, with the exception of felinae (9). This, however, does not negate the possibility that human infants have unique or different requirements for n-6 and n-3 fatty acids from that of other species for which scientific data are available.

Infant formulas in North America do not contain safflower and sunflower oil as the only source of unsaturated fatty acids. Formulas containing corn oil as the only unsaturated oil, however, were available in North America until about 1992. Corn oil contains about 55% linoleic acid and 0.8% α-linolenic acid. Formulas containing corn oil blended with a source of saturated fatty acids typically provided about 16% energy as linoleic acid and 0.2–0.3% of energy as α-linolenic acid. Premature infants fed formula with corn oil have been reported to show deficits in rod electroretinogram responses when measured at 36 weeks postconception (possibly, post-menstrual age) (53). At this time, the infants had been fed the corn oil containing formula for 20–25 days. The deficits in rod electroretinogram recordings were no longer apparent at 56 weeks (about 4 months post-term), even though feeding with the corn-oil-containing formula was continued. It was suggested that the results might indicate delayed development, rather than permanent damage to rod function. The measures of rod electroretinogram function for preterm infants fed a formula with soybean oil providing about 16% energy as linoleic acid and 1% energy as α-linolenic acid showed no statistically significant differences from those of infants fed human milk at either age (53). However, light-adapted oscillatory potentials, which reflect inner retinal function, were lower in infants fed the corn oil formula at about 4 months post-term. Some changes due to dietary α-linolenic acid deficiency may then persist, or be susceptible to change, in the post-term period. The implicit times of light-adapted oscillatory potentials of infants fed a formula containing soybean oil and those fed a formula containing fish (marine) oil to provide docosahexaenoic acid were very similar (53). This suggests the deficits in visual function were related to α-linolenic acid deficiency, with no benefit conferred by provision of dietary docosahexaenoic acid rather than an adequate supply of the n-3 fatty acid precursor. Based on these findings, it would not seem reasonable to expect differences in the measured rod electroretinogram functions in term infants fed from birth with formula containing 1% or more energy as α-linolenic acid.

Others, however, have reported higher preferential looking acuity at 2–4 months post-term in premature infants fed formula with 0.2% docosahexaenoic acid than in infants fed formula with 2% fatty acids (about 1% energy) α-linolenic acid, and no docosahexaenoic acid (54). Differences in novelty preference and in preferential looking behavior during a recognition memory test have also been described in premature infants fed a formula with docosahexaenoic acid when compared to infants fed formula without this fatty acid (55,56). These studies provide evidence that dietary docosahexaenoic acid may influence neurodevelopment in preterm infants even when the α-linolenic acid intake is presumably adequate. A review of studies with premature infants by Carlson can be found in this volume. It is reasonable then to proceed with the question: 'Does dietary docosahexaenoic acid intake influence neurodevelopment in infants after term gestation?' No studies have identified neurodevelopmental differences in animals or infants fed diets considered adequate in linoleic acid which can be potentially attributed to a lack of dietary arachidonic acid. Thus, no studies have been designed to test a hypothesis that dietary arachidonic acid deficiency impairs normal neurodevelopment in term infants fed formula.

## Studies with Term Infants

It is well known that infants fed formulas containing linoleic acid and α-linolenic acid as the only n-6 and n-3 fatty acids have lower plasma and red blood cell phospholipid levels of arachidonic acid and docosahexaenoic acid than most infants who are breast-fed (9,44–46,57–61). It is also well established that the addition of arachidonic acid and docosahexaenoic acid to formulas results in increased arachidonic acid and docosahexaenoic acid, respectively, in the blood lipids of formula-fed infants (57,58,60,62,63). This should not be surprising. Numerous studies are available to show that the composition of the dietary fatty acids determines in part the composition of plasma and red blood cell fatty acids (64–70). This relation between diet and blood lipid fatty acid composition is also seen in breast-fed infants. Infants who are breast-fed by women with higher milk levels of docosahexaenoic acid have higher blood lipid levels of docosahexaenoic acid than infants who are breast-fed by mothers with low levels of docosahexaenoic acid in their milk (71,72). No evidence has been reported to suggest that variations in docosahexaenoic acid intake, and blood lipid docosahexaenoic acid, have any measurable effect on the neurodevelopment of breast-fed infants. However, the absence of reported information may be of no meaning if appropriate studies have not been done.

Information from analyses of autopsy samples in the UK and Australia have indicated lower arachidonic acid and docosahexaenoic acid in liver, lower docosahexaenoic acid but not arachidonic acid in brain, and no difference in docosahexaenoic acid or arachidonic acid in retina of infants who had been fed formula, when compared to infants who had been breast-fed (73–76). Some of the formulas described contained amounts of α-linolenic acid similar to that (<0.5% dietary energy, <1% formula fatty acids) shown to result in reduced levels of docosahexaenoic acid in the brain of formula-fed compared to milk-fed piglets (77–79). This raises the possibility that some of the lower values were due to low levels of α-linolenic acid in the formulas given. Additional, often frequently encountered problems with analyses of autopsy samples include the small number of infants studied, differences in the ages of 'treatment' and 'control' groups, confounding effects of the circumstances surrounding death and time delays to tissue recovery, and collection of dietary information.

Studies using stable isotopes of linoleic acid and α-linolenic acid have shown that term and preterm gestation infants are able to convert linoleic acid to arachidonic acid and α-linolenic acid to docosahexaenoic acid (80–82). It is not yet clear how to translate information showing conversion of precursors to products to quantitative data on rates of synthesis in relation to the needs of growing tissues. Whether or not the plasma compartment, which reflects conversion and secretion in lipoprotein lipids (and thus presumably synthesis by liver) is relevant to organs such as the brain is not known. Similarly, it is not yet clear which plasma analyses, phospholipids, triglycerides, cholesteryl esters, or unesterified fatty acids best reflect the specific lipoprotein or fatty acid pool(s) available for uptake by the brain.

A synopsis of published studies with breast-fed and formula-fed term infants undertaken with the intent of identifying potential differences in visual or other aspects of neurodevelopment attributable to dietary n-3 fatty acid supply is set out in Table 1. The only definitive approach to establishing whether dietary docosahexaenoic acid is important for 'optimal' neurodevelopment is a randomized study of infants fed formula with or without this fatty acid. However, if the tests of neurodevelopment employed are not sensitive to postnatal dietary n-3 fatty acid deficiency (i.e. do not show a difference between breast-fed and formula-fed infants), then there is a real probability that any such randomized trial will fail to demonstrate a difference. The majority of published studies concerning n-3 fatty acid nutrition have focused on measures of visual function. This, presumably, is because of the established role of docosahexaenoic acid in the visual process in the retina and experimental evidence that dietary α-linolenic acid deficiency can alter visual function. It appears to be an inherent assumption that the measures of visual function used will provide information about the development of the infant's visual system, with little or no effect of potential confounding variables, such as home environment, birth order, or ethnic background. The assumption that confounding variables are unimportant has neither been established nor considered in the studies reported to date. Recent work has suggested an inverse association of smoking and alcohol with docosahexaenoic acid (83). Both alcohol and maternal smoking may have a negative effect on neurodevelopment (84). Thus, apparent correlations of docosahexaenoic acid and infant development may simply be a reflection of other more important maternal pre- and postnatal variables which may independently affect both infant lipids and neurodevelopment.

The potential effect of dietary n-3 fatty acid intake on visual function has been assessed using both behavioral and electrophysiological methods. Currently, the most popular behavioral method is assessment of preferential visual (preferential looking) acuity through measurement of minimum separable acuity (resolution acuity) using the acuity card procedure (85). The procedure is based on the inherent tendency of infants to orientate to a discernible patterned rather than unpatterned stimulus. The assessment involves showing the infant a series of black and white gratings (stripes) of different size and determining the smallest size grating which the infant can resolve. Simply put, the infant is shown a series of cards containing stripes on one side and a luminance-matched grey field on the other. If the infant can resolve the grating, then a preference in preferential looking at the grating rather than the grey field should be observed. Acuity is determined as the smallest grating to which the infant orientates. A review of the technique can be found in the chapter by Mayer and Dobson in this volume. Benefits of the preferential looking acuity procedure include published normative data, a highly standardized and relatively fast (3–6 min) test procedure, and excellent inter- and intraobserver reliability.

Acuity has also been estimated from measures of visual evoked potentials. The visual evoked potential is an electrophysiological measure of responses to temporal changes in visual patterns (typically, a checkerboard pattern in which white squares become black and black become white). For acuity testing, checkerboards (or gratings) ranging in size from large to small are used. The acuity threshold is estimated by extrapolation of the linear regression of the visual evoked potential amplitude versus pattern size.

Acuity estimates derived from preferential looking acuity (behavioral) methods differ from those generated from visual evoked potentials. Several factors could explain this. These include the difference in the stimuli (stationary versus moving), the neural processes involved in responding, and the method of estimating the acuity threshold. The visual evoked potential measures the neural pathway from the retina to area 17 of the occipital cortex. Preferential looking, on the other hand, measures the neural pathway from the retina to the primary visual cortex, and the ability

Table 1. Summary of published studies on polyunsaturated fatty acids and outcome in term gestation infants.

| Author | Diets | Measures | Finding |
| --- | --- | --- | --- |
| Agostoni et al. (98,101) | Human milk: NA<br>Formula: 11.1% LA, 0.7% LNA<br>Formula + 0.3% γLNA, 0.4% AA, 0.3% DHA | Brunet–Lezine psychomotor development test at 4 and 12 months | Human milk and formula + LCP > formula $p <0.01$ at 4 months, no difference at 12 months |
| Auestad et al. (90) | Human milk: NA<br>Formula: 22% LA, 2.2% LNA<br>Formula + 0.4% AA, 0.1% DHA<br>Formula + 0.2% DHA | Preferential looking acuity and VEP acuity at 2, 4, 6, 9, and 12 months | No difference in looking acuity or VEP acuity between human milk and formula, or formula + LCP |
| Birch et al. (94) | Human milk: NA<br>Formula 29.4% LA, 0.8% LNA | VEP and preferential looking acuity at 4 months. Preferential, near and distance recognition preferential looking acuity, random dot stereo acuity, letter matching, picture naming at 3 years | Human milk VEP $p <0.05$ and looking acuity $p <0.025$ > formula at 4 months. Human milk random dot stereo acuity $p <0.05$ and letter matching $p < 0.001$ formula at 3 years. Other measures not different |
| Carlson et al. (63) | Human milk: NA<br>Formula: 22% LA, 2.2% LNA<br>Formula + LCPFAs (0.4% AA, 0.1% DHA) | Preferential looking acuity at 2, 4, 6, 9, and 12 months | Human milk and formula + LCP > formula at 2, but not 4, 6, 9, and 12 months |
| Innis et al. (44) | Human milk: 13.4% LA, 1.5% LNA, 0.5% AA, 0.2% DHA<br>Formula + 17.9% LA, 2.1% LNA | Preferential looking acuity at 14 days and 3 months | No difference in looking acuity in human milk and formula-fed at 14 days or 3 months |

| Study | Composition | Measurement | Outcome |
|---|---|---|---|
| Innis et al. (45) | Human milk: 18% LA, 1.9% LNA, 0.5% AA, 0.2% DHA<br>Formula: 18% LA, 19.9% LNA<br>Formula: 34% LA, 4.7% LNA | Preferential looking acuity at 14 days and 3 months | No difference in looking acuity in human milk and formula-fed. No relation of DHA to looking acuity or of AA to growth |
| Innis et al. (87) | Human milk: NA<br>Formula: >16% LA, 1.6% LNA | Infants stratified by duration of breast-feeding. Preferential looking acuity, novelty preference at 9 months | No difference in looking acuity or novelty preference in human milk versus formula-fed, or by duration of human milk feeding |
| Jorgensen et al. (86) | Human milk: 11% LA, 1.5% LNA, 0.5% AA, 0.4% DHA<br>Formula: 14.4% LA, 1.7% LNA | Preferential looking acuity at 1, 2, and 4 months | Increase in looking acuity higher in human milk than formula fed, $p < 0.001$ |
| Makrides et al. (58) | Human milk: NA<br>Formula: 12–15% LA, 1–1.6% LNA | VEP acuity at 5 months | Human milk VEP acuity > formula, $p < 0.05$ |
| Makrides et al. (59) | Human milk: 13.9% LA, 0.9% LNA, 0.4% AA, 0.2% DHA<br>Formula: 16.8% LA, 1.6% LNA<br>Formula: 17.4% LA, 1.5% LNA, 0.27% γLNA, 0.58% EPA, 0.36% DHA | VEP at 16 and 30 weeks | Human milk and formula + LCP VEP acuity > formula at 16, $p < 0.001$ and 30 weeks, $p < 0.01$ |

AA, arachidonic acid; DHA, docosahexaenoic acid; LA, linoleic acid; LNA, α-linolenic acid; γLNA, γ-linolenic acid; LCPFAs, long chain polyunsaturated fatty acids (AA, DHA); NA, not analyzed or reported for human milk fed in the study cited; VEP, visual evoked potential.

of the infant to see and produce a motor response (fixation). It is quite reasonable then to expect that the results of studies on the effects of diet, or other variables, on visual acuity may vary between studies using behavioral and electrophysiological tests.

Jorgensen et al. (86) have reported the development of preferential looking acuity in 17 breast-fed and 16 formula-fed infants. The study was prospective, but not blinded. The formula contained 14.4% linoleic acid and 1.7% α-linolenic acid. The human milk fatty acids had about 11% linoleic acid, 1.5% α-linolenic acid, 0.4–0.5% arachidonic acid, and 0.4–0.5% docosahexaenoic acid. Two of the 17 breast-fed infants later received formula and were, therefore, excluded from the analysis for 4, but not for 1 or 2, months, of age. Among the formula-fed infants, 3 had never received breast milk; the remaining 14 infants had been breast-fed for a median of 1.5 weeks after birth. The reasons for discontinuation of breast-feeding were not given. The data were analyzed by Student's $t$-test, $\chi^2$ analysis, or two-way ANOVA. Preferential looking acuity increased over time in both groups ($p < 0.001$), with a higher increase in preferential looking acuity over time in the breast-fed than formula-fed infants ($p < 0.001$). The method for analysis of the data over time, with exclusion of two of the breast-fed infants at 4, but not at 1 and 2, months of age was not given. Inspection of the curves illustrating the change in preferential looking acuity with age suggests that the difference in acuity was primarily due to a slower rate of increase in acuity from 1 to 2 months of age, with little further difference from 2 to 4 months of age. No differences in preferential looking acuity were found between the two groups of infants at 1 month of age. Mean values for preferential looking acuity at 2 months of age were 2.6 and 1.8, and at 4 months were 6.0 and 3.8 cycles/degrees in the breast-fed and formula-fed infants, respectively. It is important to note that measures of preferential looking visual acuity are based on a logarithmic function, thus differences at low acuity thresholds have greater meaning than numerically similar differences at higher thresholds.

Carlson et al. (63) have reported the preferential looking acuity of full-term gestation infants who were breast-fed ($n = 16$ to 19) or randomized to be fed a formula containing 21.9% linoleic acid (11% energy) and 2.2% α-linolenic acid (1% energy) ($n = 17$ to 20), or 21.8% linoleic acid, 2.0% α-linolenic acid, 0.43% arachidonic acid, and 0.1% docosahexaenoic acid (from egg phospholipid) ($n = 18$–19). Preferential looking acuity was measured at 2, 4, 6, 9, and 12 months of age. Human milk samples were not analyzed. However, data on human milk fatty acids are given in the report. These are from a 1982 publication by Putnam et al. (61), which included analysis of the fatty acid composition of milk collected from nine women in Tampa, Florida. The procedure for testing preferential looking acuity deviated from current standardized procedures in that a test distance of 38 cm was used at all ages, rather than increased to 55 cm for infants of 9 and 12 months of age. Whether this could influence the ability to find differences in preferential looking acuity at 9 and 12 months is not clear. Preferential looking acuity was significantly lower in the 2-month-old infants fed the formula without docosahexaenoic acid than in those fed the formula with docosahexaenoic acid or human milk (mean (cycles/degree) ± SD (octaves): 1.85 ± 0.49, 2.69 ± 0.49, and 2.93 ± 0.49, respectively). No differences in preferential looking acuity were found at 4 months (6.08 ± 0.49, 5.35 ± 0.49, and 6.07 ± 0.49, respectively), or at 6, 9, or 12 months of age. No explanation for the apparent 'catch-up' in preferential looking acuity between 2 and 4 months of age by infants fed the formula without docosahexaenoic acid was given. Clearly, the 4-month-old formula-fed infants in this study had higher mean preferential looking acuities (6.08 cycles/degree) than the infants fed a formula containing very similar amounts of linoleic and α-linolenic acid in the study by Jorgensen et al. (3.8 cycles/degree) (86). The work of Carlson et al. (63) also shows disparity between the age at which differences in preferential looking acuity (2 months) were present and the timing of differences in blood lipid docosahexaenoic acid. The plasma phospholipid docosahexaenoic acid was not lower in the 2-month-old infants fed the formula without docosahexaenoic acid (21.6 ± 2.1%) when compared to the breast-fed infants (18.6 ± 2.8%) or infants fed the formula with docosahexaenoic acid (17.8 ± 2.0%). Similarly, the red blood cell docosahexaenoic acid values were not different among the groups at 2 months of age. Infants fed the formula with docosahexaenoic acid had higher plasma and red blood cell phospholipid docosahexaenoic acid at 4, 6, and 12 months of age than infants fed the formula without docosahexaenoic acid, but no differences in preferential looking acuity were present.

Innis and colleagues have reported the results of three separate studies on the preferential looking acuity of breast-fed and bottle-fed infants born at term (44,45,87). The first study involved a prospective measurement of preferential looking acuity in a group of 18 breast-fed and 17 formula-fed infants (44). The mother's milk was analyzed and contained 13.4% linoleic acid, 1.5% α-linolenic acid, 0.5% arachidonic acid, and 0.2% docosahexaenoic acid. The formula had 17.9% linoleic acid and 2.1% α-linolenic acid, with no arachidonic acid or docosahexaenoic acid. Preferential looking acuity was measured at 14 days and 3 months of age in all the infants. No significant differences in acuity were found at 14 days or 3 months of age, despite the lower blood lipid levels of arachidonic acid and docosahexaenoic acid in the infants fed formula. Preferential looking acuity at 3 months of age was (mean (cycles/degree) ± SD (octaves)) 3.93 ± 0.54 for the breast-fed and 4.77 ± 0.48 for the formula-fed infants. No statistical relations were found between the diet (milk or formula), or plasma or red blood cell levels of docosahexaenoic acid and visual acuity. If 'catch-up' in acuity does indeed occur, this study would have missed early (e.g. at 2 months) differences in preferential looking acuity between the breast-fed and formula-fed infants.

A subsequent multicenter study by Innis et al. (45) involved 56 breast-fed infants, 59 infants fed a formula with 18% linoleic acid and 1.9% α-linolenic acid, and 57 infants fed a formula with 34% linoleic acid and 4.7% α-linolenic acid from birth to 90 days (about 3 months) of age. Samples of milk were analyzed for 56 mothers at 3 months postpartum, with mean values of 14.6% linoleic acid, 1.2% α-linolenic acid, 0.5% arachidonic acid, and 0.2% docosahexaenoic acid. Measures of preferential looking acuity at 3 months of age again found no differences among the breast-fed infants and those fed formula. As in previous studies (44,46,57–61), the plasma and red blood cell phospholipid levels of arachidonic acid and docosahexaenoic acid were significantly lower in the infants fed formula than in those who were breast-fed. This study also explored potential relations between plasma phospholipid and red blood cell phosphatidyl-ethanolamine and phosphatidylcholine docosahexaenoic acid and preferential looking acuity, and between arachidonic acid and growth within and among the breast-fed and formula-fed groups of infants, as well as in the entire group of infants. No statistically significant relations between docosahexaenoic acid and preferential looking acuity, or between arachidonic acid and growth were found. Reports of a statistical association between growth and arachidonic acid in premature infants (88), and between birth weight and cord blood arachidonic acid (89) were the basis for the statistical analyses concerning arachidonic acid. The preferential looking acuity measures in this multicenter study (40) were lower, and had a wider variance (mean (cycles/degree) ± SD (octaves): 2.79 ± 0.77, 2.51 ± 0.50, and 2.67 ± 0.64 for the breast-fed infants, and for infants fed the formula low or high in linoleic acid, respectively) than in the prospective single-center study by Innis et al. (49) with infants of the same age. The reasons for this are not known, but could include uncontrolled variables in the multicenter test settings or procedures, or in the infant populations.

Innis et al. (87) have also measured the preferential looking acuity and novelty preference of over 400 infants of 39 ± 1 weeks of age born at term. Information on the initiation and duration of breast-feeding was collected and used to stratify the infants by duration of breast-feeding: never breast-fed, $n = 68$; breast-fed < 1 month, $n = 40$; 1–3 months, $n = 51$; 4–6 months, $n = 95$; 7–8 months, $n = 49$; or > 8 months $n = 92$. The remaining infants were breast-fed and supplemented with formula for more than 1 month at any time, or during the first month after birth. The infants fed formula were given commercial formula which, for Canadian formula at that time, all contained at least 16% linoleic acid and 1.6% α-linolenic acid, with no arachidonic or docosahexaenoic acid. No differences in preferential looking acuity were found among the infants fed formula from birth (never breast-fed) and those who were breast-fed to 39 weeks (never bottle-fed). Similarly, no relation between duration of breast-feeding and preferential looking acuity was present. Again, this study would not have detected any differences in preferential looking acuity in the first 2, or 2–4, months after birth as found in the study by Carlson et al. (63) and Jorgenson et al., respectively. It is, however, consistent with other studies by this group (44,45), and those of Carlson et al. (63), and others (90) in demonstrating no differences in preferential looking acuity attributable to breast-feeding when compared to formula feeding in infants of 3 months of age or older.

Innis et al. (87) also assessed novelty preference at 39 weeks of age using the Fagan Test of infant intelligence (91–93). This test is a measure of recognition memory and is based on the visual preference for a novel stimulus in comparison to a familiar stimulus when both stimuli are shown together. Higher novelty preference (calculated as the percentage of total looking time spent looking at the novel stimulus) is taken to indicate that information about the familiar stimulus has been retained, and that this is indicative of higher cognitive performance. A review of this procedure is provided by Singer in this volume. The infants studied by Innis et al. (87) were tested with the subset of the test specific for infants of 79 weeks postmenstrual age (40 weeks gestation + 39 weeks post-term). Analyses to explore potential relations between preferential looking behavior, including the number of looks, total looking time, average look duration (calculated from the number of looks/total preferential looking time), or time during novelty testing not attending to the stimulus (null-time), and incidence and duration of breast-feeding were also undertaken. No differences in novelty preference were found between the breast-fed and formula-fed infants, or among infants who were breast-fed for different durations. The analyses of preferential looking behavior were done because of reports that premature infants fed a formula with docosahexaenoic acid had lower novelty preference at 12 months, but a higher number of looks to the novel and the familiar stimuli at both 9 and 12 months post-term than infants fed a formula without docosahexaenoic acid (55,56). The cross-sectional study by Innis *et al.* (87) involved a large number of infants, but had several potentially important weaknesses. For example, the information on breast- and bottle-feeding provided by the mothers could be erroneous due to false or inaccurate reporting. It is also possible that novelty preference and preferential looking behavior as measured by the Fagan test do not involve processes which are sensitive to dietary docosahexaenoic acid intake from term birth to 9 months of age. The infants tested were predominantly from middle class, two-parent families, with at least some post-secondary education. Thus the study did not address the important issue of whether the absence of a dietary intake of docosahexaenoic acid has a measurable effect on neurodevelopment only in the presence of some other environmental deprivation which prevented compensation, or compounded the insult.

The results of studies on the visual evoked potential acuity of breast-fed and bottle-fed term gestation infants, and the potential relation of visual evoked potential acuity to erythrocyte docosahexaenoic acid in term infants have been published by Makrides *et al.* (59,60). An assessment of visual evoked potential acuity in infants identified at immunization and postnatal clinics found a significantly lower mean acuity for eight formula-fed than for eight breast-fed 5-month-old infants (59). Information on feeding patterns was collected by questionnaire. The formula-fed infants were fed one of three formulas, all containing 12–15% linoleic acid and 1–1.6% $\alpha$-linolenic acid. The composition of the human milk was not reported. The erythrocyte total lipid linoleic acid showed a positive relation, and docosahexaenoic acid showed an inverse relation to the acuity values measured in log MAR (minimum angle of resolution, for which lower values indicate higher acuity). In a subsequent report, the same group described the visual evoked potential acuities of 23 fully breast-fed infants (no formula prior to 16 weeks, then no more than 120 ml formula/day from 16 to 30 weeks of age), or 19 infants randomly assigned to a formula with 16.7% linoleic acid and 1.6% $\alpha$-linolenic acid; or with an additional 0.58% eicosapentaenoic acid and 0.36% docosahexaenoic acid from fish oil and 0.26% $\gamma$-linolenic acid from evening primrose oil ($n$ = 13) (60). Visual evoked potential acuity results were given for 28, 18, and 8 infants at 16 weeks, and 28, 17, and 9 infants at 30 weeks of age for the breast-fed, unsupplemented and supplemented formula groups, respectively. Visual evoked potential recordings were not made or could not be extrapolated for 15 infants at 16 weeks and for 17 infants at 30 weeks. Visual evoked potential acuities were significantly higher at 16 ($p < 0.001$) and 30 ($p < 0.01$) weeks of age for the breast-fed infants than for infants fed the formula without docosahexaenoic acid. Infants fed the formula with docosahexaenoic acid had visual evoked potential acuity results which were not different from those for the breast-fed infants. It is not clear if the lower number of infants fed the formula with docosahexaenoic acid was due to difficulties with the test, or with the infants. The same study reported that the acuity thresholds of 5 infants who were breast-fed for less than 16 weeks were not different from those of 15 infants fed the unsupplemented formula; 5 infants who were breast-fed for more than 16–30 weeks and 15 infants who were fully breast-fed had significantly better visual evoked potential acuities. It was also noted that infants fed the formula with docosahexaenoic acid had better visual evoked potential acuities than infants who were breast-fed for a mean of 23 weeks; infants who were breast-fed for 30 weeks had a mean visual evoked potential acuity between that of infants fed the formula with docosahexaenoic acid and infants who were breast-fed for 23 weeks. It was speculated that the amount and duration of docosahexaenoic acid exposure may be important to visual evoked potential acuity. These analyses, while limited by the small number of infants tested, suggest that future studies should consider whether the duration of breast-feeding may be an important variable which impacts on the ability to detect differences between breast-fed and formula-fed infants. This may be particularly important where breast-feeding is discontinued or supplemented after 3 months, but where infant assessment continues to later ages.

Birch *et al.* (94) have reported measures of visual acuity from two studies with infants born at term who were breast-fed or fed formula containing corn oil as the source of unsaturated fatty acids. One study involved 30 full-term infants who were tested at 4 months of age. A second study involved 43 infants who were fed a controlled (fat) diet to 12 months of age and then tested at 36 months. Preferential looking and visual evoked potential acuity were assessed at 4 months of age. For testing the children of 36 months, the children were taught to point to the gratings to obtain a food reward. In addition to preferential looking acuity, operant forced choice preferential looking stereo acuity was assessed at 36 months. This was done using matched pairs of random dot patterns, one with and one without a binocularly disparate stripe which, therefore, gives the illusion of depth. Letter-matching was evaluated with the letters H, O, and T on a screen by asking the child to point to the same letter on a card. Picture naming was evaluated by asking the child to name objects in pictures. In the first study, 4-month-old breast-fed infants had a significantly higher mean visual evoked potential acuity and forced choice preferential looking acuity than infants fed the formula. The number of infants tested and standard deviations of the measures were not given. When tested at 36 months of age, no statistically significant differences in preferential looking acuities, or near recognition or distance recognition acuities were found. Children who had been fed human milk, however, had better mean operant forced-choice preferential looking random dot stereo acuities and letter matching than did children who had been fed the corn-oil formula. A significant correlation was found between stereo acuity and letter matching with the ratio of docosahexaenoic acid to 22:5n-6 in red blood cell lipids collected at 4 months of age. The correlations, of course, are not evidence of cause and effect, although the red blood cell lipid docosahexaenoic acid to 22:5n-6 ratio will differentiate the breast-fed infants from those fed the corn-oil formula. The number of children in each group, and the standard deviations of the measures were not given. A symposium report by the same group described similar findings in 21 term infants fed human milk and 14 4-month-old infants fed formula with 12–18% linoleic acid and 0.5–1.0% $\alpha$-linolenic acid (95). Results of a 3-year follow-up study of three cohorts of 20–22 infants who were breast-fed or fed formula with 0.3–0.6% energy from $\alpha$-linolenic acid was also described, but did not note the number of children in each group tested at 3 years of age.

Auestad et al. (90) recently reported the results of a multicenter prospective study in which infants were randomized to be fed a formula with about 22% linoleic acid and 2.2% $\alpha$-linolenic acid, but with no arachidonic acid or docosahexaenoic acid, or with 0.43% arachidonic acid and 0.12% docosahexaenoic acid (from egg lipid), or 0.23% docosahexaenoic acid and < 0.1% eicosapentaenoic acid (from high docosahexaenoic acid fish oil), or breast-fed. The values for human milk fatty acids in this report (90) are a range taken from data on human milk fatty acids published by Innis (96) and Koletzko et al. (97). Preferential looking acuity and visual evoked potential acuity were measured at 2, 4, 6, 9 and 12 months of age. No differences in preferential looking acuity or in visual evoked potential acuity were found among the groups, including between the breast-fed group and the group fed the formula without docosahexaenoic acid. The study of preferential looking acuity involved 38, 28, 26, and 28 infants, that of visual evoked potential acuity involved 41, 26, 23, and 28 infants who were breast-fed, or fed the standard formula, formula with docosahexaenoic acid, or formula with docosahexaenoic acid and arachidonic acid, respectively. This study, while giving similar results to those found by Innis et al. (44,45,87), did not find reduced preferential looking acuity in 2-month-old formula-fed infants, or an increase in preferential looking acuity in 2-month-old infants fed formula with docosahexaenoic acid (egg lipid), as described by Carlson *et al.* (54). The study also found no differences in visual evoked potential acuity in association with either the amount or duration of docosahexaenoic acid exposure from human milk or formula, as described by Makrides *et al.* (60).

Agostoni *et al.* (98) used the Brunet–Lezine psychomotor development scale, which measures postural, motor, and social performance, to assess development in a group of 4-month-old term gestation infants randomly assigned to be fed formula with 11.1% linoleic acid, 0.70% α-linolenic acid, or a formula containing 10.8% linoleic acid, 0.73% α-linolenic acid, with 0.3% γ-linolenic acid from evening primrose oil, and 0.44% arachidonic acid, 0.05% eicosapentaenoic acid, and 0.3% docosahexaenoic acid from egg lipid (phospholipid and triglyceride), or who had been breast-fed. This Italian study did not include analysis of human milk fatty acids. The report, however, cites the composition of human milk fatty acids published by Koletzko *et al.* (97) for median levels of fatty acids in the milk of 15 German women from a study published in 1988 (99). The differences in composition of human milk fatty acids in Germany from that in other countries (96,97,99,100), leads to the question of whether the values reflect the composition of human milk fatty acids in Italy. Agostoni *et al.* (98) used randomization schedules to remove intergroup differences due to parental occupation and education. Thus, no differences in these parameters were present among the groups. The mean Brunet–Lezine developmental quotient was significantly lower in infants fed the formula without docosahexaenoic acid (mean ± SD, 96.5 ± 10.9, $n = 29$) than for infants fed the formula with docosahexaenoic acid (105.3 ± 9.4, $n = 27$) or breast-fed (102.2 ± 11.5, $n = 30$). The range of scores, however, shows considerable overlap among the groups. The lowest and highest values attained were 80 and 136, respectively, but the groups in which these values occurred were not given. When tested at 12 months of age, no differences in the developmental quotients were found among the groups (mean ± SD: 101.2 ± 8.0, 101.5 ± 9.2, and 105.4 ± 7 for infants fed the formula without or with docosahexaenoic acid, or breast-fed, respectively) (101). This would suggest that any benefit of arachidonic acid and docosahexaenoic acid supplementation is of short-term significance, or perhaps no longer measurable at 12 months of age by the tests employed in this study.

## Addition of LCPUFAs to Infant Formula: Concept of a Safe and Adequate Range

The average composition of human milk is often adopted as a minimum standard for the levels of nutrients in human milk substitutes. Exceptions occur for special nutrients such as vitamin D and iron, based on scientific evidence of a beneficial effect of providing higher levels than in human milk, or of lower bioavailability from milk substitutes. Several problems arise in extrapolating average levels of fatty acids in human milk to the minimum level for inclusion in formula. Firstly, the levels of linoleic acid, α-linolenic acid, and docosahexaenoic acid in milk are related to the amounts of these fatty acids in the maternal diet (69,71,72,96,102–105). There is no evidence of any metabolic regulation to control the secretion of these fatty acids by the mammary gland. Indeed, levels of linoleic acid in human milk in North America have increased over the last four decades from mean levels of about 10% to 16% fatty acids (99,106–108). This increase presumably reflects the decreased dietary intake of saturated fats and increased use of polyunsaturated fatty acid (linoleic acid) rich vegetable oils, such as safflower, corn, and soybean oils, which has occurred over this time (109). Similarly, levels of docosahexaenoic acid vary widely, and depend largely on the amount of docosahexaenoic acid from fish or other marine lipids in the maternal diet (71,102,104,105).

A further problem with the use of human milk as a reference for the levels of n-6 and n-3 fatty acids in breast milk substitutes is that the amount of fatty acids in human milk does not follow a normal distribution. Some consideration should, therefore, be given as to whether estimates of 'averages' should be based on mean or median values. Of course, when extrapolating average values for formula from mean or median values variability will be much greater if individual values from single studies are used rather than mean or median data from different studies. In other words, averaging mean or median values from different studies hides the variations in fatty acids in the individual (women) within each study. For example, the median level of linoleic acid in human milk in Germany was 10.8% with a range of 5.6–21.6% (97), in Spain the median was 12.0% with a range of 6.6–27.6% (100), and in Vancouver the mean content of linoleic acid was 13.4% with a range of 7.0–22.4% (44). Thus, the average level of linoleic acid in human milk, based on these data from Canada and Europe, is around 12%, but levels in the milk of individual women can be considerably different from this. A final point with regard to fatty acids is that, since fat content is fixed, increasing the proportion of any particular fatty acid, for example linoleic acid, necessarily results in a decrease in some other fatty acid(s). This may become a problem if mean or median levels of individual fatty acids are used as a minimum standard for the levels in infant formula.

Another approach to designing human milk substitutes is to provide a composition of fatty acids which will achieve a similar pattern of blood lipid fatty acids in the bottle-fed infant to that in the breast-fed infant. It is well known that the fatty acid pattern of plasma phospholipids, as well as other blood lipid fractions reflects that of the dietary fat (64–70). More specifically, the dietary intake of docosahexaenoic acid (and of fish) is positively related to the plasma and red blood cell phospholipid docosahexaenoic acid with a dose-dependent increase in response to increasing docosahexaenoic acid intake (67,70,110). Except under conditions of prolonged dietary deficiency, the brain and retina, in contrast to other tissues, are highly resistant to changes in fatty acid composition, despite wide variations in circulating fatty acids (9,111). This ability to maintain fatty acid composition despite varying dietary intakes may be partly explained by efficient conservation (recycling) (112), as well as highly selective pathways of n-6 and n-3 fatty acid uptake. Because of this, variations in plasma or red blood cell fatty acids cannot be interpreted as evidence of differences in central nervous tissue fatty acids. However, the plasma (or red blood cell) level of n-6 and n-3 fatty acids which constitute a risk for inadequate brain and retina arachidonic acid and docosahexaenoic acid accretion is not known. Thus a prudent approach is to provide a mix of fatty acids for the bottle-fed infant which will mimic as closely as possible the effects of breast-feeding on the infant's blood lipid fatty acids.

Several studies have described the effects of formulas with arachidonic acid, and docosahexaenoic acid, or docosahexaenoic acid without arachidonic acid, on the blood lipid fatty acids of infants born at term. These studies also embrace several important problems in design and interpretation. The composition of plasma and red blood cell phospholipid fatty acids changes with increasing age in the normal breast-fed infant (57,58). Thus, differences in levels of arachidonic acid and docosahexaenoic acid in infants fed formula without (or with) these fatty acids from the levels in breast-fed infants may or may not be present, depending on the age at which measurements are made. The studies by Makrides *et al.* (58) concerning formula with about 17% linoleic acid and 1.6% α-linolenic acid provide a useful illustration of this point. At 6 weeks of age, the red blood cell docosahexaenoic acid was (mean ± SD) $5.5 \pm 0.8$, $5.4 \pm 1.1$, and $4.2 \pm 0.7\%$ in breast-fed infants, infants fed a formula supplemented with 0.36% docosahexaenoic acid, and infants fed a formula without docosahexaenoic acid, respectively. The significantly lower levels of docosahexaenoic acid in infants fed the formula without, but not in those fed the formula with, docosahexaenoic acid suggest that 0.36% docosahexaenoic acid is an appropriate amount to include in formula. At 16 weeks of age, however, the level of docosahexaenoic acid was significantly higher in the red blood cell lipids of infants fed the formula with docosahexaenoic acid (6.3 ± 0.7%) than in infants fed human milk (5.4 ± 1.0%). This latter result suggests that the amount of docosahexaenoic acid in the formula was too high. Innis *et al.* (57) have described the results of a longitudinal study comparing the effects of 0%, 0.1%, or about 0.2% docosahexaenoic acid in formula on the plasma and red blood cell phospholipid fatty acids of term gestation infants. The results of this study indicate that about 0.1–0.2% docosahexaenoic acid is appropriate, if the level of docosahexaenoic acid in the blood lipids of infants currently breast-fed in North America is accepted as an appropriate reference standard. But, of course, the proportion of docosahexaenoic acid in the blood lipids of the breast-fed infant depends on the amount in the human milk fed (71,72). Therefore, conclusions with regard to appropriate levels for inclusion in infant formula can vary, depending on which group of breast-fed infants are used as the reference.

The question of which blood lipid fraction best reflects the effects of particular diets on the composition of fatty acids in the central nervous system, or on the fatty acids available for uptake and assimilation by the brain and retina, is an important issue which has not been fully resolved. Agostoni *et al.* (62) have reported the effects of feeding formula with about 11% linoleic acid and 0.7% α-linolenic acid on the plasma and red blood cell fatty acids of term gestation infants at 16 weeks of age. The plasma phospholipid arachidonic acid was significantly lower in the formula-fed infants (mean ± SD: 6.2 ± 1.3%) than the breast-fed infants (10.8 ± 2.1%), but values were not different in the red blood cell phospholipids (15.1 ± 1.8 and 14.6 ± 2.3% for the formula- and breast-fed infants, respectively). Do the formula-fed infants have a lower arachidonic acid status than the breast-fed infants or do they not? These few examples illustrate a disparate interpretation of the effects of formula-feeding, and on the appropriate levels or need for LCPUFA supplementation depending on the age of the infant, and on whether plasma or red blood cells are the basis of the assessment.

Extremes of eicosapentaenoic acid and docosahexaenoic acid intake can result in increased levels of docosahexaenoic acid and eicosapentaenoic acid, as well as reduced arachidonic acid in some lipid classes in brain and retina (43,113,114). In recent studies, these fatty acid compositional changes were found to be accompanied by reduced visual function (114). The threshold of plasma or red blood cell eicosapentaenoic acid or docosahexaenoic acid which constitute risk for increased brain accretion of these fatty acids, or at which competition with normal brain arachidonic acid metabolism occurs, is not known. It is also not known if pathways for uptake, and thus possible adverse effects of inappropriate or unbalanced n-6 and n-3 fatty acid intakes, differ (mature?) during development of the so-called blood–brain barrier from those of the more mature brain.

Consideration of the potential effects of high or unbalanced dietary intakes of LCPUFAs is different from that for many other nutrients. Dietary fatty acids, particularly polyunsaturated fatty acids, are relatively well absorbed, with no regulation of intestinal absorption, as exists, for example, for iron or calcium. Similarly, no internal excretory route, such as kidney or bile is available to dispose of intakes in excess of need. It may be generally assumed that fatty acids consumed in excess of need for energy ($\beta$-oxidation) or membrane lipid synthesis are innocuously retained in adipose tissue. There is, however, no particular evidence that arachidonic acid and docosahexaenoic acid are so conveniently disposed of. On the contrary, arachidonic acid and docosahexaenoic acid are esterified primarily in phospholipids, rather than triglycerides and cholesteryl esters, and are not good substrates for mitochondrial $\beta$-oxidation. Increasing the long-chain n-6 and n-3 fatty acid content of the diet leads to increased levels of n-6 and n-3 fatty acids in membrane phospholipids. Theoretically, this could enhance membrane susceptibility to lipid peroxidation. Whether or not this contributed to the reduced visual function in animals fed fish oil (114), or whether increased tissue levels might have untoward effects on human infants is not known. The limited available information for infants fed formula containing arachidonic acid and docosahexaenoic acid from egg lipid, however, does not suggest any adverse effect on plasma $\alpha$-tocopherol levels (115).

Several potential sources of arachidonic acid and docosahexaenoic acid are available for supplementation of infant formula. Arachidonic acid and docosahexaenoic acid are not added to formulas as unesterified fatty acids, but as components of triglycerides and/or phospholipids. These sources contain many additional fatty acids, and potential lipid or nonlipid components extracted from the lipid source or included as a result of the extraction and purification procedures. One of the most readily available sources of docosahexaenoic acid is fish oil. Fish oils high in docosahexaenoic acid and low in eicosapentaenoic acid are now generally favored over oils high in eicosapentaenoic acid. Sources of arachidonic acid and docosahexaenoic acid include egg total lipid and egg phospholipid, and single-cell oils.

There is little doubt that the long-chain n-3 fatty acids, eicosapentaenoic acid and docosahexaenoic acid are very active constituents of fish oil. Fish oil, however, contains many other saturated and unsaturated fatty acids. The possibility of contamination with heavy metals, organic residues, pollutants, and allergens also needs to be considered. Dietary α-linolenic acid is well absorbed, and is found in plasma and liver triglycerides. Eicosapentaenoic acid and docosahexaenoic acid differ in their metabolic effects, and both differ from α-linolenic acid. α-Linolenic acid appears to be preferentially oxidized rather than incorporated into membrane phospholipids (9). Eicosapentaenoic acid, but not docosahexaenoic acid, is a substrate for synthesis of eicosanoids which have either opposing or weaker effects than eicosanoids derived from arachidonic acid (10).

Docosahexaenoic acid, but not α-linolenic acid or eicosapentaenoic acid, on the other hand, is important in normal retinal and possibly other aspects of brain metabolism. Many studies are available to show that dietary fish oil alters many key resulatory steps in lipid metabolism and alters production of eicosanoids derived from the n-6 series fatty acids, leading to effects on inflammatory responses, platelet and vessel wall reactivity, and blood pressure (116–119). Whether such effects could occur in young infants fed small amounts of fish oil in formula is not known. It should not need to be noted that, while suppression of n-6 derived eicosanoid metabolism in favor of n-3 fatty acid derived eicosanoids may be beneficial in reducing the incidence or severity of chronic disease in adults, such effects are not necessarily compatible with optimum development and immune function in infants. Future studies, however, might consider including measures of functions dependent on polyunsaturated fatty acid metabolism, or their eicosanoid derivatives in tissues most likely to be susceptible to excess or unbalanced fatty acid intakes.

The demonstration of lower growth and arachidonic acid levels in premature infants fed formula with (in fatty acids) 0.3% eicosapentaenoic acid and 0.2% docosahexaenoic acid from fish oil than in infants fed a similar formula without fish oil (88) has fostered the general notion that oils containing eicosapentaenoic acid are not appropriate sources of n-3 fatty acids for infants. There is, however, no evidence to show that eicosapentaenoic acid interferes with growth, either directly or through a mechanism involving competitive interaction with arachidonic acid. Indeed, lower body weight at 6 and 9 months, and lower weight for length at 2, 6, 9, and 12 months corrected age was subsequently found in premature infants fed formula with fish oil to provide only 0.06% eicosapentaenoic acid and

0.2% docosahexaenoic acid when compared to similar infants fed a formula without fish oil (120). The less pronounced effects on growth in the latter study (120) than those found earlier with formula containing 0.3% eicosapentaenoic acid (88) could be explained by the lower total amount of n-3 fatty acids (0.5% compared to 0.26%), shorter time of feeding formula with fish oil (11 compared to 4 months), and the continued feeding of a nutrient-enriched preterm infant formula to 2 months post-term in infants fed the formula with 0.06% eicosapentaenoic acid.

The desaturation pathway of linoleic acid to arachidonic acid and of α-linolenic acid to docosahexaenoic acid, however, is sensitive to inhibition by eicosapentaenoic acid. This pathway is less sensitive to inhibition by arachidonic acid and docosahexaenoic acid. Thus a prudent approach is not to include oils containing eicosapentaenoic acid in formulas without arachidonic acid because of the possibility of compromising tissue capacity for arachidonic acid synthesis and/or acylation into membrane phospholipids. An alternative strategy is to include arachidonic acid. This latter approach requires that the triglyceride or phospholipid source of arachidonic acid is free from any potentially harmful component, and provides arachidonic acid in a manner which allows for assimilation and tissue metabolism analogous to that achieved with arachidonic acid from human milk.

Arachidonic acid is present in a variety of membrane phospholipids, but is not readily available in significant quantities on a commercial basis from traditional animal products. Egg yolk phospholipid is mainly phosphatidylcholine and is relatively low (usually < 6% fatty acids) in arachidonic acid and docosahexaenoic acid (121). As a result, rather large amounts of phospholipid are required to achieve the desired concentrations of arachidonic acid and docosahexaenoic acid in the final formula. Egg total lipid similarly contains relatively low concentrations of the fatty acids of interest, and in addition contains significant amounts of cholesterol (122). The composition of other fatty acids, sterols and the possibility of allergens also needs to be considered.

Triglycerides produced by microalgal and microfungal biomass fermentation (single cell oils) are of considerable interest because of their very high concentrations of docosahexaenoic or arachidonic acid, respectively. Major concerns with single-cell oils (i.e. the (edible) oil which can be extracted from microbial cells) include the possibility of natural toxins, biologically active components such as steroids, pigments, triglyceride components, and/or unusual fatty acids, and contaminants introduced during the fermentation or extraction processes. In 1985, Totani and colleagues discovered that the filamentous fungus *Mortierella alpina* produces large amounts of arachidonic acid (123). Species of fungi related to *M. alpina* are known to produce mycotoxins, but the strain used has not been shown to have this ability. The most common single-cell source of docosahexaenoic acid is a strain (MK8805) of the microalgae *Crypthecodium cohni*. The microalgal and microfungal masses are grown under specific (stress) fermentation conditions to facilitate synthesis of the fatty acid of interest, and force acylation towards triglycerides. The microfungal triglycerides typically contain about 40–50% arachidonic acid, and the microalgal triglycerides contain about 35–55% docosahexaenoic acid (123).

The positional distribution of fatty acids in dietary triglycerides is of importance to digestion and absorption (availability), and subsequent metabolic effects. The endogenous gastric and pancreatic lipases hydrolyze fatty acids from the sn-1 and sn-3 positions of dietary triglycerides to produce unesterified fatty acids and sn-2 monoglycerides (124,125). These products are absorbed into the enterocytes and reassembled, predominantly via the monoacylglycerol pathway. This pathway conserves the fatty acid in the sn-2 position. The unesterified fatty acids are re-esterified at random to the sn-1 and sn-3 positions (125,126). Some information is available to suggest the positional distribution of fatty acids in plasma triglycerides is important to intravascular fatty acid metabolism (125,127,128). This raises the possibility that the distribution of fatty acids in dietary triglycerides is important to normal tissue fatty acid delivery, and consequently to normal or abnormal tissue fatty acid metabolism. Docosahexaenoic acid and arachidonic acid are distributed between the sn-2 and sn-3 positions of the human milk triacylglcyerol (129). Analysis of microalgal and fungal oil triglycerides, on the other hand, has shown that arachidonic acid and docosahexaenoic acid are on the sn-1 position, as well as the sn-2 and sn-3 positions (130). Further, triglyceride species containing either two or three arachidonic acid or docosahexaenoic acid molecules are present. Whether the difference in distribution of arachidonic acid and docosahexaenoic acid between the sn-1 and sn-3 compared to sn-2 position of milk and formula lipids or lipid supplements, influences digestion and absorption (availability), subsequent delivery to the liver, or clearance by extrahepatic tissues in the capillary beds is not known.

Animal studies have shown that single-cell triglyceride sources of arachidonic acid and docosahexaenoic acid are effective in increasing tissue phospholipid levels of these fatty acids (131,132). Human adults given single oils containing arachidonic acid and docosahexaenoic acid in amounts similar to those provided by human milk (approx. 0.8 g and 0.6 g/day, respectively,) or at levels 3 or 5 times higher for 14 days also show a dose-dependent increase in plasma phospholipid, triglyceride and cholesteryl ester levels of these fatty acids (110). Comparison of the relative increase in plasma levels in response to the amounts fed indicated a greater increase in plasma arachidonic acid than docosahexaenoic acid in response to a given dietary supplement. Whether this is related to differences in pathways of absorption (availability) or regulation at the level of phospholipid synthesis or remodeling is not known. The studies with adults fed single-cell sources of arachidonic acid and docosahexaenoic acid also found a dose-dependent increase in plasma cholesterol and a decrease in serum glucose (110). The changes were of statistical significance, but not considered of clinical importance. The reason for the increase in plasma cholesterol is not known, but could involve effects of either the LCPUFAs on lipid metabolism, or possibly sterol or other components in the oils. Eructation was also noted as a significant side-effect which was also dose dependent (110). At the present time, there is no published information on the effects of single cell oils in infant nutrition, or in extended human feeding. With regard to the latter, it needs to be noted that if included in formulas, single-cell or other sources of n-6 and n-3 fatty acids may easily be fed throughout the first year of life, and to sick or otherwise compromised infants.

## Summary

Information has been presented to suggest that there is scientific basis to consider that inadequate supplies of n-6 and n-3 fatty acids during development may result in measurable, untoward changes in the central nervous system. Information collected in animal studies, and knowledge of the biochemistry and metabolism of fatty acids in the central nervous system, however, does not allow a conclusion that arachidonic acid and docosahexaenoic acid, rather than the precursors linoleic acid and α-linolenic acid, must be supplied in the diet. Studies assessing visual and other aspects of neurodevelopment in healthy term infants fed human milk in comparison to infants fed formulas without or with arachidonic acid and docosahexaenoic acid have yielded disparate findings. In some cases, evidence of reduced preferential looking acuity (63,86) or visual evoked potential acuity (59,60), or lower scores on psychometric tests (98) have been found in infants fed formula, which are corrected by supplementation with docosahexaenoic acid without or with arachidonic acid (60,63,98). Other, very similar studies, have found no differences between breast-fed and formula-fed infants (44,45,87,90), and no benefit with addition of arachidonic acid and/or docosahexaenoic acid to formula

(90). Also problematic is the lack of consistency in the timing of differences in visual function, and conclusions on whether differences in visual function or other aspects of neural development in infants fed formula without docosahexaenoic acid are transient (63) or of longer term duration. The reduced visual function in infants and children fed formula containing corn oil as the source of polyunsaturated fatty acids (94) may be explained by intakes of α-linolenic acid below requirements for n-3 fatty acids (77–79). However, one or more of the many other differences between human milk and formula, or genetic and environmental differences between groups of breast-fed and formula-fed infants may also provide the explanation. Similarly, some of the discrepancies among studies involving formulas containing apparently adequate amounts of α-linolenic acid may involve differences in the infant populations, as well as in the study designs and methodologies. The latter, of course, includes the composition of the formula and the nature and duration of feeding of any fatty acid supplement. Given the limitations of currently published information, it does not seem reasonable to attempt a conclusion as to whether inclusion of arachidonic acid and docosahexaenoic acid in formula confers any short- or long-term benefit to the neurodevelopment of term gestation infants who cannot be breast-fed. There is, however, consistent evidence that blood lipid levels of arachidonic acid and docosahexaenoic acid are lower in infants fed formula without these fatty acids than in infants fed human milk (57–61). Animal studies also consistently show that the membrane phospholipids of organs other than the brain and retina are altered to a much greater extent by diet and in a similar manner to that of plasma (111). The brain and retina, of course, are not the only organs of interest when attempting to ensure optimal dietary intakes in young infants. An approach to modify infant formula to attain similar blood lipid fatty acid transport and metabolism in formula-fed infants to that of the breast-fed infant can, therefore, be endorsed. At this time, which breast-fed infant, or group of breast-fed infants, at which age, and which blood lipid fraction(s) should be used as the reference on which to base standards for the design of infant formula fat blends has not been defined. Similarly, information on the composition of potential sources of LCPUFAs, and their pathways of digestion, absorption and metabolism are currently insufficient to allow conclusions on either appropriate and efficacious levels, or the potential for adverse effects.

## References

1. Howard CR, Weisman M. Breast or bottle: practical aspects of infant nutrition in the first 6 months. *Pediatr Ann* 1992; **21**: 619–31.
2. Fergusson DM, Beautrais AL, Silva PA. Breast-feeding and cognitive development in the first seven years of life. *Soc Sci Med* 1982; **16**: 1705–8.
3. Morrow-Tlucak M, Haude RH, Ernhart CB. Breastfeeding and cognitive development in the first 2 years of life. *Soc Sci Med* 1988; **26**: 635–9.
4. Pollock JI. Long-term associations with infant feeding in a clinically advantaged population of babies. *Dev Med Child Neurol* 1994; **36**: 429–40.
5. Rodgers B. Feeding in infancy and later ability and attainment: a longitudinal study. *Dev Med Child Neurol* 1978; **20**: 421–6.
6. Rogan WJ, Gladen BC. Breast-feeding and cognitive development. *Early Human Dev* 1993; **31**: 181–93.
7. Temboury MC, Oter OQ, Polanco I, Arribas E. Influence of breast-feeding on the infants intellectual development. *J Pediatr Gastroenterol Nutr* 1994; **18**: 32–6.
8. Grosvenor CE, Picciano MF, Baumrucker CR. Hormones and growth factors in milk. *Endocrine Rev* 1992; **14**: 710–28.
9. Innis SM. Essential fatty acids in growth and development. *Prog Lipid Res* 1991; **30**: 39–103.
10. Innis SM. Essential dietary lipids. *Int Life Sci Instrum* 1996; 58–66.
11. MacDonald JIS, Sprecher H. Phospholipid fatty acid remodeling in mammalian cells. *Biochim Biophys Acta* 1991; **1084**: 105–21.
12. Moore SA, Yoder E, Murphy S, Dutton GR, Spector AA. Astrocytes, not neurons, produce docosahexaenoic acid and arachidonic acid. *J Neurochem* 1991; **56**: 518–24.
13. Voss A, Reinhart M, Sankarappa S, Sprecher H. The metabolism of 7,10,13,16,19-docosapentaenoic acid to 4,7,10,13,16,19-docosahexaenoic acid in rat liver is independent of a 4-desaturase. *J Biol Chem* 1991; **266**: 19 995–20 000.
14. Rotstein NP, Pennacchiotti GL, Sprecher H, Aveldano MI. Active synthesis of $C_{24:5,n-3}$ fatty acid in retina. *Biochem J* 1996; **316**: 859–64.
15. Wetzel MG, Li J, Alvarez RA, Anderson RE, O'Brien PJ. Metabolism of linolenic acid and docosahexaenoic acid in rat retinas and rod outer segments. *Exp Eye Res* 1991; **53**: 437–46.
16. Flieser SJ, Anderson RE. Chemistry and metabolism of lipids in the vertebrate retina. *Prog Lipid Res* 1983; **22**: 79–131.
17. Sastry PS. Lipids of nervous tissue: composition and metabolism. *Prog Lipids Res* 1985; **24**: 169–76.
18. Hagve T-A, Sprecher H. Metabolism of long-chain polyunsaturated fatty acids in isolated cardiac myocytes. *Biochim Biophys Acta* 1989; **1001**: 338–44.
19. Sprecher H, Lutheria D, Geiger M, Mohammed BS, Reinhart M. Intercellular communication in fatty acid metabolism. *World Rev Dietet* 1994; **75**: 1–7.
20. Chen Q, Nilsson A. Desaturation and chain elongation of n-3 and n-6 polyunsaturated fatty acids in the human CaCo-2 cell line. *Biochim Biophys Acta* 1993; **1166**: 193–201.
21. Edmond J, Korsak RA, Morrow JW, Torok-Both G, Catlin DH. Dietary

cholesterol and the origin of cholesterol in the brain of developing rats. *J Nutr* 1991; **121**: 1323–30.
22. Noelle Marbois B, Ajie HO, Korsak RA, Sensharma DK, Edmond J. The origin of palmitic acid in brain of the developing rat. *Lipids* 1992; **27**: 587–92.
23. Piomelli D. Eicosanoids in synaptic transmission. *Sensitive Rev Neurobiol* 1994; **8**: 65–83.
24. Schaad NC, Magistretti PJ, Schorderet M. Prostanoids and their role in cell–cell interactions in the central nervous system. *Neurochem Int* 1991; **18**: 303–22.
25. Vazquez E, Herrero I, Miras-Portugal TM, Sanchez-Prieto J. Role of arachidonic acid in facilitation of glutamate release from rat cortical synapotosomes independent of metabotropic glutamate receptor responses. *Neurosci Lett* 1994; **174**: 9–13.
26. Delion S, Chalon S, Herault J, Guilloteau D, Besnard J-C, Durand G. Chronic dietary α-linolenic acid deficiency alters dopaminergic and serotonergic neurotransmission in rats. *J Nutr* 1994; **124**: 2466–76.
27. Nishikawa M, Kimura S, Akaitke N. Facilitatory effect of docosahexaenoic acid on $N$-methyl-D-aspartate response in pyramidial neurons of rat cerebral cortex. *J Physiol* 1994; **475**: 83–93.
28. Benolken RM, Anderson RE, Wheeler IG. Membrane fatty acids associated with the electrical response in visual excitation. *Science* 1973; **182**: 1253–4.
29. Wheeler TG, Benolken RM. Anderson RE. Visual membranes: Specificity of fatty acid precursors for the electrical response to illumination. *Science* 1975; **188**: 1312–14.
30. Neuringer M, Connor WE, Lin DS, Barstad L. Luck S. Biochemical and functional effects of prenatal and postnatal w3 fatty acid deficiency on retina and brain in rhesus monkeys. *Proc Natl Acad Sci USA* 1986; **83**: 4021–5.
31. Neuringer M, Connor WE, Van Petten C, Barstad L. Dietary omega-3 fatty acid deficiency and visual loss in infant rhesus monkeys. *J Clin Invest* 1984; **73**: 272–6.
32. Reisbick S, Neuringer M, Hasnain R, Connor WE. Polydipsia in rhesus monkeys deficient in omega-3 fatty acids. *Physiol Behav* 1990; **47**: 315–23.
33. Reisbick S, Neuringer M, Hasnain R, Connor WE. Home cage behaviour of rhesus monkeys with long-term deficiency of omega-3 fatty acids. *Physiol Behav* 1994; **55**: 231–39.
34. Chen Y, Houghton, LA, Brenna JT, Noy N. Docosahexaenoic acid modulates the interactions of the interphotoreceptor retinoid-binding protein with 11-*cis*-retinal. *J Biol Chem* 1996; **271**: 20 507–15.
35. Clandinin MT, Chappell JE, Leong S, Heim T, Swyer PF, Chance GW. Intrauterine fatty acid accretion rates in human brain: implications for fatty acid requirements. *Early Human Dev* 1980; **4**: 121–9.
36. Clandinin MT, Chappell EJ, Leong S, Heim T, Swyer PF, Chance GW. Extrauterine fatty acid accretion in infant brain: implications for fatty acid requirements. *Early Human Dev* 1980; **4**: 131–8.
37. Martinez M. Tissue levels of polyunsaturated fatty acids during early human development. *J Pediatr* 1992; **120**: S129–38.
38. Dobbing J. Vulnerable periods of brain growth. In: Elliott K, Knight J, eds. *Lipids Malnutrition and the Developing Brain*. Amsterdam: North Holland, 1972: 9–20.

39. Bourre J-M, Dumont OS, Piciotti MJ, Pascal GA, Durand GA. Dietary and linolenic acid deficiency in adult rats for 7 months does not alter brain docosahexaenoic acid content in contrast to liver, heart and testes. *Biochim Biophys Acta* 1992; **1124**: 119–22.
40. Bourre J-M, Durand G, Pascal G, Youyou A. Brain cell and tissue recovery in rats made deficient in n-3 fatty acids by alteration of dietary fat. *J Nutr* 1989; 119: 15–22.
41. Tinoco J, Miljanich P, Medwadowski B. Depletion of docosahexaenoic acid in retinal lipids of rats fed a linolenic acid-deficient, linoleic acid-containing diet. *Biochim Biophys Acta* 1988; **486**: 575–8.
42. Bourre J-M, Francois M, Youyou A et al. The effects of dietary α-linolenic acid on the composition of nerve membranes, enzymatic activity, amplitude of electrophysiological parameters, and resistance to poisons and performance of learning tasks in rats. *J Nutr* 1989; **119**: 1880–92.
43. Connor WE, Neuringer M. The effects of n-3 fatty acid deficiency and repletion upon the fatty acid composition and function of the brain and retina. In: *Biological Membranes: Aberrations in Membrane Structure and Function*. New York: AR Liss, 1988: 275–95.
44. Innis SM, Nelson CM, Rioux MF, King DJ. Development of visual acuity in relation to plasma and erythrocyte ω-6 and ω-3 fatty acids in healthy term gestation infants. *Am J Clin Nutr* 1994; **60**: 347–52.
45. Innis SM, Akrabawi SS, Dierson-Schade DA, Dobson MV, Guy DG. Visual acuity and blood lipids in term infants fed human milk or formulae. *Lipids* 1997; **32**: 63–72.
46. Ponder DL, Innis SM, Benson JD, Siegman JS. Docosahexaenoic acid status of term infants fed breast milk or infant formula containing soy oil or corn oil. *Pediatr Res* 1992; **32**: 683–8.
47. Billson FA, Fitzgerald BA, Provis JM. Visual deprivation in infancy and childhood: clinical aspects. *Austr NZ J Ophthalmol* 1985; **13**: 279–86.
48. Salomao SR, Ventura DF. Large sample population age norms for visual acuities obtained with Vistec–Teller Acuity Cards. *Invest Ophthalmol Vis Sci* 1995; **36**: 657–70.
49. Dobbing J, Sands J. Comparative aspects of the brain growth spurt. *Early Human Dev* 1979; **3**: 79–83.
50. Hershkowitz N. Brain development in the fetus, neonate and infant. *Biol Neonate* 1988; **54**: 1–19.
51. Morgane PJ, Autin-LaFrance RJ, Bronzino JD, Galler JR. Malnutrition and the developing central nervous system. In: Isaacson RL, Jensen KF, eds. *The Vulnerable Brain and Environmaental Risks*, Vol. 1, *Malnutrition and Hazard Assessment*. New York: Plenum Press, 1992: 3–44.
52. Wainwright PE. Do essential fatty acids play a role in brain and behavioural development? *Neurosci Biobehav Rev* 1996; **16**: 193–205.
53. Birch DG, Birch EE, Hoffman DR, Many RD. Retinal development in very-low-birth-weight infants fed diets differing in omega-3 fatty acids. *Invest Ophthalmol Vis Sci* 1992; **33**: 2365–76.
54. Carlson SE, Werkman SH, Rhodes PG, Tolley EA. Visual acuity development in healthy preterm infants: effect of marine-oil supplementation. *Am J Clin Nutr* 1993; **58**: 35–42.

55. Carlson SE, Werkman SH. A randomized trial of visual attention of preterm infants fed docosahexaenoic acid until two months. *Lipids* 1996; **31**: 85–90.
56. Werkman SH, Carlson SE. A randomized trial of visual attention of preterm infants fed docosahexaenoic acid until nine months. *Lipids* 1996; **31**: 91–8.
57. Innis SM, Auestad N, Siegman JS. Blood lipid docosahexaenoic and arachidonic acid in term gestation infants fed formulas with high docosahexaenoic acid, low eicosapentaenoic acid fish oil. *Lipids* 1996; **31**: 617–25.
58. Makrides M, Neumann MA, Simmer K, Gibson RA. Erythrocyte fatty acids of term infants fed either breast milk, standard formula, or formula supplemented with long-chain polyunsaturates. *Lipids* 1995; **30**: 941–8.
59. Makrides M, Simmer K, Goggin M, Gibson RA. Erythrocyte docosahexaenoic acid correlates with the visual response of healthy, term infants. *Pediatr Res* 1993; **33**: 425–7.
60. Makrides M, Neumann M, Simmer K, Pater J, Gibson R. Are long-chain polyunsaturated fatty acids essential nutrients in infancy? *Lancet* 1995; **345**: 1463–8.
61. Putnam JC, Carlson SE, DeVoe PW, Barness LA. The effect of variations in dietary fatty acids on the fatty acid composition of erythrocyte phosphatidylcholine and phosphatidylethanolamine in human infants. *Am J Clin Nutr* 1982; **36**: 106–14.
62. Agostoni C, Riva E, Bell R, Trojan S, Luotti D, Giovannini M. Effects of diet on the lipid and fatty acid status of full-term infants at 4 months. *J Am Coll Nutr* 1994; **13**: 658–64.
63. Carlson SE, Ford AJ, Werkman SH, Peeples JM, Koo WWK. Visual acuity and fatty acid status of term infants fed human milk and formulas with and without docosahexaenoate and arachidonate from egg yolk lecithin. *Pediatr Res* 1996; **39**: 882–8.
64. Asciutti-Moura LS, Guilland JC, Fuchs F, Richard D, Klepping J. Fatty acid composition of serum lipids and its relation to diet in an elderly institutionalized population. *Am J Clin Nutr* 1988; **48**: 980–7.
65. Andersen LF, Solvoll K, Drevon CA Very-long-chain n-3 fatty acids as biomarkers for intake of fish and n-3 fatty acid concentrates. *Am J Clin Nutr* 1996; **64**: 305–11.
66. Bjerve KS, Brubakk AM, Fougner KJ, Johnsen H, Midthjel K, Vik T. Omega-3 fatty acids: essential fatty acids with important biological effects, and serum phospholipid fatty acids as markets of dietary ω3-fatty acid intake. *Am J Clin Nutr* 1993; **57**: 801S-6S.
67. Brown AJ, Pang E, Roberts DCK. Erythrocyte eicosapentaenoic acid as a marker for fish and fish oil consumption. *Prostaglandins Leukotrienes Essential Fatty Acids* 1991; **44**: 103–6.
68. Innis SM. Effect of different milk or formula diets on brain, liver and blood ω-6 and ω-3 fatty acids. In Sinclair A, Gibson R, eds. *Essential Fatty Acids and Eicosanoids.* Champaign, IL: American Oil Chemists' Society, 1992; 183–91.
69. Sanders TAB, Ellis FR, Path FRC, Dickerson JWT. Studies of vegans: the fatty acid composition of plasma choline phosphoglycerides, erythrocytes, adipose tissue, and breast milk, and some indicators of susceptibility to ischemic heart disease in vegans and omnivore controls. *Am J Clin Nutr* 1978; **31**: 805–13.
70. van Houwelingen AC, Kester ADM, Kromhout D, Hornstra G. Comparison

between habitual intake of polyunsaturated fatty acids and their concentrations in serum lipid fractions. *Eur J Clin Nutr* 1989; **43**: 11–20.
71. Henderson RA, Jensen RG, Lammi-Keefe AM, Dordick KR. Effect of fish oil on the fatty acid composition of human milk and maternal and infant erythrocytes. *Lipids* 1992; **27**: 863–9.
72. Sanders TAB, Reddy S. The influence of a vegetarian diet on the fatty acid composition of human milk and the essential fatty acid status of the infant. *J Pediatr* 1992; **120**: S71–7.
73. Farquharson J, Cockburn F, Patrick WA, Jamieson EC, Logan RW. Infant cerebral cortex phospholipid fatty-acid composition and diet. *Lancet* 1995; **340**: 810–13.
74. Farquharson J, Jamieson EC, Abbasi KA, Patrick WJA, Logan RW, Cockburn F. Effect of diet on the fatty acid composition of the major phospholipids of infant cerebral cortex. *Arch Dis Child* 1995; **72**: 198–203.
75. Farquharson J, Jamieson EC, Logan RW, Patrick WJA, Howatson AG, Cockburn F. Age- and dietary-related distributions of hepatic arachidonic and docosahexaenoic acid in early infancy. *Pediatr Res* 1995; **38**: 361–5.
76. Makrides M, Neumann MA, Byard RW, Simmer K, Gibson RA. Fatty acid composition of brain, retina, and erythrocytes in breast- and formula-fed infants. *Am J Clin Nutr* 1994; **60**: 189–94.
77. Arbuckle LD, MacKinnon MJ, Innis SM. Formula 18:2(n-6) and 18:3(n-3) content and ratio influence long-chain polyunsaturated fatty acids in developing piglet liver and central nervous tissue. *J Nutr* 1994; **124**: 289–98.
78. Hrboticky N, MacKinnon MJ, Puterman ML, Innis SM. Effect of a linoleic acid rich vegetable oil 'infant' formula on brain synaptosomal lipid accretion and enzyme thermotropic behaviour in the new-born piglet. *J Lipid Res* 1989; **30**: 1173–84.
79. Hrboticky N, MacKinnon MJ, Innis SM. Effect of vegetable oil formula rich in linoleic acid on tissue fatty acid accretion in brain, liver, plasma and erythrocytes of infant piglets. *Am J Clin Nutr* 1990; **51**: 173–82.
80. Carnielli VP, Wattimean DJL, Luijendijk IHT, Boelage A, Degenhart HJ, Sauer PJJ. The very low birthweight premature infant is capable of synthesizing arachidonic and docosahexaenoic acids from linoleic and linolenic acids. *Pediatr Res* 1996; **4**: 169–74.
81. Demmelmair H, Schenck U v, Behrendt E, Sauerwalk T, Koletzko B. Estimation of arachidonic acid synthesis in full term neonates using natural variation of $^{13}C$ content. *J Pediatr Gastroenterol Nutr* 1995; **21**: 31–6.
82. Salem N Jr, Wegher B, Mena P, Uauy R. Arachidonic and docosahexaenoic acids are biosynthesized from their 18-carbon precursors in human infants. *Proc Natl Acad Sci USA* 1996; **93**: 49–54.
83. Simon JA, Fong J, Bernet JT, Browner WS. Relation of smoking and alcohol consumption to serum fatty acids. *Am J Epidemiol* 1996; **144**: 325–34.
84. Niemela A, Jarvenpaa A-L. Is breast-feeding beneficial and smoking harmful to the cognitive development of children? *Acta Paediatr* 1996; **85**: 1202–6.
85. Dobson V. Visual acuity testing by preferential looking techniques. In: Isenberg SJ, ed. *The Eye in Infancy*. New York: Mosby, 1994: 131–56.
86. Jorgensen MH, Hernell O, Lund P, Holmer G, Michaelsen KF. Visual acuity and erythrocyte docosahexaenoic acid status in breast-fed and formula-fed term infants during the first four months of life. *Lipids* 1996; **31**: 99–105.

87. Innis SM, Nelson CM, Lwanga D, Rioux FM. Waslen P. Feeding formula without arachidonic acid and docosahexaenoic acid has no effect on preferential looking acuity or recognition memory in healthy full term infants at 9 mo. of age. *Am J Clin Nutr* 1996; **64**: 40–6.
88. Carlson SE, Cook RJ, Werkman SH, Tolley EA. First year growth of preterm infants fed standard compared to marine oil (fish oil) (n-3)-supplemented formula. *Lipids* 1992; **27**: 901–7.
89. Koletzko B. *Trans* fatty acids may impair biosynthesis of long-chain polyunsaturates and growth in man. *Acta Paediatr Scand* 1992; **81**: 302–6.
90. Auestad N, Montalto MB, Hall TR et al. Visual acuity, erythrocyte fatty acid composition, and growth in term infants fed formulas with long-chain polyunsaturated fatty acids for one year. *Pediatr Res* 1997; **41**: 1–10.
91. Fagan J-F, Detterman DK. The Fagan Test of infant intelligence: a technical summary. *J Applied Dev Psychol* 1992; **13**: 173–93.
92. Fagan JF, McGrath SK. Infant recognition memory and later intelligence. *Intelligence* 1981; **5**: 121–30.
93. Thompson LA, Fagan JF, Fulker DW. Longitudinal prediction of specific cognitive abilities from infant novelty preference. *Child Dev* 1991; **62**: 530–8.
94. Birch E, Birch D, Hoffman D, Hale L, Everett J, Uauy R. Breast-feeding and optimal visual development. *J Pediatr Opthalmol Strabismus* 1993; **30**: 33–8.
95. Uauy R, Birch E, Birch D, Peirano P. Visual and brain function measurements in studies of n-3 fatty acid requirements of infants. *J Pediatr* 1992; **120**: S168–80.
96. Innis SM. Human milk and formula fatty acids. *J Pediatr* 1991; **120L**: S56–61.
97. Koletzko B, Thiel I, Abiodun PO. The fatty acid composition of human milk in Europe and Africa. *J Pediatr* 1992; **120**: S87–92.
98. Agostoni C, Trojan S, Bellu R, Riva E, Giovannini M. Neurodevelopmental quotient of healthy term infants at 4 months and feeding practice: the role of long-chain polyunsaturated fatty acids. *Pediatr Res* 1995; **38**: 262–6.
99. Koletzko B, Mrotzek M, Bremer HJ. Fatty acid composition of mature milk in Germany. *Am J Clin Nutr* 1988; **47**: 954–9.
100. de la Presa Owens S, Lopez-Sabater MC, Rivero-Urgell M. Fatty acid composition of human milk in Spain. *J Pediatr Gasteroenterol Nutr* 1996; **22**: 180–5.
101. Agostoni C, Trojan S, Bellu R, Riva E, Luotti D, Giovannini M. LCPUFA status and developmental quotient in term infants fed different dietary sources of lipids in the first months of life. In: Bindels JG, Goedhart AC, Visser HKA, eds. *Recent Developments in Infant Nutrition. Tenth Nutricia Symposium.* Boston, MA: Kluwer Academic, 1996: 212–17.
102. Harris WS, Connor WE, Lindsay S. Will dietary ω-3 fatty acids change the composition of human milk? *Am J Nutr* 1984; **40**: 786–9.
103. Chappell JE, Clandinin MT, Kearney-Volpe C. *Trans* fatty acids in human milk lipids: influence of maternal diet and weight loss. *Am J Clin Nutr* 1985; **42**: 49–56.
104. Innis SM, Kuhnlein HV. Long-chain n-3 fatty acids in breast milk of Inuit women consuming traditional foods. *Early Human Dev* 1988; **18**: 185–9.
105. Makrides M, Neumann MA, Gibson RA. Effect of maternal docosahexaenoic

acid (DHA) supplement on breast milk composition. *Eur J Clin Nutr* 1996; **50**: 352–7.
106. Glass RL, Troolin HA, Jenness R. Comparative bichemical studies of milks. IV. Constituent fatty acids of milk fats. *Comp Biochem Physiol* 1967; **22**: 415–25.
107. Insull W, Ahrens EH. The fatty acids of human milk from mothers on diets taken ad libitum. *Biochem J* 1959; **72**: 27–33.
108. Mellies MJ, Ishikawa TT, Gartside PS *et al*. Effects of varying maternal dietary fatty acids in lactating women and their infants. *Am J Clin Nutr* 1979; **32**: 299–303.
109. Slattery ML, Randall DE. Trends in coronary heart disease mortality and food consumption in the United States between 1909 and 1980. *Am J Clin Nutr* 1988; **47**: 1060–7.
110. Innis SM, Hansen JW. Plasma fatty acid responses, metabolic effects, and safety of microalgal and fungal oils rich in arachidonic and docosahexaenoic acids in healthy adults. *Am J Clin Nutr* 1996; **64**: 159–67.
111. Rioux FM, Innis SM, Dyer R, MacKinnon M. Diet-induced changes in liver and bile but not brain fatty aicds can be predicted from differences in plasma phospholipid fatty acids in formula and milk-fed piglets. *J Nutr* 1997; **127**: 370–7.
112. Bazan NG, Rodriguez de Turco EB, Gordon WC. Docosahexaenoic acid supply to the retina and its conservation in photoreceptor cells by active retinal pigment epithelium-medicated recycling. *World Rev Nutr Diet* 1994; **75**: 120–3.
113. Phibrick D-J, Mahadevappa VG, Ackman RG, Holub BJ. Ingestion of fish oil or a derived n-3 fatty acid concentrate containing eicosapentaenoic acid (EPA) affects fatty acid compositions of individual phospholipids of rat brain, sciatic nerve and retina. *J Nutr* 1987; **117**: 1663–70.
114. Weisinger HS, Vinger AJ, Sinclair AJ. The effect of docosahexaenoic acid on the electroretinogram of the guinea pig. *Lipids* 1996; **31**: 65–70.
115. Decsi T, Koletzko B. Growth, fatty acid composition of plasma lipid classes, and plasma retinol and α-tocopherol concentrations in full-term infants fed formula enriched with ω-6 and ω-3 long-chain polyunsaturated fatty acids. *Acta Paediatr Scand* 1995; **84**: 725–32.
116. Agren JJ, Hanninen O, Hanninen A, Seppanen K. Dose responses in platelet fatty acid composition, aggregation and prostanoid metabolism during moderate fresh water fish diet. *Thromb Res* 1990; **57**: 565–75.
117. Boberg M, Vessby B, Selinus I. Effects of dietary supplementation with n-6 and n-3 long-chain polyunsaturated fatty acids on serum lipoproteins and platelet functions in hypertriglyceridaemic patients. *Acta Med Scand* 1986; **220**: 153–60.
118. Cathcart ES, Gonnerman WA, Leslie CA, Hayes KC. Dietary n-3 fatty acids and arthritis. *J Int Med* 1989; **225**: 217–23.
119. Nestel PJ, Connor WE, Reardon MF, Connor S, Wang S, Boston R. Suppression by diets rich in fish oil of very low density lipoprotein production in man. *J Clin Invest* 1984; **74**: 82–9.
120. Carlson SE, Werkman SH, Tolley EA. Effect of long-chain n-3 supplementation on visual acuity and growth of preterm infants with and without bronchopulmonary dysplasia. *Am J Clin Nutr* 1996; **63**: 687–97.

121. Ide T, Murata M, Moriuchi H. Microsomal triacylglcyerol synthesis and diacylglycerol concentration in the liver of rats fed with soybean and egg yolk phospholipids. *Biosci Biotechnol Biochem* 1992; **56**: 732–5.
122. Guardiola F, Codony R, Manich A, Rafecas M, Boatella J. Stability of polyunsaturated fatty acids in egg powder processed and stored under various conditions. *J Agric Food Chem* 1995; **43**: 2254–9.
123. Kyle DJ, Ratledge C (eds). *Industrial Applications of Single Cell Oils*. Champaign, IL: American Oil Chemists' Society, 1992.
124. Rogalska E, Ransac S, Verge R. Stereoselectivity of lipases. II Stereoselective hydrolysis of triglycerides by gastric and pancreatic lipases. *J Biol Chem* 1990; **265**: 20 271–6.
125. Small DM. The effects of glyceride structure on absorption and metabolism. *Annu Rev Nutr* 1991; **11**: 413–34.
126. Yang Y-L, Kuksis A. Apparent convergence (at 2-monoacylglycerol level) of phosphatidic acid and 2-monoacylglycerol pathways of synthesis of chylomicron triacylglycerols. *J Lipid Res* 1991; **32**: 1173–86.
127. Decker E. The role of stereospecific saturated fatty acid positions on lipid nutrition. *Nutr Rev* 1996; **54**: 108–10.
128. Mortimer B-C, Kenrick MA, Holthouse DJ, Stick RV, Redgrave TG. Plasma clearance of model lipoproteins containing saturated and polyunsaturated monoacylglycerols injected intravenously in the rat. *Biochim Biophys Acta* 1992; **1127**: 67–73.
129. Martin J-C, Bougnoux P, Antoine J-M, Lanson M, Louet C. Triacylglycerol structure of human colostrum and mature milk. *Lipids* 1993; **28**: 637–43.
130. Myher JJ, Kuksis A, Geher K, Park PW, Diersen-Schade S. Stereospecific analysis of triacylglycerols rich in long-chain polyunsaturated fatty acids. *Lipids* 1996; **31**: 207–15.
131. Craig-Schmidt, Stieh KE, Lien EL. Retinal fatty acids of piglets fed docosahexaenoic and arachidonic acids from microbial sources. *Lipids* 1996; **31**: 53–62.
132. Wainwright PE, Xing H-C, Mutsaers L, McCutcheon D, Kyle D. Arachidonic acid offsets the effects on mouse brain and behaviour of a diet with a low (n-6):(n-3) ratio and very high levels of docosahexaenoic acid. *J Nutr* 1997; **127**: 184–93.

# Commentary

**Carlson**: I am skeptical about whether or not docosahexaenoic acid (and arachidonic acid) are conditionally essential nutrients for term infants, based on the virtual absence of helpful studies that can be used to address the issue. I count only three published randomized trials related to visual acuity (1–3), two of which have been widely criticized because so little docosahexaenoic acid was supplied in the experimental formula (2,3).

Several nonrandomized trials, including three of your own, and other randomized trials have included a nonrandomized reference group fed human milk. Unfortunately, there is no uniformity among these studies in 'breast-feeding' – the duration of breast-feeding, the allowance for use of supplemental formula, socio-economic factors compared with the randomized group, and milk composition are all different/highly variable – making conclusions about the effects of breast-feeding on function difficult if not impossible. Finally, and by no means least, many of the outcomes used in the nonrandomized trials of term infants have been applied at different ages or have used somewhat different versions of methods than have previously been demonstrated to have validity for detecting differences in n-3 fatty acid status in less mature infants. Thus, the absence of effects in these studies can only be taken as 'absence of evidence' rather than 'evidence of absence' that preterm infants benefit from docosahexaenoic acid.

1. Makrides M, Neumann M, Simmer K, Pater J, Gibson R. Are long-chain polyunsaturated fatty acids essential nutrients in infancy? *Lancet* 1995; **345**: 1463–8.
2. Carlson SE, Ford AJ, Werkman SH, Peeples JM, Koo WWK. Visual acuity and fatty acid status of term infants fed human milk and formula with and without docosahexaenate and arachidonate from egg yolk lecithin. *Pediatr Res* 1996; **39**: 1–7.
3. Auestad N, Montalto MB, Hall RT et al. Visual acuity, erythrocyte fatty acid composition, and growth in term infants fed formulas with long-chain polyunsaturated fatty acids for one year. *Pediatr Res* 1997; 41: 1–10.

AUTHOR'S REPLY: Your comment is consistent with my conclusion that there are discrepancies among the published studies, both for those involving comparisons of nonrandomized groups of breast-fed and formula-fed infants, and for those involving groups of infants randomized to be fed either standard formula or formula supplemented with different forms of docosahexaenoic acid or docosahexaenoic acid and arachidonic acid.

CARLSON: My recollection is that the rat studies (and those of other nonprimate animals) begat the nonhuman primate studies, which in turn begat the human preterm infant studies, which in turn begat some of the trials that have been or are being done in term infants related to n-3 fatty acids. At each stage, there was plausibility based on results that came from previous studies, and an attempt to use analogous methods to determine if the validity of the effects of these fatty acids could be extended to the new group. For example, the human preterm studies were undertaken because it was plausible to suggest that formula-fed preterm infants might not receive sufficient docosahexaenoic acid because (a) they missed the last intrauterine trimester (1), (b) they did not receive any docosahexaenoic acid after birth, (c) they had much less mature brain development at birth than the rhesus monkey infant, and (d) functional effects of inadequate docosahexaenoic acid accumulation were known to occur in primates and lower animals. Confirming the early speculation, several of these trials have now found effects of feeding docosahexaenoic acid on early visual acuity and what appears to be cognition.

The data from these trials in preterm infants have been available in at least preliminary form since 1989. Despite this, the impression one gains from reading your chapter is that the later term studies were undertaken primarily because human milk contains docosahexaenoic acid and formula does not. While I cannot speak for your own nonrandomized trials, other investigators have done and are doing randomized trials in term infants based on information available from comparable trials in preterm infants. You do indicate that randomized trials are important in the final analysis, nor do you comment on why you think so many nonrandomized trials were done after evidence became available that docosahexaenoic acid influenced function in infants studied in randomized trials.

1. Clandinin MT, Chappell JE, Leong S, Heim T, Swyer PR, Chance GW. Intrauterine fatty acid accretion rates in human brain: implications for fatty acid requirements. *Early Human Dev* 1980; **4**: 121–9.

AUTHOR'S REPLY: Each of us develop the rationale for our research for different reasons; this diversity in the long run adds strength to the data and keeps the field moving forward.

The hypothesis relating to a possible dietary need for arachidonic acid and docosahexaenoic acid in my laboratory was not based on the sequence of rat studies to nonhuman primate studies to preterm infant studies to term infant studies. I had the benefit of being at the University of Toronto in the late 1970s at a time when there was considerable interest in breast milk. The particular attributes of the amino acid composition (e.g. cysteine) and higher protein content of early than mature milk, all of which my colleagues Dr Stanley and Stephanie Atkinson worked on, were studied in the department. Dr T. Clandinin with whom I worked was at this time analyzing fetal and infant autopsy tissue, and the composition of human milk fatty acids. This work published in 1980–81 described the increase in arachidonic acid and docosahexaenoic acid in infant brain and provided mathematical estimates that human milk contains sufficient amounts of arachidonic acid and docosahexaenoic acid to at least theoretically meet the need of the developing brain (1–5). These studies and the idea of supplementing formula with egg lipids (6) to provide arachidonic acid and docosahexaenoic acid were published well before the first preterm studies by you in 1989. Thus, the hypothesis was logically born in an environment of intense interest on the benefits of human milk feeding that docosahexaenoic acid and arachidonic acid in milk play some important role in infant nutrition. The studies with rats and with monkeys fed formula extremely deficient in α-linolenic acid added credence to the possibility, because they provided information that docosahexaenoic acid has an important role in normal physiological function. Without this, of course, one could not hypothesize that human milk n-3 fatty acids are of any consequence. The animal studies have been of extreme importance in pointing to the systems which may be sensitive to changes in n-3 fatty acid intake during growth and development, but they were not the basis for formulating hypotheses and studies with infants in my laboratories.

1. Clandinin MT, Chappell JE, Heim T. Do low birthweight infants require nutrition with chain elongation–desaturation products of essential fatty acids? *Prog Lipids Res* 1981; **20**: 901–4
2. Clandinin MT, Chappell JE, Heim T et al. Fatty acid accretion in fetal and neonatal liver: implications for fatty acid requirements. *Early Human Dev* 1981; **5**: 7–14.
3. Clandinin MT, Chappell JE, Heim T et al. Fatty acid utilization in prenatal de novo synthesis of tissues. *Early Human Dev* 1981; **5**: 355–66.
4. Clandinin MT, Chappell JE, Leung S et al. Extrauterine fatty acid requirements. *Early Human Dev* 1980; **4**: 131–8.
5. Clandinin MT, Chappell JE, Leung S et al. Intrauterine fatty acid accretion rates in human brain: implications for fatty acid requirements. *Early Human Dev* 1980; **4**: 121–9.
6. Clandinin MT, Chappell JE. *US Patent 4670285*, 2 June, 1987.

CARLSON: One of the most perplexing pieces of information from my perspective is how to rationalize that human-milk-fed infants appear more likely to have higher acuity in two populations where milk levels of docosahexaenoic acid are the highest (1,2). The relationship between maternal intake of docosahexaenoic acid and the status of their infants at birth and following breast feeding has been studied enough to know that they are related. Therefore, one might infer that a woman producing milk with > 0.35% docosahexaenoic acid while consuming her 'usual diet' would also transmit more docosahexaenoic acid to her fetus, and, further, that the infant would be less likely to 'need' docosahexaenoic acid for higher visual acuity after term delivery. The fact that this does not seem to be the case could be used to argue that the critical period for docosahexaenoic acid accumulation may continue for some time after term birth. Our own studies in preterm infants suggest that for preterm infants the interval corresponding to the last intrauterine trimester and the first 2 months after term birth may be most critical.

1. Makrides M, Simmer K, Goggin M, Gibson RA. Erythrocyte docosahexaenoic acid correlates with the visual response of healthy, term infants. *Pediatr Res* 1993; **33**: 425–7.
2. Jorgensen MH, Hernell O, Lund P, Holmer G, Michaelsen KF. Visual acuity and erythrocyte docosahexaenoic acid status in breast-fed and formula-fed term infants during the first four months of life. *Lipids* 1996; **31**: 99–105.

AUTHOR'S REPLY: Dr Neuringer also commented that the studies by Jorgenson *et al.* (my reference 86) from Denmark and those of Makrides *et al.* from Australia which find differences in visual acuity may involve higher breast milk docosahexaenoic acid (> 0.35%). Studies in North America, where the reported milk docosahexaenoic acid is 0.2% fatty acids (similar to the milk in the Makrides *et al.* paper (my reference 59; and see Table 1)) have not been as consistent. Your studies, of course, do find differences in preferential looking acuity. Although you did not analyze milk samples in your study, I see no reason at this time to speculate better docosahexaenoic status in babies in Memphis, compared to Portland or Vancouver. Again, the observation of variances in human milk fatty acids is interesting and certainly merits the development of hypotheses and studies to determine if this is of any physiological relevance. It is important to recognize that these differences could be explained by other differences in formula between Europe, Australia, and North America, the populations studied, or it may simply reflect analytical differences between the laboratories.

CARLSON: What numbers do you think are needed to detect a significant difference in visual acuity between groups? The point of noting specific numbers of infants tested at each age for each diet is not clear, but individuals who are only beginning to be informed about these studies (as are most of the panel, and the ultimate readers of the book) are likely to receive the impression that the numbers are mentioned because the author thinks they are meaningful. If you do mean to imply that not enough infants were studied to reject the null hypothesis, it would be better simply to say so.

A lack of power in the smaller trials in fact does not appear to be the problem; rather, there are larger trials that find no effect on visual acuity and smaller trials that find highly significant effects. Questions of reliability necessarily arise when any trial, regardless of its size, cannot confirm the outcomes of a prospective, double-blind trial, but the burden of proof is on the trial that finds no effect rather than on the one that rejects the null hypothesis. Until we look critically at the differences in methods and design among these studies and rule them out as a possible reason (often difficult to do because the methodology is insufficiently well described), we cannot logically turn our attention to less obvious possibilities to explain the differences such as deprivation (lower socio-economic status, docosahexaenoic acid status, smoking, alcohol, etc.) mentioned by you. At any rate, these are more reasonably invoked in studies that fail to detect differences than in studies that detect differences.

AUTHOR'S REPLY: Appelbaum in his Commentary has discussed the importance of power considerations in the design and reporting of empirical results. The numbers in the text are provided for that purpose.

CARLSON: You imply that infants of 9 and 12 months cannot be tested for visual acuity using the Teller cards at 38 cm. This demonstrates a basic lack of understanding of how this procedure is used to determine visual acuity. I do appreciate that this method was ultimately put forth as a 'standardized procedure' some time after I was trained to use the cards by Dobson. Also, I appreciate that the procedure was developed to be used in the hands of trained individuals as a sensitive and specific screen for problems with grating acuity in nonverbal individuals rather than as a research tool. Over the years, I have attempted to honor the intent of the investigators who developed the procedure by carefully describing my methods and pointing out where they differ from the now 'standardized procedure'.

It does not surprise me that the results were lower in your multicenter trial because the safest way to train people who cannot go through rigorous training, including various types of reliability testing, is to teach them to make extremely conservative judgments about whether or not the infant sees the card, i.e. if they have any doubt to say that the infant does not see the card. While this method of training undoubtedly decreases variability among study sites, the sensitivity of grating cards as a research tool, e.g. their ability to detect differences related to feeding docosahexaenoic acid, may be very much diminished. It is important to remember that, within the context of a double-blind trial, one does not require a standardized procedure but only a sensitive procedure to demonstrate significant effects.

AUTHOR'S REPLY: Mayer has commented on and given a useful response to the question of whether or not the question on the test distance of 38 rather than 55 cm could have limited the ability to detect differences in visual acuity after 6 months of age. The testers in my center are well qualified and were also trained by Dr Dobson, and inter- and intratester reliabilities are routinely checked. At this point, I would be hesitant to make suggestions beyond those in the chapter as to why results of studies from different centers are not uniform in finding evidence either for or against a difference in preferential looking acuity between breast-fed infants and infants fed unsupplemented formula, or infants randomized to be fed unsupplemented and supplemented formula.

CARLSON: Your comments seem to imply that you think it is not possible for differences to occur in relevant tissues (here the brain and possibly the retina) before they are observed in the red blood cell or, for that matter, to continue to be present in the brain after they are no longer apparent in the red cell. While I am aware that relationships between circulating docosahexaenoic acid and visual function have been reported, such associations cannot reasonably be extrapolated to include all situations, methods, populations, etc. For example, in our second supplementation trial, all preterm infants were fed the identical formula after 2 months adjusted age. By 12 months, there was no detectable evidence of early formula feeding in the plasma and red blood cells (11), but infants fed DHA until 2 months adjusted age had shorter look duration and more looks compared with controls fed α-linolenic acid as the sole source of n-3 fatty acids (2).

1. Carlson SE, Werkman SH, Tolley EA. The effect of long chain n-3 fatty acid supplementation on visual acuity and growth of preterm infants with and without bronchopulmonary dysplasia. *Am J Clin Nutr* 1996; **63**: 687–97.
2. Carlson SE, Werkman SH. A randomized trial of visual attention of preterm infants fed docosahexaenoic acid until 2 months. *Lipids* 1996; **31**: 85–90.

AUTHOR'S REPLY: My position with regard to red blood cell lipids is that they do not give a reliable index of the composition of fatty acids in the developing (or adult) brain.


CARLSON: You do not describe the Fagan Test method used in your reference 87, but the numbers you report suggest that this is the newer version that was put out for IBM sometime after 1991. In our hands, the version sold (and the program provided to measure direction and duration of looks) for the old Commodore computer yields significantly lower novelty preference scores, lower numbers of looks, and longer look durations compared with the newest version sold on or before 1991 to be used with the IBM. More importantly, the older version, but not the newer version, was sensitive to the differences in docosahexaenoic acid intake in preterm infants fed diets with and without docosahexaenoic acid. If you have been given the older version, I would be convinced that there are no effects of breast-feeding as it exists in your center on look duration. However, if you have administered the newer version, I would maintain that again we have an 'absence of evidence' because the test is not sensitive to the effects of docosahexaenoic acid supplementation that have been observed in two trials of preterm infants.

Jacobson may have used a third version of the Fagan Test in her studies of alcohol and polychlorobiphenyl exposure (1,2), because with Fagan's permission she adapted the software for her research. These comments are not meant to cause concern about the significant data that have been observed within individual studies, but to emphasize that not all versions of the test are the same. The results of published studies cannot be compared directly without first confirming that the methods are equivalent, and this is not possible to do from most of the published data. This is one example where a 'standardized procedure' was far from standardized.

1. Jacobson SW, Jacobson JL, Sokol RJ, Martier SS, Ager JW. Prenatal alcohol exposure and infant information processing ability. *Child Dev* 1993; **64**: 1706–21.
2. Jacobson SW, Fein GG, Jacobson JL, Schwartz PM, Dowler JK. The effect of intrauterine PCB exposure on visual recognition memory. *Child Dev* 1985; **56**: 853–860.

AUTHOR'S REPLY: We use the newer IBM program. You wrote in your comments, subsequently not included in this text, that in your studies with term infants, visual acuity at 2 months was the only parameter to show an effect. You stated that visual acuity, visual recognition memory, and look duration at 6.5, 9, and 12 months and the Bayley MDI and PDI at 12 months were not different. This is similar to the results which we have published, as well as for the longitudinal data to 18 months which we have yet to publish. Whatever the reason, it would seem that our results for term infants are the same. Whether or not the tests were not sufficiently sensitive to find differences is another issue.

CARLSON: It is worth noting also that Willatts and co-workers have shown some interesting effects of both arachidonic acid and docosahexaenoic acid supplementation on look duration of infants who achieve their peak look after the first look in habituation trials (1). In a later preliminary report, the supplemented infants also performed significantly better on an infant play measured at 10 months (2). Unfortunately, these authors did not state the amount of α-linolenic acid in the control diet, and so it is unclear if the effects they have reported occur because the control group received too little α-linolenic acid or despite good intakes of α-linolenic acid.

1. Forsyth JS, Willatts. Do LCPUFA influence infant cognitive behavior? In, Bindels JG, Goedhart AC, Visser HKA, eds. *Recent Developments in Infant Nutrition*. Dordrecht: Kluwer Academic, 1995: 225–34.
2. Willatts P, Forsyth JS, DiModugno MK, Varma S, Colvin M. Improved means–end problem solving at ten months by infants fed a formula supplemented with long-chain polyunsaturated fatty acids. In: *American Oil Chemists' Conference on PUFA in Infant Nutrition: Consensus and Controversies, Program and Abstracts*. Champaign, IL: American Oil Chemists' Society, 1996; 43.

AUTHOR'S REPLY: There are certainly a number of abstracts and chapters which indicate beneficial effects of docosahexaenoic acid. These reports also include some apparent 'negative' effects, for example the lower language scores in 14-month-old infants fed formula with docosahexaenoic acid.

CARLSON: The example you give of reduced visual function in rats fed fish oil is analogous to other studies that have shown changes in acoustic responses in rats (1). I have also been told that the offspring of rodents fed fish oil as the only source of fat looked and behaved in an unusual manner at birth. There is little doubt in my mind that there can be problems with a diet that contains only fish oil fat. It should be emphasized, however, that the experimental diets that have been fed to human infants contain many orders of magnitude less n-3 LCPUFAs on a body weight basis. Judging from the comments of some of the vision and behavior experts, they are clearly impressed with the concerns about long-chain fatty acids and safety. For balance, it should be emphasized that virtually all essential nutrients have side-effects of various severity from both under- and overconsumption. Moreover, there are tests available to detect pesticides and other contaminants in novel oil sources (many of which are known to be in far higher amounts in human milk than in formulas in North America!) before they are fed to infants.

Your questions related to what happens to docosahexaenoic acid and arachidonic acid consumed in excess of physiological need and stores should be taken to heart. Clearly these compounds are stored or it would not take months for them to decline in the circulation once their feeding is discontinued, but there is a need for study of how much is too much. I anticipate trials in children with attention deficit hyperactivity disorder (ADHD) following a recent report that these children have lower than normal levels of long-chain n-3 and n-6 fatty acids (2), so the need to resolve this issue takes on more urgency. I am very concerned about parents medicating their ADHD children with LCPUFAs based on these data.

1. Stockard JE, Carver JD, Register B, Benford VJ, Chen L, Phelps C. Dietary fatty acid (FA) effects upon auditory brainstem responses (ABR). *Pediatr Res* 1996; **39**: 320A.
2. Stevens LJ, Zentall SS, Deck JL et al. Essential fatty acid metabolism in boys with attention-deficit hyperactivity disorder. *Am J Clin Nutr* 1995; **62**: 761–768.

**Neuringer and Reisbick**: This chapter spells out many critical issues regarding LCPUFA supplementation of term infant formulas, and the various studies of the effects of supplementation are described comprehensively and in detail. Both this chapter and Lucas' chapter stress the discrepancies in the results among the studies, as well as their important methodological differences. Both note that those studies in which the unsupplemented formula contained 0.8% α-linolenic acid (ALA) found visual acuity differences, compared with either breast-fed or LCPUFA-supplemented groups, which may be attributed to insufficient ALA. However, both chapters overlook a second consistent and fairly straightforward pattern in the outcomes of both randomized and nonrandomized term infant studies: regardless of the ALA content of the unsupplemented formula, those studies in which either the human milk or the LCPUFA-supplemented formula contained 0.3% or more docosahexaenoic acid (DHA) found differences in acuity (see Innis' references 53, 54, and 80) (1) and the Brunet–Lezine (reference 90), whereas those providing 0.2% or less DHA did not (references 38, 39, 81, and 84; and reference 57, except for a small acuity difference at 2 months only). The data at this point are limited, but given the differences noted in populations, types of control formula, and study methods, this consistency appears fairly striking.

'Current information, however, does not provide a mechanism by which dietary n-6 fatty acid deficiency could lead to immediate or long-term deficits in neural function.' This statement appears to be completely at odds with the list of powerful potential mechanisms which precedes it. Certainly total essential fatty acid deficiency has significant effects on brain development and behavior, much more so than specific n-3 fatty acid deficiency.

The correct reference for the monkey plasma values is your reference 28, and for cerebral cortex and retina is reference 27. The values shown for monkey retina DHA are not correct but are actually the values for frontal cortex. In addition, all the values shown are for total n-3 fatty acids rather than for DHA. The correct values for occipital cortex and retina ethanolamine phosphoglycerides are, respectively: 5.8% and 7.1% in deficient animals, 34% and 36.4% in controls. In addition, these are values for 2-year-old monkeys, which are the same as in adults. The more relevant comparison would be the perinatal monkey values, which are, for occipital cortex and retina, respectively: 4.0 and 8.6% in deficient animals, 15.0 and 17.3% in controls.

The term 'sensitive period' is more appropriate than the term 'critical period' for primates (monkey and human) where such periods are less tightly defined, and there tends to be a gradual and progressive decrease in vulnerability to deprivation and in the potential for reversal. In addition, the specification of critical periods depends on the outcome. There is no one 'critical period' for all aspects of visual development or even all aspects of visual image deprivation. It depends, for example, whether one is talking about the sensitive period for induction or reversal of changes in cell size in the lateral geniculate nucleus (LGN), which is different for the magnocellular and parvocellular systems, or for shifts in ocular dominance columns in primary visual cortex, which are different for the different cortical layers, or for acuity and contrast sensitivity loss, or for failure to develop binocular vision. Shifts in ocular dominance and loss of binocularity apply primarily to monocular deprivation or interocular asymmetries, and therefore are less relevant to the possible effects of n-3 fatty acid deficiency, whereas changes in LGN cell size and acuity/contrast sensitivity loss can occur with both monocular and binocular deprivation of pattern vision.

It may be helpful for us to spell out more clearly the hypothesis underlying comparisons of the effects of visual deprivation to acuity loss due to n-3 fatty acid deficiency. The hypothesis is (i) that the primary effect of the deficiency is to alter retinal development in ways (as yet unknown) which result in slower acuity development, and (ii) that poorer spatial vision due to this altered retinal development affects cortical development in ways similar to loss of sharp spatial vision induced optically (by cataracts or other ocular media defects, by artificial blur or refractive errors). In both cases, the input of visual information (specifically high spatial frequency information) to the LGN and visual cortex are restricted. There are no primate studies which provide an optimal model for binocular mild acuity loss, so we can only extrapolate from the available data for (i) mild monocular acuity loss induced by optical blur (3), and (ii) binocular pattern deprivation (e.g. (4)).

Although safflower oil diets are (deliberately) very low in $\alpha$-linolenic acid (about 0.3%), the level is not very extreme compared with the old standard corn oil formulas, which were often as low as 0.5–0.6%. The differences in the n-6/n-3 ratio between these two are greater than the absolute differences in $\alpha$-linolenic acid.

It is not appropriate to compare the percentage fat in our maternal monkey diets with the percentage fat in human infant formulas. The percentage fat in our infant monkey formulas has been 30–49%.

The human autopsy data indicated higher n-6 LCPUFAs (AA: 22:4 and 22:5 in your reference 67, and only 22:4 and 22:5 in reference 70) in the cerebral cortex of formula-fed infants: a consistent finding in $\alpha$-linolenic acid deficiency. Statements here and elsewhere in your chapter, and in several other chapters, treat the potential importance of dietary DHA and AA as equivalent issues, when they are different in important ways.

1. Birch EE, Birch DG, Hoffman DR et al. Visual maturation of term infants fed omega-3 long-chain polyunsaturated fatty acid (LCPUFA) supplemented formula. *Invest Ophthalmol Vis Sci* 1996; **37**: S1112 (abstr).
2. Boothe RG, Dobson V, Teller DY. Postnatal development of vision in human and non-human primates. *Ann Rev Neurosci* 1985; **8**: 495–545.
3. Movshon JA, Eggers HM, Gizzi MS, Hendrickson AE, Kiorpes L, Boothe RG. Effects of early unilateral blur on the macaque's visual system: III. Physiological observations. *J Neurosci* 1987; **7**: 1340–51.
4. Harwerth RS, Smith EL III, Paul AD, Crawford MLJ, von Noorden GK. Functional effects of bilateral form deprivation in monkeys. *Invest Ophthalmol Vis Sci* 1991; **32**: 2311–27.

AUTHOR'S REPLY: This is an interesting observation which several investigators have made. One of the problems in interpreting the data at the present time is that many of the studies have not reported the composition of human milk fed in their own studies (for example, my references – Makrides et al. (59), Agostoni et al. (98), Carlson et al. (63), and Auestad et al. (90)). Makrides et al. (60) used a DHA content of 0.35% in the supplemented formula, but had 0.2% DHA in breast milk (breast-fed infants here did better than those fed unsupplemented formula). Thus, the data are incomplete and there remains inconsistency in what is available.

Another important point to raise here is a note of caution in comparison of fatty acid data, not only from different laboratories, but also from different countries. In milk (and plasma, red blood cell, etc.) analysis, the total fatty acids identified are summed, then the individual components are calculated as a percentage of the total. This can be problematic. For example, if one investigator identifies 40 fatty acids, and another 25 fatty acids, a difference of 0.1% (e.g. 0.2 vs 0.3) can readily be accounted for by fatty acid methodologies (and mathematics). Fatty acid analysis is not standardized across laboratories. Thus, I think some degree of caution is needed in viewing differences of the order of 0.1% fatty acids and their potential significance to meaningful differences in biological function. I would, however, agree that information on the development of breast-fed infants in relation to varying intakes of DHA (i.e. the natural history study) would be very helpful.

**Mayer and Dobson**: Regarding Innis' discussion of critical, or sensitive periods in visual development, stimulus deprivation does adversely affect development of the human visual system before age 4 months. For example, Birch and Stager (1) review their own data and other evidence that the extent of acuity loss in childhood due to a unilateral congenital cataract depends upon the age of onset of treatment and is more severe if treatment is not instituted before age 4 months, with the optimal age of treatment being before age 2 months. However, amblyopia due to stimulus deprivation may have a different neurophysiological basis than amblyopia due to strabismus (see review in (2)).

Overt signs and manifestations of strabismic amblyopia, such as fixation preference for one eye or unequal visual acuity, are not observed until after age 4 months. This fits nicely with the hypothesis that strabismic amblyopia is due to abnormal binocular competition and with the finding that stereopsis (fine binocular vision) develops rapidly at about age 4 months.

Visual acuity development under conditions of bilateral stimulus deprivation is more relevant to studies of early nutrition than is unilateral stimulus deprivation. The few studies of this condition in humans and monkeys suggest that spatial resolution deficits are mild; behaviorally, visual responsiveness is more affected than visual acuity, and spatial resolution of individual neurons is unaffected until the cortical level (see Boothe *et al.* (3) and also chapter by Shaw, this volume).

The upper limit of the amblyopiagenic sensitive period in visual system development, as evaluated in terms of treatment of amblyopia, extends to age 8–10 years in humans. Interestingly, the upper limit of amblyopia reversal and the neurophysiological sensitive period in a monkey model is about 2 years (4). This follows the 4:1 rule for monkey versus human infant grating acuity development (3,5). That is, the time course of development of grating acuity in monkeys and humans can be made to superimpose by multiplying monkey age by a factor of 4 (e.g. acuity of human infants at age 4 months equals acuity of monkey infants at age 1 month).

The most rapid improvement of visual acuity in human development is within the first 6–8 months after birth, and thus beyond the 4 months stated by Innis, and continues to improve more slowly to adult acuity levels by ages 10–12 years (see the review in (6)). Rapid improvement of grating acuity to age 6 months is also shown in the preferential looking acuity data reported in the reference Innis cites (7), as well as in many other behavioral acuity studies. Acuity measured using the swept visual evoked potentials asymptotes at about 8 months for the stimulus conditions used (8).

1. Birch EE, Stager DR. Prevalence of good visual acuity following surgery for congenital unilateral cataract. *Arch Ophthalmol* 1988; **106**: 40–3.
2. von Noorden GK. *Binocular Vision and Ocular Motility*. 5th edn. St Louis: Mosby, 1996: 249–54.
3. Boothe RG, Dobson V, Teller DY. Postnatal development of vision in human and nonhuman primates. *Ann Rev Neurosci* 1985; **8**: 495–545.
4. Boothe RG. Amblyopia. In: Albert DM, Jakobiec FA, eds. *Principles and Practice of Ophthalmology*. Philadelphia: WB Saunders, 1994: 663–82.
5. Teller DY. The development of visual acuity in human and monkey infants. *Trends Neurosci* 1981; **4**: 21–4.
6. Hamer RD, Mayer DL. The development of spatial vision. In: Albert DM, Jakobiec FA, eds. *Principles and Practice of Ophthalmology*. Philadelphia: WB Saunders, 1994: 578–608.
7. Salomao SR, Ventura DF. Large-sample population age norms for visual acuities obtained with Vistech/Teller Acuity Cards. *Invest Ophthalmol Vis Sci* 1995; **36**: 657–70.

8. Norcia AM, Tyler CW. Spatial frequency sweep VEP: visual acuity during the first year of life. *Vision Res* 1985; **25**: 1399–408.

AUTHOR'S REPLY: In the paper cited, Birch and Stager report a prospective study of preferential looking acuity in infants following surgery for cataract. The paper and abstract conclude that best outcomes are achieved with surgery in the first 2 months after birth. However, the authors state in the paper that those infants with later surgeries were also less likely to tolerate the occlusion therapy necessary for rehabilitation. A figure to illustrate the high rate of noncompliance of occlusion therapy for infants over 5 months is given. The authors note in the discussion that the benefit of early surgery involves better compliance with therapy, whereas poor outcome with late surgery involve high noncompliance. I do not agree that this paper provides specific evidence that stimulus deprivation in the first 4 months does adversely affect development of the human visual system.

With regard to monkey versus human data, the issue in my chapter is simpler, and I believe more generic. It is simply a cautionary note to be careful in extrapolating data from long-term deficiency studies in animals, for which sensitive periods differ, and for which the duration and severity of deficiency differs from the human. The intent is not to invalidate or lessen the importance of the work which has been done in monkeys and rats regarding dietary $\alpha$-linolenic acid, brain and retina docosahexaenoic acid and visual function.

MAYER AND DOBSON: Innis' claim that a 38 cm rather than a 55 cm test distance used the acuity card testing of the Carlson *et al.* study (1) 'limit[s] the ability to find differences in acuity at 9 and 12 months', is not warranted. As long as the grating spatial frequencies used are high enough to encompass threshold visual acuity at these ages, acuity would be as accurately tested at a 38 cm distance as at a 55 cm distance. The grating stimuli used by Carlson *et al.* (Teller Acuity Cards) range in spatial frequency up to 26 c/d at the 38 cm distance, which is at or outside the 95% tolerance limits for binocular acuity of 9- and 12-month-old infants (reference (8) above).

1. Carlson SE, Ford AJ, Werkman Sh, Peeples JM, Koo WWK. Visual acuity and fatty acid status of term infants fed human milk and formulas with and without docosahexaenoate and archidonate from egg yolk lecithin. *Pediatr Res* 1996; **39**: 882–8.

AUTHOR'S REPLY: I think that you have misunderstood the question I raised. It was not intended to infer that one cannot measure visual acuity at 38 cm, of course one can. Simply, the question is whether the test is equally sensitive to detect differences among treatment groups when the acuity cards are held closer than in the standardized test procedure. When the card is moved from the infant, both the distance and the angle over which stimulation is derived is increased. In reviewing information on the test, I was not able to find information to tell me that this is not a possibility.

MAYER AND DOBSON: The statement that mean acuities of 4 month olds in the Carlson *et al.* study (reference (1) above) were 'much higher' than those in the study by Jorgensen *et al.* (1) is not supported by the data. In Carlson *et al.*, the highest acuities were shown in the formula groups without bronchopulmonary dysplasia (BPD) (6 c/d and 5.7 c/d for control and fish oil formulas, respectively) and the mean acuity of the breast-fed group in the Jorgensen *et al.* study was equally high (6 c/d). Indeed, it appears as if the formula-fed group in Jorgensen *et al.* had mean acuity (3.8 c/d) in the lower range of mean acuities in studies of normal infants (see Table 1 in our chapter).

1. Jorgensen MH, Hernell O, Lund P, Holmer G, Michaelsen KF. Visual acuity and erythrocyte docosahexaenoic acid status in breast-fed and formula-fed term infants during the first four months of life. *Lipids* 1996; **31**: 99–105.

AUTHOR'S REPLY: The 4-month-old formula-fed babies reported by Carlson *et al.* had a preferential looking acuity of 6.08 (vs 6.07 in the breast-fed); Jorgensen *et al.* reported 3.8 c/d for formula-fed infants (and 6 for the breast-fed). The numbers from Jorgensen *et al.* are lower than those reported by Carlson *et al.*

MAYER AND DOBSON: The larger standard deviations of preferential looking acuity obtained in the multicenter study by Innis and colleagues (1) could be due, at least in part, to tester differences, as we discuss in our chapter. However, in fact, only in one group of the three reported by Innis was the standard deviation in the higher end of the range found in other nutrition studies (SD 0.77 octave in the human milk group) (see Tables 2 and 3 in our chapter).

1. Innis SM. Visual acuity and blood lipids in term infants fed human milk or formulae. *Lipids* 1997; **32**: 63–72.

AUTHOR'S REPLY: The larger standard deviations in preferential looking acuity in the multicenter study by Innis could be due to tester differences. However, the standard deviation is the high end of that found in other nutrition studies in only one of the three groups, that being the breast-fed group (SD 0.77 octave).

MAYER AND DOBSON: Innis' argument that average, circulating levels of any one fatty acid should not be used as a minimum standard for public health policy is difficult to follow. First, it would be useful for those of us not in the nutrition field to understand why we should conclude that the data sets from three different countries on median levels of linoleic acid in human milk show wide individual differences. Interestingly, these data sets suggest no evidence for differences among countries in mean levels and ranges. The second point Innis makes, that the total amount of circulating fatty acids is constant and therefore that the ratio of different fatty acids must vary proportionally, is good. But how does the conclusion follow that levels of one fatty acid should not be used to guide policy? The argument needs information on the other fatty acid constituents, how they vary proportionally, etc. Perhaps if Innis gave examples of variability in ratios as a consequence of variability in linoleic acid and other fatty acids, the argument would be clearer.

AUTHOR'S REPLY: If minimum standards are set at average levels, then by definition the levels in formulas are at the high end of the normal. This raises issues of where to set the maximum, bearing in mind that a certain range must be given for practical purposes of product production. This is somewhat secondary to the nutrition issue though. For fatty acids it is important to remember that total fat is constant, so if you pin the mimimum for one fatty acid at the average amount, then by necessity the content of some other fatty acids must be below average. Of course, you could set the minimum for all fatty acids at the average in milk; if you do this you will have too much fat.

MAYER AND DOBSON: Innis states that variations in circulating fatty acids (measured in plasma and red blood cells) are not necessarily representative of variations in fatty acids in the central nervous system. Why then would we expect a correlation between behavioral or electrophysiological measures (of vision) and concurrent measures of fatty acids in plasma and red blood cells?

AUTHOR'S REPLY: I do not suggest we would.

MAYER AND DOBSON: Further, Innis states that there are age-related changes in circulating fatty acids. Innis further points out that plasma and red blood cell phospholipids may disagree. Can the discrepancies among studies finding and not finding a correlation between direct measures of fatty acids and outcome measures in the nutrition studies be explained by these or any factors or are they uninterpretable on this basis?

**Lucas**: Innis' critique of dietary polyunsaturated fatty acids in term infant nutrition makes it abundantly clear that the biology of long-chain lipids and the theoretical problems involved in achieving physiologically appropriate supplementation are of extraordinary complexity. For this reason, theoretical bases for intervention in such complex systems could be flawed and, indeed, if acted on could even do more harm than good. The question, then, of immediate public health and clinical relevance is what quality of support is needed to satisfy regulatory bodies and nutrition experts that action should be taken to modify the lipid content of existing, modern infant formulas? Innis tentatively takes the position that she could endorse designing breast milk substitutes that result in fatty acid contents in the blood similar to those found in the breast-fed baby (though she does go on to raise caveats about that approach).

To put this in context, I shall consider more generally the type of support that could be used to defend the incorporation of long-chain polyunsaturated fatty acids (LCPUFAs) into infant formula. We could rank such support into six categories of ascending robustness:

(i) Nonrandomized, 'hypothesis-generating' observations on, say, cognitive or visual developmental advantage in breast versus formula fed babies. In the light of these:
(ii) formula designed to 'mimic' breast milk in LCPUFA content;
(iii) formula designed to induce a similar pattern of blood lipids as found in the breast fed infant ('biochemical efficacy');
(iv) small 'pilot' randomized studies (as currently found in the literature) examining the possible effects of LCPUFAs on selected efficacy outcomes (e.g. visual and cognitive development);
(v) full-scale randomized trials that examine efficacy and safety with adequate power;

(vi) formal evaluation of numerous large studies covering a broad range of efficacy and safety outcomes.

For a number of reasons that I have considered in my own chapter in this volume, it would seem prudent for us to wait until data emerge from studies in at least category (v) above (full-scale randomized efficacy and safety trials) before official recommendations are conceived.

I would suggest there are good reasons why we should not be satisfied with data in category (iii) (biochemical efficacy) to dictate any immediate change in policy with regard to the addition of LCPUFA docosahexaenoic acid (DHA) or arachidonic acid (AA) to infant formula.

Firstly, the biochemical pathways of LCPUFAs and related eicosanoids are so complex that it would be unrealistic to assume that 'correcting' DHA and AA to breast-fed-infant values would ensure either efficacy or safety – indeed current data are quite inconsistent on the relationship of blood biochemistry to functional outcomes such as visual or cognitive development and growth (1–7).

Secondly, as pointed out by Innis herself, the breast-fed baby is not necessarily a gold standard. Considerable variation in both function and in blood lipid biochemistry is seen in the breast-fed population. Interestingly, Makrides (unpublished) finds that cognitive scores in breast-fed infants relate to the mother's breast milk DHA content and to the baby's red cell DHA content, though these associations could be confounded and not necessarily causal. The important endpoint, however, is the safe, optimization of functional outcome in formula-fed babies, regardless of the range of performance in breast-fed infants.

Finally, it is clear from randomized LCPUFA supplementation studies performed on formula-fed babies to date, particularly preterm babies, that detrimental effects of LCPUFA supplementation including those on growth, cognitive function, motor function, and possibly sepsis can occur (for a review see my own chapter in this volume). These findings should make us nervous about recommending manipulation of key biochemical pathways with biologically potent lipids without good efficacy and safety data.

Thus, returning to Innis' remarks about the possible desirability of bringing the blood lipid biochemistry of formula-fed babies more in line with that of breast-fed infants, it would seem reasonable that the formula-fed infants' blood biochemistry could be used as a guide to LCPUFA precursor intakes ($\alpha$-linolenic acid (LNA) and linoleic acid (LA)) – and perhaps this is what Innis intended, but in my view it is too soon to recommend that we routinely manipulate blood biochemistry further, using dietary LCPUFAs themselves. Given modern formulas, now providing n-3 and n-6 precursors, it seems we are in a satisfactory 'holding position' until large-scale studies convince us one way or the other on the safety and efficacy of further changes. Until then, it may be prudent to accept some biochemical differences between breast-fed and formula-fed infants, and indeed differences amongst breast fed infants themselves. I do accept, however, that formulas with added LCPUFAs designed to produce patterns of blood LCPUFAs within the range of those seen in breast-fed babies should be amongst those tested in intervention studies.

1. Makrides M, Simmer K, Goggin M, Bigson RA. Erythrocyte docosahexaenoic acid correlates with the visual response of healthy, term infants. *Pediatr Res* 1993; **33**: 425–27.
2. Jorgensen MH, Hernell O, Lund P et al. Visual acuity and erythrocyte docosahexaenoic acid status in breast-fed and formula-fed term infants during the first four months of life. *Lipids* 1996; **31**: 99–105.
3. Makrides M, Neumann M, Simmer K, Pater J, Gibson R. Are long-chain polyunsaturated fatty acids essential nutrients in infancy? *Lancet* 1995; **345**: 1463–68.
4. Innis S, Nelson CM, Lwanga D, Rioux FM, Waslen P. Feeding formula without arachidonic acid and docosahexaenoic acid has no effect on preferential looking acuity or recognition memory in healthy full-term infants at 9 months of age. *Am J Clin Nutr* 1996; **64**: 40–6.
5. Innis SM, Nelson CM, Rioux FM, King DJ. Development of visual acuity in relation to plasma and erythrocyte ω-6 and ω-3 fatty acids in healthy term gestation infants. *Am J Clin Nutr* 1994; **60**: 347–52.
6. Auestad N, Montalto MB, Hall RT et al. Visual acuity, erythrocyte fatty acid composition and growth in term infants fed formula with long chain polyunsaturated fatty acids for one year. *Pediatr Res* 1997; **41**: 1–10.
7. Hartmann EE, Neuringer M. Longitudinal behavioural measures of visual acuity in full-term human infants fed different dietary fatty acids. *Invest Ophthalmol Vis Sci* 1995; **36**: S869.

AUTHOR'S REPLY: Dr Lucas ranks support for addition of LCPUFA to infant formulas into six categories in ascending order of robustness:

(i) nonrandomized, hypothesis generating observations, e.g. on cognitive and/or visual development;
(ii) formula designed to mimic breast-milk LCPUFAs;
(iii) formula designed to mimic blood lipid patterns of breast-fed infants;
(iv) small 'pilot-sized' randomized trials;
(v) large randomized trials which examine safety and efficacy;
(vi) evaluation of numerous large studies covering a broad range of efficacy outcomes.

Firstly, I do not agree with this rank order. Categories (i)–(iii) are done for different reasons. When starting the work with addition of LCPUFAs to formula (ii), it is reasonable to start in the range of that in milk; the measure of interest is (iii) – blood lipids. In the first instance, is it absorbed? I will come back to this. The final point, in practical terms, is that it may, to summarize your final point, i.e. it would be foolhardy to add LCPUFAs to formulas in general use either for term or premature infants until the safety data from large randomized trials (point (v)) are available. I would add, an adverse outcome resulting from changes in infant formula compositions is a mistake which would be difficult to overcome, either for the industry concerned or for the regulatory agency which endorses it. Because of this it may be argued that (v) is a prerequisite. As discussed in my chapter, a worthwhile point to emphasize is that docosahexaenoic acid and arachidonic acid are not added as single fatty acids. They are components of oils which contain many other components and with which there is no experience of feeding infants (history of use). The study of safety is by no means simple. Some oils contain components which have not been fully identified, and thus their biological activity or potency is not known. One might astutely ask: 'How does one define safety, and how should it be tested?' However, this is not done in infants. I suggest then, that your point (v) is not valid; you can test efficacy but should not have an endpoint of safety in a human study.

Dr Lucas notes that studies to date have found detrimental effects on growth, cognitive function, motor function, and possibly sepsis. These changes have occurred despite manipulation of blood biochemistry, giving 'biochemical efficacy' but not safety. He suggests then that point (iii) – mimicking blood lipid patterns by adding LCPUFAs – is not justification for LCPUFA addition until the large randomized trials (point (v)) have been completed.

I suggest a potential approach may be to acheive similar patterns of lipid metabolism in formula-fed infants to that in breast-fed infants, i.e. point (iii) in Dr Lucas' hierarchy. The question is whether this is appropriate.

It is possible to achieve a similar percentage of arachidonic acid and docosahexaenoic acid in total plasma phospholipids in formula-fed infants to that in breast-fed infants by the addition of small amounts of arachidonic acid and docosahexaenoic acid to formula. Do we conclude that a similar percentage of these fatty acids is evidence of biochemical equivalence (as many indeed do)? This is not only simplistic, it is incorrect. Lipoprotein concentrations remain completely different (remember that phospholipids are distributed in chylomicron, VLDL, LDL, and HDL) between babies fed breast (or expressed) milk and formula fed babies, and the levels of many other fatty acids, e.g. linoleic acid, remain different. The pathways by which fatty acids enter brain have not been identified. Similar percentage values for individual fatty acids in a heterogenous mix of total phospholipids is meaningless from a biochemical perspective. For example, what if fatty acid ratios (e.g. EPA/AA) (as for amino acids) have important effects on brain fatty acid uptake. What is the relevance of plasma phospholipid docosahexaenoic acid, if it is lysophospholipid that delivers n-3 fatty acids to brain. Equivalence of blood lipids has not been achieved in any study to date; my interpretation of what is needed for point (iii) in your hierachy has not been accomplished.

**Shaw and McEachern**: The increase in looking acuity over the first few months of postnatal life reflects maturation of the retina. Developmental 'plasticity' of the visual system, particularly in the visual cortex, appears to be most prominent only after such retinal development is complete. As such, the period of retinal maturation should not be confused with critical periods for activity-dependent neuronal plasticity.

**Colombo**: The inconsistency of human studies of LCPUFA supplementation studies on perceptual–cognitive function is in some contrast with animal studies in which LCPUFA levels are severely deprived. While I realize that the level of deprivation in animal studies very likely constitutes a qualitatively different phenomenon than is studied in the human studies, I am still looking to somehow reconcile the inconsistencies in the human clinical trial. From an outsider's view of this literature, it would appear that positive and null outcomes of LCPUFA supplementation studies are consistently and respectively associated with particular laboratories and/or sites at which the research has been run. When this occurs across laboratories in developmental psychology, one of the first considerations brought to the table are the characteristics of the samples that comprise the contradicting studies. Is there any basis for the possibility that demographic characteristics (socio-economic status, race, education, parental characteristics, etc.) may mediate whatever effects might (theoretically or empirically) be engendered by LCPUFA supplementation?

**Singer**: This chapter summarizes the biochemical rationale for the supplementation of infant formula with LCPUFAs, as well as the animal and human studies in term infants to date which have examined whether supplementation has produced beneficial effects in the two areas targeted as outcomes, i.e. visual acuity and cognition. It appears that benefits of supplementation observable using animal models may not be generalizable to human term infants because, in animal models, severe deficiencies have been induced which are not likely to occur in human infants. The largest and most well-controlled studies of human infants have not documented beneficial effects. However, these studies have not focused on lower socio-economic status samples which might have nutritional deficiencies and might differentially benefit from supplementation. The possibility of such an interaction/moderator effect should be considered in future studies. This chapter also emphasizes the lack of knowledge regarding the correlates of variability in levels of LCPUFAs in breast milk across time, cultures, and in conjunction with specific diets, making it difficult to specify an ideal supplementation level.

**Heird**: Do the concerns about the safety of single-cell oils and other sources of LCPUFAs for supplementing infant formulas represent real or potential concerns? Surely, since formulas containing the products are available in Europe and trials of formulas containing them have been and are being conducted in the USA, the presence or absence of potential toxins has been documented. Are you aware of any data? Also, how serious are the concerns about the position and number of LCPUFAs in the triacylglycerol molecule?

AUTHOR'S REPLY: It is difficult to assess the potential for adverse effects of new oil sources (this is generic and not just relevant to single-cell oils) without knowledge of all the components of the oil itself, the presence of any other components which may be present as a result of the extraction and purification processes, the concentration of these components, and their potency. Without this knowledge it is difficult to know if 'safety' studies have addressed the systems which may be sensitive to adverse effects, or if particular groups of infants, for example certain groups of high-risk premature infants, are more vulnerable.

I would be hesitant to conclude that the availability of certain foods or supplements in other countries is in itself evidence of safety.

I have not seen any published information on the digestion and absorption of fatty acids relating to single-cell triglycerides with two or three LCPUFAs per triglyceride, and with LPCUFAs in the 1 and 3 positions of the triglcyeride. Whether or not arachidonic acid could be released as a free fatty acid in the lumen, and whether this could have efffects on intestinal eicosanoids or immune function has not to my knowledge been addressed. Eructation was a significant side-effect in the adult studies with single-cell oils. Considerations in clinical studies of parameters such as time to attain full enteral feeds, days on which feeds are withheld and other evidence of feed intolerance, sepsis and necrotizing enterocolitis certainly seem reasonable parameters to include in clinical studies.

**McCall and Mash**: Is it possible that a minimum amount of DHA and AA and perhaps LCPUFAs is required but that, after that minimum is attained, more is not necessarily better and potentially harmful? The result of such a system is that individual differences in levels of these substances within the infant's body may not be very meaningful (i.e. may not correlate much with other functional and behavioral outcomes) as you suggest. Further, the system might be structured to be highly tolerant to great variations in dietary intake of such substances, since only a minimum is needed. Further, your discussion of the ability or inability of the system to rid itself of excesses may also be a component of a system that needs only a minimum amount and more is not better, unless excesses lead to deficits or other disorders and problems, a possibility that is at least suggested to be the case in this literature. I think an integrated discussion in one place of this possibility might be highly relevant to this literature, since most benefits are observed among low birth weight or preterm infants who may not get the minimum amount under certain circumstances and in which at least some evidence of poorer outcomes is present among term infants who may be getting too much of these substances. This possibility, plus your intriguing discussions of the issues of supplementation (e.g. how much supplement is appropriate at one age versus another age, and which form of supplementation is most appropriate given the other components that come along with DHA and AA in different forms of supplementation) would seem crucial. More specifically, your discussion is rather biochemically technical for us nonbiochemists. Could you add to this discussion what would or might happen functionally or behaviorally if an excess of these substances were present and persistent?

AUTHOR'S REPLY: This is a very reasonable suggestion and very consistent with other dietary 'lipids'; for example, vitamin A, vitamin D, total fat, cholesterol, and many other nutrients also have clear 'adverse' effects when consumed in excess of need.

McCALL AND MASH: The above issue pertains to a second consideration. You present a very interesting discussion of other substances that accompany DHA and AA in different sources of those LCPUFAs that may have positive or negative effects either on the targeted outcomes or other biochemical, neurological, or behavioral outcomes. However, you do not consider whether the variation in results that is observed within subdomains of this literature could be attributed to different sources used for supplementation. Granted, the literature may be highly fragmented, but I was impressed with all the possible 'side-effects' associated with different sources of supplementation, and I simply wonder to what extent these additional biochemical characteristics might produce the rather pronounced variation in results that have been observed in some subdomains. For example, the differences in results in Carlson's studies between fish oil I and fish oil II may illustrate the differential consequences of different types of supplementation.

AUTHOR'S REPLY: A response to this can only be speculation based on animal and in vitro studies, and diseases related to high n-6-derived eicosanoid metabolism.

Excess intake of LCPUFAs will result in increased amounts of these fatty acids in membrane phospholipids, largely at the expense of linoleic acid. The longer chain length and greater number of double bonds (kinks in the fatty acid chain) can be expected to alter the membrane microenvironment, which could alter integral or associated membrane proteins, ion channels, or receptors. Increased amounts of LCPUFAs in a membrane also increase the potential for membrane damage due to oxidation. This could be of particular significance for tissues exposed to higher oxygen and might well have long-term significance. There is considerable and growing evidence that LCPUFAs, particularly n-6 LCPUFAs, play a role in gene expression, and this opens the door to a suggestion that many fundamental processes ranging from cell division, to hormone release, to intermediary metabolism could be changed. Although there is some evidence that docosahexaenoic acid and arachidonic acid may be involved with neurotransmitter function, it is not clear if this is through a role in membrane lipids (e.g. influencing receptor uptake) or if there are direct effects after release of the LCPUFA from the membrane lipids in response to the stimuli. Another, and probably most widely appreciated, mechanism, would be through changes in eicosanoid metabolism; however, this again is usually in response to some specific signal.

These are simply suggestions, for which there is no evidence for infants or animals fed high levels of LCPUFAs in the context which we are discussing.

McCall and Mash: To what extent could general nutritional enrichment (e.g. in the form of the special preterm formulas) accomplish the same thing relative to some of the outcomes that we are focusing on as could human milk or, more specifically, LCPUFAs? For example, in some of Lucas' comparisons, human milk produced better developmental scores than standard formula in low birth weight infants, despite the fact that human milk was lower in general nutrition. Furthermore, human milk was not more beneficial relative to an enriched preterm formula. This suggests that at low levels of nutritional substances contained in human milk (LCPUFAs) may convey some benefits, but that generally enriched nutrition can compensate for a deficit in those specific substances. Is there any biochemical reason or rationale that might support this possibility?

Author's reply: Human milk contains many nutrient and non-nutrient components that are absent from formula which could have beneficial effects on the premature infant (1). An example is thyroid hormone, and indeed there has been speculation that neurological and developmental problems in premature infants are related to thyroid function (2,3). Even though the measured outcome indicates benefit, I see no reason to speculate that the biochemical mechanism would be the same as any effect which LCPUFAs may have.

1. Grosvenor CE, Picciano MF, Baumrucker CR. Hormones and growth factors in milk. *Endocrine Rev* 1992; **14**: 710–28.
2. Meijer WJ, Verloove-Varhorick SP, Brard R, van den Brande JL. Transient hypothyroxenaeria associated with developmental delay in very preterm infants. *Arch Dis Child* 1992; **67**: 944–7.
3. Vulsma T, Kok JH. Prematurity-associated neurologic and developmental abnormalities and neonatal thyroid function. *N Engl J Med* 1996; **334**: 857–8.

McCall and Mash: Is there any biochemical rationale to the possibility that there is something special to be gained by an infant if that infant is fed milk from his or her own mother versus human milk from an unrelated donor? That is: (i) Are there differences in the characteristics of human milk from mother to mother? (ii) Are there differences in the biochemical characteristics of individual infants? I assume the answer to both these questions is profoundly 'yes'. More to the point, however, (iii) is there something to be gained by the infant by a certain match in the individual differences of the milk received from his or her own mother with that particular infant's biochemistry? I ask this question, because some of Lucas' results might be explained if such a match were beneficial.

AUTHOR'S REPLY: At this point, I do not think there is any reason to speculate this would be the case. I think that it is more reasonable that losses in activity of one or more of the biologically active components in milk for e.g. growth-hormone releasing hormone, TRH, TSH, PTH-related peptide, IGFs, or EGF (1) are inactivated by processing or storage conditions. These conditions can also affect nutrient quality, for example by changes to the milk fat globule structure, and by inactivation of some of the enzymes in milk. A secondary issue is that for the mother who expresses milk there is the potential that the infant may be breast-fed. This is not a potential for infants fed donor milk and whose mothers do not establish a milk supply. This now introduces variables of breast-feeding, and potentially secondary variables of physiological/psychological differences in lactating/nonlactating women. However, you note (point (vi)) that Jacobson and Jacobson suggest breast- versus bottle-feeding is not important when characteristics of the mother and home environment are included as covariates.

1. Grosvenor CE, Picciano MF, Baumrucker CR. Hormones and growth factors in milk. *Endocrine Rev* 1992; **14**: 710–28.

McCALL AND MASH: The Jacobson and Jacobson (1) data suggest that it is possible that, when characteristics of the mother and the home environment assessed after birth and at a time relevant to the actual rearing of the child are included as covariates, the breast versus formula distinction in developmental and mental performance washes out.

1. Jacobson SW, Jacobson JL. Breastfeeding and intelligence. *Lancet* 1992; **339**: 926.

McCall and Mash: Since these infants were tested at 36 months of age and the tasks used were more cognitive than those necessarily used at younger ages, is it possible that the observed difference is associated with rearing circumstances rather than feeding, in that we know that parents who choose to breast feed tend to be higher educated and provide a more stimulating home environment?

You seem to endorse supplementing infant formulas 'to attain similar patterns of blood lipids fatty acid transport and metabolism in formula-fed infants to that of breast-fed infants.' There must be a biochemical distinction here that I am missing.

Throughout this paper you have generally argued for the uncertainty of evidence regarding the functional, if not biochemical, benefits of supplementation, but now it appears to a biochemically naive reader that you are endorsing such supplementation. Perhaps you should make any distinctions that are relevant clearer to the reader. However, if indeed you are endorsing supplementation, then I think the argument to bolster that conclusion needs to be stronger than you have made it. Would you be making such an argument based solely on mimicking the biochemical outcome in breast-fed infants in the absence of firm functional and behavioral benefits and in the presence of some risks?

Author's reply: Yes, this is certainly a possibility.

# Statistically Significant Versus Biologically Significant Effects of Long-Chain Polyunsaturated Fatty Acids on Growth

WILLIAM C. HEIRD

*Children's Nutrition Research Center, Department of Pediatrics, Baylor College of Medicine, Texas Medical Center, 1100 Bates Street, Houston, TX 77030, USA*

## Introduction

Many individuals (1,2) as well as groups (3–6) have advocated supplementation of infant formulas with long-chain polyunsaturated fatty acids (LCPUFAs), i.e. fatty acids of a chain length greater than 18 carbon atoms and with more than two double bonds, particularly n-3 LCPUFAs. Indeed, many formulas containing only n-3 LCPUFA or a mixture of n-3 and n-6 LCPUFAs are available in Europe and Asia (7). To date, however, the major North American dietary advisory groups have not advocated such supplementation, and supplemented formulas are not available in North America.

Even if it becomes clear that supplementation of formulas with n-3 and n-6 LCPUFAs is desirable for visual and/or neural development, several questions concerning the safety of such supplementation must be addressed before supplemented formulas are likely to be available in North America. One of the major questions concerns the effect of n-3 LCPUFAs on growth.

Three studies (8–10) have now been published showing that infants fed n-3 LCPUFA-supplemented formulas or formulas with a high content of α-linolenic acid (ALA; 18:3n-3) the precursor of n-3 LCPUFAs, experience lower rates of growth than do infants fed unsupplemented formulas or formulas with a low ALA content. Two additional studies, as yet published only as abstracts (11,12), show similar effects.

In one of the first studies of n-3 LCPUFA supplementation (8,13), preterm infants fed a formula supplemented with fish oil (0.3% of total fatty acids as eicosapentaenoic acid (EPA; 20:5n-3) and 0.2% of fatty acids as docosahexaenoic acid (DHA; 22:6n-3) weighed less from 40 weeks postmenstrual age throughout the first year of life than infants fed an unsupplemented formula. The mean weight of the supplemented infants at one year of age (corrected for prematurity), 3 months after supplementation was stopped, was approximately 750 g less than that of unsupplemented infants. Moreover, across groups, there was a direct relationship between weight and the arachidonic acid (AA; 20:4n-6) content of plasma and erythrocyte phospholipids (13). Since the supplement consisted of EPA as well as DHA but no AA, the growth failure was attributed to depression of the AA content of plasma and erythrocyte phospholipid secondary to the high EPA content of the supplement. A precise mechanism, however, was not identified.

Uauy et al. (14), interestingly, found no statistically significant differences in growth from shortly after birth through 57 weeks postmenstrual age (PMA) among groups of preterm infants fed a formula supplemented with a greater amount of fish oil (0.65% of total fatty acids as EPA and 0.35% as DHA), two unsupplemented formulas (either 24.2% of total fatty acids as linoleic acid (LA; C18:2n-6) and 0.5% as α-linolenic acid (ALA; 18:3n-3), or 20.8% LA and 2.7% ALA) and human milk. However, supplemented formulas were fed only until discharge from the nursery (about 37 weeks PMA). Moreover, only 10–12 infants per group were evaluated at 57 weeks PMA.

Subsequently, less marked, but nonetheless statistically significant, differences were observed between the growth of preterm infants fed an unsupplemented formula versus a formula supplemented with low-EPA fish oil (0.06% vs 0.3% EPA; 0.2% DHA). In this study (9), the mean weight of the supplemented group was 200 g less at 2 months corrected age (48 weeks PMA), when the supplemented formula was stopped, and differences of at least this magnitude between supplemented and unsupplemented infants persisted through 12 months corrected age. However, only the differences at 9 (470 g) and 12 months corrected age (290 g) were statistically significant.

In another study, so far reported only as an abstract (11), the mean head circumference, length, weight and fat-free mass (measured by bioelectric impedance) of low birth weight (LBW) infants who received a formula supplemented with low-EPA fish oil (0.06% EPA, 0.2% DHA) from the time full enteral feedings were tolerated through 168 days of age were lower at 4 and 6 months of age than observed in infants fed an unsupplemented formula for the same period.

In both studies of the low-EPA fish oil supplement, the plasma and erythrocyte phospholipid content of DHA was higher in supplemented versus unsupplemented infants and, in one (9), weight-for-length at 2, 6, 9, and 12 months PMA was inversely related to the DHA content of erythrocyte phosphatidylethanolamine at 2 months PMA. However, in neither study was there a statistically significant difference between groups in the AA content of erythrocyte phospholipids. Further, there was no straightforward relationship between any aspect of growth and the AA content of erythrocyte phospholipids. Carlson *et al.* (9) concluded that an 'imbalance' between DHA and AA might explain the apparent effects of DHA on growth. Again, however, no specific mechanism was proposed.

Jensen *et al.* observed an apparent effect of ALA intake, or the ratio of LA/ALA, on growth of both preterm (12) and term (10) infants. Preterm infants fed a formula with about 16% of total fatty acids as LA and 3.2% of total fatty acids as ALA (LA/ALA ratio about 5) from shortly after birth to 56 weeks PMA weighed approximately 550, 650, and 800 g less, respectively, at 48, 56, and 74 weeks PMA than infants fed a formula with the same LA content but only 1% of total fatty acids as ALA (LA/ALA ratio about 16) for the same period. Although the mean birth weight of the group that received the higher ALA intake was about 150 g less than that of the other group, normalized weight of the group that received the higher ALA intake also was significantly lower at 48, 56, and 74 weeks PMA (about 0.7. 0.8, and 0.85 SD, respectively).

Term infants fed a formula with the same high ALA or low LA/ALA fat blend (about 16% of total fatty acids as LA and 3.2% as ALA) from birth to 4 months of age (56 weeks PMA) weighed 760 g less at 4 months of age than infants fed a formula with the same LA content but only 0.4% of total fatty acids as ALA (LA/ALA ratio about 40) for the same period, and about 400 g less than infants fed formulas with the same LA content and 1% or 2% of total fatty acids as ALA (10). The mean weight of the group fed the highest ALA (or lowest LA/ALA ratio) intake for the first 4 months of life also was less at 8 months of age than that of the group fed the lowest ALA intake (or highest LA/ALA ratio); however, the difference between groups at 8 months of age (about 520 g) was not statistically significant.

Across groups of term infants fed formulas with different contents of ALA, there was a positive correlation between weight at both 4 and 8 months of age (56 and 74 weeks PMA) and the plasma phospholipid content of AA at 4 months of age. The weight of preterm infants at these ages correlated with the plasma phospholipid content of other n-6 LCPUFAs, but not consistently with the content of AA. There was no correlation between the weight of either term or preterm infants who received different ALA intakes or different LA/ALA ratios and the DHA content of the plasma phospholipid fraction at any age.

It should be noted that the high ALA content of the formulas studied by Jensen et al. (10,12) was achieved with canola oil, which is not a conventional oil for formulas marketed in the USA. This raises the question of whether the observed effects on growth are related to canola oil, rather than ALA content or the LA/ALA ratio of the formula. Currently, LA and ALA intakes of 16% and 3.2% of total fatty acids, respectively, cannot be achieved without use of canola oil or another unconventional oil (e.g. flax seed or linseed oil). Thus, it is difficult to address this concern. However, canola oil is a component of many formulas currently marketed outside the USA and, to our knowledge, these formulas do not interfere with growth. On the other hand, these formulas do not contain as much canola oil as the high ALA formula studied by Jensen et al. (10,12).

If the apparent effect of n-3 LCPUFA supplementation and high ALA intake on growth is secondary to a low AA content of plasma and, presumably, tissue lipids and/or an imbalance between n-3 and n-6 LCPUFA content, it seems likely that the effect can be overcome by concurrent AA supplementation. In fact, because of this likelihood, many assume that the effect of n-3 LCPUFA on growth is no longer a relevant issue. Unfortunately, data concerning the growth of either preterm or term infants fed formulas supplemented concurrently with n-3 LCPUFA and AA are scarce.

In one relatively large study (15), there was no statistically significant difference in growth of term infants fed formulas supplemented with DHA alone (0.2% of total fatty acids) versus DHA and AA (0.12% and 0.43% of total fatty acids, respectively), despite the fact that the group fed the formula supplemented with only DHA had lower plasma and erythrocyte phospholipid contents of AA. Moreover, growth of these two groups of infants also did not differ from that of either a group of breast-fed infants or a group of infants fed a standard unsupplemented formula.

Interestingly, the DHA supplement used in this study was the same low-EPA fish oil that, in two studies (9,11), resulted in at least modest effects on growth of preterm infants. Since the effects of n-3 LCPUFA on growth are likely to be more relevant for preterm versus term infants, data concerning growth of preterm infants fed unsupplemented formulas versus formulas supplemented with DHA alone *as well as* DHA plus AA are needed to clarify the possibility that the effects of n-3 LCPUFA on growth can be prevented by concurrent supplementation with AA. Such studies currently are underway but, to date, no data are available.

From the above summary of the published sets of data, it is clear that there are statistically significant effects of n-3 fatty acids on growth of both term and preterm infants. A more relevant issue, however, concerns the biological significance of this effect, and to evaluate this issue several questions must be addressed. The first is the general question of why the observed effects of n-3 LCPUFA on growth have generated so much concern. Another concerns the magnitude of the effect that has been observed and, specifically, whether it is sufficient to justify the concern that it has generated. Yet another concerns the mechanism(s) of the observed effect of n-3 fatty acids on growth. Each of these questions is addressed in the sections that follow. The final section includes the author's conclusions concerning the biological significance of the observed effects of n-3 LCPUFAs on growth.

## Why be Concerned about Growth?

The growth and development of infants and children versus adults is the factor that distinguishes pediatrics from all other medical specialties. Thus, pediatricians have always been concerned about growth and development, and any practice and/or condition that interferes with either growth or development is considered undesirable. Part of the concern, perhaps, can be attributed to the fact that most chronic and many acute childhood diseases are associated with or accompanied by poor growth. In fact, the development of pediatrics as a specialty around the turn of the century was precipitated by the prevalence at that time of growth disorders secondary to a variety of nutritional deficiencies (e.g. rickets) and diseases resulting in growth failure (e.g. diarrhea).

The majority of the disorders that precipitated the development of pediatrics as a specialty have now been conquered, at least in developed countries. Nonetheless, emphasis on growth and development continues to be a hallmark of pediatrics. The appropriateness of this emphasis is supported by a number of studies from both developed and underdeveloped countries showing an association between nutritional status, as identified by disordered growth, and subsequent mortality. As summarized by Cooper and Heird (16), mortality is about three-fold greater in children from underdeveloped countries whose weight-for-age is < 60% versus > 75% of expected. A similar difference in mortality is associated with weight-for-length of < 70% versus > 90% of expected as well as height-for-age of < 85% versus > 95% of expected. Relationships between poor nutritional status and subsequent mortality also have been noted in hospitalized children and adults in more developed countries (17,18). This greater subsequent mortality of severely malnourished versus reasonably well nourished infants/children may be related, in part, to the known adverse effects of malnutrition on immune function and, hence, decreased resistance to infection.

For the past 30–40 years, the impact of a discrete period of subnormal growth, or malnutrition, on subsequent growth and development, particularly development of the central nervous system, has received considerable attention. The initial studies in rats suggested that a discrete period of malnutrition during a critical period of central nervous system growth and/or development, unless corrected during the critical period of growth and/or development, resulted in irrecoverable deficits in brain structure and function (19,20). Since the same biochemical deficits observed in the brains of previously malnourished versus well-nourished rats were observed also in malnourished versus well-nourished human infants who died suddenly during early life (21), it was assumed that this was true also for the human infant.

Currently, it is recognized that the situation in the human infant is more complex. Nonetheless, there appears to be an association, albeit a complex one, between inadequate early growth and subsequent central nervous system development as assessed by standard behavioral tests. A major problem in understanding this relationship more fully is the difficulty in distinguishing between the effects of inadequate growth or malnutrition, per se, and the myriad medical and/or environmental conditions resulting in inadequate growth or malnutrition. It is particularly difficult, for example, to determine if poor neurodevelopmental function at school age is secondary to an earlier period of inadequately treated malnutrition and, hence, inadequate growth during early life, to the effects of the usual environment in which malnutrition or inadequate growth is usually encountered (i.e. poverty, emotional deprivation) or to a combination of inadequately treated malnutrition and environmental factors.

The hundreds of published studies concerning the subsequent effects of early malnutrition on neurodevelopmental outcome have been reviewed recently by a number of authorities (22–31). Virtually all conclude that malnutrition produces acute neurobehavioral effects but that these effects are at least partially reversible, provided the postrecovery environment is satisfactory. However, if the postrecovery environment is not satisfactory, many of the changes incident to the early period of inadequate growth seem to persist despite adequate nutritional rehabilitation. Most also allow for the possibility that maintenance of adequate nutrition during childhood may protect against or attenuate the effects of poor environmental factors.

Two relatively recent studies, one conducted in Jamaica (32) and the other in Indonesia (33), provide further insight into this issue. In the first, 9- to 24-month-old infants whose length was at least 2 SD below that expected for age were assigned randomly to a control group, a nutrition supplementation group, a psychosocial stimulation group or a group that received both nutritional supplementation and stimulation. During the 2 years of intervention, developmental indices of the three latter groups improved, but improvement was greatest in the combined intervention group. In the other study (33), nutritionally 'at-risk' infants between 6 and 20 months of age were assigned randomly to receive, or not receive, a nutrition supplement for 90 days. Bayley Scales of infant development scores improved considerably in the supplemented versus the control group. In both studies, nutritional supplementation improved indices of motor development much more dramatically than indices of mental development.

The majority of studies have focused on the subsequent effects of an early period of severe malnutrition or growth failure. Much less emphasis has been placed on the degree of growth failure that is likely to result in long-term effects. Findings of a recent multinational study suggest that even mild to moderate growth retardation secondary to a variety of specific or mixed nutrient deficiencies results in acute as well as longer term deficits in neurodevelopmental outcome (34). A number of other correlational studies suggest that attentiveness, school performance, and standard neurobehavioral indices are less in smaller versus larger infants and children as well as in those growing more slowly versus those growing at normal rates (27). However, as in most such studies, the effect of the nutrient deficiencies, per se, on subsequent outcome was not differentiated from the effects of environmental factors. Moreover, it is not clear that the effects of a specific nutrient deficiency (e.g. iron) is as amenable to correction as a more general deficiency of protein and/or energy.

A major issue concerns the period during which the central nervous system is vulnerable to the effects of malnutrition and, hence, the period during which nutritional rehabilitation might prevent the effects of malnutrition on subsequent function. This period is usually thought to encompass the period of the brain growth spurt. In the human infant, this begins during early prenatal life and continues to at least 2 years postnatal age

(35). Most of the neuronal growth spurt occurs during the first half of gestation, while the late prenatal and postnatal growth spurt includes astrocyte proliferation and growth, deposition of myelin, dendritic arborization, and formation of synapses. Thus, it is unlikely that malnutrition will affect neuronal proliferation and growth, as was once suggested (19). Nonetheless, it is clear that malnutrition during the latter period of the brain growth spurt, if not corrected, can result in lower neurobehavioral indices later in life. It also is clear that these effects of early malnutrition can be at least partially reversed by nutritional supplementation as well as psychosocial stimulation (32). However, neither the extent of malnutrition likely to interfere with neurobehavioral development nor the precise period during which nutritional rehabilitation must be accomplished in order to prevent or attenuate the later neurobehavioral effects is entirely clear.

In developed countries, the population of infants for which these considerations are most relevant is the increasing number of surviving infants who were born weighing < 1500 g. These very-low-birth-weight (VLBW) infants have a high incidence of neonatal illnesses and most probably receive suboptimal nutrition for the first several days of life. For example, these infants lose 10–15% of body weight during the first several days of life and, on average, do not regain this lost weight until about 15 days of age (36). Thus, while only 22% weigh less than the 10th percentile for intrauterine weight at birth, 69% weigh less than this at discharge (36). Moreover, only a slightly smaller percentage weigh less than the 10th percentile at 40 weeks PMA (37) and many continue to be small throughout life (37,38). In addition, these infants have a reasonably high incidence of neurodevelopmental handicaps (e.g. cerebral palsy) and, as a group, score lower on standardized tests of mental and psychomotor development.

In these infants, it is very difficult to dissect the contributions, if any, of poor intrauterine growth, neonatal illness, and poor neonatal growth to either neurodevelopmental status or size later in life. However, recent studies by Hack *et al.* (37,39,40) have shed some light. Hack and her colleagues have followed a cohort of VLBW infants born during the late 1970s and early 1980s and have published follow-up data through 8 years of age. At 3 years of age, the Bayley Mental Development Index (MDI) of infants whose weight at birth was appropriate for gestational age (AGA), after controlling for a variety of biologic and socio-demographic variables, was related to a subnormal head circumference at 8 months of age (39).

Intrigued by the fact that the early childhood outcomes of AGA and SGA (small for gestational age) VLBW infants appeared to be similar, despite lower rates of catch-up brain growth in the latter, they investigated the differential effect of small head circumference at 8 months of age on outcome of AGA versus small for gestational age (SGA) VLBW infants (40). At 20 months corrected age, neither the MDI scores nor the incidence of neurologic impairment differed between the two groups. In addition, subnormal head size at 8 months corrected age was a statistically significant predictor of 20-month MDI score for both groups. However, after adjusting for neonatal illness, neurologic impairment, socio-economic status, and race, the relationship between head circumference at 8 months of age, while remaining a statistically significant predictor of the 20-month MDI score of AGA infants, was no longer a statistically significant predictor of the 20-month MDI score of SGA infants. The investigators interpreted these findings as suggesting that a subnormal head circumference in AGA infants at 8 months of age results primarily from neonatal illness and/or poor neonatal nutrition and has adverse effects on outcome. In SGA infants, on the other hand, a subnormal head circumference at 8 months of age results from many causes, including intrauterine growth failure, and does not, by itself, appear to affect outcome. Thus, it appears that intrauterine growth failure, unless accompanied by neonatal illness or poor neonatal nutrition, may not have an adverse effect on outcome, whereas postnatal growth failure secondary to illness and/or poor nutrition is likely to have a more marked effect on outcome.

In this population, abnormal head circumference at 8 months of age also had an independent adverse effect on overall IQ as well as receptive language, speech, reading, and spelling scores at 8–9 years of age (41). Other statistically significant predictors of these scores included maternal socio-economic status, neonatal risk score, and neurologic status. Size for gestational age was not a significant predictor of outcome at 8–9 years of age, but the number of SGA infants in the population was quite small.

At one time, it was assumed that catch-up growth did not occur in VLBW infants after 2–3 years of age (42,43). While the studies of Hack et al. (37,40,41) seem to indicate that catch-up in brain growth does not appear to occur beyond 8 months of age in either AGA or SGA VLBW infants, several recent studies show that catch-up in weight and length growth occurs at least through 8 years of age (38,44,45). Hack et al. (37) have shown that the mean weight as well as the mean length of a group of 249 former VLBW infants were at or only slightly below the NCHS means at 8–9 years of age. However, both weight and length were somewhat lower than observed in a group of matched normal weight infants followed concurrently. Multiple regression analysis showed that the mother's height and race, the infant's birth weight, and the presence of a neurologic abnormality were statistically significant determinants of the 8-year height percentile. Maternal height, being small for gestational age, and having a neurologic abnormality were statistically significant predictors of subnormal height at 8 years of age.

Recent epidemiological data suggesting a relationship between size at birth and/or 1 year of age and the incidence of a number of adult diseases also may be relevant to the question, 'Why be concerned about early growth?' Lower versus higher birth weights and/or weights at 1 year of age appear to be associated in adulthood with higher death rates from coronary heart disease and stroke, higher systolic and diastolic blood pressure, and a higher incidence of insulin resistance or non-insulin-dependent diabetes mellitus (46). Some of the adult outcomes appear to be associated with birth weight and others with weight at 1 year of age. Moreover, the associations appear to be continuous but independent of length of gestation; for example, the coronary heart disease death rates of men who weighed 8.2 kg or less at 1 year of age were three times greater than those of men who weighed 12.3 kg or more at 1 year of age.

Also relevant is the finding of significantly lower 18-month Bayley psychomotor development index (PDI) scores in infants who were randomly assigned to be fed a standard term formula versus a nutrient-enriched preterm formula for a discrete period during hospitalization as preterm infants (47). Further, Lucas (this volume) states that this difference persists through 7–8 years of age. These differences at 18 months and 7–8 years of age presumably reflect lower rates of growth secondary to the lower nutrient intakes of the infants fed standard term formula. However, illustrating the complexity of this issue, Lucas *et al.* (48) found no difference in developmental outcomes at 18 months of age between infants fed the nutrient-dense preterm formula versus banked human milk. Since the banked human milk was even less nutrient-dense than the term formula, infants fed this regimen undoubtedly experienced even lower rates of growth. While both studies were randomized controlled trials, it should be noted that they were performed at different sites and involved totally different populations, although both included a group fed the same preterm formula and the scores of these two groups were similar.

As illustrated by this brief summary of a huge body of literature, the pediatrician's concern with maintaining adequate growth and assuring optimal development is both understandable and supportable. However, as also illustrated, it is not easy to differentiate the effects of inadequate growth on poor subsequent outcome from the effects of the variety of other factors giving rise to inadequate growth. Another major problem concerns the definition of adequate (or inadequate) growth; there being no straightforward definition that is likely to be acceptable to all concerned, any deviation from 'normal growth' is usually considered to be undesirable.

## Observed Effects of ω-3 LCPUFA on Growth

With this background, we return to the effects of n-3 LCPUFAs on growth and the specific question of whether the magnitude of the observed effect is sufficient to warrant the concern that it has generated. To do so, it is necessary to consider in more detail the effects that have been observed. To this end, the findings of the two published studies showing an effect of n-3 LCPUFA supplementation on growth of preterm infants and the one published study showing an effect of a high ALA intake on growth of term infants are summarized in Tables 1 and 2.

At some or all ages, there are statistically significant differences between the mean body weight of preterm infants who received n-3 LCPUFA supplemented versus unsupplemented formula and term infants who received formula with high versus low ALA contents or low versus high LA/ALA ratios (see Table 1). Moreover, the magnitude of the maximum difference in mean body weight of preterm infants who were fed the fish oil supplemented versus unsupplemented formulas (8), i.e. 750 g at 12 months corrected age, and the term infants who were fed formulas with an ALA content of 3.2% versus 0.4% of total fatty acids (10), i.e. about 760 g at 4 months of age (56 weeks PMA), is considerable. The magnitude of the differences in mean body weight between preterm infants who received a formula supplemented with low-EPA fish oil versus an unsupplemented formula, i.e. about 200–250 g at 2, 4, and 6 months corrected age, 470 g at 9 months corrected age, and 290 g at 12 months corrected age (9), are somewhat less.

While the maximum difference in body weight between term infants fed formula with high versus low ALA contents (10), i.e. about 750 g at 56 weeks PMA, is of roughly the same magnitude observed by Carlson *et al.* (8) between preterm infants fed formula with versus without the high-EPA fish oil supplement, i.e. 750 g at 93 weeks PMA, the difference between the two groups of term infants at other ages is not always statistically significant. This, most likely, is because the term infant study was designed to have sufficient power to detect a difference between groups of only one standard deviation, whereas the two preterm infant studies were designed to have sufficient power to detect smaller differences.

In one study (8) of groups who received supplemented versus unsupplemented formulas, statistically significant differences in length also were

Table 1. Reported mean weights, lengths, and head circumferences of LCPUFA supplemented versus unsupplemented preterm infants and of term infants fed formula with 3.2% versus 0.4% fatty acids as ALA. (Data from Carlson et al. (8,*9) and Jensen et al (10).)

| Weeks (PMA) | Control (8) | Suppl. (8) | Control (9) | Suppl. (9) | 0.4% ALA (10) | 3.2% ALA (10) |
|---|---|---|---|---|---|---|
| *Weight (kg)* | | | | | | |
| 40 | 3.10 | 2.91 | 2.81 | 2.83 | 3.377 | 3.231 |
| 48 | 4.64 | 4.49 | 4.94 | 4.74 | 5.556 | 4.984 |
| 57 | 6.22 | 5.95 | 6.42 | 6.23 | 6.972 | 6.214 |
| 68 | 7.57 | 7.11 | 7.84 | 7.58 | – | – |
| 74 | – | – | – | – | 8.660 | 8.144 |
| 79 | 8.51 | 7.91 | 8.66 | 8.19 | – | – |
| 93 | 9.43 | 8.68 | 9.31 | 9.02 | – | – |
| *Length (cm)* | | | | | | |
| 40 | 48.1 | 47.3 | 46.6 | 46.7 | 49.8 | 49.6 |
| 48 | 54.3 | 53.8 | 54.6 | 54.8 | 56.7 | 57.0 |
| 57 | 60.6 | 60.2 | 60.5 | 60.3 | 62.9 | 61.9 |
| 68 | 66.0 | 64.7 | 65.7 | 65.6 | – | – |
| 74 | – | – | – | – | 69.9 | 69.3 |
| 79 | 69.6 | 68.7 | 69.6 | 69.2 | – | – |
| 93 | 73.9 | 72.6 | 73.2 | 73.1 | – | – |
| *Head circumference (cm)* | | | | | | |
| 40 | 35.2 | 34.9 | 34.8 | 34.9 | 34.2 | 34.0 |
| 48 | 38.8 | 38.4 | 39.4 | 39.2 | 39.0 | 38.3 |
| 57 | 41.8 | 41.2 | 42.2 | 41.5 | 41.7 | 40.8 |
| 68 | 43.9 | 43.3 | 44.1 | 43.7 | – | – |
| 74 | – | – | – | – | 44.5 | 43.4 |
| 79 | 45.2 | 44.6 | 45.3 | 44.7 | – | – |
| 93 | 46.3 | 45.9 | 46.3 | 45.9 | – | – |

* Weighted means for males and females combined were calculated from reported means for males and females.

observed. However, differences in length were more marked in female versus male infants. Interestingly, in all studies the normalized length of both groups improved from 40 to 56 weeks PMA and then remained essentially unchanged for the remainder of the first year of life (see Table 2), a pattern that appears to be characteristic of the growth of infants who are born prematurely or are growth-retarded at birth (42,49).

Table 2. Reported weight for age, length for age, weight for length, and head circumference for age Z-scores of preterm infants fed ω-3 LCPUFA supplemented formulas and term infants fed formulas with 3.2% versus 0.4% fatty acids as ALA. (Data from Carlson et al. (8,[*]9) and Jensen et al. (10).[†]

| Weeks (PMA) | Control (8) | Suppl. (8) | Control (9) | Suppl. (9) | 0.4% ALA (10) | 3.2% ALA (10) |
|---|---|---|---|---|---|---|
| *Weight for age Z-scores* | | | | | | |
| 40 | −0.29 | −0.65 | −0.42 | −0.38 | 0.08 | −0.19 |
| 43 | − | − | − | − | 0.19 | −0.04 |
| 48 | 0.32 | −0.15 | 0.19 | −0.04 | 0.68 | 0.08 |
| 57 | 0.25 | −0.32 | 0.15 | −0.02 | 0.53 | −0.10 |
| 68 | −0.12 | −0.62 | −0.03 | −0.38 | − | − |
| 74 | − | − | − | − | 0.04 | −0.32 |
| 79 | −0.28 | −0.84 | −0.27 | −0.76 | − | − |
| 93 | −0.45 | −1.00 | −0.59 | −0.77 | − | − |
| *Length for age Z-scores* | | | | | | |
| 40 | −0.91 | −1.18 | −1.22 | −1.08 | −0.36 | −0.43 |
| 43 | − | − | − | − | −0.33 | −0.01 |
| 48 | −0.43 | −0.94 | −0.91 | −0.81 | −0.42 | −0.14 |
| 57 | −0.37 | −0.76 | −0.79 | −0.74 | −0.15 | −0.29 |
| 68 | −0.46 | −0.96 | −0.87 | −0.87 | − | − |
| 74 | − | − | − | − | −0.21 | −0.24 |
| 79 | −0.44 | −0.81 | −0.74 | −0.77 | − | − |
| 93 | −0.38 | −0.77 | −0.84 | −0.71 | − | − |
| *Weight for length Z-scores* | | | | | | |
| 43 | − | − | − | − | 0.49 | −0.16 |
| 48 | 0.83 | 0.71 | 1.16 | 0.76 | 1.30 | 0.24 |
| 57 | 0.70 | 0.33 | 0.92 | 0.67 | 0.74 | 0.09 |
| 68 | 0.38 | 0.14 | 0.77 | 0.37 | − | − |
| 74 | − | − | − | − | 0.26 | 0.12 |
| 79 | 0.24 | 0.18 | 0.37 | −0.17 | − | − |
| 93 | 0.04 | −0.53 | 0.04 | −0.30 | − | − |
| *Head circumference for age Z-scores* | | | | | | |
| 40 | 0.78 | 0.19 | 0.78 | 0.96 | 0.71 | 0.57 |
| 48 | 1.04 | 0.40 | 0.80 | 0.75 | 1.04 | 0.43 |
| 57 | 0.80 | 0.09 | 0.64 | 0.40 | 0.82 | 0 |
| 68 | 0.50 | 0.03 | 0.39 | 0.10 | 0.36 | −0.63 |
| 79 | 0.36 | −0.15 | 0.22 | −0.21 | − | − |
| 93 | −0.25 | −0.23 | −0.13 | −0.25 | − | − |

* Weighted mean Z-scores for males and females combined were calculated from published mean values for males and females.
† Mean head circumference for age Z-scores were calculated from normative data of Karlberg et al. (50).

Despite the greater effect of n-3 LCPUFAs and high-ALA intakes on weight gain than on length gain, normalized weight-for-length of all groups in all studies was above the mean NCHS standard for most of the first year of life. In fact, normalized weight-for-length was appreciably below the mean NCHS standard for term infants in only one study (8) and, in this study, by only 0.5 SD (see Table 2). In preterm infants, this most likely reflects the lower normalized length at the beginning of the study.

Statistically significant effects of n-3 LCPUFA supplementation on head circumference also were observed consistently in one study (8). However, this effect, like the apparent effect of n-3 LCPUFAs on length observed in the same study, was more marked in female infants. No consistent effect on either absolute or normalized head circumference was found in either of the other studies, although in one (9) there was a statistically significant difference between supplemented and unsupplemented infants in absolute, but not normalized, head circumference at 9 months of age. Interestingly, in all studies, the mean head circumference of supplemented infants, although sometimes lower than that of unsupplemented infants, was very close to the 50th percentile value for normal term infants (50).

Fomon and Nelson (51) have proposed that comparing actual rates of increase in weight and length during various intervals to expected rates of normal infants is a more sensitive method for evaluating growth than comparing the observed weight and/or length of a specific infant or group of infants to the weight and/or length expected for age. This is likely to be particularly true for preterm infants who begin most studies with weights and lengths considerably below those of the normal term infant and still have a lower weight and length at 40 weeks PMA (37).

The mean rates of increase in weight and length of the groups of infants discussed above (calculated, by necessity, as differences between the mean weight of groups at different ages divided by the interval between ages), plotted against the mean rates of increase of normal infants during each interval (52), are shown in Fig. 1. When viewed in this way, the effects of n-3 LCPUFAs and/or ALA intake on growth are less impressive. For example, despite statistically significant effects of n-3 LCPUFA supplementation and high-ALA intake on both absolute and normalized weight and length, the mean rates of increase in both weight and length of the n-3 LCPUFA supplemented and high-ALA groups, at least for the first 6 months of life (or from 40 to roughly 66 weeks PMA), are very close to the mean rates of normally growing term infants (52). The mean rate of increase in weight of the control infants is usually higher than that of the infants who received the n-3 LCPUFA supplemented or high-ALA formulas, but the rates of increase in length differ minimally between groups

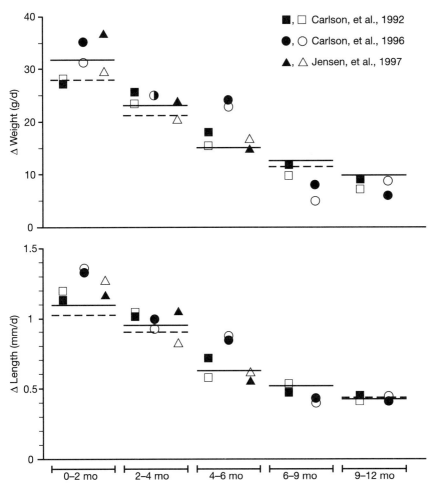

Fig. 1. Mean rates of weight gain (g/day) and length gain (mm/day) of preterm infants fed n-3 LCPUFA supplemented versus unsupplemented formulas and of term infants fed formulas with 3.2% versus 0.4% total fatty acids as ALA in comparison with mean rates of gain of normal term infants. Mean rates of gain were calculated from mean weights and lengths at the different study times. Filled symbols depict mean rates of unsupplemented preterm infants and term infants fed formula with 0.4% ALA; open symbols depict mean rates of n-3 LCPUFA supplemented preterm infants and term infants fed a formula with 3.2% ALA. Solid horizontal lines depict the mean rates of gain of normal term male infants (52); broken horizontal lines depict the mean rates of normal female term infants (52). The mean rates shown for supplemented and unsupplemented infants reported in 1992 by Carlson et al. (8) are the weighted means of males and females combined, as calculated from the mean data reported for males and females separately (see Table I). The 4–6 month rates shown for infants reported by Jensen et al. (10) are actually rates for 4–8 months.

and/or are lower in the control groups. This suggests that a portion of the deficits in weight and length at various ages was sustained prior to 40 weeks PMA.

In summary, three studies (two published (8,9) and one published only as an abstract (11)) have shown that preterm infants fed formulas supplemented with either a high (0.3% EPA and 0.2% DHA) or a low EPA fish oil (0.06% EPA and 0.2% DHA) weigh significantly less at some or all times during the first year of life than preterm infants fed formulas without n-3 LCPUFA supplementation. Another published study (14) evaluating the effect of a somewhat larger supplement of high EPA fish oil (0.65% EPA and 0.35% DHA) did not show a statistically significant effect of the supplement on growth of preterm infants. However, the period of supplementation in this study was short in comparison with the period of supplementation in the other studies. Moreover, data beyond 40 weeks PMA were available for only a limited number of infants. In yet another published study, there were no effects of either the low EPA fish oil or a mixture of AA and DHA (0.43% and 0.12% of total fatty acids) on growth of term infants (15).

One published study (10) shows that term infants fed a formula with 3.2% of total fatty acids as ALA, or a LA/ALA ratio of about 5, weigh significantly less at 4 months of age than infants fed a formula with 0.4% of total fatty acids as ALA, or a LA/ALA ratio of about 40. An additional unpublished study (12) shows that preterm infants fed a formula with the same fat blend (about 16% of total fatty acids as LA and 3.2% as ALA) weigh less at 48, 56, and 74 weeks PMA than infants fed a formula with about 16% of total fatty acids as LA and 1% as ALA.

In some of the published studies, length and head circumference of the n-3 LCPUFA supplemented and high-ALA groups also were significantly lower, but effects on length and head circumference were much less marked than the effects on weight. In fact, there was a consistent statistically significant effect of n-3 LCPUFA supplementation on length (approximately 0.4–0.5 SD) in only one study, and in all studies, normalized length tended to improve over time. As a result of the greater effect of n-3 LCPUFA and ALA on the rate of increase in weight versus length, the normalized weight-for-length of the supplemented versus the unsupplemented infants or the infants who received a higher versus a lower ALA intake was usually lower, but by no more than approximately 0.5 SD and only at the latter times of study. More important, weight-for-length was greater than the mean NCHS standard at most times in all studies.

The rates of increase in weight of both n-3 LCPUFA supplemented and unsupplemented groups of preterm infants and term infants fed both the higher and the lower ALA intakes were close to, and sometimes above, the mean rate of normal infants at 0–2, 2–4, and 4–6 months of age. However, despite the apparently 'normal' or 'near-normal' mean rates of weight gain of all groups, the mean rates of weight gain of the control groups were as much as 7 g/day greater during some intervals in some studies. In general, differences between rates of weight gain of n-3 LCPUFA supplemented versus unsupplemented preterm infants were somewhat less than the difference between the rates of weight gain of term infants who received higher versus lower intakes of ALA (or lower versus higher LA/ALA ratios).

The rates of increase in length were near the mean rate of normal term infants in all groups during most intervals of all published studies. Further, differences in rates of increase in length between n-3 LCPUFA supplemented versus unsupplemented groups and groups who received higher versus lower ALA intakes were minimal. During some study intervals, in fact, the rate of increase in length was greater in supplemented versus unsupplemented or high versus low ALA groups.

From the above, it appears that the mean rates of increase in both weight and length of all groups were close to, or actually greater than, the mean rates of normal term infants. However, the mean rates of increase in weight of the n-3 LCPUFA supplemented groups and the groups fed the high-ALA intakes tended to be very near the mean rate of increase of normal term infants, whereas the mean rates of increase in weight of the control groups tended to be greater, resulting in statistically significant differences in absolute and normalized weight between supplemented versus unsupplemented and higher versus lower ALA groups. This raises the question of whether statistically significant differences in absolute as well as normalized weight, length, and/or head circumference between two groups that are both growing at near normal rates are also biologically significant. In considering this question, several issues are of importance.

One concerns the rates of increase in weight and length that are appropriate, or optimal, particularly for LBW and/or preterm infants. While it is difficult to argue that the mean rates of increase observed in normally growing term infants are not appropriate standards against which to judge the growth of term infants, these rates may not be appropriate for preterm and/or LBW infants whose weight and length are not at or near the 50th percentile for normal term infants at 40 weeks PMA. In fact, most smaller preterm and/or LBW infants are considerably smaller at 40 weeks PMA than the normal term infant. Thus, for these infants, some catch-up growth may be desirable. If so, achievement of the desired 'catch-up' necessitates achievement of rates of increase in weight and length greater than the mean rates of normally growing term infants. Nonetheless, it is difficult to argue that even preterm and/or LBW infants who are growing, after 40 weeks PMA, at the mean rate of normal infants born at term are likely to be at any great disadvantage other than, perhaps, for not realizing their 'growth potential'.

In this regard, Fomon and Nelson (51) have proposed the hypothesis that all infants have a genetically determined potential for growth of fat-free body mass and that healthy infants who receive sufficient intakes of all required nutrients will meet their growth potential. If true, it follows that the ideal rate of growth is the rate that allows maximum gain in fat-free mass without excessive gain in fat, and that any lower rate of growth is suboptimal. Since it is difficult, if not impossible, to measure rates of increase in fat-free versus total mass, it is difficult to prove or disprove this hypothesis. Nonetheless, it is reasonable and implies that achievement of a 'normal' rate of increase in length may be more important than the rate of increase in weight that it accompanies. Obviously, however, there are limits to this implication; thus, this definition of ideal growth rate is not very helpful.

Another issue concerns the fact that the above discussion is based on the mean rates of growth of groups of infants. Some members of each group, of course, increase in weight and length at rates above the mean rate of the group whereas others increase at rates lower than the mean. Thus, while the overall mean effect of n-3 LCPUFA and high-ALA intakes on growth may not be problematical, this conclusion cannot be applied to all infants. Any factor that lowers the mean rate of growth of a population will result in fewer members being above a fixed upper limit standard for achieved size and more members being below a fixed lower limit standard. Thus, the lower rates of growth incident to n-3 LCPUFA and/or high-ALA intakes are likely to result in more infants being below any arbitrarily determined lower standard. For these infants, effects of n-3 LCPUFA on growth are likely to be biologically important. The biological importance of these effects on growth for the remainder of the group is unclear.

Since the primary effect of n-3 LCPUFA and high ALA intakes on growth appears to be inhibition of the rate of weight gain rather than the rate of increase in either length or head circumference, this effect may not necessarily be an undesirable one. If the lower rate of weight gain results solely from a lower rate of fat deposition, some would argue that the effect is, in fact, desirable. This, of course, may depend upon the mechanism whereby n-3 fatty acids affect rates of weight gain, which is discussed in the following section.

## Mechanism of the Effect of n-3 Fatty Acids on Growth

At the outset, it should be noted that none of the studies showing an effect of n-3 fatty acids on growth provides insights into the specific mechanism of the observed effects. In general, however, differences in rates of growth between/among groups of similar infants reflect differences in nutrient intake, nutrient absorption, and/or nutrient utilization/deposition. All the studies cited included assessment of intake and difference between groups were minimal. None included assessment of nutrient absorption, but there is little reason to expect differences. How n-3 fatty acids affect nutrient utilization and/or deposition is not known with certainty but there are some clues to possible mechanisms.

In two (8,10) of the three published studies, statistically significant correlations between weight and/or length and the plasma and/or erythrocyte phospholipid content of AA were observed. Moreover, in both of these studies, the plasma/erythrocyte phospholipid content of EPA was considerably higher in the supplemented versus unsupplemented group and the high versus low ALA group. In the third published study (9), there were statistically significant inverse correlations between growth and the DHA content of plasma/erythrocyte phospholipid. In this study there was a strong correlation between the AA content of plasma/erythrocyte phospholipid at 2 months corrected age and maternal height; further, substitution of the AA content of plasma phosphatidylcholine at 2 months corrected age for maternal height in a multiple regression model explained 8% of the variance in length at all ages. In the final model, which did not include the AA content of plasma phosphatidylcholine at 2 months corrected age, a combination of maternal height, birth weight, and the DHA content of erythrocyte phosphatidylcholine explained 10–30% of the variance in weight for length at different ages.

LA is known to be important for growth and this may be related to its role as a precursor of AA. For example, the growth effects of classical n-6 essential fatty acid deficiency can be overcome with AA as well as LA. In addition, correlations between birth weight and the AA content of cord plasma lipids (53) as well as cord endothelial lipids have been reported (54). However, the specific role of AA in growth has not been elucidated.

Since the initial observation of an effect of n-3 LCPUFAs on growth was made in infants who received a conventional fish oil supplement, it was assumed that the effect was related to the high EPA content of the supplement. Possibilities include inhibition of AA synthesis from linoleic acid (by EPA) and/or an imbalance of eicosanoids derived from EPA versus

AA. Since the high versus low-ALA intake or the low versus high-LA/ALA ratio results in higher plasma lipid contents of EPA and lower plasma lipid contents of AA, the same general mechanism proposed to explain the effects of the high-EPA fish oil could conceivably explain the effects of the high-ALA intake on growth. However, it is more difficult to attribute the effects of low-EPA fish oil on growth to a mechanism involving a higher EPA content of plasma/erythrocyte (and presumably other tissue) lipids. Although the AA content of plasma/erythrocyte phospholipids was lower in at least some of the studies in which the low-EPA fish oil supplement was used (9,12), there was no straightforward relationship between growth and AA status. This raises the possibility that n-3 LCPUFA and/or ALA may both decrease the AA content of plasma/erythrocyte lipid and inhibit growth, but that these effects, although correlated, are independent.

Inhibitory effects of n-3 LCPUFAs on growth of both rats (55,56) and mice (57,58) have been reported. One of the most informative such studies is that of Wainwright *et al.* (58) demonstrating that pups suckled by dams fed a very low n-6/n-3 ratio diet (i.e. 0.32), with or without half of the n-6 fatty acids as AA, weighed 12% less after 15 days of age than pups suckled by dams fed an n-3 deficient diet (n-6/n-3 = 49) or diets with an n-6/n-3 ratio of 4, either with or without AA. Energy expenditure of the rats fed a high n-3 fatty acid diet also has been shown to be greater than that of a control group (59). This was attributed to the effect of n-3 fatty acids on the structure of cellular and subcellular membranes resulting in a requirement for more energy to maintain crucial intracellular/extracellular gradients. Other direct effects of the n-3 fatty acids, however, are possible.

Both n-3 and n-6 fatty acids have been shown to inhibit transcription of genes coding for lipogenic enzymes, e.g. fatty acid synthase (60–62). This appears to be an immediate effect. The mRNA encoding the fatty acid synthase gene decreases before a change in the fatty acid pattern of cellular and subcellular membranes could have occurred. Although LA and ALA also reduce the abundance of hepatic fatty acid synthase mRNA, the actual intracellular modulators appear to be the $\Delta^6$ desaturation products of these fatty acids.

Insulin sensitivity of adult humans (63,64) as well as several animal species (65,66) appears to be related to the muscle membrane content of a variety of LCPUFAs. Specifically, positive correlations between insulin sensitivity and the muscle phospholipid content of AA, as well as the sum of $C_{20}$ and $C_{22}$ n-6 and n-3 fatty acids and the unsaturation index of all muscle phospholipid fatty acids have been demonstrated. Greater insulin sensitivity means that normoglycemia can be maintained with a lower concentration of insulin; hence, the effects of insulin on muscle (i.e. to impede breakdown) and adipose tissue (i.e. to prevent lipolysis) are minimized. With insulin resistance, of course, plasma insulin concentration is increased, resulting in greater effects on muscle and adipose tissue.

It is difficult to relate these associations between muscle membrane fatty acid pattern and insulin sensitivity in adult humans and rats to the effects of n-3 fatty acids on the growth of infants. For example, a higher AA content of plasma and/or erythrocyte phospholipid appears to be related to higher rates of weight gain in infants, whereas a higher content of AA in muscle phospholipid is associated with greater insulin sensitivity and, presumably, lower rates of fat deposition and, hence, weight gain. However, since the correlation between insulin sensitivity and the muscle phospholipid content of all $C_{20}$ and $C_{22}$ polyunsaturated fatty acids appears to be as strong as that between insulin sensitivity and the muscle phospholipid content of AA, the correlation with AA content of muscle phospholipid may simply reflect the fact that AA is the most predominant LCPUFA of muscle phospholipid. Thus, it is probably best to conclude that insulin sensitivity is associated with a higher content of polyunsaturated fatty acids in muscle phospholipid rather than a higher content of either n-3 or n-6 LCPUFAs. In addition, it must be noted that the fatty acid pattern of muscle phospholipid may not necessarily reflect the fatty acid pattern of plasma phospholipid.

From the foregoing, it appears that there are a number of mechanisms by which n-3 LCPUFAs might affect the rate of growth of infants. First, a higher phospholipid content of EPA versus AA might lead to an imbalance in eicosanoids derived from EPA versus AA, and this imbalance, in turn, could conceivably inhibit growth, although a specific mechanism by which the imbalance might do so has not been identified. Alternatively, and/or additionally, the n-3 LCPUFAs might decrease the levels of various lipogenic enzymes. It also is likely that changes in the characteristics of cellular and subcellular membranes incident to dietary intake of n-3 LCPUFAs (and perhaps n-6 LCPUFAs as well) alters attachment of hormone receptors to cellular and subcellular membranes as well as second messenger signaling, either or both of which could result in lower rates of weight gain. Changes in the characteristics of cellular and subcellular membranes might also increase the energy required to maintain ionic and other gradients thus leading to an increased resting metabolic rate and, hence a lower rate of weight gain.

Interestingly, all these mechanisms, theoretically, result in lower rates of fat deposition. Thus, emphasis has been placed on the possibility that n-3 LCPUFAs may be beneficial in preventing obesity in adults, thereby decreasing the incidence of the many modern diseases with which it is associated. While such an effect is clearly desirable for adults, it is not clear that it is desirable for infants. However, if inhibition of fat deposition during infancy proves to be desirable (e.g. if higher rates of fat deposition during infancy are shown to 'program' for higher rates of fat deposition later in life, as some epidemiological observations suggest (67)) and if n-3 fatty acids result in lower rates of fat deposition, the effects of n-3 LCPUFAs on the growth of infants also might be considered desirable. This concept is, of course, counter to the possibility, based on other epidemiological data (46), that a smaller versus a larger size at 1 year of age may be associated with a greater risk of death from coronary heart disease.

It should be noted that a large part of this discussion of mechanisms whereby n-3 LCPUFA might affect the growth of infants, although based on concrete observations in human adults and other animal species, may not be applicable to infants. Unfortunately, there also is no information concerning body composition of infants fed n-3 LCPUFA supplemented versus unsupplemented formulas or infants fed high versus low ALA intakes. In this regard, although the difference was not statistically significant, the lower skinfold thicknesses of term infants fed formulas with higher versus lower contents of ALA (10) may be biologically significant. Moreover, the lower skinfold thickness of preterm infants fed the formula with the higher versus the lower content of ALA was statistically significant (12). Nonetheless, the lack of information concerning the effects of n-3 fatty acids on energy expenditure of human infants makes it very difficult to distinguish a specific mechanism for the apparent inhibitory effect of n-3 fatty acids on growth of infants.

## Conclusions

From the foregoing, it is clear that n-3 fatty acids affect growth of both preterm and term infants. To date, statistically significant differences between growth of infants fed n-3 LCPUFA supplemented versus unsupplemented formulas and term as well as preterm infants fed formulas with a higher versus a lower content of ALA have been observed in two as yet unpublished (11,12) and three published studies (8–10). Moreover, statistically significant differences in length and/or head circumference also were observed in some of these studies; however, these differences generally were less impressive than the differences in weight, and in all studies, normalized length of all groups tended to improve over time. In addition, normalized weight-for-length, although greater in the control groups of all studies, tended to be greater than the mean weight-for-length of normal term infants in all groups. The rates of increase in weight and length of the study groups were lower during most intervals of all studies, but these rates tended to be near the mean rates of increase observed in normal term infants (52).

In assessing the biological significance of these statistically significant differences in growth between infants fed n-3 LCPUFA supplemented formulas and formulas with a high content of ALA or a low LA/ALA ratio versus infants fed unsupplemented formulas and formulas with a lower content of ALA or a higher LA/ALA ratio, it is clear that the supplemented and high ALA formula groups are not severely malnourished. On the other hand, it also is clear that they are smaller than the control groups. However, whether they are sufficiently smaller to justify the concern that has been generated is not clear. Further, it is likely that the magnitude of difference necessary to cause concern is different for preterm versus term infants.

As discussed above, it is difficult to argue that a group of term infants growing at or near the mean rate of normal term infants, although at a lower rate than a control group, is 'at risk'. For example, it is unlikely that a difference of 750 g between the mean body weights of a group of breast-fed versus a group of formula-fed infants would cause concern. Indeed, it is likely that most would consider the lower mean weight of the breast-fed group to be more appropriate. However, as illustrated in Table 3 and Fig. 2, it is not clear that this conclusion is justified. For example, the normalized weight of all groups in comparison to recent growth standards of breast-fed infants (68), although greater after 6 months of age, is less through 6 months of age (see Tables 2 and 3).

It is easier to argue that different rates of growth between two groups of preterm infants might impose some risk for the slower growing group but, even in these infants, there are few data to support this argument. The major support for the argument that a lower rate of growth may be problematical for the preterm infant is the fact that most LBW infants are considerably lighter and shorter at 40 weeks PMA than the infant who is born after a normal term pregnancy. Thus, to achieve the same subsequent size as the term infant, the LBW infant must experience more rapid rates of growth. In this situation, any dietary factor that inhibits growth for any period of time decreases the likelihood of the LBW infant's achieving the same size as the infant born at term. On the other hand, it is not clear that failure to achieve the same size as the infant born at term is a major handicap, particularly if the major difference in size represents primarily a difference in weight as suggested by the greater effects of n-3 fatty acids on the rate of weight gain versus the rate of length gain. The relative lack of an effect of n-3 fatty acids on normalized head circumference also argues

Table 3. Reported weight for age, length for age, head circumference for age, and weight for length Z-scores (calculated from WHO growth standards for breast-fed infants (68)) of preterm infants fed ω-3 LCPUFA supplemented versus unsupplemented formulas and term infants fed formulas with 3.2% versus 0.4% fatty acids as ALA. (Data from Carlsen et al. (8,9) and Jensen et al. (10).)

| Age (months) | Control (8) | Suppl. (8) | Control (9) | Suppl. (9) | 0.4% ALA (10) | 3.2% ALA (10) |
|---|---|---|---|---|---|---|
| *Weight for age Z-scores* | | | | | | |
| 0 | −0.95 | −1.44 | −1.70 | −1.65 | −0.23 | −0.61 |
| 2 | −1.24 | −1.51 | −0.69 | −1.05 | 0.43 | −0.61 |
| 4 | −0.67 | −1.07 | −0.375 | −0.65 | 0.44 | −0.68 |
| 6 | −0.09 | −0.68 | 0.26 | −0.08 | − | − |
| 8 | − | − | − | − | 0.39 | −0.21 |
| 9 | −0.10 | −0.79 | 0.07 | −0.48 | | |
| 12 | −0.13 | −0.70 | 0 | −0.32 | | |
| *Length for age Z-scores* | | | | | | |
| 0 | − | − | − | − | − | − |
| 2 | −1.95 | −2.22 | −1.78 | −1.67 | −0.65 | −0.49 |
| 4 | −1.29 | −1.50 | −1.34 | −1.45 | −0.08 | −0.60 |
| 6 | −0.32 | −1.00 | −0.47 | −0.53 | − | − |
| 8 | − | − | − | − | 0.30 | 0 |
| 9 | −0.44 | −0.88 | −0.44 | −0.63 | | |
| 12 | −0.26 | −0.83 | −0.56 | −0.83 | | |
| *Weight for length Z-scores* | | | | | | |
| 2 | +0.5 | +0.5 | +1.25 | +0.50 | +1.00 | −0.20 |
| 4 | +0.25 | 0 | +0.75 | +0.40 | +0.40 | −0.25 |
| 6 | 0 | 0 | +0.75 | +0.20 | − | − |
| 8 | − | − | − | − | +0.25 | −0.25 |
| 9 | 0.25 | −0.25 | +0.40 | −0.20 | | |
| 12 | 0 | −0.25 | +0.10 | −0.30 | | |
| *Head circumference for age Z-scores* | | | | | | |
| 2 | −0.67 | −1.05 | −0.10 | −0.29 | −0.48 | −1.14 |
| 4 | −0.24 | −0.81 | 0.14 | −0.52 | −0.33 | −1.19 |
| 6 | 0.09 | −0.45 | 0.27 | −0.09 | − | − |
| 8 | − | − | − | − | −0.52 | −1.48 |
| 9 | −0.35 | −0.87 | −0.26 | −0.78 | | |
| 12 | −0.46 | −0.79 | −0.46 | −0.79 | | |

against the likelihood that the lower rates of growth incident to these fatty acids is problematical.

Clearly, the consequences of the lower rates of growth of preterm infants fed n-3 fatty acid supplemented versus unsupplemented formulas are not

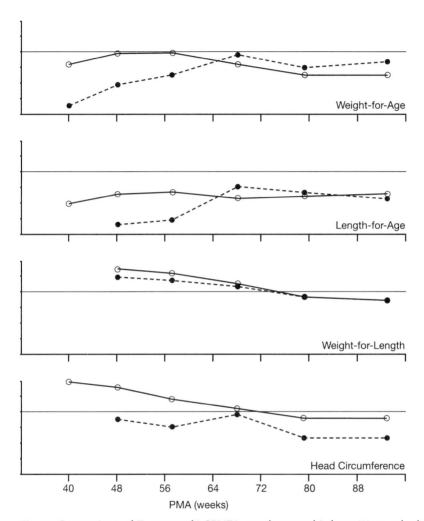

Fig. 2. Comparison of Z-scores of LCPUFA supplemented infants (9) as calculated from WHO breast-fed infants (68) versus NCHS standards (69). (- - - -) Z-scores based on WHO standards calculated from published mean weights, lengths, and head circumferences (9); (———) Z-scores based on NCHS standards are the published Z-scores of the LCPUFA supplemented group at the designated ages.

known. Moreover, despite the argument that these lower rates of growth may not be problematical, the lower 12-month Bayley PDI score of the high EPA fish oil supplemented group versus the control group studied by Carlson et al. (70) cannot be ignored. Nor can the lower 18 month and 7–8 year Bayley PDI score reported by Lucas et al. (47; see also Lucas, this volume) in infants who were fed standard term infant formulas versus nutrient-enriched preterm formulas for a discrete period during hospitalization as LBW infants. The epidemiological evidence linking the incidence of adult diseases to a lower birth weight or weight at 1 year of age (46) also must be considered.

Considering these findings of an apparent relationship between early feeding, and hence growth and subsequent neurobehavioral outcomes as well as the incidence of adult-onset diseases, it seems reasonable to conclude that anything which interferes with growth of infants during early infancy is a cause for concern. This is because brain growth during infancy and, perhaps, adult health appear to be linked in some complex way with growth in weight and length as well as a number of other factors. Thus, since normal brain growth during infancy and adult health are desirable, and since the non-nutritional factors which affect these outcomes are difficult to modify, promotion of optimal somatic growth is the only way we can hope to promote optimal brain growth. Interestingly, this conclusion paraphrases the conclusion reached also by Dobbing (71) over a decade ago. Data on which to base a more specific conclusion remain unavailable.

## Acknowledgements

This work is a publication of the USDA/ARS Children's Nutrition Research Center, Department of Pediatrics, Baylor College of Medicine, Houston, Texas, and has been funded, in part, with federal funds from the US Department of Agriculture, Agriculture Research Service, under Cooperative Agreement No. 38-6250-1-003. The contents of this publication do not necessarily reflect the views or policies of the US Department of Agriculture, nor does the mention of trade names, commercial products, or organizations imply endorsement by the US Government. The author's research cited herein was supported also by Grants EY00871 and EY04149 and by grants from Mead-Johnson Nutrition Group, The Foundation Fighting Blindness, Research to Prevent Blindness, Inc., and Retina Research Foundation. Study formulas as well as regular formulas were provided by Mead Johnson Nutrition Group and weaning foods were provided by Gerber Products Company.

# References

1. Simopoulos AP. Omega-3 fatty acids in health and disease and in growth and development. *Am J Clin Nutr* 1991; **54**: 438–63.
2. Connor WE. Omega-3 essential fatty acids in infant neurological development. *Backgrounder* 1996; **1**: 1–6.
3. Food and Agriculture Organization/World Health Organization. *Fats and Oils in Human Nutrition.* Rome: FAO/WHO, 1994: 6–7, 49–55.
4. The British Nutrition Foundation. Unsaturated fatty acids. Nutritional and physiological significance. In: *The Report of the British Nutrition Foundation's Task Force.* Padstow, Cornwall: TJ Press, 1992: 63–7.
5. ESPGAN Committee on Nutrition. Committee Report: Comment on the Content and Composition of Lipids in Infant Formulas. *Acta Paediatr Scand* 1991; **80**: 887–96.
6. International Society for the Study of Fatty Acids and Lipids (ISSFAL) board statement: recommendations for the essential fatty acid requirement for infant formulas. *Nutrition Today* 1995; **30**: 46.
7. Koletzko B. Long-chain polyunsaturated fatty acids in infant formulae in Europe. *ISSFAL News* 1995; **2**: 3–5.
8. Carlson SE, Cooke RJ, Werkman SH, Tolley EA. First year growth of preterm infants fed standard compared to marine oil (fish oil) n-3 supplemented formula. *Lipids* 1992; **27**: 901–7.
9. Carlson SE, Werkman SH, Tolley EA. Effect of long-chain n-3 fatty acid supplementation on visual acuity and growth of preterm infants with and without bronchopulmonary dysplasia. *Am J Clin Nutr* 1996; **63**: 687–97.
10. Jensen CL, Prager TC, Fraley JK, Anderson RE, Heird WC. Functional effects of dietary linoleic/α-linolenic acid ratio in term infants. *J Pediatr* 1997 (in press).
11. Montalto MB, Mimouni FB, Sentipal-Walerius J, Bender CV, Groh-Wargo S. Reduced growth in hospital discharged low birth weight (LBW) infants fed formulas with added marine oil (fish oil) (MO). *Pediatr Res* 1996; **39**: 316A.
12. Jensen CL, Chen HM, Prager TC, Anderson RE, Heird WC. Effect of 18: 3ω-3 intake on plasma fatty acids, growth and visual development of preterm infants. *Pediatr Res* 1995; **37**: 311A.
13. Carlson SE, Werkman SH, Peeples JM, Cooke RJ, Tolley EA. Arachidonic acid status correlates with first year growth in preterm infants. *Proc Natl Acad Sci USA* 1993; **90**: 1073–7.
14. Uauy RD, Hoffman DR, Birch EE, Birch DG, Jameson DM, Tyson JE. Safety and efficacy of omega-3 fatty acids in the nutrition of very-low-birth-weight infants: soy oil and marine oil supplementation of formula. *J Pediatr* 1994; **124**: 612–20.
15. Austad N, Montalto MB, Hall RT et al. Visual acuity, erythrocyte fatty acid composition, and growth in term infants fed formulas with long chain polyunsaturated fatty acids for one year. *Pediatr Res* 1997; **41**: 1–10.
16. Cooper A, Heird WC. Nutritional assessment of the pediatric patient. *Am J Clin Nutr* 1982; **35**: 1132–41.
17. Merritt RJ, Suskind RM. Nutritional survey of hospitalized pediatric patients. *Am J Clin Nutr* 1979; **32**: 1320–5.

18. Bistrian BR, Blackburn GL, Vitale J, Cochran D, Naylor J. Prevalence of malnutrition in general medical patients. *JAMA* 1976; **235**: 1567–70.
19. Winick M, Noble A. Cellular response in rats during malnutrition at various ages. *J Nutr* 1966; **89**: 300–6.
20. Dobbing J. The later development of the brain and its vulnerability. In: Davis JA, Dobbing J, eds. *Scientific Foundations of Paediatrics*, 1st edn. Philadelphia, WB Saunders.
21. Winick M, Rosso P. The effect of severe early malnutrition on cellular growth of human brain. *Pediatr Res* 1969; **3**: 181–4.
22. Dobbing J, ed. *Early Nutrition and Later Achievement*. London: Academic Press, 1987.
23. Levitsky DA, Strupp BJ. Malnutrition and the brain: changing concepts, changing concerns. *J Nutr* 1995; **125**: 2212S–20S.
24. Levitsky DA, Strupp BJ. Enduring cognitive effects of early malnutrition: a theoretical reappraisal. *J Nutr* 1995; **125**: 2221S–32S.
25. Grantham-McGregor S. A review of studies of the effect of severe malnutrition on mental development. *J Nutr* 1995; **125**: 2233S–8S
26. Gorman KS. Malnutrition and cognitive development: evidence from experimental/quasi-experimental studies among the mild-to-moderately malnourished. *J Nutr* 1995; **125**: 2239S–44S.
27. Wachs TD. Relation of mild-to-moderate malnutrition to human development: correlational studies. *J Nutr* 1995; **125**: 2245S–54S.
28. Schürch B. Malnutrition and behavioral development: the nutrition variable. *J Nutr* 1995; **125**: 2255S–62S.
29. Golub MS, Keen CL, Gershwin ME, Hendrickx AG. Developmental zinc deficiency and behavior. *J Nutr* 1995; **125**: 2263S–71S.
30. Pollitt E. Functional significance of the covariance between protein energy malnutrition and iron deficiency anemia. *J Nutr* 1995; **125**: 2272S–7S.
31. Brown JL, Sherman LP. Policy implications of new scientific knowledge. *J Nutr* 1995; **125**: 2281S–4S.
32. Grantham-McGregor SM, Powell CA, Walker SP, Himes JH. Nutritional supplementation, psychosocial stimulation and mental development of stunted children: the Jamaican study. *Lancet* 1991; **338**: 1–5.
33. Husaini MA, Karyadi L, Husaini YK, Karyadi D, Pollitt E. Developmental effects of short-term supplementary feeding in nutritionally at risk Indonesian infants. *Am J Clin Nutr* 1991; **54**: 799–804.
34. Allen LH. The nutrition CRSP: what is marginal malnutrition, and does it affect human function? *Nutr Rev* 1993; **51**: 255–67.
35. Dobbing J, Sands J. Quantitative growth and development of human brain. *Arch Dis Child* 1973; **48**: 757–67.
36. Hack M, Horbar JD, Malloy MH, Tyson JE, Wright E, Wright L. Very low birth weight outcomes of the National Institute of Child Health and Human Development Neonatal Network. *Pediatrics* 1991; **87**: 587–97.
37. Hack M, Weissman B, Borawski-Clark E. Catch-up growth during childhood among very low-birth-weight children. *Arch Pediatr Adolesc Med* 1996; **150**: 1122–9.
38. Kitchen WH, Doyle LW, Ford GW, Callanan C. Very low birth weight and growth to age 8 years. 1: Weight and height. *Am J Dis Child* 1992; **146**: 40–5.
39. Hack M, Breslau N. Very low birth weight infants: effects of brain growth

during infancy on intelligence quotient at 3 years of age. *Pediatrics* 1986; **77**: 196–202.
40. Hack M, Breslau N, Fanaroff AA. Differential effects of intrauterine and postnatal brain growth failure in infants of very low birth weight. *Am J Dis Child* 1989; **143**: 63–8.
41. Hack M, Breslau N, Weissman B, Aram D, Klein N, Borawski E. Effect of very low birth weight and subnormal head size on cognitive abilities at school age. *N Engl J Med* 1991; **325**: 231–7.
42. Casey PH, Kraemer HC, Bernbaum J, Yogman MW, Sells JC. Growth status and growth rates of a varied sample of low-birth-weight, preterm infants: a longitudinal cohort from birth to three years of age. *J Pediatr* 1991; **119**: 599–605.
43. Hack M, Merkatz IR, McGrath SK, Jones PK, Fanaroff AA. Catch-up growth in very low-birth-weight infants: potential and clinical correlates. *Am J Dis Child* 1984; **138**: 370–5.
44. Ross G, Lipper EG, Auld PAM. Growth achievement in low-birth-weight premature children at school age. *J Pediatr* 1990; **117**: 307–12.
45. Ovigstad E, Verloove-Vanhorick SP, Ens-Dokkum MH et al. Prediction of height achievement at five years of age in children born very preterm or with very low birth weight: continuation of catch-up growth after two years of age. *Acta Paediatr* 1993; **82**: 444–8.
46. Barker DJP, Fall, CHD. Fetal and infant origins of cardiovascular disease. *Arch Dis Child* 1993; **68**: 797–9.
47. Lucas A, Morley R, Cole TJ et al. Early diet in preterm babies and developmental status at 18 months. *Lancet* 1990; **335**: 1477–81.
48. Lucas A, Morley R, Cole TJ, Gore SM. A randomised multicentre study of human milk versus formula and later development in preterm infants. *Arch Dis Child* 1994; **70**: F141–6.
49. Binkin NJ, Yip R, Fleshood L, Trowbridge FL. Birth weight and childhood growth. Pediatrics 1988; **82**: 828–34.
50. Karlberg P, Engström I, Lichtenstein H et al. The development of children in a Swedish urban community. A prospective longitudinal study. III. Physical growth during the first three years of life. *Acta Paediatr Scand* 1968; **48**(Suppl 187): 48–66.
51. Fomon SJ, Nelson SE. Size and growth. In: Fomon SJ, ed. *Nutrition of Normal Infants*. St Louis, MI: Mosby, 1993: 36–84.
52. Guo S, Roche AF, Fomon SJ et al. Reference data for gains in weight and length during the first two years of life. *J Pediatr* 1991; **119**: 355–62.
53. Koletzko B, Braun M. Arachidonic acid and early human growth: is there a relation? *Ann Nutr Metab* 1991; **35**: 128–31.
54. Leaf AA, Leighfield MJ, Costeloe KL, Crawford MA. Long-chain polyunsaturated fatty acids and fetal growth. *Early Human Dev* 1992; **30**: 183–91.
55. Clarke SD, Benjamin L, Bell L, Phinney SD. Fetal growth and fetal lung phospholipid content in rats fed safflower oil, menhaden oil, or hydrogenated coconut oil. *Am J Clin Nutr* 1988; **47**: 828–35.
56. Pan DA, Storlien LH. Dietary lipid profile is a determinant of tissue phospholipid fatty acid composition and rate of weight gain in rats. *J Nutr* 1993; **123**: 512–9.

57. Cunnane SC, McAdoo KR, Horrobin DF. n-3 Essential fatty acids decrease weight gain in genetically obese mice. *Br J Nutr* 1986; **56**: 87–95.
58. Wainwright PE, Xing X-C, Mutsaers L, McCutcheon D, Kyle D. Arachidonic acid offsets the effects on mouse brain and behaviour of a diet with a low (n-6):(n-3) ratio and very high levels of docosahexaenoic acid. *J Nutr* 1997; **127**: 184–93.
59. Pan DA, Hulbert AJ, Storlien LH. Dietary fats, membrane phospholipids and obesity. *J Nutr* 1994; **124**: 1555–65.
60. Clarke SJ, Jump DB. Regulation of gene transcription by polyunsaturated fatty acids. *Prog Lipid Res* 1993; **32**: 139–49.
61. Jump DB, Clarke SD, MacDougald O, Thelen A. Polyunsaturated fatty acids inhibit S14 gene transcription in rat liver and cultured hepatocytes. *Proc Natl Acad Sci* 1993; **90**: 8454–8.
62. Clarke SD, Jump DB. Dietary polyunsaturated fatty acid regulation of gene transcription. *Annu Rev Nutr* 1994; **14**: 83–98.
63. Borkman M, Storlien LH, Pan DA, Jenkins AB, Chisholm DJ, Campbell LV. The relationship between insulin sensitivity and the fatty-acid composition of skeletal-muscle phospholipids. *N Engl J Med* 1993; **328**(4): 238–44.
64. Pan DA, Lillioja S, Milner MR et al. Skeletal muscle membrane lipid composition is related to adiposity and insulin action. *J Clin Invest* 1995; **96**: 2802–8.
65. Storlien LH, Jenkins AB, Chisholm DJ, Pascoe WS, Khouri S, Kraegen EW. Influence of dietary fat composition on development of insulin resistance in rats: relationship to muscle triglyceride and 3 fatty acids in muscle phospholipid. *Diabetes* 1991; **40**: 280–9.
66. Storlien LH, Pan DA, Kriketos AD, Baur LA. High fat diet-induced insulin resistance. Lessons and implications from animal studies. *Ann NY Acad Sci* 1993; **683**: 82–90.
67. Dietz WH. Critical periods in childhood for the development of obesity. *Am J Clin Nutr* 1994; **59**: 955–9.
68. WHO Working Group on Infant Growth. *An Evaluation of Infant Growth.* Geneva: Nutrition Unit, World Health Organization, 1994.
69. Hamill PVV, Drizd TA, Johnson CL, Reed RB, Roche AF, Moore WM. Physical growth: National Center for Health Statistics percentiles. *Am J Clin Nutr* 1979; **32**: 607–29
70. Carlson SE. Lipid requirements of very-low-birth-weight infants for optimal growth and development. In: Dobbing J, ed. *Lipids, Learning, and the Brain: Fats in Infant Formulas, Report of the 103rd Ross Conference on Pediatric Research.* Columbus, OH: Ross Laboratories, 1993: 188–207.
71. Dobbing J. Infant nutrition and later achievement. *Am J Clin Nutr* 1985; **41**: 477–84.

## COMMENTARY

**Lucas**: Heird's review summarizes evidence that n-3 fatty acids do indeed suppress growth. The evidence from preterm studies that n-3 LCPUFA can suppress growth is compelling. The evidence that α-linolenic acid (LNA) suppresses growth will be surprising to many readers, and a broad examination of the extensive experience of formulas containing higher LNA would be desirable. The mechanism by which n-3 LCPUFA could suppress growth is unknown, but the very widespread actions of LCPUFAs, summarized in my own and in Heird's chapters in this volume, provide numerous hypotheses as to possible mechanisms, some of which are well discussed by Heird.

The question is whether the described effects on growth with n-3 LCPUFA supplementation are biologically significant, and, of practical relevance, whether these effects should influence regulatory bodies in their recommendations for the addition, or not, of n-3 LCPUFAs to infant formula.

The potential impact of early growth on long-term outcome is currently an issue of major interest and could be expanded further by the author. For instance low body weight in early life (from birth to 1 year) appears to have major predictive value for morbidity in later life (1), including death form ischemic heart disease, raised blood pressure, impaired glucose tolerance, and adverse changes in blood lipids. The impact of size in early life on long-term bone health is also of interest. I do not suggest by any means that causation has been proven in these epidemiological associations, but they should be discussed, given the intensity of current interest in them. For instance, body weight at 1 year is one of the strongest predictors of death from ischemic heart disease in men; in Barker's study body weight at 1 year was linearly related to decreasing mortality from ischemic heart disease. Thus a weight at 1 year of <18 lb was associated with a mortality from ischemic heart disease of about three times that seen in individuals whose weight was >26 lb at 1 year. If this relationship was a causal one, a 750 g weight difference induced by n-3 LCPUFA could equate to, say, a 40% increase in ischemic heart disease risk. The validity of Barker's data to modern cohorts remains to be evaluated.

The relationship between early growth and later neurodevelopment is a complex one and, as identified by Heard, often confounded in epidemiological studies of malnourished and underprivileged cohorts. Experimental studies are, therefore, most informative. It would be unbalanced for Heird just to quote our experimental study on the impact on the 18-month Bayley score of standard versus preterm formula given to preterm neonates. The advantage seen here in the preterm formula-fed group in psychomotor development index (PDI) was not seen in a parallel randomised study comparing the same preterm formula with banked breast milk (2), yet the banked breast milk group showed the poorest neonatal growth performance of any group studied. These data provide evidence that diet (i.e. the pattern of nutrients supplied) can be of critical importance to later neurodevelopment; but these later effects cannot necessarily be related to early growth per se.

Heird discusses the potential impact of n-3 LCPUFAs on catch-up growth in preterm infants. This could, theoretically, affect size in later childhood and beyond; indeed our own (3) and Hack's data show preterm infants to be at risk of long-term growth deficits; Heird does not consider the impact of childhood final size on psychological outcomes.

Finally, a less central point: Heird's rather strong assertion about the relationship between malnutrition and infection/immunity, often uncritically accepted in articles on nutrition, can be challenged (see Gareth Morgan's recent review article (4)). A more moderate stance here would be justified.

Overall, however, I agree with Heird that, despite the uncertain significance of growth reduction due to n-3 LCPUFA supplementation, it would be prudent, in the light of our current knowledge, to regard this as a potentially adverse consequence and hence a safety issue.

1. Barker DJP, ed. *Fetal and Infant Origins of Adult Disease*. London: BMJ 1992.
2. Lucas A, Morley R, Cole TJ, Gore SM. A randomised multicentre study of human milk versus formula and later development in preterm infants. *Arch Dis Child* 1994; **70**: F141–6.
3. Lucas A. Nutrition, growth and development of postdischarge preterm infants. In: *Post-hospital Nutrition of the Preterm Infant. Report of the 106th Ross Conference on Pediatric Research*. Columbus, OH: Ross Products, 1996: 81–9.
4. Morgan G. What, if any, is the effect of malnutrition on immunological competence? *Lancet* 1997 (in press).

**Singer**: As noted in this summary of a complex literature, understanding the determinants of growth outcome in very low birth weight (VLBW) infant cohorts is a formidable task. Heird's perspective on the relative clinical importance of the growth differences observed in LCPUFA-supplemented cohorts suggests that they may not be of great concern. A major shortcoming of the studies cited which have investigated growth outcomes of VLBW infants to school age, however (1,2), is that these cohorts were studied prior to the routine use of cranial ultrasonography, and were unable to assess the influence of neurologic factors such as intraventricular hemorrhage and periventricular hemorrhage on growth outcomes.

Other factors to consider in assessing growth which are often not routinely evaluated include: the use of medication, such as steroids and antibiotics, affecting appetites with and without weight gain; infections and rehospitalizations; and pulmonary status which can be affected by the stress of feedings. For example, infants with bronchopulmonary dysplasia who had been considered clinically normoxic with pulmonary assessment while alert and at rest, were found to become hypoxic during oral feedings (3). Mothers of VLBW infants are more likely to be severely depressed post partum, affecting their feeding interactions, and the VLBW infant elicits different, more active maternal care-giving than do term infants (4–6). Feeding practices prevalent because mothers of VLBW infants tend to view their infants based on their chronological rather than conceptional ages have also been identified which could result in compromised nutrition, e.g. introduction to cow's milk at an early age.

1. Hack AM, Merkatz IR, McGrath SK, Jones PK, Fanaroff AA. Catch up growth in very low birthweight infants: potential and clinical correlates. *Am J Dis Child* 1984; **138**: 370–5.
2. Kitchen WH, Doyle LW, Ford GW, Callanan C. Very low birthweight and growth to age 8 years. 1: Weight and height. *Am J Dis Child* 1992; **146**: 40–5.
3. Singer L, Martin RJ, Hawkins SW, Benson-Szekely LJ, Yamashita TS, Carlo WA. Oxygen desaturation complicates feeding in infants with bronchopulmonary dysplasia after discharge. *Pediatrics* 1992; **90**: 380–4.
4. Singer LT, Davillier M, Bruening P, Hawkins S, Yamashita TS. Social support, psychological distress, and parenting strains in mothers of very low birthweight infants. *Family Relations* 1996; **45**: 343–50.
5. Singer LT, Davillier M, Pruess L *et al*. Feeding interactions in infants with very low birth weight and bronchopulmonary dysplasia. *Dev Behav Pediatr* 1996; **17**: 69–76.
6. Ernst JA, Bull MJ, Rickard KA, Brady MS, Lemons JA. Growth outcome and feeding practices of the very low birthweight infant (less than 1500 grams) within the first year of life. *J Pediatr* 1990; **117**: S156–66.


**Carlson**: From your article, I gather that you believe there are real effects of n-3 fatty acids on growth, but that we do not know if these effects have any relevance to short- or long-term development. I would only add to this that it could be helpful to put the data in some perspective. We have noted in our first study (your reference 8), that the first year growth achievement of the preterm infants we studied (mean birth weight 1074 g) was equivalent to that of the larger preterm infants (mean birth weight 1690 g) in a very large eight-center study by Casey *et al.* (1). Similarly, you note in your *Journal of Pediatrics* article (your reference 10) that the growth of the group fed the highest ALA was similar to the group fed human milk. It seems to me that this kind of information puts the growth effects that have been seen in a somewhat different light.

I am coming to believe that these small (albeit statistically significant) effects of n-3 fatty acids on growth within the normal range of variability have little meaning. If, however, we decide they must be avoided, the same scrutiny needs to be applied to the high levels of α-linolenic acid that are being added to some formulas (far above what is generally found in human milk). In this context, it is worth noting that some commercially available formulas have ALA exceeding the 1.6% energy from α-linolenic acid that led to the slowest growth in your study (your reference 10).

The effects of these supraphysiological n-3 fatty acid intakes from α-linolenic acid are not necessarily benign and, one could argue, are even more of a theoretical concern than the inclusion of small amounts of n-3 LCPUFA for diseases of the preterm (Lucas) and oxidation (Mayer and Dobson). For example, the inclusion in formulas of 5% of total fatty acids from linolenic acid (three double bonds) to avoid adding n-3 LCPUFA certainly raises the potential for oxidation much more than the inclusion of 0.2% of total fatty acids from DHA (six double bonds) with a more physiological amount of α-linolenic acid (about 1%).

You refer to statements about growth in the Uauy *et al.* paper (your reference 14). I believe that growth data from this study were subsequently published in one of the pediatrics journals for 0, 2, and 4 months past expected term (2). The numbers of infants with growth measurements at 4 months were not enough to draw conclusions about growth, as you indicated for the earliest age, but there were data for two other ages.

You refer to the Montalto *et al.* abstract (your reference 11). It is important to note that the effects on growth were not present in the group as a whole or among females.

Your reference 8 is our first study. It is not correct to say that the differences were limited to female infants, because, while the same proportion of males were studied in each diet group, too few males were studied to draw conclusions about them. The most reliable comparison (and the planned design) dictated a comparison between diets with normalized scores for all infants included. This analysis shows a significantly lower growth in the n-3 LCPUFA-supplemented group. You are correct that infants fed n-3 LCPUFA were shorter than those fed α-linolenic acid.

A more pronounced effect of high n-3 fatty acid intake on weight compared with length gain cannot be the cause for a normalized weight-for-length above the mean NCHS references. At least in preterm infants, we find that weight percentiles increase dramatically between 40 and 57 weeks postmenstrual age (PMA) with length increasing much less so. This results in a quite disproportionately higher weight-for-length by 57 weeks. During the remaining 8 months of infancy, normalized weight declines and normalized lengths increase somewhat so that by 92 weeks PMA (12 months) the infants are quite well-proportioned weight-for-length.

It might be worth pointing out that conclusions based on head circumference in preterm infants can be a problem because head shape influences occipital circumference. Given the same brain mass, an occipital circumference that is farther from round (a frequent problem in preterm infants who are hospitalized for several months), will be larger than the circumference if the head is closer to round. This has been thought to be at least part of the reason for the decline in normalized head circumference in preterm infants during the first year, as illustrated in your Table 2.

1. Casey PH, Kraemer HC, Bernbaum J, Yogman MW, Sells JC. Growth status and growth rates of a varied sample of low-birth-weight, preterm infants: a longitudinal cohort from birth to three years of age. *J Pediatr* 1991; **119**: 599–605.
2. Uauy R, Hoffman DR, Birch EE, Birch DG, Jameson DM, Tyson J. Safety and efficacy of omega-3 fatty acids in the nutrition of very low birth weight infants soy oil and marine oil (fish oil) supplementation of formula. *J Pediatr* 1994; **124**: 612–20.


**Bornstein**: Heird concludes: 'From the above summary of published sets of data, it is clear that there are statistically significant effects of ω-3 fatty acids of growth of both term and preterm infants.' This conclusion may be at variance with the preponderance of literature available. Heird moves us from statistical significance to functional or biological significance of potential differences in groups of n-3 deficient versus n-3 adequate LCPUFA babies. Heird also argues, on Fomon's analysis, that the effects of n-3 LCPUFAs and/or ALA intake on growth are 'less impressive' than heretofore thought. Thus, deficiencies in n-3 LCPUFAs or lower ALA intakes may not result in 'significant' differences from normal or near-normal mean rates of weight gain. Moreover, what accounts for (any) differences (deficiency in AA, an imbalance DHA and AA, or some other mechanism) is apparently unknown.

There is the possibility that n-3 LCPUFAs are beneficial in preventing obesity in adults, thereby decreasing the incidence of many modern diseases with which it is associated; here is another example of looking beyond acuity and IQ for potential new dependent measures (Bornstein, this volume).

A repeated safety concern is whether n-3 LCPUFAs negatively affect child growth (in different studies, head circumference, length, weight, and fat-free mass).

**Shaw and McEachern**: The issue of the possible effects of DHA on infant growth raises a broader concern. A number of chapters, including our own (Shaw and McEachern, Lucas, and Innis, all this volume) have touched on various areas where DHA may have deleterious effects, particularly in cases where AA decreases simultaneously. In a previous review (1), we have discussed the question of whether any molecule added to a complex system out of context to its normal constraints will be beneficial. For neurons at least, our conclusion was that such molecules would, at best, be ineffective. At worst, they may induce pathological activity. Based on this, how likely is it that DHA in formula will be beneficial, especially where normal constraints provided by the myriad other molecules in breast milk are missing? We also raise the following questions for general discussion: (i) At what point does science decide that the cumulative risk of DHA in formula exceeds its possible benefits? (ii) Even if possible negative effects of DHA can be neutralized by AA, can any formula possibly be tailored to fit the probable wide range of variations across infants (and possibly within infants, at different ages) in the uptake, use, and degradation of such molecules?

1. Shaw CA, Lanius RA, van den Doel K. The origin of synaptic neuroplasticity: crucial molecules or a dynamical cascade? *Brain Res Rev* 1994; **19**: 241–63.

**Wainwright and Ward**: This chapter addresses important issues with respect to the biological significance of the effects of supplementation with n-3 fatty acids on growth. Ultimately, dietary supplementation, if implemented, will be at the population level. Thus one of the concerns you have raised is particularly salient, namely, that any factor that changes the mean growth rate of the population will result in a shift in the distribution, such that more members of that population will now be below the standard lower limit and therefore 'at risk'. But, as Heird documents, it also depends on how growth is assessed, with n-3 supplementation affecting weight more than length. In light of this, future studies that provide more data on body composition in these infants will be informative. As he says, it may well be that a reduction in fat mass will be beneficial at the population level. This raises the question of whether long-term health outcomes with respect to outcomes such as non-insulin-dependent diabetes mellitus are optimum in the population on which the NICHD norms are based?

We have recently obtained data in a study conducted on growth and development in mice that speaks to the issue of whether the reduction in growth consequent on DHA supplementation may be related to the reduction/unavailability of AA (1). These data show that extremely high levels of dietary DHA with relatively low levels of n-6 FA as LA do result in growth retardation, but that this can be offset after weaning (but not before) by partial replacement of LA with AA. We have subsequently looked at glucose regulation in these animals (unpublished data) and found no effects.

1. Wainwright PE, Xing H-C, Mutsaers L, McCutcheon D, Kyle D. Arachidonic acid offsets the effects on mouse brain and behaviour of a diet with a low (n-6):(n-3) ratio and very high levels of docosahexaenoic acid. *J Nutr* 1997; **127**: 184–93.

**Neuringer and Reisbick**: As noted here, neural cell division is complete, or almost complete, by the end of the second trimester in humans if not by mid-second trimester (1–3). However, cell growth (including increases in size, dendritic arborization, and synaptic proliferation) does not peak until 8 months from term birth (1,3). Of specific relevance to this workshop, these processes impose a large demand for DHA for incorporation into newly forming neural membranes, and their timing closely overlaps formula feeding for many infants (also see comments on Shaw).

Might the ratio of DHA to AA or the ratio of total n-6 to n-3 LCPUFA be a better measure than absolute levels when considering effects on growth? Especially if the effects may be mediated by metabolites of these competing families of fatty acids?

1. Dobbing J, Sands J. The quantitative growth and development of the human brain. *Arch Dis Child* 1973; **48**: 757–67.
2. Rakic P. Corticogenesis in human and non-human primates. In: Gazzaniga MS, ed. *The Cognitive Neurosciences*. Cambridge: MIT Press, 1995: 127–45.
3. Jacobson M. *Developmental Neurobiology*. New York: Plenum Press, 1991.

# Methodology: Clinical Studies

# Methodological Considerations in Longitudinal Studies of Infant Risk

LYNN T. SINGER

*General Pediatrics, Stop 1038, Rainbow Babies & Children's Hospital, 2101 Adelbert Road, Cleveland, OH 44106, USA*

## Introduction

Visual and cognitive developmental advantages have been reported for term and preterm infants fed human breast milk or formula fortified with docosahexaenoic acid (DHA) and arachidonic acid (AA) in comparison to infants fed formula without these long-chain polyunsaturated fatty acids (LCPUFAs), but these findings remain unclear, controversial, and contradictory (1–12). These studies raise the possibility that specification of, and delivery of, optimal nutrition in the later states of fetal development or in early postnatal life, can have significant, positive, long-term impact on brain development, manifest in enhanced intellectual competence. Further studies, however, are necessary to establish rigorously if there are long-term benefits of the addition of LCPUFAs to infant formulas, substances which occur naturally in human breast milk.

Large-scale, randomized longitudinal clinical trials are critical to assess the benefits and potential risks of a nutritional intervention in infancy (13).

While this is standard practice in clinical drug trials, and in child and adult psychological interventions, such designs have only recently been applied to research in early human development (14), but their utility and complexity have been established. Current studies can benefit from knowledge obtained in the past two decades from a number of sources: e.g. (i) basic developmental research on the continuity and discontinuity of cognitive and developmental processes (15–17); (ii) longitudinal follow-up and intervention studies of normal, high risk and preterm infants (18–22); (iii) behavioral teratologic studies of fetal drug exposure (23–25) and environmental contaminants (26,27); and (iv) studies of malnutrition and nutritional deficit (28,29).

Establishing the efficacy of an intervention demands documentation of a reliable association between the intervention and outcome, documenting causality through this reliable association and through a temporal cause–effect relationship, as well as through proof that the relationship cannot be explained by some other factor (13,30). In order to achieve this end, research trials need appropriate representative cohorts for study, accurate information about other factors of importance to the intervention model, high cohort maintenance, and accurate and valid assessments of the intervention and outcome variables which are coherent with the theoretical intervention model (31). While these requirements may appear straightforward, they are particularly difficult to achieve in infant developmental research, which poses special methodologic and procedural dilemmas. Specific problems include the rapid developmental shifts and discontinuities of the infancy period, the relatively recent identification of biobehavioral correlates of individual differences in infancy, and the lack of extensive databases establishing predictive relationships from infant behavioral measurements or risk conditions (unless extreme) to later intellectual outcome, which are independent of social/environmental conditions. In this chapter, I discuss selected methodologic issues in infancy research important in their application to the extant research on breast feeding, LCPUFA supplementation, and infant developmental outcomes.

Central nervous system development occurs rapidly during infancy and is determined by precisely regulated genetic processes in interaction with a variety of environmental influences to which the infant brain may be particularly vulnerable, dependent on timing, nature, and degree (32). Because of transactional processes (33) the developing infant brain, especially in the preterm infant, may be sensitive to deficiencies in LCPUFAs. Early alterations in brain functioning due to deficiencies might result in the decrement in school age IQ reported in bottle versus breast fed children (4–8). Several factors underlie the complications of measuring the developmental changes potentially caused by an early nutritional intervention. These include the special responsiveness between the infant and the care-giving environment, promoting the organizational plasticity of the brain. The process of developmental canalization or self-righting is also important. There is also evidence that parental educational and socio-economic factors are more predictive of childhood IQ than many biologic risk conditions of infancy, suggesting a strong environmental influence on the development of infant skills.

## Assessment Instruments

Unlike the measurement of physical properties, such as height and weight, which can be more easily and unequivocally defined and operationalized, and which maintain their validity across ages, psychological research attempts to measure hypothetical constructs which are less clear, less directly measured, and which need strict definition to avoid confusion (34). IQ is a prized psychological construct because of its stability when measured after 3 years of age, and because it has been meaningfully related to so many other important characteristics which are societally valued. Among the significant problems in assessing the effects of an intervention in early infancy are the long time lapse between the first manifestation of benefits and stable effects, the difficulties in defining IQ in early stages of development in a theoretically or empirically consistent way, and the continuous reorganization of behavior during development. Because of these factors, choice of assessment instrument and timing of assessment are critical aspects of research design in early intervention studies (35).

An assessment instrument must have adequate basic psychometric properties, the integrity of which must be maintained during the research trial, if the clinical trial is not to be compromised by measurement error (36,37). In general, this requires that the assessment be standardized (i.e. having a uniform set of procedures for administration and scoring), be reliable (i.e. measure a cohesive construct in a similar fashion across multiple administrations and time), and be valid (i.e. measure the construct it purports to). Because reliability is a necessary, but insufficient, factor in validity, it is incumbent on researchers to evaluate both characteristics of a measure carefully before using it in research protocols.

In this regard, the most commonly used assessments of overall developmental functioning in infancy have been the Bayley Scales (38,39), which can be administered from 1 to 42 months of age. The Bayley is considered a 'broad-based' measure because it provides a global score which reflects the infant's functioning across diverse abilities in comparison to its age-peer group. Because of its excellent psychometric properties, the Bayley is nearly a gold standard in infant research. Its predictive validity for childhood IQ is its property most relevant to the study of LCPUFA supplementation, but at the same time the property for which it has received the most criticism. Bayley Mental Scale scores in the first year of life have not been shown to be predictive of later IQ for nonrisk infants (17,40), but those after 18–24 months of age do provide useful predictions (40). More recent research, however, indicates that the Bayley is sensitive to a wide range of teratogenic substances and risk conditions even in the first year of life, including alcohol exposure (25,41), polychlorinated biphenyls (PCBs) (42), lead (43), and iron deficiency (29). Moreover, when administered between 7 and 12 months of age, the Bayley Scales have discriminated known risk groups, including failure-to-thrive infants (28), preterm and term infants (44) and between high- and low-risk preterm groups (21). Jacobson and Jacobson (45) note that the Bayley Scales appear capable of detecting population effects, even when prediction of an individual score to later functioning is not as reliable.

A significant problem with the Bayley, because it yields a conglomerate score representing the sum of the infant's performance on a number of diverse skills, is that it does not provide quantifiable information on more discrete processing skills, such as language, attention, or rate of information processing. Because it is an apical test, i.e. providing a summary score of achievement of a number of diverse skills, the Bayley summary score may dilute and mask subtle benefits or detrimental effects of an intervention.

Because of the lack of predictive validity of the traditional infant developmental assessments in the first year of life, developmentalists have continued to seek an assessment instrument which would have more diagnostic (predictive) validity. A number of promising paradigms and alternative assessments have been comprehensively reviewed by Colombo (16), two of which have been utilized in human studies of LCPUFA supplementation, i.e. novelty preference and fixation duration. Infant preferences for novel visual stimuli have been manipulated in habituation tasks and in paired-comparison or visual recognition memory (VRM) tasks. The infant's tendency to fixate a novel versus a previously seen target is an operational definition of VRM. Infant performance on these tasks is presumed to reflect infant perceptual orientation, encoding, discrimination, and memory of presented stimuli (17). Significant predictive validity of childhood IQ measures have been found using both habituation and recognition memory (paired-comparison) tasks in the first year of life, with such measures largely independent of socio-economic status, and with prediction from measures administered at 2–7 months of age (16,46–48). Moreover, moderate levels of prediction have been derived for both risk and non-risk samples, despite significant methodological and procedural differences across studies, which include differing sample sizes, paradigms, stimuli, number of tasks, stimulus exposure duration, and predictor variables (16,22,28; see McCall and Carriger (48) for a recent review and meta-analysis). Visual recognition memory task performance has also been discriminated between groups of high- and low-risk infants, such as those with failure to thrive (28) and PCB exposure (27). A specific aspect of information processing which can be derived from both habituation and visual recognition memory tasks is that of duration of visual fixation. Although habituation tasks can be administered through all sensory modalities, vision has been the most widely studied. Fixation duration refers to the duration of the infant's individual fixations of the visual stimulus, and can be computed in various ways in both habituation and VRM tasks. Across both paradigms, this measure has shown significant negative concurrent and prediction relationships to measures of sensorimotor and cognitive processing, suggesting that 'shorter looking times' are reflective of better information processing on the part of the infant (16, pp. 54–9). Faster information processing is also intuitively a component of successful performance on visual recognition memory tasks, since novelty preferences must be demonstrated with specific exposure time limitations to familiar stimuli.

Because some studies have suggested that LCPUFAs may be important in early visual and later intellectual development, and because visual habituation and recognition memory tasks have proven to be predictive of later intellectual functioning, such tasks are plausible candidates for investigation as outcome measures in intervention trials of LCPUFA supplementation. Of the infant cognitive tasks available to date, however, only Fagan's test, which utilizes novelty preferences to assess infant visual recognition memory (49), meets minimal standards of the following requirements for psychometric adequacy to allow in a clinical trial. Such requirements include: development and publishing of standardized equipment, stimuli, and test items; administration and scoring procedures for test items; training requirements for examiners; development of tasks on a representative, normative sample; and provision of normative, reliability, and validity data. Reliability data include intra-rater or test–retest reliability, chronological age stabilities, inter-rater agreements, and internal consistency. Validity is derived, in part, as a function of the number of established relationships documented for a particular assessment over time along a variety of parameters. In this latter regard, only Fagan's (49) test has had significant demonstration of relationships with predictive/diagnostic categories suggesting that it relates to general intellectual status for a variety of populations. It should be noted, however, that some degree of predictive validity has been established for only one outcome measure on Fagan's test, i.e. percentage of infant visual fixation to the novel stimulus as averaged over all test items (mean percentage novelty). While other measures from the test are obtainable, the meaning and utility of derived measures await further documentation. One such measure which can be extracted from the Fagan test, used by Jacobson (23), averages fixation duration during the familiarization phase, and has been conceptualized as a measure of information processing speed. Such derived measures are statistically weak, however, because they cannot be assumed to have any of the psychometric characteristics of tests from which they were taken (36,50). They should be used only for heuristic or exploratory purposes, and should be interpreted with extreme caution.

Even with its relative strengths as a psychometric instrument, however, it should be noted that Fagan's test is not comparable in psychometric properties to standard IQ tests, which, after 86 years of refinement, are among the most well validated of psychological measures. Current deficiencies in Fagan's test include: poor short-term reliabilites which reduce validity, and which require multiple chronological age administrations to produce stability; lack of standard score indices based on distribution in infant performance, which preclude comparison of scores to those of an age reference group or to other tests; and difficulty of interpreting group findings if they are within normal ranges. For example, discrepancies on Bayley Scale scores within normal ranges can be interpreted to reflect population differences because of the scaling system of the Bayley. When differences on VRM performance on the Fagan Test are significant, but in average ranges, they are more difficult to interpret.

## Timing of Assessments

Scheduling data collection points in infancy research is dependent on several factors, including goals of the research trial, age spans of appropriate test instruments, and practical considerations. Practical considerations include expense, the need to avoid burdening families, integration of the research into schedules compatible with medical visits for preterm and term infants, and the avoidance of redundancy. Another consideration is compatibility with the general developmental research literature in the field, i.e. to facilitate comparisons across studies and expand inferences from the available data. As noted previously, infancy is a time period in which the developmental process particularly complicates measurement of function as the age of the child interacts with the factors that can be measured at the time, as well as with test construction artifacts (15,50). Moreover, even when an individual test is designed to span a wide age range and includes several skill domains, such as the Bayley, factors affecting its validity may differ depending on the age of the child. For example, the Bayley Scales may function quite well as a measure of gross motor, but not language, skills at 12 months as they are just emerging, or as a predictor of IQ at 24, but not 6 months. Similarly, Fagan's Test (and other measures of habituation/VRM) have been reported to be most predictive of later IQ at 2–7 months (45). To predict school age IQ with a reasonably high degree of confidence would entail a standard IQ assessment at 3 years of age. Even at later ages, however, beneficial or detrimental effects which may be quite important, e.g. learning disabilities, which are relatively independent of IQ by their definition, may not be discernible when only IQ is measured. For example, higher rates of learning disabilities have been found in normal IQ preterms at school age (20), suggesting that specific domains of cognitive function may be differentially affected by an intervention. 'Sleeper effects' are well known in teratologic studies, while some intervention effects have not been apparent in preterm studies until school age (51–53). Risk conditions, teratogens, or interventions which significantly affect specific domains of cognitive functioning require multidimensional assessments at school age to detect effects.

## Confounding Variables

Inferring that the effects of a nutritional intervention are causally related to the specific intervention under study requires that the demonstrated relationship between the independent and dependent variables cannot be attributed to any other factors, thus rendering the inferred relationship as spurious. Indeed, avoidance or minimization of erroneous relationships is one of the significant strengths of the randomized clinical trial, as randomization is theoretically the only method of controlling all possible extraneous variables (13: p. 288).

Confounding variables can exert an influence on both the intervention (independent) and outcome (dependent) variables. Thus, when early studies of term and low birthweight infants indicated that breast-feeding was associated with better child intellectual functioning later in life, careful analysis discerned that social, educational, and demographic factors known to be related to cognitive development also discriminated mothers who chose to breast-feed from those who did not. Randomization to treatment increases the probability of group equivalence, but does not necessarily remove all biases. For one, mothers who choose to breast-feed cannot be randomized and are likely to continue to differ in important ways from those who agree to enter the randomized trial. Additionally, those mothers whose infants are randomized to the clinical trial are likely to differ from those who refuse to enter the trial, even when they are not breast-feeders. Both maternal and infant characteristics may bias the study groups. For example, mothers who are more invested in their child's development may be more motivated, better educated, and more health conscious than those who do not have the energy or interest to perform the record and appointment keeping required by a longitudinal study. Their infants may also differ in salient characteristics in unexpected ways. For example, in our longitudinal studies of very low birth weight (VLBW) infants with and without bronchopulmonary dysplasia (BPD) (21,54,55), we originally feared that mothers of the sicker infants would not wish to enter the study. To our surprise, we found that few mothers of the sick infants with BPD declined the study. In contrast, mothers of healthier, VLBW infants without complications often declined to enter the study until a later time point, stating that they wished to forget the upsetting hospital experience and 'normalize' their experiences with their infants. While biases related to entrance to study are not necessarily destructive to a study's validity, especially if treatment is randomized, they do affect generalizability of findings. As much as possible, characteristics of subjects who refuse to enter a clinical trial should be documented in order to aid interpretation of results.

Even with randomization, confounding can occur either by chance, due to smaller sample sizes that are inadequate to control for all potential confounders, or through some unknown systematic error in subject selection. In order to infer with as high a degree of confidence as possible that statistical relationships between nutritional supplementation and outcome variables reflect the influence of supplementation per se, it is recommended that potential confounders be identified and documented so that balance for these factors can be determined in randomization (56–58).

As many potential confounders as possible should be identified prior to data collection, not after statistical analysis, and should be determined based on their known theoretical or empirical relationships to the study variables. Data from longitudinal studies of infant risk over the past two decades suggest a number of known influences on the most frequently used measures of child outcome in longitudinal infancy studies, i.e. developmental quotient (DQ) or IQ.

Once identified, confounders can be managed in several ways, i.e. through exclusion, matching, stratification, randomization, or data analytic strategies (58). Jacobson and Jacobson (57) have presented a list of control variables tested routinely in their studies of infant cognition and prenatal alcohol exposure. Because a variable cannot be a confounder unless it is related to both the intervention and outcome variables, and because the randomized intervention should reduce bias associated with the independent variable, control variables are generally assessed for their influence on outcome.

Socio-economic status is probably the most important confounder in developmental studies due to its strong and pervasive relationship to child intellectual outcome, usually manifest by 2 to 3 years of age and correlating at about 0.50 with school age IQ (15). Socio-economic status measures generally reflect some amalgam of parental educational level, occupational status, and economic resources. In the USA, the Hollingshead Index (59,60) is the most widely used. Maternal IQ is another confounder which, because it reflects both genetic and environmental influences, should be assessed, if possible. Widely used, well validated, measures are available which are quick and easy to administer. Such measures should be used cautiously, however, in populations where there is significant ethnic and socio-economic status diversity. In our studies of VLBW and cocaine exposed infants, for example, we attempted to use the Peabody Picture Vocabulary Test (61) to exclude mothers with mental retardation more reliably. We found, however, that among low socio-economic status, African–American mothers, almost 50% of the sample would have been excluded based on test norms (62). Nonetheless, there is reported validity to use of this test as a correlate of relevant demographic and personality variables within lower class, African–American subjects (57). Thus, factors that contribute to between population variability are not necessarily those that contribute to within population variability.

Specific aspects of the home environment have also been found to be strongly related to cognitive and language competence through age 5 years (63,64). The HOME Inventory (65), an observation/interview technique that assesses the quality of the cognitive, social, and emotional stimulation the infant receives, is composed of six subscales: parental responsivity, acceptance of child, organization of the environment, provision of appropriate materials, parental involvement, and variety of stimulation. In one study, total HOME score at 6 months of age was demonstrated to correlate highly ($r = 0.50$) with 3 year IQ (66). Its validity has also been established for African–American children (67). Constructs assessed through the HOME Scale, such as warmth and responsivity, are likely to be particularly relevant to nutritional intervention studies, such as those which provide supplements found in breast-milk, because breast-feeding may be associated with subtle differences in the quality of mothering not wholly explained by socio-economic status (3,8). Although designed to be administered on a home visit, the HOME Scale has also been validated with laboratory administration via interview (57).

Other routinely assessed confounders included maternal age, parity, ethnicity, marital status, child gender, birth order, and total number of children in the home.

A growing body of developmental research has also implicated certain maternal psychological characteristics, especially depression and other psychological distress symptoms, as important correlates of poorer child cognitive and behavioral outcomes through their negative impact on maternal child interactions (68,69). Clinically significant levels of maternal post partum depression are common, and may be particularly relevant to assess in studies of feeding practices and supplementation. In a study of VLBW and term infants, 20–34% of mothers self-reported clinically significant symptoms of depression at a visit 1 month post partum (54). Greater severity of depressive symptoms was associated with fewer maternal verbal interactions with her infant during feeding (54). Psychological distress symptoms were also associated with maternal report of higher parenting stress, including a lower sense of parenting competence, and lower perceived spousal and social supports, even after control for socio-economic status and race (55). Using the Nursing Child Assessment Feeding Scale at 1 month post partum, maternal depression was related to decreased sensitivity to infant cues, and to decreased cognitive growth fostering, but only for term mothers, indicating that infant risk or prematurity may also affect how maternal depression impacts on mother–child interactions (70).

These data all suggest that maternal depression or psychological distress symptoms post partum have complex relationships to mother–child interactions, including feeding interactions, maternal and infant hormonal levels, and maternal feelings of competence and perceived social supports. At least one study (10) found, in a UK sample, in a retrospective survey of mothers of 5-year-old children, that those who had a high score on a standardized inventory indicating psychiatric disturbance were less likely to report having breast-fed their infants. Follow-up studies of lower socio-economic infants in the USA have found maternal psychological distress or depression post partum to be negatively related to child Bayley Mental Development Index (MDI) at 17–18 months of age, after control for other relevant factors (71,72). Similarly, maternal postnatal depression has been related to poorer 3 year IQ in a UK study (73).

Several useful self-report instruments are widely available and have been found to be sensitive to maternal depressive affect across racial and socio-economic status groups. These include the Beck Depression Inventory (74), the Brief Symptom Inventory (75), and the Center for Epidemiologic Studies Depression Scale (76).

Maternal use of several non-nutritional substances during pregnancy should also be considered as a potential confounder. Tobacco, alcohol, marijuana, cocaine, and opiates have recently been recognized for their potentially harmful effects on fetal and infant growth and development (24,25,52,77–79). Estimates of the most recent United States Household Drug Survey (80) indicates that pregnant women aged 15–44 years reported use of substances during the previous year at the following rates: tobacco (31%); alcohol (64%); marijuana (8%); cocaine (2%); and other 'hard' drugs (0.2%). Maternal use of substances during pregnancy is differentially associated with lower socio-economic status, older maternal age, poor maternal nutritional status, higher rates of infectious disease, prematurity and low birth weight, and ethnicity. Moreover, varying combinations of polydrug use may have their own deleterious effects on outcome, and may interact with maternal nutritional factors. Animal studies have found that the combined metabolism of cocaine and alcohol produces a third substance, cocaethylene, with its own neurochemical properties (81). For example, in a case control study, we found that the combination of cocaine and alcohol predicted lower birth lengths in lower social class, African–American infants who had been exposed in utero to cocaine and alcohol, while there were no deleterious effects for infants who were exposed to either drug in isolation (78,79). Because they are likely to have serious social and psychological problems which can affect infant outcome, heavy alcohol and marijuana users, and cocaine and opiate users, should be excluded from longitudinal studies of infant supplementation unless severity and chronicity of substance use can be reliably documented.

Examiner effects are confounding factors which should be routinely monitored, but which are often ignored. There is considerable evidence that test results may vary systematically from examiner to examiner (36). For example, the Brazelton Neonatal Behavioral Assessment Scales have been well studied with regard to examiner effects and found to be highly susceptible (82). Infant tests are likely to be more vulnerable to examiner effects given the propensity of infants to be highly variable in mood, requiring greater efforts from the examiner.

## Confounders in Preterm Populations

Preterm populations pose special problems for intervention studies because of rapidly evolving changes in neonatal intensive care unit technology and medical practices, and the large number of medical and neurological complications which must be considered as possible confounders in longitudinal developmental studies. Among the most widely documented known confounders bearing a relationship to early developmental outcomes are gestational age, birth weight, intrauterine growth retardation, multiple birth status, Apgar scores, patent ductus arteriosus, septicemia, peak bilirubin level, apnea, retinopathy of prematurity, necrotizing enterocolitis, tracheotomy, hearing test abnormalities, number of days spent on ventilator, number of days with supplemental oxygen, number of days hospitalized, and diagnosis of BPD. Numerous neurologic sequelae of VLBW bear direct relationships to cognitive outcome, including intraventricular hemorrhage and periventricular leukomalacia, seizures, meningitis, porencephaly, hydrocephalus, ventriculoperitoneal shunting, and echodense lesions (21,83).

As various medical complications may cluster nonrandomly, interpretation of studies may be difficult if clear, reliable information about as many medical complication variables as possible is not given regarding exclusion of subjects, or incidence within study populations. For example, our studies of BPD and VLBW infants indicated that, after control for race, socioeconomic status, and multiple birth status, BPD infants are more likely to have: necrotizing enterocolitis; lower birth weight and gestational age; more septicemia and retinopathy of prematurity; lower peak bilirubin; more visual abnormalities; higher incidence of tracheotomy; greater neurologic risk, reflected in higher incidence of seizures and both mild (grades I–II) and severe (grades III–IV) intraventricular hemorrhage; more oxygen desaturations; and feeding problems (21,54,84). Given these possible confounders, interpretation of findings of effects of long chain fatty acid supplementation on growth, visual acuity and information processing outcomes of small groups of BPD and non-BPD VLBW infants in prior studies becomes exceedingly problematic (1).

Because studies of preterm infants usually involve multiple sites in order to increase sample size, they are more likely to be affected by significant variation in patient populations and obstetric and neonatal care practices across sites. For example, use of surfactant to prevent respiratory distress, administration of steroids or magnesium sulfate to mothers during pregnancy, or use of steroids to wean infants from the ventilator are examples of care practices likely to vary nonrandomly which can influence both medical complications and developmental outcomes in unforseen ways. These rapidly changing practices are also likely to produce strong cohort effects.

The importance of accurate measurement of control or confounding variables has been emphasized because, in multivariate analysis, unreliable measurement will lead to an underestimate of the confounder and a subsequent spurious inflation of the variance attributable to the intervention (57,58). While many of the perinatal risk variables noted above can be measured in relatively straightforward, objective fashion, measurement of gestational age (GA) merits special scrutiny. GA is particularly important because of its potentially confounding effects, but also because inaccurate estimation of GA will have an inordinate effect on infant developmental outcome measures administered. Since it is general practice to administer developmental assessments based on infant age after correction for prematurity, inaccurate estimates will have larger impact on smaller, more premature infants and when tests are administered at earlier ages in infancy. Even within the normal full term range, however, differences in GA estimates could conceivably influence test findings. As full term infants can range from 37 to 42 weeks GA, tests based on chronological age can include infants varying by as much as 5–6 weeks.

DiPietro and Allen (85) have reviewed findings on GA estimation, noting that measurement standards are often inadequate, determination by postnatal examination is less accurate for preterm infants, and that GA assessments are inherently confounded by factors of developmental risk under study, leading to substantial variability in measurement. Their recommendation is to review both maternal and neonatal chart data, combining last menstrual period date based on maternal report, sonogram results at < 18 weeks, prenatal examination at < 12 weeks, and Dubowitz or Ballard postnatal assessment, and excluding cases where there are widely discrepant findings. An alternative strategy currently used by many researchers is to use only one type of GA estimate, usually the postnatal examination, a practice which can be defended as at least introducing a consistent source of bias. Another alternative is to average assessment results if obstetric and pediatric assessments differ by more than 2 weeks.

## Confounding versus Moderating versus Mediating Variables

An important consideration in intervention studies is determination of whether variables other than the intervention and dependent variables should be conceptualized as confounding, moderating, or mediating variables. While lack of control for confounders may lead to a distorted estimate of the impact of an intervention, failure to consider moderating or mediating variables may be equally misleading. Moderating variables are contextual, demographic, medical, or other characteristics which differentially affect the efficacy of an intervention. Mediating variables are actually part of the causal pathway between intervention and outcome variables. If a mediator is treated as a confounder, it can obscure the true relationship of the intervention to outcome, as variance shared by the intervention and the mediator will be attributed to the mediator (57). Baron and Kenny (86) have presented a thorough discussion of the distinction between, and statistical handling of, these constructs.

An interesting example of the need for greater specification of these constructs in study design can be found in the emerging studies of developmental effects of LCPUFA supplementation and DHA and AA levels in preterm infants (1,9,12). As noted correctly by Carlson et al. (1), while many perinatal risk variables precede or are independent of the dietary intervention (e.g. birth weight), several significant medical complications can occur, or are diagnosable, after introduction of the dietary change. These include development of BPD, septicemia, and necrotizing enterocolitis, as well as intraventricular hemorrhage (IVH). Indeed, in some studies of neonatal intensive care unit behavioral interventions, reduction of incidence of BPD and IVH has been specifically targeted as an outcome (87). Studies available on addition of LCPUFAs to preterm infants' diets that have found differential effects for infants with BPD are hard to interpret in that (i) it is not clear if BPD was conceptualized as a confounder which escaped randomization; (ii) or if BPD was evaluated post hoc as an outcome of intervention or (iii) as a moderator of the intervention's effects. Other medical complications in these same studies, such as occurrence of sepsis or necrotizing enterocolitis, were not assessed as outcomes, suggesting that BPD was considered post hoc as an outcome. Additionally, some perinatal complications which were not excluded do not appear to have been evaluated, such as retinopathy of prematurity and grades I–II IVH. In studies in which subgroup sizes are quite small, lack of randomization of such complications can bias results. We have found, for example, that, of over 20 perinatal medical complications assessed in a sample of VLBW infants, mild IVH was the only complication which discriminated cocaine-exposed from those not exposed (78). Other studies have found that healthy preterms with only mild IVH performed more poorly than nonaffected preterms at 5-year follow-up on school readiness tasks (88), suggesting that grade I–II hemorrhage cannot be ignored as a confounder.

## General Considerations

As one investigator has noted (89), data collection is one of the most time consuming and important aspects of a clinical trial, but often receives little attention since it is done by the 'soldiers', while design is done by 'generals' and analysis by their 'lieutenants'. Morley (56) and others (90) have written excellent overviews of data management factors relevant to clinical trials with urges to investigators to attend to these details in advance. This chapter will conclude with attention to several specific issues relevant to data collection and subject management in infant studies.

### Training of Examiners

Attention to measurement issues in training examiners in reporting procedural details in published studies is an important component of conducting randomized clinical trials. Even when assessment instruments, such as the Bayley Scales and the Fagan Test, are well standardized, valid, and reliable, ensuring that standard administration and scoring procedures are followed consistently across examiners and sites over the longitudinal trial is a major task for the investigator (89). Surprisingly, few published studies report on these procedural issues. Frequently, even when inter-rater reliabilities are reported for a measure, they are taken from the normative data, not the researcher's own setting, and they may not reflect the actual properties of the data reported.

Well known examiner influences on the data include examiner expectations or self-fulfilling prophecy (13) which occurs when a tester's beliefs about the efficacy of an intervention or obvious characteristics of the subject affect the tester's ratings of infant performance. For example, the phenomenon of 'prematurity stereotyping' (91) has been documented in which infants labeled premature have been perceived less favorably than those not so labeled, despite identical characteristics. Attempts at masking 'examiners' are often inadvertently undermined. For example, in studies in which breast-feeders are a separate group, testers who have beliefs about the positive benefits of breast-feeding may rate these infants more favorably. Despite masking, because most infant assessments require maternal involvement, the breast- or bottle-fed status of the infant may be determinable by the bottles protruding from an infant's diaper bag, or the infant's behavior towards the mother's breasts during the research visit.

Examiner characteristics, such as warmth and enthusiasm, have been demonstrated to affect test performance (36), and some examiners who are naturally good with babies may systemically elicit more co-operative, and thus, higher level performances than others. In general, children are more susceptible to examiner and behavioral characteristics than adults, and Anastasi (36) notes that: 'in the examination of preschool children, the role of the examiner is especially crucial' (p. 38). Similarity/dissimilarity of the examiner in ethnic, gender, or other personal characteristics may also affect maternal comfort level, indirectly affecting infant performance.

Examiner 'drift' over the course of a longitudinal study is also commonly reported (92). 'Drift' refers to subtle deviations from standard administration or scoring of procedures which can occur over time, or it can refer to the effects of greater expertise in examiner functioning over time, which might be discrepant from function at an earlier test point (93). Situational variables associated with site and examiner may also create bias. Infants seen for follow-up in the same medical offices where they have also received injections, or who are seen in conjunction with a medical visit, may be more fearful and uncooperative, providing less valid assessment data (36).

Investigators and examiners must be constantly on guard to detect the possible operation of these extraneous factors in order to minimize their influence. Common documented training procedures should be described, implemented across examiners and sites, and acceptable inter-rater reliabilities established for both item administration and scoring. Assumption of adequate reliability based on normative or previously established levels is inappropriate, and threatens the validity of results. Throughout the study, periodic, regular assessments should be made to assure that procedures do not 'drift' over time. Continued monitoring and repeated calculations of inter-rater reliabilities for each assessment point should be conducted. Independent scoring of videotapes can facilitate cross-site monitoring.

Sample attrition can be a serious problem in longitudinal studies, since even with moderate dropout at each time point, the number of subjects with incomplete data will accumulate (94). Usually attrition is not random, and reassessment of samples at each time point with regard to relevant study and confounding factors is necessary to assess how attrition bias may affect findings. In a large follow-up study of preterm infants (95), dropout rates were higher for lower social class Black and Hispanic infants in urban areas.

Stipends and incentives are often important to consider in managing recruitment as well as attrition biases. Incentives are given to defray time and economic expenses entailed in a family's participation in a study which might lead to refusal or dropout. Incentives may be more or less meaningful to subjects' participation based on a variety of factors. For example, in our own longitudinal studies of preterm infants in a large urban center, socio-economic status is not evenly distributed, and our sample includes a large proportion of poor, urban families. Although all families received the same monetary stipend to cover time and transportation costs, we initially noted a high dropout rate among poor, urban subjects. With investigation, we determined that stipends were mailed by the University by check several months after the research visit, a factor which did not deter socially advantaged subjects, but which differentially affected poorer families who moved residences more frequently and who had to pay a fee to cash the check. Further, although bus transportation tickets may be provided by a study, the difficulties of finding a reliable bus service within an urban area, particularly with small and physically vulnerable children in tow, and, frequently, in bad weather, can make transportation a significant problem for lower socio-economic status subjects in the USA. Removal of these barriers through immediate cash stipends at the time of visit, co-ordination of the visit with other medical appointments, and provision of cab transportation when needed, significantly increased visit compliance. In studies of nutritional supplementation, free formula is often a benefit of participation, but formula may be of little benefit for breast-feeding mothers or for low social class mothers who already receive free nutritional supplementation through government programs. These study subgroups may have differing participation rates based on these extraneous factors.

A significant issue in infancy research concerns the sharing of assessment results with parents over the course of the study. One benefit of clinical developmental research commonly noted in Human Subjects' Institutional Review Board protocols is that assessments may allow early detection of problems and delays that can lead to earlier intervention than would have occurred if the infant had not participated. When pronounced subaverage performance is detected in clinically relevant assessments, it is incumbent upon the researcher to notify parents and to help make appropriate referrals. Some discussion of the limitations of predicting from infant tests should accompany any communication, especially in light of infant variability of performance from day to day and the greater variation in attainment of developmental milestones across infants. Working with the parents to include the pediatric or medical caregiver in assessing the infant's needs is essential. It should be stressed to parents that the research visit is not equivalent to a clinical assessment and is used primarily to evaluate grouped data.

Preterm or other medically at risk groups pose special problems since rates of detectable developmental delay are higher than in term populations, and because infant assessments are scored based not on a child's chronologic age, but on age corrected for prematurity, which, as has been previously noted, may reflect an unreliable estimate of gestational age. In addition to their unreliability, age corrected scores may significantly obscure real developmental delays in preterm infants. The extent of this masking of deficits varies depending on the degree of prematurity and the age at which an individual infant is tested. DiPietro and Allen (86) report that one study found that the mean difference between corrected versus uncorrected scores for Bayley outcomes for a sample of preterms was 36 points, with a range of 0 to 61 points. Recent data from preterm studies using the 1969 version of the Bayley Scales found high- and low-risk VLBW infants performing well within normal limits on Bayley MDI outcomes at 12 months, but their scores were 7 (low risk) to 21 (high risk) points lower than a race, gender, and socio-economic status equivalent control group (47). Thus, parents need to be particularly cautioned about the uncertainty of infant scores when infants are very premature.

## Acknowledgements

Preparation of this paper was supported by grants R01–DA07957 from the National Institute on Drug Abuse, and MCJ-390592 from the Bureau of Maternal and Child Health Services.

## References

1. Carlson S, Werkman S, Trolley E. Effect of long chain n-3 fatty acid supplementation on visual acuity and growth of preterm infants with and without bronchopulmonary dysplasia. *Am J Clin Nutr* 1996; **63**: 689–97.
2. Carlson S, Werkman S. A randomized trial of visual attention of preterm infants fed docasohexaenoic acid until two months. *Lipids* 1996; **31**: 85–90.
3. Jacobson SW, Jacobson JL. Breast feeding and intelligence [letter to the editor]. *Lancet* 1992; **339**: 926.
4. Lucas A, Cole TJ, Morley, R et al. Factors associated with a mother's choice to provide breast milk for a pre-term infant. *Arch Dis Child* 1988; **63**: 48–52.
5. Lucas A, Morley R, Cole TJ, Gore SM. A randomized multicentre study of human milk versus formula and later development in preterm infants. *Arch Dis Child* 1994; **70**: 141–6.
6. Lucas A, Morley R, Cole T, Lister G, Leesom-Payne C. Breast milk and subsequent intelligence quotient in children born preterm. *Lancet* 1992; **339**: 261–4.
7. Morley R, Cole TJ, Powell R, Lucas A. Mother's choice to provide breast milk and developmental outcome. *Arch Dis Child* 1988; **63**: 1382–5.

8. Pollock JI. Mother's choice to provide breast milk and developmental outcome. *Arch Dis Child* 1989; **64**: 763–4.
9. Rioux F, Innis S, Lupton B, Nelson C, Waslen P, Whitefield M. Cognitive and visual development, psychomotor performance and growth in term compared to preterm infants. *Pediatr Res* 1995; **38**: 1889 (abstr).
10. Taylor B, Wadsworth J. Breast feeding and child development at five years. *Dev Med Child Neurol* 1984; **26**: 73–80.
11. Janosky J, Scott D, Wheeler R, Auestad N. Fatty acids affect early language development. *Pediatr Res* 1995; **38**: 1847 (abstr).
12. Werkman SH, Carlson S. A randomized trial of visual attention of preterm infants fed docosahexaenoic acid until nine months. *Lipids* 1996; **31**: 91–7.
13. Kerlinger F. *Foundations of Behavioral Research*, 3rd edn. New York: Holt, Rinehart & Winston, 1986.
14. Richmond J. Low birthweight infants: can we enhance their development? *JAMA* 1990; **263**: 3069–70.
15. McCall RB. Nature nuture and the two realms of development: a proposed integration with respect to mental development. *Child Dev* 1981; **55**: 1–12.
16. Colombo J. *Infant Cognition: Predicting Later Intellectual Functioning*. Newbury Park: Sage, 1993.
17. Fagan JF, Singer LT. Infant recognition memory as a measure of intelligence. *Adv Infancy Res* 1983; **2**: 31–78.
18. Broman S, Nichols PL, Kennedy W. *Preschool IQ: Prenatal and Early Developmental Correlates*. Hillsdale, NJ: L. Erlbaum, 1975.
19. Werner E, Bierman JM, French FE. *The Children of Kauai: A Longitudinal Study from the Prenatal Period to Age Ten*. Honolulu: University of Hawaii Press, 1971.
20. Klein NK, Hack M, Breslau N. Children who were very low birthweight: developmental and academic achievement at nine years of age. *J Dev Behav Pediatr* 1989; **10**: 32–7.
21. Singer LT, Yamashita TS, Lilien L, Collin M, Baley J. Three year outcome of infants with BPD and VLBW. *Pediatrics* 1997; (in press).
22. Rose S, Wallace I. Visual recognition memory: a predictor of later cognitive development in preterms. *Child Dev* 1985; **56**: 843–52.
23. Jacobson SW, Jacobson JL, Sohol RJ, Martier SS, Ager JW. Prenatal alcohol and infant information processing. *Child Dev* 1993; **64**: 1706–21.
24. Fried PA, Watkinson B. 12 and 24 month neurobehavioral follow-up of children prenatally exposed to marijuana, cigarettes, alcohol. Neurotoxicol Teratol 1990; **11**: 49–58.
25. Streissguth AP, Barr HM, Martin DC, Herman CS. Effects of maternal alcohol, nicotine, and caffeine use during pregnancy on infant mental and motor development at 8 months. *Alcohol Clin Exp Res* 1980; **4**: 152–64.
26. Needleman HL, Gunnoe C, Leviton A et al. Deficits in psychologic and classroom performance of children with elevated dentine lead levels. *N Eng J Med* 1979; **300**: 689–95.
27. Jacobson JL, Jacobson SW. Intellectual impairment in children exposed to polychlorinated biphenyls in utero. *N Eng J Med* 1996; **335**: 783–9.
28. Singer LT, Fagan JF. The cognitive development of the failure-to-thrive infant: a three year longitudinal study. *J Pediatr Psychol* 1984; **9**: 363–84.
29. Lozoff B. Has iron deficiency been shown to cause altered behavior in infants?

In: Dobbing J, ed. *Brain Behavior and Iron in the Infant Diet*. London: Springer-Verlag, 1990.
30. Gordon R. The design and conduct of randomized clinical trials. *IRB – A Review of Human Subjects Research* 1985; **7**: 1–3, 12.
31. Zetterstrom R. Data in pediatric longitudinal research. In: Magnusson D, Bergman L, eds. *Data Quality in Longitudinal Research*. New York: Cambridge University Press, 1990: 72–84.
32. Dobbing J. Vulnerable periods in developing brain. In Dobbing J, ed. *Brain Behavior and Iron in the Infant Diet*, New York: Springer-Verlag, 1990: 1–17.
33. Sameroff AJ, Chandler MJ. Reproductive risk and the continuum of caretaking casualty. *Rev Child Dev Res* 1975; **4**: 187–244.
34. Bergman L, Magnusson D. General issues about data quality in longitudinal research. In: Magnusson D, Bergman L, eds. *Data Quality in Longitudinal Research*. New York: Cambridge University Press, 1990: 1–31.
35. Kalverboer A. Follow-up of biological high risk groups. In: Rutter M, ed. *Studies of Psychosocial Risk*. New York: Cambridge University Press, 1988: 114–37.
36. Anastasi A. *Psychological Testing*. New York: MacMillan, 1988.
37. Cronbach L. *The Essentials of Psychological Testing*, 4th edn. New York: Harper & Row, 1984.
38. Bayley N. *Bayley Scales of Infant Development*. New York: Psychological Corporation, 1969.
39. Bayley N. *The Bayley Scales of Infant Development*. San Antonio, TX: Psychological Corporation, 1993.
40. Kopp CB, McCall RB. Stability and instability in mental performance among normal, at-risk, and handicapped infants and children. In: Baltes PB, Brim OG, eds. *Lifespan Development and Behavior*, Vol. 4. New York: Academic Press, 1980: 33–61.
41. O'Connor MJ, Brill NJ, Sigman M. Alcohol use in primiparous women older than 30 years of age: relation to infant development. *Pediatrics* 1986; **78**: 444–50.
42. Jacobson SW, Fein GG, Jacobson JL, Schwartz PM, Dowler JK. The effect of PCB exposure on visual recognition memory. *Child Dev* 1985; **56**: 853–60.
43. Bellinger D, Leviton A, Waternaux C, Needleman H, Rabinowitz M. Longitudinal analyses of prenatal and postnatal lead exposure and early cognitive development. *N Engl J Med* 1987; **316**: 1037–43.
44. Ross G. Use of the Bayley Scales to characterize abilities of preterm infants. *Child Dev* 1985; **56**: 835–42.
45. Jacobson J, Jacobson S. Assessment of teratogenic effects on cognitive and behavioral development in infancy and childhood. In: Kilbey M, Asghar K, eds. *Methodological Issues in Controlled Studies on Effects of Prenatal Exposure to Drug Abuse*. Rockville, MD: NIDA, 1991: 248–61.
46. Bornstein M, Sigman M. Continuity in mental development from infancy. *Child Dev* 1986; **57**: 251–74.
47. O'Connor MJ, Cohen SL, Parmelee, AH. Infant auditory discrimination in preterm and full-term infants as a predictor of 5-year intelligence. *Dev Psychol* 1984; **20**: 159–65.
48. McCall R, Carriger M. A meta-analysis of infant habituation and recognition memory performance as predictors of later IQ. *Child Dev* 1993; **64**: 57–9.

49. Fagan JF, Shepherd P. *The Fagan Test of Infant Intelligence Training Manual.* Cleveland, OH: Infantest Corp, 1989.
50. Kenny T, Holden EW, Santilli L. The meaning of measures: pitfalls in behavioral and developmental research. *J Dev Behav Pediatr* 1991; **12**: 355–60.
51. Fried PA, Watkinson B. 36 and 48 month neurobehavioral follow-up of children prenatally exposed to marijuana, cigarettes, alcohol. *J Dev Behav Pediatr* 1990; **11**: 49–58.
52. Fried PA, O'Connell CM, Watkinson B. 60- and 71-month follow-up of children prenatally exposed to marijuana, cigarettes, and alcohol: cognitive and language assessment. *J Dev Behav Pediatr* 1992; **13**: 383–91.
53. Achenbach T, Phares V, Howell C. Seven year outcome of the Vermont Intervention Program for low birthweight infants. *Child Dev* 1990; **61**: 1672–81.
54. Singer LT, Yamashita T, Davillier M et al. Feeding interactions in infants with very low birthweight and BPD. *J Dev Behav Pediatr* 1996; **17**: 69–76.
55. Singer LT, Bruening P, Davillier M, Hawkins S, Yamashita T. Social support, psychological distress and parenting strains in mothers of very low birthweight infants. *Family Relations* 1996; **45**: 343–50.
56. Morley R. Data collection and management for research studies. *Arch Dis Child* 1995; **73**: 364–7.
57. Jacobson JL, Jacobson SW. Strategies for detecting the effects of prenatal drug exposure: lessons from research on alcohol. In: Lewis M, Bendersky M, eds. *Mothers, Babies, and Cocaine: the Role of Toxins in Development.* Hillsdale, NJ: Lawrence Erlbaum Associates, 1995: 111–127.
58. Neuspiel D. The problem of confounding in research on prenatal cocaine effects on behavior and development. In: Lewis M, Bendersky M, eds. *Mothers, Babies and Cocaine: the Role of Toxins in Development.* Hillsdale, NJ: Lawrence Erlbaum Associates, 1995: 95–110.
59. Hollingshead AB. *Two Factor Index of Social Position.* New Haven, CT: Department of Sociology, 1957 (unpublished paper).
60. Hollingshead AB. *Four Factor Index of Social Status.* New Haven, CT: Yale University, Department of Sociology, 1975 (unpublished paper).
61. Dunn LM, Dunn LM. *Peabody Picture Vocabulary Test – Revised.* Circle Pines, MN: American Guidance Service, 1981.
62. Singer LT, Farkas K, Arendt R, Minnes S, Yamashita T, Kliegman R. Increased psychological distress in post partum, cocaine using mothers. *J Substance Abuse* 1995; **7**: 165–74.
63. Bradley R, Caldwell B. Home environment, cognitive competence and IQ among males and females. *Child Dev* 1980; **51**: 1140–8.
64. Siegal LA. Home environmental influences on cognitive development in preterm and full term children during the first 5 years. In: Gottfried AW, ed. *Home Environment and Early Cognitive Development: Longitudinal Research.* Orlando, FL: Academic Press, 1990 (197–233).
65. Caldwell B, Bradley R. *Home Observation for Measurement of the Environment.* Little Rock, AK: University of Arkansas, 1984.
66. Bradley R, Caldwell B, Rock S. Home environment and school performance: a ten year follow-up and examination of three models of environmental action. *Child Dev* 1988; **58**: 852–67.

67. Bradley R, Caldwell B. The HOME Inventory: a validation of the preschool scale for black children. *Child Dev* 1981; **52**: 708–10.
68. Field T. Infants of depressed mothers. *Infant Behav Dev* 1995; **18**: 1–3.
69. Cohn JF, Tronick EZ. Three month old infants' reaction to simulated maternal depression. *Child Dev* 1983; **54**: 185–93.
70. Singer L, Yamashita T, Dorsey P, Baley J. Effects of BPD and very low birthweight birth on maternal infant interaction, depression, and coping mechanisms. *Pediatr Res* 1995; **38**: 273A.
71. Lyons-Ruth K, Connell DB, Grunebaum HU. Infants at social risk: maternal depression and family support services as mediators of infant development and security of attachment. *Child Dev* 1990; **61**: 85–98.
72. Singer L, Arendt R, Farkas K, Minnes S, Huang J, Yamashita T. The relationship of prenatal cocaine exposure and maternal postpartum distress to child developmental outcome. *Dev Psychopathol* 1997 (in press).
73. Cogill S, Caplan H, Alexandra H et al. Impact of postnatal depression on cognitive development in young children. *Br Med J* 1986; **292**: 1165–7.
74. Beck AT, Ward CH, Mendelson M, Mock F, Erlbaugh J. An inventory for measuring depression. *Arch Gen Psychiatry* 1961; **4**: 387–402.
75. Derogatis L. *The Brief Symptom Inventory: Administration, Scoring, and Procedures Manual* 2nd edn., Baltimore, MD: Clinical Psychometric Research, Inc., 1992.
76. Radloff LS. The CES-D Scale: a self-report depression scale for research in the general population. *Appl Psychol Measurement* 1977; **1**: 385–401.
77. Zuckerman B, Frank DA, Hingson R et al. Effects of maternal marijuana and cocaine use on fetal growth. *N Eng J Med* 1989; **320**: 762–8.
78. Singer LT, Arendt R, Song L, Warshawsky E, Kliegman R. Direct and indirect interactions of cocaine with child birth outcomes. *Arch Pediatr Adolescent Med* 1994; **148**: 959–64.
79. Singer LT, Yamashita T, Hawkins S, Cairns D, Baley J, Kliegman R. Increased incidence of intraventricular hemorrhage and developmental delay in cocaine-exposed very low birthweight infants. *J Pediatr* 1994; **124**: 765–71.
80. US Department of Health and Human Services. *Preliminary Estimates from the 1995 National Household Survey on Drug Abuse (Report No. 18)*. Rockville, MD: Mental Health Services Administration, 1995.
81. Church MW, Holmes D, Overbeck G, Tieak J, Zajac C. Interaction effects of prenatal alcohol and cocaine exposure on postnatal mortality, development, and behavior in the Long–Evans rat. *Neurotoxicol Teratol* 1991; **13**: 377–86.
82. Coles CD, Platzman KA, Smith I, James ME, Falek A. Effects of cocaine and alcohol use in pregnancy on neonatal growth and neurobehavioral status. *Neurotoxicol Teratol* 1992; **14**: 23–33.
83. Landry S, Chapieski L, Fletcher J, Denson S. Three year outcomes for low birth weight infants: differential effects of early medical complications. *J Pediatr Psychol* 1988; **13**: 317–27.
84. Singer LT, Martin RJ, Hawkins S, Benson-Szekely L, Yamashita T, Carlo W. Oxygen desaturation complicates feeding of bronchopulmonary dysplasia infants in the home environment. *Pediatrics* 1992; **90**: 380–4.
85. DiPietro J, Allen M. Estimate of gestational age: implications for developmental research. *Child Dev* 1991; **61**: 1184–99.
86. Barron RM, Kenny DA. The moderator–mediator variable distinction in social

psychological research: conceptual, strategic, and statistical considerations. *J Personality Social Psychol* 1986; **51**: 1173–82.
87. Als H, Lawhon G, Duffy FH, McAulty GB, Gibes-Grossman R, Blickman JG. Individualized developmental care for the very low birthweight preterm infant. *JAMA* 1994; **272**: 853–8.
88. Sostek AM. Prematurity, as well as intraventricular hemorrhage, influences developmental outcome at five years. In: Friedman SL, Sigman M, eds. *The Psychological Development of Low Birthweight Children*. Norwood, NJ: Ablex, 1993: 259–74.
89. Ban T, Guy W, Wilson W. Organizing and conducting clinical trials. *Neuropsychology* 1983; **10**: 137–40.
90. Bielefelt R, Yamashita T, Kerekes E, Ercanli E, Singer LT. A research database for improved data management and analysis in longitudinal studies. *MD Comput* 1995; **12**: 200–6.
91. Stern M, Hildebrandt KA. Prematurity stereotyping: effects on mother–infant interaction. *Child Dev* 1986; **57**: 308–15.
92. Berk RA. Generalizability of behavioral observations: a clarification of interobserver agreement and interobserver reliability. *Am J Mental Deficiency* 1979; **83**: 460–3.
93. Black M, Holden EW. Longitudinal intervention research in children's health and development. *J Clin Psychol* 1995; **24**: 163–72.
94. Flick SN. Managing attrition in clinical research. *Clin Psychol Rev* 1988; **8**: 499–515.
95. Aylward G, Hatcher R, Stripp B, Gustofson N, Levitt L. Who goes and who stays: subject loss in a multicenter, longitudinal follow-up study. *J Dev Behav Pediatr* 1985; **6**: 3–8.

## Commentary

**Lucas, Heird, and Innis**: Innis and Lucas expressed concerns about reduced growth in n-3 LCPUFA-supplemented infants, whereas Heird remained less concerned. These authors' views are in fact quite similar and have the following in common.

Firstly, there is consensus that the cumulative evidence that n-3 PUFAs may reduce growth rate (even beyond the period that they are administered) is sufficiently compelling to warrant further investigation. Secondly, there is consensus that, while it is difficult to prove that lower growth rates (of the magnitude observed in DHA supplementation) have adverse consequences, early growth, and notably brain growth, have been identified as potential markers of important outcomes. For instance, reduced head growth in preterm infants at 8 months has been related to impaired later neurodevelopmental outcome. Finally, given that growth is such a fundamental process in early life and that reduced brain growth during potentially vulnerable stages in brain development is a reasonable concern, it would seem prudent to regard reduced growth rate as a safety issue in future LCPUFA trials.

**Innis**: What are transactional processes? Why do transactional processes make the brain susceptible to LCPUFA deficiency?

What is the evidence that correcting for gestational length in term infants (37–42 weeks) has a significant influence on developmental test results?

AUTHOR'S REPLY: Strictly speaking, correlation cannot be equated with causation. Correlations obtained under controlled conditions of an experimental trial in which subjects are randomized to a condition prior to the measurement of an outcome provide better evidence of a causal effect (1).

1. Kerlinger F. *Foundations of Behavioral Research*, 2nd edn. 1973, New York: Holt, Riverhart & Winston, 1973: 314–16.

INNIS: Full-term can range from 37 to 42 weeks, infants varying by 5–6 weeks. This issue has been raised previously in the context that variability in gestational age among and within treatment versus nontreatment groups might explain the inability to find significant group differences. You state that full-term varies from 37 to 42 weeks. Thus there is an apparent difference of 5–6 weeks. Do you mean full-term based on last menstrual period is 37–42 weeks? This is not the same as the actual length of gestation. Since ovulation can be 7–45 days after the last menstrual period in normal women, this amounts to a 5 week variance. This section should be more specifically written as there is a potential for misunderstanding and misquotation. Please state specifically, for the infant born after spontaneous onset of labor 37 weeks after the last menstrual period, the true variability in length of gestation (i.e. take away the variable of individual differences in length of cycle).

What is the evidence that correcting for gestational length in term infants (37–42 weeks) has a significant influence on developmental test results?

AUTHOR'S REPLY: My understanding is that full term is 37–42 weeks based on last menstrual period dates. I am unaware of evidence related to correction of gestational age on the outcomes of infant assessments for full term infants. For the Bayley Scales, there is usually not a correction for infants of ≥ 37 weeks gestation. Thus, their scores are computed from chronologic age. The Fagan Test computes test ages for full term infants based on actual gestational age, so that there is correction within this 5 week period. I know of no studies, however, which have reported whether this correction affects outcome scores.


INNIS: Since BDP occurs after initiation of feeding (at least in these types of studies) you cannot randomize for BPD, i.e. option is not possible.

What do you do about infants who are withdrawn for reasons of sepsis, etc., since these infants cannot proceed to succumb to BPD. How can you have an outcome variable which not all infants are equally able to attain, i.e. removal from study.

What happens if you have a treatment effect which selectively removes infants from the study. For example, in the LCPUFA studies, more infants are lost due to necrotizing enterocolitis (NEC) and sepsis from the LCPUFA group, with those who succumb to adverse effects being the less mature more vulnerable infants. The group left, i.e. those who proceed to measurement, have now been selected to be the most robust of the preterm infant population. The nonintervention group has not been so stratified as to remove the weaker individuals. Some discussion of this would be useful. In addition, what is your opinion on the validity of proceeding to developmental measures on infants in groups of infants where there has been unbalanced dropout among the groups?


AUTHOR'S REPLY: Treatment effects which selectively remove subjects from a study are a commonly recognized occurrence in clinical trials. The intention to treat principle necessitates that the preferable analysis of study results is based on including *all* participants and their follow-up results in the intervention groups as initially assigned, since analysis of only subjects who have completed a trial rather than the intervention to which they were originally assigned may produce invalid results. Complementing this principle is also the need for the investigator to provide data about reasons for dropout from a trial, since adverse reactions are one important cause of dropout which should be considered in evaluation of intervention risk/ benefits (1). With regard to the issue that a potential adverse effect of LCPUFAs to affect NEC/sepsis outcomes, may obscure the assessment of the effects of the intervention on BPD, no conclusions can be drawn regarding BPD outcomes, and trials assessing the outcome of NEC/sepsis need to be designed as the next step in the investigative process.

In clinical trials of behavioral/psychotropic interventions, it is not uncommon to find interventions which may have adverse effects for some individuals, but be quite beneficial for other specific groups of patients.

Proceeding to developmental measures in the face of unbalanced dropout should be considered using both the intent to treat analysis noted, which would include all subjects who had been randomized originally as the basis from which to determine a positive/negative outcome. In some cases, however, an efficacy analysis of outcomes of infants who were able to be maintained on formula can also be useful, but both types of analyses should be reported. The intent to treat principle would have dictated retention of infants who have developed NEC/sepsis for the developmental outcomes, since they are not true 'dropouts'.

1. The Standards of Reporting Trials Group. A proposal for structured reporting of randomized controlled trials. *JAMA* 1994; **272**: 1926–31.

INNIS: Age-corrected scores may significantly obscure real developmental delays. This is not clear. Why would age-corrected scores obscure delays? Should or should not test results for premature infants be corrected?

AUTHOR'S REPLY: Age-corrected scores should be used. While some developmental functions may be more biologically determined, it is likely that the additional postnatal experience of the very preterm infant also affects some functions; however, I do not believe we have much data as yet with which to model such differential endpoints.

**Wainwright and Ward**: The thoughtful discussion provided by this paper on the multitude of variables potentially associated with intervention and outcome in longitudinal clinical trials of development during childhood should prove invaluable to investigators working in this area, as well as to those wishing to have a basis from which to evaluate published studies.

The author notes that the aim of such trials is to obtain accurate and valid assessments of the intervention and outcome which are coherent with the theoretical intervention model. A weakness of much of the research endeavor on LCPUFAs is the lack of such a theoretical approach based on an understanding of putative mechanistic relationships between dietary lipids, brain and behavior (as discussed in the chapter by Shaw and McEachern).

The paragraph on the distinction among confounding, moderating, and mediating variables and how these should be handled statistically is extremely important, particularly that related to avoiding treating mediators as confounders.

Why does the scaling system of the Bayley allow it to reflect population differences whereas the Fagan Test does not?

The statement that 'interventions which significantly affect specific domains of cognitive functioning require multidimensional assessments at school age to detect effects' seems to me to be a very important one. This is related to an earlier point made in this paper, namely, that apical tests such as the Bayley's, by averaging across items, may mask effects on specific response categories. It seems that by focusing on straight cognitive outcomes such as IQ we may overlook effects on other aspects of behavior that may be of considerable importance to overall quality of life, mood disorders, or affiliative behavior, for example.

AUTHOR'S REPLY: I agree with you. The study of learning disabilities, as well as populations of head-injured adults, illustrate that an individual can have significantly impairing deficits in specific areas of functioning, such as memory, attention, or reading skills, in the context of average IQ.

**Lucas:** Singer provides a most welcome review of methodological considerations in conducting randomized intervention studies of early nutrition in infants. This is an area of prime concern to our MRC Childhood Nutrition Research Centre in the UK. Randomized trials of nutrition have been conducted sporadically for decades. However, in the early 1980s we elected to pioneer the concept of using the randomized nutritional intervention trial with long-term follow-up in a manner similar to that used for pharmaceutical interventions – that is with targeted efficacy and safety outcomes, adequate power and sample size, and formal procedures in line with good clinical practice. Given the potential importance of early nutrition (based on animal data) as an 'intervention' that could potentially influence or program (1) long-term health and neurodevelopmental outcomes, we considered that a series of such randomized 'therapeutic' intervention trials, with follow-up into childhood and beyond, could give new insight into the consequences of infant nutrition in humans. Our first trial, involving 926 preterm infants, was commenced in 1982 (the cohort is still being actively followed up), and since then 15 interventional trials on term and preterm infants with long-term follow-up, involving thousands of infants, are currently in progress. Four of these trials are large studies involving the use of LCPUFAs.

In the light of this cumulative experience, I shall add a few points, of relevance to LCPUFA trials, to the considerations laid out by Singer. A modern trial requires a committed multidisciplinary team. In our UK trials all staff are employed specifically for the project. We virtually never rely on busy clinicians or service nurses for essential trial work, since in our experience clinical staff invariably have patches of time in which they must, correctly, put their clinical duties before their research ones, often with serious and sometimes disastrous consequences for the trial. In a major trial we would allocate specific staff for the following key elements: (i) overall trial leader; (ii) trial scientific co-ordinator, responsible for ethical submissions, data collection form design, day-to-day trial management, and monitoring for good clinical practice; (iii) data input; (iv) data management (management of databases, data cleaning, etc.); (v) statistician (trial design, randomization, analyses, etc.); (vi) scientific staff for (a) clinical assessments, (b) biological assessments (e.g. neurodevelopment), and (c) laboratory assessments; (vii) field staff, usually research nurses; (viii) training staff for field workers and scientists; and (ix) administrative staff, particularly important for follow-up tracing. Because of the economics of scale, such multidisciplinary core staffing can be best maintained in a centre conducting and co-ordinating a rolling program of multisite trials. It is often difficult for centers unspecialized in clinical trials to stage manage and staff large multisite trials – and this factor alone probably accounts most for scientists' choice to conduct inappropriately undersized trials.

As Singer points out, sample size is indeed a major issue in randomized trials, particularly in preterm infants where the common multiple confounders (identified in Singer's review) can be unequally distributed, by chance, in small sized groups. In our first preterm infant trials comparing donor breast milk with a new preterm formula (1), the first planned interim analysis of short-term findings revealed a near-significant increase in the death rate in the preterm formula group. A couple more deaths in this group might have stopped the trial and resulted in withdrawal of the formula. By the time our large target sample size had been recruited, however, death rate was almost identical between groups. Looking retrospectively, out of interest, at our databases in chronological order of subject recruitment, we can identify numerous outcomes that would have reached spurious significance had we curtailed our studies with small sizes. As discussed in my own chapter in this volume, most reported LCPUFA trials have insufficient sample sizes to detect realistic neurodevelopmental differences between groups with adequate power; but it should be recognized that testing for safety and ensuring that high risk infant groups are likely to be balanced for significant confounders may require even larger sample sizes still than those needed for efficacy testing.

Singer raises the point about the importance of adequate follow-up and the possibility of selective dropouts. This is another major consideration. To cite an example from one of our own studies, in 1991 we reported the neurodevelopmental outcome at 18 months post-term of 424 preterm infants randomly assigned to term or preterm formula in the neonatal period (2). Follow-up rate was only 89% of survivors, largely relating to a temporary follow-up staffing problem in one center. We found a major difference in Bayley Scale scores between groups, as hypothesized, but no significant differences in neuromotor impairment. Currently we have analysed for publication our data on the same cohort at 7.5–8 years. My colleague, Morley, resolved to achieve a near-complete follow-up and we examined 96% of survivors (98% of survivors still remaining in Britain – a few emigrated to the USA). With this relatively small increase in sample size (from 89% to 96%) we traced a number of children with neuromotor impairment who had selectively not turned up at 18 months during the temporary staffing problem period (with full staffing we could have encouraged their parents to attend, as we did at 7.5–8 years). The consequence is that we have now found a significant, and possibly causal difference in neuromotor impairment between groups at 7.5–8 years that could have been detected at 18 months. This example illustrates the importance of near-complete ascertainment in the neurodevelopmental follow-up of high risk infants. Others have also noted the dangers of incomplete follow-up (3). Follow-up rates as low as 60–80% (or even 90%), as often reported in developmental studies of high risk infants, could yield highly misleading findings. The implication of this is that there are many centers, particularly in some parts of the USA, that may be unsuitable for outcome studies of LCPUFA interventions when there is high population mobility and poor compliance. The lack of a national health service that can be used to log and trace patients wherever they are is a serious problem in the USA, and invokes the need for new, inventive techniques to maintain follow-up by effective tracing.

1. Lucas A. Role of nutritional programming in determining adult morbidity. *Arch Dis Child* 1994; **71**: 288–90.

2. Lucas A, Morley R, Cole TJ et al. Early diet in preterm babies and developmental status at 18 months. Lancet 1990; **335**: 1477–81.
3. Wariyar U. Morbidity and preterm delivery: importance of 100% folow-up. Lancet 1989; **i**: 387–8.

**Heird**: You state that, 'As much as possible, characteristics of subjects who refuse to enter a clinical trial should be documented in order to aid interpretation of results.' I agree that this is desirable, but doesn't recording and subsequently using such data from a subject who does not consent to participate raise some ethical issues? Presumably, every subject who is asked to participate in a study meets at least minimal criteria for inclusion, or consent would not have been sought. Is it sufficient to document simply that the subject met enrollment criteria? Or, should information concerning socio-economic status, etc., also be collected? Finally, are you aware of data indicating the extent of the problem of failing to account for differences in the characteristics of subjects who agree versus those who refuse to participate in a clinical trial?

AUTHOR'S REPLY: Generally, institutional human investigation review boards exempt anonymous, patient chart review data from consent requirements. These data can provide valuable information about socio-economic, ethnic, and medical factors that may be related to entrance into a clinical trial.

I do not know of any particular study in which conclusions might have differed if such information were available. As an example in our studies of infant cocaine exposure, however, we had some concerns that mothers who agreed to enter the study might differ in their severity of addiction from those who consented. If there were a high rate of refusal, and if refusers tended to be the more severely addicted women, the study population might not reflect the real extent of the effects of fetal cocaine exposure, perhaps underestimating effects due to a preponderance of 'light' drug users. Medical chart review and birth information regarding prenatal care, birth weight, infection, and maternal drug history can shed some light on whether there is a systematic bias in the samples, as an aid to data interpretation. Any information available about characteristics of refusers, exclusions, or dropouts can aid in judgements regarding how confident one can feel about results.

HEIRD: I am impressed by the overlap in outcomes between or among groups that have been reported in this area of research. This appears to be true for biochemical as well as functional outcomes. This being the case, should regression analysis not be a reasonably powerful tool? Or, does the demonstration that there is a strong correlation between a biochemical finding and a functional finding, as hypothesized pre hoc, merely support, but not necessarily prove the hypothesis?

AUTHOR'S REPLY: Correlation cannot be equated with causation. Correlations obtained under controlled conditions of an experimental trial in which subjects are randomized to a condition prior to the measurement of an outcome provide better evidence of a causal effect (1).

1. Kerlinger F. *Foundations of Behavioral Research*, 2nd edn. New York: Holt, Riverhart & Winston, 1973; 314–16.

HEIRD: Your chapter and the one by Wainwright should be required reading for all who are involved in or plan to become involved in clinical studies, particularly clinical studies concerning nutritional issues in low birth weight infants. The discussions of confounding variables, mediating variables, etc., are particularly relevant and important. On the other hand, these variables are inherent in any low birth weight infant population, and a broader discussion of how to deal with them in realistic clinical studies is desirable. For example, should study populations be highly selective or inclusive of all members of the population.

With respect to formulas for feeding low birth weight infants, it is unlikely that multiple formulas will ever be available for specific subgroups of this total population. Further, even if multiple formulas were available, most medical practitioners would lack the expertise to decide which infants should receive which formula. Thus, as is now the case, availability of multiple formulas for this diverse population is unlikely. Considering this, further discussion of the merits and/or pitfalls of studying well-defined subgroups of the total population versus a group that is more representative of the total population would be desirable. For example, while the variance of outcome variables is likely to be much less if the study population includes only infants with a narrow range of birth weights, gestational ages, etc., and with few, if any, confounding and/or mediating variables, the results obtained in such a study will only be applicable to the type of infant studied. Despite a greater variance of outcome variables and, therefore, the necessity for studying more infants, the results obtained in a population that includes representatives of all infants likely to receive the intervention under study are more likely to be applicable to the entire population. Moreover, this approach should permit detection of confounding and/or mediating factors which contribute to the variance in the results obtained. A broader discussion of the relative merits versus the disadvantages of these two approaches would be helpful.

AUTHOR'S REPLY: I would agree that larger samples, more representative of heterogeneous preterm populations, will provide more generalizable data, and contribute to our understanding of potentially mediating/moderating variables. Understanding which are the relevant variables to consider, however, often stems from data from smaller, better controlled studies.

**Carlson**: You might mention the large number of potential confounders in preterm populations for early developmental outcomes. It is worth mentiooning that many of these factors cluster in the same infants, i.e. are highly interrelated. (You do mention this when speaking specifically about BPD). Many of these confounders prevent enteral feeding according to protocol so that infants required by protocol to receive sustained enteral feeding are effectively removed (along with the confounding variables). Of course, small studies that do not permit retention of all infants cannot then be generalized to all preterm infants as you point out in your chapter. For this reason, there is a need for larger randomized trials of n-3 LCPUFA supplementation in preterm infants. We will likely have quite a bit of discussion on the issue of trial size, who should be excluded, and so on, in future LCPUFA feeding trials. In my view, the first trials in preterm infants necessarily had to exclude as many confounders as possible, while not being so restrictive as to prevent completion of the studies in a reasonable length of time.

Intuitively, the inclusion of infants at risk for developmental delay in even very large 'take-all-comers' studies opens the door for false issues and problems. For example, infants have poor development as the expected outcome of their disease. However, the effects of disease on development are not linearly or normally distributed. If more risk for developmental delay is placed in the experimental group, they could appear to do worse than the controls even though the intervention has no or a beneficial effect. Alternatively, if the intervention improved performance this wouldn't necessarily appear to be so if the increase in performance occurred from a lower baseline, in effect canceling out the negative effects of the disease.

These outcomes could lead to conclusions of (i) harm or (ii) no effect and delays in implementing a change that might benefit infant health.

Another possible scenario of a 'take-all-comers' trial is that the population as a whole may not benefit because the risk factors overwhelm or preclude any benefit on development from an intervention such as n-3 LCPUFAs. Even if there was general agreement that healthy infants benefited from n-3 LCPUFAs, in such a circumstance some might argue from a public health standpoint that it is not worth the cost to include n-3 LCPUFAs in formula, as only a subset of the population benefits.

In the broadest sense, the comments above could just as easily apply to the healthiest preterm infants, because they (compared with healthy term infants) are at risk for many adverse consequences, including developmental delay. In the final analysis, no single study can be accepted as final proof or used as the basis to implement changes in policy (to state what is obvious to all).

My own preference would be to study gradually some of these potential confounders of development in a very controlled way. We tried to do this in our second trial for infants with BPD who, except for their continued oxygen requirement and lung changes at 28 days, were still able to be fed enterally according to nursery protocol. Compared to most BPD infants such as you describe, our BPD infants were undoubtedly at far lower risk for developmental delay (e.g. their 12-month Bayley MDI scores were identical with our center investigator norm. However, for this very reason, the study allowed us to obtain a better picture of the effects of BPD unconfounded by many of its co-linear variables. Even though the number of BPD infants studied was small, they had lower early visual acuity development and poorer information processing or ability to disengage at 12 months (whatever longer look duration means). I think this type of design has the potential to narrow the focus of possible variables that produce an insult to development, and that it would be valuable to study larger numbers of relatively unaffected infants with BPD in this way.

In response to the issue of design, BPD was conceptualized as a potential confounder of virtually all the study outcomes (growth, grating acuity, and the developmental outcomes of visual attention and the Bayley MDI and PDI). The expected incidence in the infants enrolled in the birth weight range of the study (725–1275 g) in 1989 was 40%. Although we did not anticipate that the disease would not occur randomly in the diet groups, we planned to analyze infants by diet–BPD group. Other medical complications were in fact assessed as outcomes. Only sepsis and necrotizing enterocolitis (NEC) occurred with any frequency, as indicated (1). These were not significantly influenced by diet, but they also did not appear to randomize equally by diet. Lucas (this volume) has pointed out that, while the numbers were too small to detect differences related to diet, larger studies are needed to rule out problems with n-3 LCPUFA supplementation.

Other medical complications were not mentioned by name (1), because they occurred with exceedingly low frequency and there was not even any suggestion that they did not randomize by diet. There was never any intent to analyze by NEC–diet or sepsis–diet because these diseases interfere with enteral feeding, leading to loss from the study when enteral feeding is not tolerated for more than 7 days.

Only a preliminary report of the 12-month Bayley MDI data from our second trial has been written (2), referred to by McCall and Mash (this volume). When infants are analyzed by diet–BPD, these are the results: control–no BPD, 95.6 ($n = 14$); control–BPD, 97.1 ($n = 11$); experimental–no BPD, 114.5 ($n = 15$); experimental–BPD, 98.0 ($n = 7$); center investigator norm, 97.2 ($n = 253$, $< 1500$ g infants studied prior to our studies). Although the number of infants in each diet–BPD group is small, only infants without BPD appear to benefit from receiving n-3 LCPUFA. Would you consider BPD to be a moderating variable in this example?

1. Carlson SE, Werkman SH, Tolley EA. Effect of long-chain n-3 fatty acid supplementation on visual acuity and growth of preterm infants with and without bronchopulmonary dysplasia. *Am J Clin Nutr* 1996; **63**: 687–97.
2. Carlson SE, Werkman SH, Peeples JM, Wilson WM III. Growth and development of premature infants in relation to $\omega$-3 and $\omega$-6 fatty acid status. *World Rev Nutr Diet* 1994; **75**: 63–9.

AUTHOR'S REPLY: Moderator effects are tested by testing the main effects of diet and BPD, and then testing the interaction term BPD $\times$ diet, which must be significant beyond the main effects (1).

1. Barron R, Kenny D. The moderator–mediator variable distinction in social psychological research. *J Personality Social Psychol* 1986; **51**: 1173–82.

**Mayer and Dobson**: Singer's discussion of various confounding factors in neurodevelopmental studies of infants includes examiner effects in measuring infant behavior. In our chapter, we discuss several studies showing systematic differences among testers' estimates of infant acuities tested both with the psychophysically objective, forced-choice preferential looking (FPL) method, and with the more subjective, acuity card procedure

(ACP). Two points should be made: (i) inter-rater or between-tester reliabilities should be obtained for the examiners who assess data in each study and from tests of subjects from the same population; and (ii) correlation coefficients will not indicate if there are systematic differences among testers, and therefore some scaled measure of between-tester differences should also be reported (e.g. mean difference between acuities or novelty preference scores). We applaud Singer's recommendation for periodic, regular assessments of tester reliability to guard against 'drift'.

Sample attrition was noted by Singer to be a serious problem in longitudinal studies because attrition is usually not random. Equally serious is the problem of studies in which only part of the sample provides data on the dependent variable. For example, in a nutrition study comparing visual evoked potential (VEP) acuity of infants fed formula versus breast milk, 17% in the 16 week age group and 21% in the 30 week age group did not contribute data to the average acuities (1). Although the authors of this study claim that there was no association between dietary grouping and subject loss due to inability to obtain a VEP acuity, this is not borne out by the number of subjects in each dietary group that contributed to VEP data. Other nutrition studies can be faulted because the number of subjects for each dietary group and age point are not given (e.g. 2,3) and, therefore, this potential confounding factor cannot be evaluated.

1. Makrides M, Neumann M, Simmer K, Pater J, Gibson R. Are long-chain polyunsaturated fatty acids essential nutrients in infancy? *Lancet* 1995; **345**: 1463–8.
2. Birch EE, Birch DG, Hoffman DR, Uauy R. Dietary essential fatty acid supply and visual acuity development. *Invest Ophthalmol Vis Sci* 1992; **33**: 3243–53.
3. Birch E, Birch D, Hoffman D *et al*. Breast-feeding and optimal visual development. *J Pediatr Ophthalmol Strabismus* 1993; **30**: 33–8.

AUTHOR'S REPLY: Attrition is often not random, and may even be the result of adverse effects of the intervention (see also my response to Innis regarding attrition).

MAYER AND DOBSON: It seems incumbent upon investigators to delineate the relevant subject variables (confounders as described by Singer) and demographic characteristics of the infants' parents for each dietary group, including those who were lost to follow-up, and from whom data were not obtainable, as well as for the subjects who did not participate. This seems particularly important in studies in which feeding conditions include breast-feeding, given the apparent overriding mediating influence of maternal factors (education, intelligence, home environment), as discussed by others in this volume as well.

Some of the contributors to this symposium conclude their discussions of discrepancies among findings of various nutrition studies by attributing the discrepant findings to 'population differences' between study samples. However, although these differences are discussed among investigators in private, no one in this symposium has provided data on population differences in, for example, fatty acid ratios, metabolism, breast-feeding, and other relevant factors. Although the study designs are not comparable, it is interesting to compare racial distributions of subjects in nutrition studies conducted by two investigators who report sample demographics. In Carlson *et al.*'s two studies of preterm infants, 83% and 88% of the infants were African–American (1,2), while in Innis *et al.*'s study of the effect of duration of breast-feeding on visual behavior (3), 57% were Canadian or European, 39% were Chinese, East Indian or other Asian, and 4% were classified 'other'. These populations differed on socio-economic status characteristics as well. Other nutrition studies have not reported the racial or demographic characteristics of their samples. It appears that most writers agree that socio-economic status is an important factor in this area of research. Is there any evidence as to whether the infant's race/ethnicity is relevant to the results of nutrition studies?

1. Carlson SE, Werkman SH, Tolley EA. Effect of long-chain n-3 fatty acid supplementation on visual acuity and growth of preterm infants with and without bronchopulmonary dysplasia. *Am J Clin Nutr* 1996; **63**: 687–97.
2. Carlson SE, Werkman SH, Rhodes PG, Tolley EA. Visual-acuity development in healthy preterm infants: effect of marine-oil supplementation. *Am J Clin Nutr* 1993; **58**: 35–42.
3. Innis SM, Nelson CM, Lwanga D, Rioux FM, Waslen P. Feeding formula without arachidonic acid and docosahexaenoic acid has no effect on preferential looking acuity or recognition memory in health full-term infants at 9 months of age. *Am J Clin Nutr* 1996; **64**: 40–6.

AUTHOR'S REPLY: The issue of socio-economic status/ethnic differences as potential moderators of effects in nutrition studies has not been assessed, as has been noted by Colombo in this volume. Such influences may occur in light of nutritional deficiencies often found in lower socio-economic status groups, or because of well known socio-economic status/ethnic influences on some of the outcomes now being assessed in the LCPUFA studies such as language and intellectual functioning.

**Neuringer and Reisbick**: It seems odd to consider the Bayley Scales the gold standard for excellent psychometric properties when they have such poor predictive validity during infancy. However, the distinction made here between individual prediction and detection of group differences is a very important one which merits more consideration at this workshop. As noted in our general commentary, the detection of group differences is the much more relevant issue for the design of future studies of LCPUFAs.

A paragraph is needed here, parallel to the paragraph on fixation duration, which briefly describes concurrent and predictive validity, theoretical basis and interpretation (why they measure visual recognition memory (VRM)) of visual paired comparison tests. This would serve as an introduction to the discussion of the Fagan Test. It is confusing to follow the paragraph on look duration with a description of the advantages of the Fagan Test, but only mention on the next page that all the psychometric advantages apply only to novelty preference and not to look duration.

AUTHORS' REPLY: The Fagan Test of Infant Intelligence (FTII) (1) evolved from over 20 years of developmental research on infant perceptual and visual memory processes. These studies capitalized on the seminal methodology of Robert Fantz, which documented the tendency of the infant to fixate some stimuli more than others, i.e. visual preferences (2). By 1–2 months of age, infants tend to look preferentially at a novel, in comparison to a previously exposed, target, thus operationally defining VRM (3). Numerous studies by Fagan, Rose, and others have used the paired comparison procedure to explore developmental progressions in infant visual perceptual and memory capabilities and their predictive validity for later intellectual outcome (4–6).

In the paired comparison procedure, novelty problems are presented to the infant through pictures presented on a pivoting stimulus presentation stage. Usually two stimulus targets are placed on the stage, and an observer views infant corneal reflection of the targets through a 0.64 cm peephole in the center of the screen, recording looking time on a computerized program. For each problem, the infant is exposed to the stimulus until he or she has looked at it for a standard period of time (familiarization phase) which can vary from 6 to 60 s, depending on the age of the infant and the nature of the stimulus. After familiarization, the previously seen picture is paired with a novel picture, and presented simultaneously for a standard length of time (test phase), with right–left positions switched midway to control for side preferences. A 'novelty' score is then computed, consisting of the amount of fixation during the test phase devoted to the novel picture divided by the total fixation time to both the novel and familiar, multiplied by 100 (percentage novelty score). The FTII consists of 10 novelty problems, yielding a mean novelty score (7) used to assess the infant between 6.5–12 months of age. Rose (8) has also developed an assessment using paired comparisons, using differing stimuli consisting of both faces and abstract targets to comprise test items, while the FTII uses only faces.

The short-term test–retest reliability of VRM scores has been noted to be low (4–6), typically between 0.30 and 0.45, and remains a persistent problem for construction of tests using VRM items (9). McCall and Carriger (5) in their meta-analysis of 31 samples of the predictive relationships between habituation and VRM assessments and later IQ in both risk and non-risk samples in the first year of life found that they predicted later IQ between 1 and 8 years of age with a weighted average normalized correlation of 0.36, and a raw median correlation of 0.45. They note also, however, that the largest correlations are derived from the smallest samples (some as small as $n = 13$), a difficulty which has recently been underscored in a study by Anderson (10), which found significant, but low, correlations of 0.18 and 0.21, in separate samples of 110 and 86 infants, between FTII scores and 5 year verbal IQ. Predictive validity for individual risk, i.e. sensitivity and specificity, also continue to be less than optimal for VRM assessments (8,9,11).

1. Fagan J, Shepherd PA. *The Fagan Test of Infant Intelligence*. Cleveland, OH: Infant's Corporation, 1990.
2. Fantz RL. A method for studying early visual development. *Percept Motor Skills* 1956; **6**: 13–15.
3. Fagan JF. Memory in the infant. *J Exp Child Psychol* 1970; **9**: 217–26.
4. Fagan JF, Singer LT. Infant recognition memory as a measure of intelligence. *Adv Infancy Res* 1983; **2**: 31–78.
5. McCall RB, Carriger MS. A meta-analysis of infant habituation and recognition memory performance as predictors of later IQ. *Child Dev* 1993; **64**: 57–9.
6. Bornstein MH, Sigman MD. Continuity in mental development from infancy. *Child Dev* 1986; **57**: 251–74.
7. Fagan JF, Detterman D. The Fagan Test of Infant Intelligence: a technical summary. *J Appl Dev Psychol* 1992; **13**: 173–93.
8. Rose SA, Feldman JF, Wallace IF. Individual differences in infant information processing: reliability, stability, and prediction. *Child Dev* 1988; **59**: 1177–97.
9. Benasich AA, Bejar II. The Fagan Test of Infant Intelligence: a critical review. *J Appl Dev Psychol* 1992; **13**: 153–71.
10. Anderson HW. The Fagan Test of Infant Intelligence: predictive validity in a random sample. *Psychol Rep* 1996; **78**: 1015–26.
11. Fagan JF, Singer LT, Montie JE, Shepherd PA. Selective screening device for the early detection of normal or delayed cognitive development in infants at risk for later mental retardation. *Pediatrics* 1986; **78**: 1021–6.

NEURINGER AND REISBICK: The fact that the Fagan Test is standardized only for novelty preference is an important point. However, if the novelty preference measure is insensitive to effects of LCPUFAs (as has been true to date for both human and monkey infants), then its psychometric properties become irrelevant, unless the only issues of interest are performance on this measure and its potential prediction of later IQ. If differences are found on derived and nonstandardized measures such as look duration (which again is true for both human and monkey infants), then it would seem appropriate to follow up those exploratory findings with other tests more specifically designed to assess look duration and for which the predictive validity of look duration has been evaluated.

No infant tests exist which are comparable to standard IQ tests in their psychometric properties or which meet all the desired criteria as assessment tools. What conclusion is to be made from this fact? That we should simply give up on any attempt to assess infant development, even if we have some preliminary data and some clear hypotheses? Are there any constructive alternatives to offer? Please see our general commentary, page 517.

AUTHOR'S REPLY: It is quite possible that look duration may be differentially sensitive to the effects of LCPUFAs while novelty preference may not, as has been noted by Colombo (this volume). The important point you have made is that such measures are only exploratory at this time. Other tests should be designed to assess the measure of look duration, and will need to be standardized, normed, and their psychometric properties explored, before drawing conclusions about the efficacy of an intervention using such measures. We should not give up on attempts to assess infant development, but tests must be constructed first.

NEURINGER AND REISBICK: The Carlson study is a useful example here. We interpret the study as treating BPD as a moderating factor. However, all the alternatives – BPD as a confounder, moderator, or differential outcome – could be true simultaneously. Often in studies such as this, one does not have data a priori to know which is the most valid interpretation. What then is the best strategy for data analysis? IVH provides another relevant example which could fit any of the three categories. One might reasonably hypothesize that n-3 LCPUFA supplementation might increase incidence or severity of IVH by increasing bleeding time, as this is a known effect of larger doses in adults and was demonstrated, although not considered clinically significant, in the original preterm trial of Uauy *et al.* (1),

1. Uauy R, Hoffman DR, Birth EE, Birch DG, Jameson DM, Tyson J. Safety and efficacy of ω-3 fatty acids in the nutrition of very low birth weight infants: soy oil and marine oil supplementation of formula. *J Pediatr* 1994; **124**: 612–20.

**McCall and Mash**: While the purpose of the chapter is to review clinical trials methodology and while such experimental approaches provide valuable information that more clearly identifies cause–effect relations than other approaches, it may be prudent to acknowledge that clinical trials methodology has potential limitations when it comes to generalizing results to applied situations. Such limitations may apply to the case of LCPUFAs and the use of supplemented formulas in the general population. For example, whenever individual choice is likely to be involved in who actually receives, can afford, will use, or otherwise selects even a public health treatment, such as the use of special supplemented formulas, one must be concerned about the external validity or generality of clinical trials research to the realities of applied use or even enthusiasm for use of a publicly available treatment. For example, suppose it was shown in clinical trials research that LCPUFAs had beneficial effects, but also suppose that such effects could be produced by enriched nutrition alone, as Lucas has found in some of his human milk versus standard/enriched formula studies. Further, suppose also, as some research suggests, that too high a level of LCPUFAs could be deleterious. Now, if a supplemented formula is widely available, even with appropriate cautions, individuals likely to select to use the formula are also ones who are likely to be more conscientious about the feeding of their infants, perhaps giving them nutritionally enriched diets that exceed the levels at which the LCPUFAs are safe. Even restricting the supplementation to prescription by a physician might not eliminate such a hazard. Only nonrandomized, naturalistic studies of who uses the supplemented formula and the outcomes for their infants under conditions that are identical to the public health distribution of supplemented formulas will be able to answer such questions. Randomized trial research is exceedingly valuable, but it has limitations, especially with respect to its generality to applied situations; nonrandomized studies can contribute valuable information that clinical trials research often cannot, because in the applied context treatments are typically not 'randomly assigned' or 'uniformly delivered'. Both strategies are often necessary in a total program of research on a public health intervention.

AUTHOR'S REPLY: You have raised an important point with regard to the contribution of nonrandomized studies to understanding the risks and benefits of an intervention. Like child development, general health practices likely reflect transactional processes in which both randomized clinical trials inform both the public and healthcare professionals, and their adaptation of interventions provides further refinement of the beneficial and harmful parameters of the intervention.

McCALL AND MASH: Most of the infant attention measures that have been found to predict later IQ, including shorter looking times, have been widely interpreted to reflect faster information processing. Although Colombo may wish to debate the point, I would argue that there is precious little direct evidence that duration of looking is actually very directly related to the speed of information processing, at least processing the information in stimuli that are routinely shown infants in the infant assessment context.

Is depression specifically related to failure to breast-feed successfully? You report that depression is related to the breast versus bottle difference, but is it related specifically to mothers who want to breast-feed but who are unable to do so successfully by their own standards? I ask because some of Lucas's results for mothers who chose to provide milk but whose infants needed substantial supplementation performed more poorly on later assessments, a result that might be partly explained by maternal depression at their infant's inability to be sustained primarily on human milk.

AUTHOR'S REPLY: The relationships among maternal depression during pregnancy, post-partum depression, breast feeding, and child developmental outcome are complex and, to date, poorly understood. Post-partum depression is quite common. Major depression is the most common diagnosis among psychiatric disorders beginning in the first 3 months post partum (1), with controlled studies indicating a 10–15% incidence of minor–major depression post partum (2). Other factors, such as lower socio-economic status, fewer social supports (3), lower infant birthweight (3,4), and delivery complications (4) may elevate these rates in particular populations.

Post-partum depression has been related to less optimal motor tone, more fussiness, less consolability, lower activity, elevated norepinephrine levels, and indeterminate sleep patterns in offspring neonatally (5). Lower levels of maternal verbal and behavioral responsiveness during maternal–infant interactions have also been reported in mothers with depressive symptomatology, and may also affect child developmental outcome (5–8). Prior to infant birth, maternal depression during pregnancy may have affected maternal eating behaviors, nutrition, medical care, and sleep patterns, which directly affect fetal nutrition and development (9), perhaps leading to the neonatal physiologic differences in infants of depressed mothers noted by Field (5). An additional complicating factor is whether or not, and how, the mother is treated for her depression.

In the developmental studies cited above, it is not clear whether mothers were treated for their depressive symptoms. Drug treatment of depression is now common, and a 1990 review (10) noted that all major classes of psychotropic drugs have been shown to pass into breast milk following maternal ingestion, although the doses to which infants are exposed are small. A recent controlled study of pregnancy outcomes of 228 women taking fluoxetine (Prozac), the most commonly used antidepressant drug in the USA, found that those taking fluoxetine in the third trimester are at increased risk for prenatal complications, including premature delivery, infant admission to the neonatal intensive care unit, and poor neonatal adaptation, including respiratory difficulty, cyanosis on feeding, and jitteriness (11). Whether symptoms are due to the drug or the underlying medical condition of the mother has not been determined.

Further complicating our understanding of the relationship between breast/bottle-feeding and maternal depression are current clinical recommendations and practice within the psychiatric and pediatric fields. Wisner *et al.* (12) summarized the clinical dilemmas which have ensued, given the known negative effects of maternal depression on child development, the transfer of drugs from maternal breast milk to the infant, and a sparse and inconclusive database on effects on the infant. Thus, it is possible that failure at breast-feeding or inability to breast-feed because of infant very low birth weight may affect maternal depressive symptoms, but it is also equally possible that maternal depression may contribute to a mother's choice not to breast-feed, due to either low motivation or energy secondary to depression, or to concerns related to potential negative long-term effects of drug transmission through breast-feeding on child development.

1. Kendell RE, Wainwright S, Hailey A et al. The influence of childbirth on psychiatric morbidity. *Psychiatr Med* 1976; **6**: 297–302.
2. O'Hara MW, Zekoski EM, Phillips LH et al. A controlled prospective study of postpartum mood disorders: comparison of childbearing and non-childbearing women. *J Abnormal Psychol* 1990; **99**: 3–15.
3. Singer LT, Bruening P, Davillier M, Hawkins S, Yamashita T. Social support, psychological distress and parenting strains in mothers of very low birthweight infants. *Family Relations* 1996; **45**: 343–50.
4. Hannah P, Adams D, Lee A et al. Links between early postpartum mood and postnatal depression. *Br J Psychol* 1992; **160**: 777–80.
5. Field T. Infants of depressed mothers. *Infant Behav Dev* 1995; **18**: 1–14.

6. Cohn J, Tronick E. Three month infants reaction to simulated maternal depression. *Child Dev* 1983; **154**: 185–93.
7. Singer LT, Yamashita T, Davillier M et al. Feeding interactions in infants with very low birthweight and BPD. *J Dev Behav Pediatr* 1996; **17**: 69–76.
8. Singer LT, Arendt R, Farkas K, Minnes S, Huang J, Yamashita T. The relationship of prenatal cocaine exposure and maternal psychological distress to child developmental outcome. *Dev Psychol* 1997 (in press).
9. Zuckerman B, Amaro H, Bauchner H, Cabral H. Depressive symptoms during pregnancy: relationship to health behaviors. *Am J Obstet Gynecol* 1989; **160**: 1107–11.
10. Buist A, Norman T, Dennerstein L. Breast feeding and the use of psychotropic medication: a review. *J Affective Disorders* 1990; **19**: 197–206.
11. Chambers C, Johnson K, Dick LM, Felix R, Jones KL. *N Engl J Med* 1996; **335**: 1010–15.
12. Wisner K, Perl J, Findling R. Antidepressant treatment during breast feeding. *Am J Psychol* 1996; **153**: 1132–7.

# Grating Acuity Cards: Validity and Reliability in Studies of Human Visual Development

D. LUISA MAYER

*Department of Ophthalmology, Boston Children's Hospital, Harvard Medical School, 300 Longward Avenue, Boston 02115, MA, USA*

VELMA DOBSON

*College of Medicine, Department of Ophthalmology, The University of Arizona, Health Services Center, 1801 N. Campbell Avenue, Tucson, AZ 85719, USA*

---

Introduction

The purpose of this chapter is to evaluate critically the acuity card procedure (ACP), a preferential looking (PL) measure of grating acuity, that has become a major outcome measure in studies of the effects of dietary fatty acids on human infant development. The impetus for visual acuity measures in human studies has come directly from research with rhesus monkeys made deficient in n-3 fatty acids (1,2). The n-3 deficient monkey infants showed abnormalities in visual acuity development, as measured with forced-choice preferential looking (FPL) and the visual evoked potential (VEP). The deficient monkeys also showed abnormalities in retinal function as measured with the electroretinogram (ERG) (see review in (3)). As reviewed in this chapter, the findings from monkeys have been replicated in some, but not all, studies of the effects of dietary fatty acids in human infants.

As background, the history and methodology of the ACP are presented.

The validity of the ACP specifically, and grating acuity generally, is discussed, and we devote considerable space to a discussion of measurement precision, or reliability, of the ACP. We attempt to sort out possible sources of variability in grating acuity measured by the ACP. We briefly discuss the maturation of binocular grating acuity and possible substrates for acuity development. In order to evaluate acuity data obtained in the infant nutrition studies, we present data from all published studies on binocular acuities obtained by ACP in normal, full-term infants (see Table 1). We limit the age range of the acuity data presented to between term and age 1 year because this is the age range included in most nutrition studies. Finally, grating acuity data from published nutrition studies are presented in Tables 2 and 3 and discussed.

In the last section we summarize major points regarding the validity and reliability of the ACP, and the across-study variability in binocular acuity norms. This section also discusses the variability and effect size in nutrition studies, and we suggest ways to minimize variability in ACP measurements of grating acuity in future studies. The section ends with a discussion of an alternative method, the VEP, which has been used to measure grating acuity in infants in nutrition studies, and summarizes the advantages and disadvantages of using a VEP rather than a behavioral (FPL or ACP) method to assess grating acuity in infants in nutritional outcome studies.

## Acuity Card Testing: History and Methodology

The acuity card procedure was developed to allow rapid assessment of visual acuity in infants and other individuals who cannot be tested with standard adult letter acuity charts. The procedure measures grating acuity, or the ability to resolve a pattern consisting of repeating black-and-white stripes. The feature of the procedure that makes it useful for testing infants is that it does not rely on verbal responses; rather, the test is constructed such that the infant's eye and head movements can be used as indicators of the infant's ability to resolve the stripes in the grating.

### Preferential Looking

The roots of the acuity card procedure lie in 25 years of laboratory research in which a technique termed 'preferential looking' (PL) was used to study the visual capabilities of infants. In the earliest PL studies, conducted in the late 1950s and early 1960s, Fantz and his colleagues presented infants with pairs of stimuli, e.g. a two-dimensional ball versus a three-dimensional ball, and watched to see if infants showed differential amounts of looking toward the two stimuli (4) (reviewed in (5)). If they did, the researchers concluded that the infants could discriminate between the stimuli.

Fantz and his colleagues obtained PL estimates of visual acuity by pairing a black-and-white grating with a homogeneous gray field of the same average luminance ('brightness') (6). When gratings were composed of large stripes, infants looked more at the grating than at the gray field, indicating that they could resolve the grating. When presented with gratings composed of small stripes, infants looked equally at the grating and gray stimuli, suggesting that they could not tell that one of the two stimuli was composed of stripes. Younger infants showed evidence of being able to resolve only fairly coarse gratings (wide stripes), while older infants could resolve finer gratings, suggesting an age-related improvement in grating acuity.

## Forced-choice Preferential Looking

The PL technique underwent a major change in the early 1970s, when Teller introduced the forced-choice preferential looking (FPL) procedure (7) (reviewed in (8)). In Fantz' PL procedure, the task of the tester was to monitor the right or left direction and duration of the infant's looks to two different stimuli. The tester in Fantz' PL procedure used the corneal reflex of the stimulus on the infant's cornea to determine the infant's look direction and thus was aware of the positions of the grating and gray stimuli. In contrast, Teller introduced a procedure in which the tester was masked to the relative positions (left versus right) of the grating and gray stimuli, and the tester's task was to use the infant's looking behavior to come to a decision as to whether the grating was on the left or right. The tester's judgment as to grating location on each trial was scored objectively ('correct' or 'incorrect'), and the results of many trials, usually 20 per grating spatial frequency (stripe width), were used to determine the finest grating on which the tester's score was significantly above chance (50%). The logic behind the FPL procedure is that, because the tester does not know where the grating is located, the only way that the tester can be correct on more than 50% of stimulus pairings is to get information *from the infant* as to the location of the grating. Thus, any grating on which the tester scores significantly above 50% correct must be visible to the infant.

The term 'forced-choice' preferential looking refers to the requirement that the tester must choose either the right or left stimulus position on each paired presentation of grating and gray stimuli. Forced-choice procedures have a long tradition in sensory psychophysical experiments and are considered objective and unbiased because the observer's response ('stripes on the right' versus 'stripes on the left') is based on an objective, physical referent (right versus left grating position) only, and not on subjective criteria (e.g. 'I see it' versus 'I don't see it').

The FPL procedure permitted a great leap forward in our knowledge about visual development in human infants, not just in the area of visual acuity (reviewed in (9)), but also other areas, including color vision and vision under conditions of dim illumination (10). Despite its success in the laboratory, however, the FPL procedure never achieved widespread acceptance or use as a clinical tool for assessment of acuity in infants (11,12), primarily because of the large number of trials and considerable amount of time required to measure an acuity threshold.

## The Acuity Card Procedure

The acuity card procedure (ACP) was developed by Teller and colleagues in the early 1980s, in response to requests from clinicians for a behavioral method for measurement of visual acuity in infant patients (see review in (11)). The critical difference between the ACP and FPL is in the tester's task. In the FPL procedure, the tester's only task on each trial is to make a forced-choice judgment as to whether the grating is on the left or on the right. This yields scorable (correct/incorrect) data from which a scientifically rigorous measurement of acuity threshold can be obtained. In contrast, in the ACP, the tester's task is to make a decision as to whether the infant can resolve each grating spatial frequency. Instead of requiring a set number of stimulus trials, as in the FPL procedure, the ACP permits the tester to determine the number of presentations needed to indicate whether the infant can resolve the grating. Furthermore, the ACP permits the tester to incorporate all aspects of the infant's response to each grating into the estimate of acuity threshold, whereas much of this information is discarded in the forced-choice judgment required of the FPL tester.

### *The acuity cards*

*Stimulus configuration.* Teller and colleagues designed the acuity cards to be a streamlined, easy-to-use adaptation of FPL stimulus presentation methods. Initially, cards were 28 × 71 cm pieces of gray cardboard, containing two 9-cm diameter stimulus apertures, one located to the left and one located to the right of a 4-mm peephole, through which the tester viewed the infant. On each card, a black and white grating was mounted behind one 9-cm aperture and a piece of gray cardboard, or a very high spatial frequency grating (that appeared to be a uniform gray field), was mounted behind the other 9-cm aperture. This arrangement was similar to Teller et al.'s (7) FPL stimulus arrangement.

Following the successful incorporation of the prototype acuity cards into ten clinical settings (13), commercial production of the acuity cards was undertaken. Commercial production resulted in a change in the stimulus configuration. No longer were the grating and gray stimuli presented behind apertures; instead, the grating was printed on the gray card, to one side of the peephole. This had the advantage of removing the distracting edges around the stimuli, but the disadvantage of removing the distinct gray comparison stimulus. Dobson and Luna (14) compared acuity results in 30 4-month-olds, 30 12-month-olds, and 30 36-month-olds, each of whom was tested with both prototype and commercially produced acuity cards. No differences were found between acuity results obtained with the two types of cards, suggesting that differences in stimulus configuration between prototype and commercially produced cards do not affect acuity results in infants and young children.

*Differences in grating spatial frequency between successive acuity cards.* In the first acuity card study, grating spatial frequency (stripe width) varied from card to card in one-octave* steps (15,16). That is, the width of the stripes on each successive card was half that of the stripes on the preceding card. Subsequently, prototype cards were constructed that varied in 0.5-octave steps (e.g. 17,18), and in 0.3-octave steps (19). Commercially produced cards (Vistech, Inc., Dayton, OH) have a fixed step size of 0.5 octave.

## Testing procedure

*Standard clinical procedure.* The standard method for assessment of visual acuity with the ACP is the procedure described in the *Teller Acuity Card Handbook* (20). Prior to testing, the acuity cards are stacked face down on a table behind the acuity card testing stage. Cards are arranged in order, from lower to higher spatial frequency (wider to narrower stripes), beginning with a card containing a grating spatial frequency chosen to be above threshold, based on the child's age. This initial card is referred to as the 'start card'.

When testing begins, the infant is held by an adult at a specified distance

---

* An octave is a halving or doubling of spatial frequency, e.g. from 10 to 20 cycles/degree (c/d).

in front of the acuity card stage. The distance used depends on the child's age. The adult is cautioned to maintain the specified test distance, and to avoid looking under the eye shield at the acuity cards. After the infant is positioned, the tester sits behind the stage and attracts the infant's attention by talking to the infant through the large opening in the stage. As soon as the infant is looking, the tester picks up the start card, which is face down on the top of the stack of cards, and puts it up to the opening. The tester watches the infant's response through the peephole, then removes the start card, rotates it by 180° without looking at the front of the card, and presents it again. If the infant gives a strong response to one side of the card on the first presentation and the other side of the card on the second presentation, the tester will judge that the infant can resolve the grating and that the grating is on the side of the card to which the infant showed the strong response. The tester is now permitted to look at the front of the card, to confirm the location of the grating.

If the infant shows a clear response to the start card, the tester presents successive cards in order of increasing spatial frequency, showing each card as many times as is necessary to determine if the infant can resolve the grating on the card. When a card is reached that contains a grating that the tester judges to be below threshold for the infant, the tester will re-present the start card or another card containing a coarse grating, to ensure that the infant is still attentive to the task. After re-establishing the infant's interest in the procedure, the tester will again present the below-threshold card, to confirm that the infant does not see it, and will also present the last card that was judged to contain a suprathreshold grating, to confirm that the infant can, indeed, resolve the gratings on this card. Acuity threshold is judged to be the spatial frequency of the finest grating that the tester judges that the infant can resolve. As the interval between adjacent acuity cards is 0.5 octave, the ACP provides an estimate of grating acuity to the nearest 0.5 octave.

If an infant has low vision, he or she may have difficulty resolving the gratings on the start card. In this case, the tester will begin testing with the card containing the lowest spatial frequency grating available in the set of acuity cards. If the child can resolve this grating, the tester will present finer and finer gratings until acuity threshold is determined.

*Random start card procedure.* While testing in clinical and in most research settings is typically conducted using the standard clinical procedure, described above, Dobson and colleagues used a protocol best described as a 'random start card' procedure to study visual acuity in several clinical populations (21,22). The purpose of this procedure was to mask testers to the spatial frequencies on the acuity cards, in order to allow an unbiased examination of within-tester and between-tester test–retest reliability. As discussion of within- and between-tester reliability is an integral part of this chapter, we will describe the essential differences between random start card and standard clinical procedures.

There are two basic differences between the random start card procedure and the standard clinical acuity card procedure. First, in the random start card procedure, the start card for each infant is chosen at random from among seven possible start cards, ranging from slightly above the expected acuity threshold for the infant's age to well above the expected acuity threshold. Second, in the random start card procedure, the tester remains masked to the spatial frequency of the gratings on all cards throughout the testing procedure. After the tester has presented a card to the infant a sufficient number of times to decide whether the infant can resolve the grating and where the grating is located, the tester may ask an assistant to indicate the left–right location of the grating, but the tester may not look at the front of the card. These modifications in the procedure prevent the tester from being biased by knowledge of the spatial frequency of the gratings on the acuity cards, and ensure that the tester remains masked to the acuity threshold obtained on the first test when conducting the second test required for assessment of within-tester reliability.

All other aspects of the random start card procedure, such as presentation of individual cards and judgment of acuity threshold, are identical to testing in the standard clinical procedure.

## Validity of the Acuity Card Procedure

Validity refers to the extent to which a measurement represents the true state of the individual on that measure. In terms of visual acuity, validity refers to the extent to which an individual's measured visual acuity represents that individual's actual ability to resolve the pattern used during testing. For measures such as visual acuity, in which no physical standard of the true state exists, the validity of the measure must be confirmed indirectly, through examination of the content, construct, and criterion validity of the measure.

### Content Validity

The content validity of a measure is the extent to which it includes all dimensions of the function being measured and nothing more. For example, the estimate of an infant's visual acuity obtained in a preferential looking test is a function of both the infant's resolution ability and the infant's general attentional state. Young infants who have normal acuity but who are sleepy or ill may provide low acuity values due to poor attention. Similarly, acuity scores of infants and children with brain damage or pervasive developmental disabilities often exhibit day-to-day variability that is related more to fluctuations in the child's attentional state than to fluctuations in visual acuity (23,24).

*Construct Validity*

The construct validity of a measure is the extent to which it varies with other established indicators of the function measured. Grating acuities measured using the ACP do not differ systematically from acuities measured with more objective FPL procedures, either as tested within studies (18,25) or as shown in between-study comparisons (see Dobson (9), p. 139, Figure 12–7). Thus, the data indicate that the ACP is a valid behavioral measure of grating acuity in infants.

Another issue is whether the ACP, or any measure of grating acuity, is correlated with or predictive of recognition acuity (acuity for letter optotypes), the type of acuity that is most frequently measured in adults. Grating acuity is a type of resolution acuity. Resolution and recognition acuity tests differ both in terms of the visual stimulus (e.g. grating versus letter) and the task (i.e. stimulus seen versus stimulus identified). Data from older children and adults, who can be tested using both grating acuity and recognition acuity tests, indicate that grating acuity is not necessarily identical to recognition acuity, and that the relation between the two is complex and depends upon the visual pathology. In individuals with normal acuity or mild acuity deficits, grating acuity and recognition acuity results are in good general agreement (26). However, agreement between the two measures worsens with increasing acuity deficits, that is, deficits in recognition acuity are greater than deficits in grating acuity in many ocular and vision disorders. Discrepancies between grating and recognition acuity are greatest in patients with damage to the central retina (macula, fovea) and in those with abnormal central neural pathways (amblyopia) (27–32). Grating stimuli are large in comparison with the small size of recognition acuity targets. For example, at 55 cm, the Teller acuity card grating subtends 12°, whereas the 20/200 letter on a Snellen letter chart subtends 50 minutes of arc. Patients with central retinal pathologies can use peripheral retina to detect and resolve large-field grating stimuli, whereas recognition acuity targets at the same acuity level cannot be resolved by the same area of peripheral retina. This stimulus size factor cannot account entirely for discrepancies between grating and recognition acuities in amblyopia, however (26,28,29,31).

*Criterion Validity*

The criterion validity of a measure is the extent to which it predicts a directly observable phenomenon. In terms of visual acuity, it would be expected that reduced acuity would be found in eyes with optical and/or ocular disorders known to impair vision.

Important factors that would be expected to reduce visual acuity include abnormalities of the ocular media (corneal opacities, cataract), refractive error (high myopia, hyperopia, astigmatism), damage to the posterior pole of the retina (retinal residua of retinopathy of prematurity, retinal degeneration, chorioretinal coloboma, toxoplasmosis scarring), and strabismus (horizontal or vertical misalignment of the eyes) and visual pathway damage due to central nervous system injury. As reviewed by Dobson (9), all the factors listed in the preceding sentence have been shown to be associated with reduced acuity in infants, as measured with either FPL or the ACP. Thus, acuity results from infant pediatric ophthalmology and infant pediatric neurology patients support the criterion validity of behavioral assessment of grating acuity in infants.

## Reliability of the Acuity Card Procedure

The reliability, or precision, of a measure is the extent to which repeated observations agree. In an ideal world, if an infant's grating acuity were tested twice by the same tester, the same acuity score would result from both tests. Furthermore, in the ideal world, an infant tested by two different testers of equal skill would also show the same acuity score on both tests. However, in reality, there are several factors that introduce variability into acuity scores, with the result that test–retest acuity scores often do not show perfect agreement. One source of variability is the infant's biological fluctuation in vision and attention between the time of the first test and the time of the second test. Other sources of variability include across-time fluctuation in a tester's ability to conduct the test, and differences across testers in the skill with which they conduct the test and the criteria they use to judge acuity threshold. Final sources of variability lie within the test procedure itself.

In this section, we first discuss tester-related factors that can influence measurement precision in the ACP. This is followed by a section on methodological sources of variability in the ACP. Finally, data are presented on within- and between-tester reliability results obtained with the ACP.

## Tester-related Factors that can Influence Reliability

### Tester's knowledge of grating spatial frequency

In the random start card variant of the ACP, cards are presented in order from lower to higher spacial frequencies, as in the standard clinical procedure, but the tester is unaware of the spatial frequencies of the gratings used during testing. In contrast, testers who use the standard clinical ACP *are* aware of the grating spatial frequencies used during testing.

In clinical patients, where acuity level is unknown prior to testing, knowledge of grating spatial frequency is unlikely to influence the tester's judgment concerning acuity threshold. However, in studies of normal full-term and healthy preterm infants, in which testers are often familiar with the expected acuity scores for infants of a particular age, knowledge of the spatial frequency of the gratings on the acuity cards could bias the tester's judgment concerning acuity threshold, thereby reducing variability of acuity results. Consistent with this hypothesis, somewhat higher variability in test–retest comparisons has been found in studies using the random start card procedure (21,22) than in studies using the standard clinical procedure (33,34). However, other factors that differed between the two types of studies, such as the number of testers and the effect of start card spatial frequency in the random start card procedure, could also explain the difference in variability.

### Tester differences

In any psychophysical procedure, sensory threshold is affected by subjective aspects of the tester's response, including the criterion that the tester uses to decide whether the stimulus is present. In the ACP, the tester's estimate of an infant's acuity threshold is based on subjective judgments made by the tester on the basis of the infant's behavior. Therefore, it should not be surprising if testers differ in their judgments of the same infant's acuity. Until recently, tester differences were assumed to have a random effect on acuity results, increasing variability but not having a systematic effect. However, several studies have now shown that there can be systematic differences between testers.

Direct, pairwise comparisons of ACP testers' results from the same infant have been reported in two studies. In the first study, in which three testers used the standard clinical ACP to assess acuity both clinical patients and healthy control subjects, one tester's acuity in scores averaged 0.4 octave lower than those of another tester and 0.3 octave lower than those of the third tester for tests of the same subjects (33). In the second study, in which two testers used the standard clinical procedure to test acuity in 1- to 48-month-old normal children, one tester's acuity scores were significantly higher than those of the other tester at three of the twelve ages tested, with a maximum average difference of 0.3 octave (at the 1- and 2.5-month test ages) (34).

Systematic tester differences have also been shown in two studies that reported mean acuities from different individuals in the same population. In the first study, in which the standard clinical ACP was used to test acuity in a large population of pediatric ophthalmology patients (35), mean acuities obtained by three testers showed significant differences for right eye tests but not left eye tests. The maximum difference between the tester with the highest and the tester with the lowest acuity scores was 0.4 octave; other differences ranged from 0.06 to 0.31 octave. In a study of between-tester reliability (22), in which the random start card procedure was used by up to six testers to test infants at four test ages (term, 4, 8, and 11 months), differences between the mean acuity scores of two testers were less than 0.5 octave in 89% of the comparisons.

Two conclusions can be drawn from these studies. First, systematic differences between testers, i.e. where one tester tends to get higher acuity values than another tester in the same infant or population of infants, have been found in studies using the ACP. Second, the magnitude of the tester differences is not large, and the majority of the differences between testers are less than 0.5 octave, or less than the difference between adjacent acuity cards.

One final note concerning tester differences is that they have also been reported in the rigorous, laboratory-based FPL procedure. In a study by Teller *et al.* (36), two testers each measured acuity in 20 2-month-old infants, using a protocol in which the tester presented 500 stimulus trials to each infant. Although acuity thresholds measured by the two testers did not differ significantly, there were significant differences between testers in the upper asymptotes and slopes of the psychometric functions generated. This suggests that tester differences can increase variability of acuity results in both ACP and FPL studies.

## Methodological Sources of Variability

### Interval between stimuli

The magnitude of measurement error in a psychophysical test depends upon the size of the interval between stimuli. Larger intervals between stimuli result in greater variability. The first acuity card studies utilized 1-octave steps between adjacent grating stimuli, while later studies used 0.5- or 0.3-octave steps. Variability of acuity scores in the studies that used 1-octave steps larger than variability of scores in studies that used 0.5-octave steps (e.g. (15,16) versus (17)). However, there were no differences in standard deviations between studies that used 0.5-octave steps (e.g. 17,25,34) and one that used 0.3-octave steps (19) over the same age range (12–48 months).

### Number of alternative stimulus locations

The magnitude of measurement error in a psychophysical test also depends on the number of alternative locations in which the stimulus can be presented. Variability is lower in procedures that use three or more alternative stimulus locations than in two-alternative procedures. However, higher-alternative procedures are difficult to use with infants (8). Therefore, nearly all PL studies of infant vision have been conducted using a two-alternative procedure. The high variability inherent in the two-alternative procedure means that accurate acuity measurements will be obtained only with large numbers of trials (37). It is this feature of the FPL procedure that made it impractical for clinical testing and led to the development of the ACP (11).

### The spatial frequency of the initial stimulus

Measurement error can also arise from a bias in threshold measurement that occurs in some psychophysical procedures when the value of the stimulus that starts the series of presentations (e.g. start card in the ACP) is either too far from or too near the threshold value. This effect was shown in computer simulations of a particular staircasing sequence used with an FPL procedure (38–40). The effect was also shown in an ACP study conducted by Mash *et al.* (22) that used the random start card procedure, in which each infant was tested with one of seven possible subsets of acuity cards. Acuity scores obtained were found to be correlated with the spatial frequency of the start card, i.e. higher spatial frequency start cards tended to produce better acuity scores and lower start cards tended to produce lower acuity scores. The start card bias found in this study was associated with increased between-tester variability at two of the eight test ages.

*Variability related to poorly controlled test distance and lighting*

Although the acuity card manual specifies test distance and lighting (20), it is sometimes necessary to use nonstandard methods of testing to obtain acuity results from a patient in a clinical setting (41). The nonstandard methods include presenting the cards without using the acuity card stage and presenting the cards to patients in the supine position. As a result, distance and light levels may vary, contributing to variability in acuity results.

*Studies of Test–Retest Reliability of the ACP*

In studies of the ACP, reliability has been evaluated by comparing test–retest acuity results obtained by one tester on one infant (within-tester reliability), and by comparing test–retest acuity results obtained by two testers on one infant (between-tester reliability). As within-tester reliability depends on within-infant variability, within-tester variability, and variability related to measurement procedures, whereas between-tester reliability depends on these three factors *plus* between-tester differences, it would be expected that test–retest reliability would be better for within- than for between-tester comparisons.

*Within-tester reliability*

Only three studies have investigated within-tester reliability of the ACP in infants. The first study was the original acuity card study, in which 4-, 8-, and 16-week-old and 6-month-old infants were tested with cards differing in spatial frequency by 1 octave (15). The results indicated a difference of 1 octave or less between test–retest acuity scores obtained by one tester in 88% (56/64) of test–retest pairs.

In the second study, in which two testers used the standard clinical procedure to assess monocular acuity in 460 children in 12 age groups between 1 and 48 months of age, each tester retested half of the sample on a second day (34). The results indicated that 99% of within-tester differences were 1 octave or less and 95% were 0.5 octave or less. In 51% of the test-retest pairs, identical acuity scores were obtained.

In the third ACP study, within-tester reliability of monocular acuity results was assessed in 91 infants, aged 4 to 17 months, who had been treated in a neonatal intensive care unit (42). All testing was completed in one session, and the tester was masked to the results of testing through the use of the random start card procedure. For the eye tested first for each subject, within-tester differences were 1 octave or less in 93% of test–retest pairs, and 0.5 octave or less in 71% of test–retest pairs. For the eye tested second, within-tester differences were 1 octave or less in 89% of test–retest pairs, and 0.5 octave or less in 67% of test–retest pairs.

Within-tester reliability has also been reported in two studies that used the FPL procedure and staircase methods to estimate acuity in normal infants (43,44). In a longitudinal study of normative binocular and monocular acuity, 105 subjects in ten age groups between 1 and 60 months were retested on the same day (this was possible because the tester was kept unaware of the result of the first acuity test). In the age range 1–12 months, the 99% upper confidence limit for binocular acuity was 1.0–1.3 octaves, and the 99% upper limit for monocular acuity was 0.8–1.2 octaves (data from Figure 1B in (43)). In the second FPL study, which used 0.3-octave intervals between grating spatial frequencies, 95% of binocular test–retest comparisons by a single tester ($n = 20$) agreed to within 0.6 octave (44). These within-tester reliability data using FPL and staircase methods are similar to those reported in the studies in which testing was conducted with the ACP.

*Between-tester reliability*

In ACP studies in which between-tester reliability for binocular acuities of healthy preterm and full-term infants were reported, the median percentage of test–retest pairs in which agreement was 1 octave or less was 92% (range 90–97%) (15,16,19,33,45). For monocular testing, the median percentage of between-tester test–retest pairs in which agreement was 1 octave or less was 95% (range 86–98%) (16,33,34). Among studies of binocular or monocular acuity that reported between-tester test–retest differences broken down into smaller bins, the median percentage of test–retest pairs in which agreement was 0.5 octave or better was 85% (range 72–90%) (17,19,33,34), and the median percentage of test–retest pairs with no difference between acuity scores was 40% (range 34–45%) (binocular and monocular comparisons in Getz *et al.* (33); monocular in Mayer *et al.* (34)).

When clinical populations are tested, between-tester reliability is lower than that reported above for normal infants. Among studies of binocular acuity in clinical populations, the median percentage of test–retest pairs in which agreement was 1 octave or better was 87% (range 75–100%) (18,21–24,33,46). Among studies of monocular acuity in clinical populations, the median percentage of test–retest pairs showing agreement of 1 octave or better was 84% (range 80–95%) (18,21,22,33).

*Comparison of within- and between-tester reliability*

It is possible to estimate the contribution of multiple testers to variability in acuity card testing by comparing reliability within testers to that between testers. Only two studies have examined both within-tester and between-tester reliability in the same infants. In the first study, in which a coarse (1 octave) step size was used, there was slightly better agreement for between-tester comparisons (92% differed by 1 octave or less) than for within-tester comparisons (88% differed by 1 octave or less) (15).

The other study in which both within-tester reliability and between-tester reliability were examined in the same subjects is the ACP study of normative monocular acuity (34). In this study, within-tester variability was lower than between-tester variability: 99% of within-tester differences versus 98% of between-tester differences were 1 octave or less; 95% versus 90% were 0.5 octave or less; and 51% versus 45% showed no difference.

Mash and colleagues examined within- and between-tester reliability in different subjects within the same population of infants who had been treated in the neonatal intensive care unit (22,42). As in the Mayer *et al.* study (34), within-tester variability was slightly lower than between tester variability: 91% of within-tester differences versus 87% of between-tester differences were 1 octave or less; and 69% versus 67% were 0.5 octave or less.

The differences between the test–retest results of one tester versus two testers are relatively small, suggesting that most of the variability in test–retest comparisons is due to biological differences within individual infants and to variability arising from the measurement procedure. Nevertheless, the results from the studies by Mayer and colleagues (34) and Mash and colleagues (22,42) suggest that increasing the number of testers who assess acuity in a population of infants can contribute increased variability to the acuity estimates obtained.

## Binocular Acuity Maturation and Norms

This section describes the course of grating acuity maturation and briefly summarizes possible substrates for acuity development. Then, we examine the variability of binocular acuity values obtained in studies of normal infants tested with an acuity card procedure between term and 1 year of age.

## Maturation of Binocular Acuity and Possible Substrates

The maturation of grating acuity measured by preferential looking methods can be described by a growth curve with acuity plotted on a logarithmic scale* against linear age (47). There is a rapid early phase of improvement in behavioral acuity, from approximately 1 c/d at age 1 month to approximately 6 c/d at age 6 months. Acuity improves more slowly thereafter, to about 30 c/d by age 4–5 years, an improvement that is about one-tenth of the rate between ages 1 and 6 months.

Several models have been proposed to account for early human visual development (e.g. 48–50). These models differ somewhat in their assumptions regarding retinal receptoral and postreceptoral processes and all have been limited by the paucity of relevant anatomical data. Most of these models agree that prereceptoral factors (optics of the eye) and the development of the cone photoreceptor mosaic (changes in density, relative distribution, and morphology) can account for some but not all the differences between infant and adult visual acuity. Changes in postreceptoral factors, including neural development of the lateral geniculate nucleus and striate cortex, are required to account for behavioral maturation of acuity, at least in monkeys (51).

## Binocular Norms

Table 1 summarizes the results of 11 published studies of binocular acuity in infants born full-term. Results are broken into the eight age groups for which four or more studies report data. For each age group, the median and the highest and lowest number of subjects tested per study are reported, as are the median acuity value across studies and the median standard deviation value across studies. Also shown are the range of mean acuity values

---

* Visual acuity measurement is scaled and plotted logarithmically (e.g. log base 10 or log base 2, an octave scale) because the visual system, like other sensory systems, processes physical stimuli as a logarithmic function. That is, the 'output' of the system – a summary measure of the subject's responses to visual stimuli – is linear with log 'input' (47). This means that 'just-noticeable differences' are equal anywhere on a logarithmic scale, or, in simpler terms, a 1-octave difference in acuity in the low spatial frequency range is equal perceptually to a 1-octave difference in acuity in the high spatial frequency range.

Table 1. Summary of studies of binocular ACP acuity in infants born at term.

| Age (weeks) | No. of studies (refs) | No. of subjects per study | | | Grating acuity | | | Standard deviation | | |
|---|---|---|---|---|---|---|---|---|---|---|
| | | Median | Range | Total | Median (c/d) | Range (c/d) | Diff.* (oct) | Median (oct) | Range (oct) | Diff.† (oct) |
| Term | 6 (25,45,52,54,55,84) | 20 | 13–60 | 150 | 0.93 | 0.59–1.11 | 0.91 | 0.54‡ | 0.30–0.78 | 0.48 |
| 4–6 | 5 (15,17,25,52,84) | 11 | 6–54 | 99 | 1.10 | 0.80–2.40 | 1.60 | 0.68 | 0.10–1.10 | 1.00 |
| 8–10 | 5 (15,17,25,52,54) | 9 | 6–54 | 88 | 2.10 | 2.00–3.56 | 0.83 | 0.69‡ | 0.06–0.90 | 0.84 |
| 12–14 | 4 (52,54,84,85) | 21 | 20–194 | 246 | 3.74 | 2.60–5.30 | 1.03 | 0.60‡ | 0.03–1.00 | 0.97 |
| 16–19 | 7 (15,17,25,52–54,86) | 25 | 6–54 | 194 | 5.48 | 3.70–6.50 | 0.81 | 0.56‡ | 0.03–0.70 | 0.67 |
| 26–31 | 9 (15,17,25,52–54,84–86) | 20 | 6–206 | 544 | 5.90 | 4.70–10.30 | 1.13 | 0.55‡ | 0.02–0.80 | 0.78 |
| 38–45 | 6 (17,25,52,54,85,86) | 27 | 6–194 | 318 | 7.53 | 5.00–14.20 | 1.50 | 0.50‡ | 0.02–1.00 | 0.98 |
| 48–58 | 8 (17,25,52–54,84–86) | 30 | 6–248 | 592 | 10.00 | 6.30–17.90 | 1.50 | 0.50‡ | 0.02–0.74 | 0.72 |

Values of mean acuity in the tables are in linear c/d (cycles per degree), but were calculated using $\log_2$ c/d. Standard deviations and differences between acuities are in $\log_2$ (octaves).
* Difference between highest and lowest mean acuity in the age group, calculated from $\log_2$ values.
† Difference between highest and lowest standard deviation in age group.
‡ Standard deviations not provided in study by Salomao and Ventura (54); median derived from $n-1$ studies.

and the range of standard deviations reported across studies, and the difference (in octaves) between the highest and lowest value in each of the ranges. Data were derived from tables, text, or, where necessary, extrapolated from graphs in the published studies.

*Variability in mean grating acuity across studies*

Notable in Table 1 is the large range of mean acuity values published for infants at each age. The smallest range was 0.81 octave in 16- to 19-week-olds and the largest range was 1.6 octaves in 4- to 6-week-olds. Some of this between-study variation in mean acuities may be due to the relatively generous age window for some age groups, but this is insufficient to account for the large range of mean acuities found for all age groups. Other possible sources of variability among studies include sampling error due to small sample sizes in some studies (6–11 in four studies), tester differences, population differences, and methodological differences, including a large step size between grating spatial frequencies in some studies.

*Variability in standard deviation across studies*

As shown in Table 1, standard deviations of acuity results also vary among studies within age groups. The difference between the largest and smallest standard deviations ranges from 0.67 octave in 16- to 19-week-olds to 1.0 octave in 4- to 6-week-olds. Across age groups, however, the median standard deviation was relatively consistent, ranging from 0.5 to 0.69 octave. The number of testers who obtained acuity data in these studies may account for differences among standard deviations across studies. The study showing the lowest standard deviations (0.02–0.10 octave) in every age group shown in Table 1 was conducted longitudinally on 11 subjects by a single tester (52). Larger standard deviations were found in studies in which three or more testers obtained acuity data (15,16,25,53,54). Methodological explanations for lower variability in the normative monocular study, for example the tester's knowledge of card spatial frequencies, are less likely as only one study of infants tested at term (55) and one study of older age groups (54) did not use the clinical ACP. Also, the standard deviations in these two studies were not different from those of the other studies in Table 1.

## Grating Acuity Results in Studies of Nutrition in Infants

In recent years, binocular grating acuity has been used in a number of studies as an outcome measure of the effect of dietary fatty acids on infant development. In these studies, grating acuity was assessed using both behavioral (FPL or ACP) and electrophysiological (VEP) methods. The studies reviewed here include nonrandomized studies of infants who were breast-fed versus formula-fed, and randomized trials in which infants were fed formulas with different fatty acid compositions. We do not attempt to evaluate the potential explanations for significant group differences on the basis of differences in fatty acid exposure or duration. These issues are examined in other chapters and commentaries in this volume.

Tables 2 and 3 summarize acuity results obtained from full-term (Table 2) and preterm (Table 3) subjects in these studies, reporting both mean acuities and standard deviations, and also the magnitude of difference in mean acuity values between dietary groups within individual sites.

The first point to note is that the mean FPL and ACP acuities obtained from many dietary groups (Tables 2 and 3) are within the range of mean ACP acuity values in normative studies of binocular acuity in infants of the same age (Table 1). Exceptions are found in the results from 2-month-olds in four studies (56–60) and 3-month-olds in another study (61), in which mean acuities of some of the dietary groups are slightly below the range of mean acuities of normal infants shown in Table 1. In a fifth study (62), of 3-month-olds, one nutrition group showed mean acuity slightly above the range for normal infants. As is well established (63), VEP acuities are higher than acuities obtained with behavioral (FPL and ACP) procedures, and this is reflected in acuity values from studies using the VEP measure (Tables 2 and 3).

Standard deviations for acuity estimates obtained with FPL, ACP, and VEP procedures in nutrition studies are generally in the range or are slightly below the range in normative binocular ACP studies of infants of the same age (Table 1). Notable exceptions are the large standard deviations reported in VEP studies of infants aged 4 (64), 5 (65), and 7 (66) months, and in 1- and 2-month-olds tested with the ACP in one study (60).

With behavioral (FPL or ACP) acuity testing, significant differences in mean acuity between dietary groups were found in term infants in two ACP studies (57,60) and one FPL study (67) (Table 2). In three other studies of the effects of fatty acids in the diet of term infants, no significant differences between dietary conditions were found (56,61,62). As shown in Table 3, all four studies in which behavioral methods were used to measure

Table 2. Dietary studies in infants born at term: binocular acuities.

| Study | No. of subjects | Age (months) | Measure | Conditions | Mean grating acuity (c/d) BM | Other | Difference* (octaves) | SD (octaves) BM; Other |
|---|---|---|---|---|---|---|---|---|
| Auestad et al. 1997 (56) | 120 | 2<br>4, 6, 9, 12<br>2, 4, 6, 9, 12 | ACP<br>ACP<br>VEP | BM vs 3 formulas<br>BM vs control<br>BM vs control<br>BM vs control | 1.88 | 1.43 | 0.40†<br>0.13 to 0.25<br>−0.26 to −0.04 | 0.56; 0.85<br>0.36–0.60"<br>0.26–0.62" |
| Birch et al. 1993 (67) | 30 | 4<br>4 | VEP<br>FPL | BM vs corn oil<br>BM vs corn oil | 9.27<br>5.59 | 7.19<br>4.65 | 0.37‡<br>0.27‡ | Not given<br>Not given |
| Carlson et al. 1996 (57) | 58 | 2<br>4, 6, 9, 12 | ACP<br>ACP | BM vs 2 formulas<br>BM vs formula (−)<br>BM vs formula (+; −) | 2.93 | 1.85 | 0.66‡<br>−0.16 to 0.18 | 0.49; 0.49<br>0.49–0.50" |
| Innis et al. 1994 (62) | 35 | 14 days<br>3 | ACP<br>ACP | BM vs formula<br>BM vs formula | 0.95<br>3.93 | 0.91<br>4.77 | 0.06<br>−0.28 | 0.51; 0.49<br>0.54; 0.48 |
| Innis et al. 1996 (87) | 160 | 9 | ACP | Duration of BM<br>>8 months vs never | 9.10 | 8.60 | 0.08 | 0.42; 0.47 |
| Innis et al. (in press) (61) | 172 | 3<br>3 | ACP<br>ACP | BM vs formula 1<br>BM vs formula 2 | 2.79<br>2.79 | 2.51<br>2.67 | 0.15<br>0.06 | 0.77; 0.50<br>0.77; 0.64 |
| Jorgensen et al. 1996 (60) | 35 | 1<br>2<br>4 | ACP<br>ACP<br>ACP | BM vs formula<br>BM vs formula<br>BM vs formula | 0.82<br>2.59<br>6.15 | 0.86<br>1.85<br>3.78 | −0.07<br>0.49‡<br>0.70‡ | 0.20; 0.24<br>0.65; 0.53<br>1.45; 0.81 |

| | | | | | | | |
|---|---|---|---|---|---|---|---|
| Makrides et al. 1993 (65) | 16 | 5 | VEP | BM vs formula | 8.07§ | 3.89§ | 1.13‡ | 0.96; 0.90 |
| Makrides et al. 1995 (66) | 66 | 4 | VEP | BM vs placebo | 6.41§ | 3.68§ | 0.78‡ | 0.66; 0.37 |
| | | 4 | VEP | BM vs supplemented | 6.41§ | 5.84§ | 0.11 | 0.66; 0.50 |
| | 62 | 7 | VEP | BM vs placebo | 16.47§ | 7.37§ | 1.13‡ | 1.03; 0.66 |
| | | 7 | VEP | BM vs supplemented | 16.47§ | 15.02§ | −0.13 | 1.03; 1.16 |

ACP, acuity card procedure; BM, breast milk; FPL, forced-choice preferential looking; VEP, visual evoked potential.
* Difference between mean acuities by order of condition (BM − Other).
† Largest difference between mean acuities.
″ Range of SDs across age and diet groups.
‡ Statistically significant (based on study).
§ Based on the fundamental spatial frequency of the checkerboard stimulus used.

Table 3. Dietary studies in preterm infants: binocular acuities.

| Study | No. of subjects | Age (months) | Measure | Conditions | Mean grating acuity (c/d) | | Difference* (octaves) | SD (octaves) | |
|---|---|---|---|---|---|---|---|---|---|
| | | | | | Group 1 | Group 2 | | Group 1 | Group 2 |
| Birch et al. 1992 (64) | 73 | | | 3 formulas (oils): | | | | | |
| | | 36 weeks | VEP | soy/fish vs corn | 3.08 | 2.42 | 0.35† | 0.34–0.49‡ | |
| | | 4 | VEP | soy/fish vs corn | 11.94 | 6.40 | 0.90† | 0.63–1.13‡ | |
| | | 4 | VEP | soy/fish vs soy | 11.94 | 7.14 | 0.74† | 0.63–1.13‡ | |
| | | 4 | FPL | soy/fish vs corn | 4.89 | 3.62 | 0.43† | 0.49–0.63‡ | |
| Birch et al. 1993 (67) | 30 | 4 | VEP | BM vs corn oil | 10.40 | 5.85 | 0.83† | Not given | |
| | | | FPL | BM vs corn oil | 5.21 | 3.78 | 0.47† | Not given | |
| Carlson et al. 1993 (68) | 77 | | | Fish oil vs commercial formula | | | | | |
| | | 2 | ACP | | 3.15 | 2.13 | 0.56† | 0.96 | 0.63 |
| | | 4 | ACP | | 6.53 | 5.07 | 0.37† | 0.47 | 0.48 |
| | | 0, 6.5, 9, 12 | ACP | | | | −0.27 to 0.37 | 0.30–0.58§ | |
| Carlson et al. 1996 (58) | 59 | | | Fish oil vs control formula: | | | | | |
| | | 2 | ACP | Without BPD | 2.90 | 2.15 | 0.43† | 0.54 | 0.53 |
| | | 2 | ACP | With BPD | 1.45 | 2.08 | −0.52† | 0.53 | 0.54 |
| | | 0, 4, 6, 9, 12 | ACP | Without BPD | | | −0.24 to 0.24 | 0.55–0.61§ | |
| | | 0, 4, 6, 9, 12 | ACP | With BPD | | | −0.78 to −0.23 | 0.54–0.61§ | |

ACP, acuity card procedure; BM, breast milk; BPD, bronchopulmonary dysplasia; FPL, forced-choice preferential looking; VEP, visual evoked potential.
* Difference between mean acuities by order of condition (Group 1 − Group 2).
† Statistically significant (based on study).
‡ Derived from the range of SEM provided in the text for the age group and measure; standard deviations for nutrition groups are not provided in text.
§ Range of SDs across age and diet groups.

acuity in preterm infants found significant differences between different diet groups in at least one age group. It is noteworthy, however, that dietary supplementation resulted in *decreased* acuity in a subgroup of infants in one study, the group of 2-month-old infants who developed bronchopulmonary dysplasia (BPD) (58).

Across both full-term and preterm infant studies, the significant differences obtained with FPL and ACP methods ranged from 0.27 to 0.66 octave. Two other studies showed a difference in mean acuities between groups that was within this range (0.40 octave at 2 months (56); and 0.37 octave at 4 months (68)), but neither difference was statistically significant.

Notable methodological differences between the studies that obtained significant differences and those that did not are the number of testers and the number of sites in the study. In the three ACP studies in which significant differences were found, a single tester obtained most or all of the acuities (58,60,68). The two FPL studies, both of which found significant differences between dietary groups, probably also used one tester, as has been done in previous FPL studies from the same laboratory (e.g. (43)). In contrast, two of the four studies that reported no significant differences between dietary groups used more than one tester and multiple sites (56,61). However, no significant differences in acuities between the three sites in one study (56) nor the seven sites in another study (61) were reported. One study averaged acuity scores from two testers on each infant, which would be expected to reduce variability of acuity results (62). Nevertheless, this study found no significant differences between dietary groups.

A final point concerning the use of the ACP in nutrition studies is that, because the ACP relies on the subjective judgment of the tester, it is extremely important to keep the tester masked to the dietary group of the infant being tested. One group of investigators noted that it was not possible to mask the tester as to whether an infant was breast-fed or bottle-fed (60), and it is likely that difficulties in masking ACP or FPL testers to dietary condition are pervasive, although unacknowledged, in studies comparing breast-fed to bottle-fed babies. Therefore, it is possible that results of studies comparing ACP grating acuity in breast milk versus formula groups may have been influenced by tester bias concerning the value of breast- versus formula-feeding.

How should one interpret the significant differences between mean acuities of breast-fed versus formula-fed infants and between acuities of infants fed formulas with different fatty acid compositions in some but not all ages? In some studies infants fed diets low in long-chain polyunsaturated fatty acids (LCPUFAs) or precursors show lower acuity than breast-fed infants and infants fed diets containing these fatty acids at 2 months, or 2 and 4 months, but not at earlier or later ages. One cannot conclude from the data presented in Tables 2 and 3 that the LCPUFA-rich diets cause an acceleration of acuity development or supernormal acuity. Rather, together, these data suggest only that a transient decrement in acuity occurs in visual development in infants with deficient fatty acid and that acuity normalizes later in these groups. To argue an acceleration in acuity development in one group, the rate of change of acuity with age would have to be different for different groups. There are no data from which to argue this. Delayed acuity maturation is not a very satisfactory model either, because in the few studies in which acuity was tested earlier than 2 months, no difference between groups was found.

## Summary and Implications for Clinical Research

As indicated previously in this chapter, the ACP can be considered a valid test of grating acuity in infants. This conclusion is based upon the good agreement between acuities obtained with the ACP and acuities obtained using the more rigorous FPL procedure. However, grating acuity is not always equivalent to recognition acuity, the type of acuity that is most frequently measured in adults. Research suggests that measures of grating acuity in infants may underestimate true acuity deficits (i.e. deficits in recognition acuity) in infants with certain types of pathology, for example, amblyopia and disorders selectively affecting the macula/fovea (reviewed by Dobson (9)). In contrast, the data suggest that measures of grating acuity provide accurate estimates of acuities in infants with severe abnormalities of the optical and visual pathway (reviewed by Dobson (9)).

The reliability data presented earlier in this chapter indicate that the majority of test–retest differences in ACP acuity tests of most infant populations are within two card steps (one octave). Not surprisingly, the test–retest reliability of two testers' results is somewhat lower than the test–retest reliability of a single tester. This is undoubtedly related to systematic differences among testers in the criteria used to estimate acuity threshold. It should be noted, however, that studies that have investigated differences among testers have reported that these differences are relatively small, less than 0.5 octave (or smaller than a step size between the spatial frequencies of the acuity cards). We have also presented a discussion of factors likely to contribute to variability in ACP results. These factors include biologic variations in an infant's acuity and attention, differences in acuity among infants in the population, variability due to tester-related factors, and variability related to methodological factors (measurement error). Based on data from Birch and Hale (43) and from Mayer et al. (34), we estimate that measurement error, individual infant factors, and single tester factors contribute about half of the total variance, while between-tester differences and biologic differences amongst infants in the population contribute the remaining variability.

We have described acuity results and variability in published studies of binocular acuity in full-term infants. As shown in Table 1, the range of mean acuities reported for infants within one age group varies from 0.81 to 1.6 octaves. Between-study differences could be due to sampling error related to small sample size in some studies, to differences in criteria used by testers between sites to estimate acuity, to methodological factors, and/or to population differences among studies. The considerable variability across studies in mean binocular acuity values suggests that prospective clinical research will be less variable if conducted with a small number of testers at each site.

Our review of published studies of the effects of early diet on grating acuity in infants shows that mean acuities and standard deviations obtained in the majority of studies are within or close to the ranges found in studies of normal infants. Significant effects of diet on FPL and ACP acuity are reported only in studies in which a single tester obtained most of the acuity results. This suggests that some multiple-site/multiple-tester studies may have had insufficient power to detect dietary effects and/or that bias in the results of the single tester in other studies may have influenced acuity results, for example, in studies in which the tester was not masked as to whether an infant was breast- or formula-fed.

The smallest significant differences between the mean acuities of two dietary groups in the nutrition studies was 0.27 octave (67). In all other studies the significant differences between mean acuities were $\geq$ 0.40 octave. These results suggest that care should be taken to design studies to detect differences of 0.3–0.4 octave. Table 4 shows the sample sizes

Table 4. Number of subjects required to detect two effect sizes with a power of 0.8 and a significance level of 0.05.

| Standard deviation (octaves) | Difference between mean acuities (octaves) | |
|---|---|---|
| | 0.30 | 0.50 |
| 0.50 | 45 | 17 |
| 1.00 | 176 | 64 |

required to detect a difference between groups of 0.3 and 0.5 octave, with standard deviations of acuity results of 0.5 and 1.0 octave. Standard deviations lower than 0.4–0.5 octave are unlikely in studies of infant grating acuity (Table 1), and in clinical trials with multiple sites and many testers, standard deviations as high as 1.0 octave would not be unexpected. Thus, for example, if a significant difference of 0.4 octave and standard deviations between 0.5 and 1.0 octave are predicted, approximately 50 subjects in each group would be required.

Finally, mention should be made of the effect size (difference between mean acuities divided by the standard deviation) found in studies with significant differences between dietary groups. For studies using ACP or FPL, effect sizes were a median 0.84 standard deviation ($n = 16$ comparisons, range 0.48–1.35 SD). For studies using VEP methods, effect sizes were a median 1.14 standard deviation ($n = 12$ comparisons, range 0.65–1.43 SD). Thus, effect sizes of about one standard deviation are obtained in studies of dietary fatty acids in human infants using behavioral and VEP measures of grating acuity.

## How can Clinical Research Studies using ACP Minimize Variability?

### Careful control and monitoring of testing conditions

The luminance of the grating cards should be maintained at a constant value across days and across test sites, and the distance between the infant's eyes and the front of the acuity card stage should be carefully monitored during each test. There should be a trained person who holds the infant during ACP testing, especially when infants are between term and 2 months of age. This is because grating acuities may be higher in infants who are actively turned toward each stimulus position, compared to acuities of infants who are seated passively in an adult's lap (see discussion in Dobson (9), pp. 150–1). Infants should be tested only when they are alert and attentive, and not when they are sleepy, fussy, or ill.

### Consistent training of testers and monitoring of testers' results

Ideally, the study should use a small number of testers, each of whom is trained by the same trainers in the identical ACP protocol. We recommend that multisite studies use two testers per site. Testers should be experienced in testing infants in the age range of the study and should maintain their skills by frequent testing of infants. Between-tester reliability should be monitored throughout the study, and mean acuities obtained by each tester should be checked periodically during the study to detect any systematic tester biases. Testing sites should be kept to a minimum, and quality control measures should be implemented to maintain consistency across testers in the testing techniques used. One study of nutritional effects on ACP grating acuity (56) brought all testers to a single site for across-site tester reliability three times during the study, a commendable practice. We recommend monitoring data among sites during the study, and analyzing and correcting multisite data for among-site differences.

## Excluding infants who are at risk of reduced acuity

In studies of preterm infants, those with threshold stage 3 or worse retinopathy of prematurity (ROP) should be excluded, since visual acuity deficits may arise from retinal residua of ROP (69,70). Similarly, in studies of either preterm or full-term infants, any infant with an ocular problem identified or confirmed by an ophthalmologist should be excluded. Because visual deficits can accompany neurodevelopmental abnormalities, for example, intraventricular hemorrhage or periventricular leukomalacia, infants with these problems and infants with confirmed neuromotor abnormalities should be excluded. Also, data from Carlson's nutrition study (58) suggest that inclusion of infants with chronic perinatal complications (e.g. BPD) may mask dietary effects in a group of healthy infants.

The only way to be sure that there are no infants with optical or ocular problems in a study is to conduct a complete eye examination in all infants. This may be more important in older infants (i.e. > 6 months), when acuity has improved to values likely to be affected by refractive error and small cataracts, and when strabismic amblyopia may first appear. Eye examinations are more important in studies of preterm infants than in studies of full-term infants, since ocular, refractive, and strabismic disorders are more prevalent in preterm than in full-term infants (71).

## Correcting for gestational age at birth

Studies that have compared grating acuity development of healthy preterm infants with acuity development of full-term infants have shown that from birth to age three years, preterm and full-term infants show similar acuity if the results are compared based on the infant's age relative to term (45,72–76). After age 3 years, however, healthy preterm children tend to show grating acuity scores at the bottom of the normal range for full-term children (74,77). Thus when preterm infants are tested, results should be analyzed and compared with full-term norms based on age relative to full-term, that is, age relative to due date.

If accurate estimates of gestational age are available, it is also advisable to use age relative to due date, rather than age relative to birth, when testing full-term infants, in order to minimize maturity as a potential source of variability of acuity data. For example, if 'term' birth is defined as $\geq 37$ weeks gestation at birth, the maturational age of a group of infants tested at 1 month postnatal age could range from 1 week (for the infant born at 37 weeks, or 3 weeks prior to a full-term gestation of 40 weeks) to 6 weeks (for the infant born at 42 weeks, or 2 weeks after term). As grating acuity is more closely related to post-term age than to postnatal age, considerable variability could arise from including a 5-week age span of maturation in a group of infants tested nominally at 1 month of age.

In practice, however, correction for age relative to term in studies of full-term infants may have little effect on the variability of acuity results, as the method used to estimate term gestation in most full-term infants is not age from conception, but rather, date of last menstrual period, which introduces more variability into gestational age estimates. In contrast, correction for age relative to term should be carried out in studies of preterm infants because the difference between post-term and postnatal age is substantial, and because estimation of gestational age is typically conducted with more accurate methods (e.g. ultrasound or perinatal examination).

## Is the VEP a Better Measure of Grating Acuity for Clinical Research Studies?

There are two possible reasons for considering VEP measures of acuity to be superior to PL measures of acuity in infants. The first concerns population variability. Clearly, for clinical research studies, one would prefer to use the measure that has the least variance. However, as shown in Tables 2 and 3, data from VEP studies of the effect of diet on acuity provide no evidence that VEP acuities are less variable than acuities obtained from PL studies (FPL and ACP). Nor do normative acuity data obtained with the swept VEP method provide evidence for lower variability than data obtained with FPL and ACP. For example, Norcia and Tyler (78) tested binocular grating acuities in infants between 1 and 12 months of age using the swept VEP method. The median standard deviation of the VEP acuities from 12 age groups was 0.43 octave (range, 0.31–0.65 octave). This value is identical to the median standard deviation of 0.43 octave (range 0.29–0.47 octave) reported in an ACP study of monocular grating acuities in seven age groups between 1 and 12 months (34). Thus, VEP acuity measures appear to be no less variable than PL measures of acuity in infants.

The other argument for using a VEP rather than PL measures of acuity is that the VEP may be more sensitive to optical and visual pathway disorders. In part, this can be argued on the basis that VEP acuities are systematically higher than behavioral acuities, develop to adult levels more rapidly and reflect central retinal rather than peripheral retinal function in adults (see 79). However, the basis for higher VEP than FPL acuity is still unresolved (see discussion in Dobson and Teller (80) for possible explanations for this discrepancy). The nearly constant difference between VEP and behavioral acuities in normal human infants in the first year (see Auestad *et al.* (56) for one of the few within-study comparisons of VEP and PL acuities), argues for methodological or scoring criterion differences to explain at least part of the difference between acuity results obtained with the two procedures.

Higher sensitivity of VEP compared to behavioral grating acuity measures is suggested in a comparison between dietary groups in one study of the effects of fatty acids on visual development in human infants (66). In a rhesus monkey model of n-3 fatty acid deficiency, however, Neuringer found significant differences between deficient and control monkeys in both VEP and FPL acuities at all ages tested (1).

In closing, we leave the reader with a list of our view of advantages and disadvantages of current VEP measures of acuity versus behavioral measures of acuity for use in studies of the effects of fatty acids on infant visual development.

*Advantages of VEP measures of visual acuity:*

(i) The pattern VEP in adults is dominated by projections from the central retina (fovea/parafovea), consistent with the large area of striate cortex devoted to foveal projection and the location of the recording electrodes (79). Although the specific retinal area responsible for the pattern VEP in infants is unknown, VEP acuity is more likely to represent more central retinal function than is behavioral acuity.

(ii) Contrast thresholds and resolution acuity estimated using VEP measurements correlate well with psychophysical thresholds in adults (see reviews in 63 and 81).

(iii) Binocular acuity norms have been published for the swept VEP, a method that has considerable promise for rapid and reliable assessment of acuity in infants (78).

(iv) VEP methods usually estimate higher acuity than behavioral methods in infants, and thus may be more sensitive to precortical and cortical perturbations of the developing visual system than are behavioral methods.
(v) Overt eye movements and orienting behavior of the infant are not required for VEP testing, although attention directed toward the stimulus display must be maintained for varying periods of time, depending upon the VEP paradigm.
(vi) In theory, examiner subjectivity is minimized using an electrophysiological measure such as the VEP. However, in some VEP paradigms, the examiner must select critical parameters that are used to estimate acuity, thus potentially introducing some subjectivity. The examiner also has to judge when the infant is fixating the stimulus display in order to ensure that the infant's central retina is being stimulated.

*Disadvantages of VEP measures of visual acuity*

(i) The course of maturation of VEP acuity depends upon the VEP paradigm and stimulus parameters used (82; see discusson in 63). Thus, normative acuities must be developed for each VEP method and set of stimulus conditions.
(ii) Variability due to within-individual differences and information on population differences have been reported for the swept VEP method in normal infants (78,83) tested within one laboratory. No data on individual and population variability are available for other VEP methods, or for the swept VEP method from other laboratories or for clinical populations.
(iii) Infant state affects the VEP.
(iv) Untestability of a proportion of infants is a problem with VEP technologies (see 66).
(v) VEP technologies require complicated and expensive equipment and a high level of technical expertise to operate and maintain the equipment as well as to record and score the results. At present, equipment and support for the one VEP acuity paradigm on which there are infant norms, the swept VEP, are unavailable, even for instruments already in use. Another commercial company that produced equipment for measuring pattern VEPs in infants has gone out of business. This is likely to be a continuing problem for any VEP technology.

*Summary of behavioral and VEP measures of grating acuity in studies of the effects of fatty acids on visual development*

A VEP measure of grating acuity has theoretical advantages over a behavioral measure of grating acuity in testing infants, primarily because the pattern VEP is selective for central retinal function, while there is no guarantee that a behavioral measure of acuity reflects central retinal activity. Reduced variability would weigh strongly in favor of a VEP measure; however, thus far, evidence for a difference in variability between VEP and behavioral acuities obtained under optimal conditions is lacking. Higher sensitivity to abnormalities in central retinal to cortical projections would be expected for a VEP measure of acuity, but, overall, the results of studies to date in human infants and monkey infants suggest that both behavioral and VEP methods of grating acuity are sensitive to the effects of fatty acids on visual development.

## References

1. Neuringer M, Connor WE, Van Petten C, Barstad L. Dietary omega-3 fatty acid deficiency and visual loss in infant rhesus monkeys. *J Clin Invest* 1984; **73**: 272–6.
2. Neuringer M, Connor WE, Lin DS *et al*. Biochemical and functional effects of prenatal and postnatal omega-3 fatty acid deficiency on retina and brain in rhesus monkeys. *Proc Natl Acad Sci USA* 1986; **83**: 285–94.
3. Neuringer M. The relationship of fatty acid composition to function in the retina and visual system. In: *Lipids, Learning, and the Brain: Fats in Infant Formulas. Report of the 103rd Ross Conference on Pediatric Research.* Columbus, OH: Ross Laboratories, 1993: 134–63.
4. Fantz RL. Pattern vision in young infants. *Psychol Record* 1958; **8**: 43–7.
5. Fantz RL, Fagan JF III, Miranda SB. Early visual selectivity. In: Cohen LB, Salapatek P, eds. *Infant Perception: From Sensation to Cognition*, Vol. 1, *Basic Visual Processes*. Orlando, FL: Academic Press, 1975: 249–345.
6. Fantz RL, Ordy JM, Udelf MS. Maturation of pattern vision in infants during the first six months. *J Comp Physiol Psychol* 1962; **55**: 907–17.
7. Teller DY, Morse R, Borton R, Regal D. Visual acuity for vertical and diagonal gratings in human infants. *Vision Res* 1974; **14**: 1433–9.
8. Teller DY. The forced-choice preferential looking procedure: a psychophysical technique for use with human infants. *Infant Behav Dev* 1979; **2**: 135–53.
9. Dobson V. Visual acuity testing by preferential looking techniques. In: Isenberg SJ, ed. *Eye in Infancy*. St Louis, MO: Mosby, 1994: 131–56.
10. Brown AM. Development of visual sensitivity to light and color vision in human infants: a critical review. *Vision Res* 1990; **30**: 1159–88.
11. Teller DY, McDonald M, Preston K, Sebris SL, Dobson V. Assessment of visual acuity in infants and children: the acuity card procedure. *Dev Med Child Neurol* 1986; **28**: 779–89.

12. Dobson V. Visual acuity testing in infants: from laboratory to clinic. In: Simons K, ed. *Early Visual Development. Normal and Abnormal*. New York: Oxford University Press, 1993: 318–334.
13. Sebris SL, Dobson V, McDonald M, Teller DY. Acuity cards for visual acuity assessment of infants and children in clinical settings. *Clin Vision Sci* 1987; **2**: 45–58.
14. Dobson V, Luna B. Prototype and Teller acuity cards yield similar acuities despite stimulus differences. *Clin Vision Sci* 1993; **8**: 395–400.
15. McDonald M, Dobson V, Sebris SL, Baitch L, Varner D, Teller DY. The acuity card procedure: a rapid test of infant acuity. *Invest Ophthalmol Visual Sci* 1985; **26**: 1158–62.
16. McDonald M, Sebris SL, Mohn G, Teller DY, Dobson V. Monocular acuity in normal infants: the acuity card procedure. *Am J Optom Physiol Opt* 1986; **63**: 127–34.
17. McDonald M, Ankrum C, Preston K, Sebris SL, Dobson V. Monocular and binocular acuity estimation in 18- to 36-month-olds: acuity card results. *Am J Optom Physiol Opt* 1986; **63**: 181–6.
18. Preston KL, McDonald MA, Sebris SL, Dobson V, Teller DY. Validation of the acuity card procedure for assessment of infants with ocular disorders. *Ophthalmology* 1987; **94**: 644–53.
19. Heersema DJ, van Hof-van Duin J. Age norms for visual acuity in toddler using the acuity card procedure. *Clin Vision Sci* 1990; **5**: 167–74.
20. *Teller Acuity Card Handbook*. Dayton, OH: Vistech Consultants, 1989.
21. Dobson V, Carpenter NA, Bonvalot K, Bossler J. The acuity card procedure: interobserver agreement in infants with perinatal complications. *Clin Vision Sci* 1990; **6**: 39–48.
22. Mash C, Dobson V, Carpenter N. Interobserver agreement for measurement of grating acuity and interocular acuity differences with the Teller acuity card procedure. *Vision Res* 1994; **35**: 303–12.
23. Hertz BG, Rosenberg J. Acuity card testing of spastic children: preliminary results. *J Pediatr Ophthalmol Strabismus* 1988; **254**: 139–44.
24. Hertz BG, Rosenberg J. Effect of mental retardation and motor disability on testing with visual acuity cards. *Dev Med Child Neurol* 1992; **34**: 115–22.
25. Mohn G, Van Hof-van Duin J. Rapid assessment of visual acuity in infants and children in a clinical setting, using acuity cards. *Documenta Ophthalmol Proc Ser* 1988; **45**: 363–72.
26. Gstalder RJ, Green DG. Laser interferometric acuity in amblyopia. *J Pediatr Ophthalmol* 1971; **8**: 251–6.
27. Dobson V, Quinn GE, Tung B, Palmer EA, Reynolds JD. Comparison of recognition and grating acuities in very-low-birth-weight children with and without retinal residua of retinopathy of prematurity. *Invest Ophthalmol Vis Sci* 1995; **36**: 692–702.
28. Friendly DS, Jaafar MS, Morillo DL. A comparitive study of grating and recognition visual acuity testing in children with anisometropic amblyopia without strabismus. *Am J Ophthalmol* 1990; **110**: 293–9.
29. Levi DM, Klein S. Differences in vernier discriminations for gratings between strabismic and anisometropic amblyopes. *Invest Ophthalmol Vis Sci* 1982; **23**: 398–407.

30. Mayer DL, Fulton AB, Rodier D. Grating and recognition acuities of pediatric patients. *Ophthalmology* 1984; **91**: 947–53.
31. Mayer DL. Acuity of amblyopic children for small field gratings and recognition stimuli. *Invest Ophthalmol Vis Sci* 1986; **27**: 1148–53.
32. White JM, Loshin DS. Grating acuity overestimates Snellen acuity in patients with age-related maculopathy. *Optom Vision Sci* 1989; **66**: 751–55.
33. Getz LM, Dobson V, Luna B, Mash C. Interobserver reliability of the Teller acuity card procedure in pediatric patients. *Invest Ophthalmol Vis Sci* 1996; **37**: 180–7.
34. Mayer DL, Beiser AS, Warner AF, Pratt EM, Raye KN, Lang JM. Monocular acuity norms for the Teller acuity cards between ages 1 month and 4 years. *Invest Ophthalmol Vis Sci* 1995; **76**: 671–85.
35. Quinn GE, Berlin JA, James M. The Teller Acuity Card procedure: three testers in a clinical setting. *Ophthalmology* 1993; **100**: 488–94.
36. Teller DY, Mar C, Preston KL. Statistical properties of 500-trial infant psychometric functions. In: Werner LA, Rubel EW, eds. *Developmental Psychoacoustics*. Washington, DC: American Psychological Association, 1992: 211–27.
37. McKee SP, Klein SA, Teller DY. Statistical properties of forced-choice psychometric functions: implications of probit analysis. *Perception Psychophys* 1985; **37**: 286–98.
38. Gwiazda J, Wolfe JM, Brill S, Mohindra I, Held R. Quick assessment of preferential looking acuity in infants. *Am J Optom Physiol Opt* 1980; **23**: 420–7.
39. Nachmias J. Starting-point bias of a recent psychophysical method. *Am J Optom Physiol Opt* 1982; **59**: 845–7.
40. Wolfe JM, Held R, Gwiazda J. Reply, to Nachmias, J. Starting-point bias of a recent psychophysical method. *Am J Optom Physiol Opt* 1982; **59**: 848.
41. Trueb L, Evans J, Hammel A, Bartholomew P, Dobson V. Assessing visual acuity of visually impaired children using the Teller acuity card procedure. *Am Orthopt J* 1992; **42**: 149–154.
42. Mash C, Dobson V. The Teller Acuity Card procedure: intraobserver agreement among a sample of infants treated in a neonatal intensive care unit (NICU). *Invest Ophthalmol Vis Sci* 1995; **36**: S369.
43. Birch EE, Hale LA. Criteria for monocular acuity deficit in infancy and early childhood. *Invest Ophthalmol Vis Sci* 1988; **29**: 636–43.
44. Saunders KJ, Westall CA, Woodhouse JM. Longitudinal assessment of monocular grating acuity. *Neuro-ophthalmology* 1996; **16**: 15–25.
45. Brown AM, Yamamoto M. Visual acuity in newborn and preterm infants measured with grating acuity cards. *Am J Ophthalmol* 1986; **102**: 245–53.
46. Hertz BG, Rosenberg J, Sjo O, Warburg M. Acuity card testing of patients with cerebral visual impairment. *Dev Med Child Neurol* 1988; **30**: 632–7.
47. Westheimer G. Scaling of visual acuity measurements. *Arch Ophthalmol* 1979; **97**: 327–30.
48. Banks MS, Bennett PJ. Optical and photoreceptor immaturities limit the spatial and chromatic vision of human neonates. *J Opt Soc Am A* 1988; **5**: 2059–79.
49. Brown AM. Intrinsic contrast noise and infant visual contrast discrimination. *Vision Res* 1994; **34**: 1947–64.

50. Wilson HR. Development of spatiotemporal mechanisms in infant vision. *Vision Res* 1988; **28**: 611–28.
51. Jacobs DS, Blakemore C. Factors limiting the postnatal development of visual acuity in the monkey. *Vision Res* 1988; **28**: 947–58.
52. Vital-Durand F. Acuity card procedures and the linearity of grating resolution development during the first year of human infants. *Behav Brain Res* 1992; **49**: 99–106.
53. Vital-Durand F, Ayzac L, Pinzaru G. Acuity cards and the search for risk factors in visual development. In: Vital-Durand F, Atkinson J, Braddick OJ, eds. *Infant Vision*. Oxford: Oxford University Press, 1996: 186–200.
54. Salomao SR, Ventura DF. Large-sample population age norms for visual acuities obtained with Vistech/Teller Acuity Cards. *Invest Ophthalmol Vis Sci* 1995; **36**: 657–70.
55. Dobson V, Schwartz TL, Sandstrom DJ, Michel L. Binocular visual acuity of neonates: the acuity card procedure. *Dev Med Child Neurol* 1987; **29**: 199–206.
56. Auestad N, Montalto MB, Hall RT, et al. Visual acuity, erythrocyte fatty acid composition, and growth in term infants fed formulas with long-chain polyunsaturated fatty acids for one year. *Pediatr Res* 1997; **41**: 1–10.
57. Carlson SE, Ford AJ, Werkman SH, Peeples JM, Koo WWK. Visual acuity and fatty acid status of term infants fed human milk and formulas with and without docosahexaenoate and arachidonate from egg yolk lecithin. *Pediatr Res* 1996; **39**: 882–8.
58. Carlson SE, Werkman SH, Tolley EA. Effect of long-chain n-3 fatty acid supplementation on visual acuity and growth of preterm infants with and without bronchopulmonary dysplasia. *Am J Clin Nutr* 1996; **63**: 687–97.
59. Birch EE, Spencer R. Visual outcome in infants with cicatricial retinopathy of prematurity. *Invest Ophthalmol Vis Sci* 1991; **32**: 410–15.
60. Jorgensen MH, Hernell O, Lund P, Holmer G, Michaelsen KF. Visual acuity and erythrocyte docosahexaenoic acid status in breast-fed and formula-fed term infants during the first four months of life. *Lipids* 1996; **31**: 99–105.
61. Innis SM, Akrabawi SS, Diersen-Schade DA, Dobson MV, Guy DG. Visual acuity and blood lipids in term infants fed human milk or formulae. *Lipids* 1997; **32**: 63–72.
62. Innis SM, Nelson CM, Rioux MF, King DJ. Development of visual acuity in relation to plasma and erythrocyte ω-6 and ω-3 fatty acids in healthy term gestation infants. *Am J Clin Nutr* 1994; **60**: 347–52.
63. Hamer RD, Mayer DL. The development of spatial vision. In: Albert DM, Jakobiec FA, eds. *Principles and Practice of Ophthalmology*. Philadelphia, PA: WB Saunders, 1994: 578–608.
64. Birch EE, Birch DG, Hoffman DR, Uauy R. Dietary essential fatty acid supply and visual acuity development. *Invest Ophthalmol Vis Sci* 1992; **33**: 3242–53.
65. Makrides M, Simmer K, Goggin M, Gibson RA. Erythrocyte docosahexaenoic acid correlates with the visual response of healthy, term infants. *Pediatr Res* 1993; **34**: 425–7.
66. Makrides M, Neumann M, Simmer K, Pater J, Gibson R. Are long-chain polyunsaturated fatty acids essential nutrients in infancy? *Lancet* 1995; **345**: 1463–8.

67. Birch E, Birch D, Hoffman D, Hale L, Everett M, Uauy R. Breast-feeding and optimal visual development. *J Pediatr Ophthalmol Strabismus* 1993; **30**: 33–8.
68. Carlson SE, Werkman SH, Rhodes PG, Tolley EA. Visual-acuity development in healthy preterm infants: effect of marine-oil supplementation. *Am J Clin Nutr* 1993; **58**: 35–42.
69. Dobson V, Quinn GE, Saunders RA, et al. Grating visual acuity in eyes with retinal residua of retinopathy of prematurity. *Arch Ophthalmol* 1995; **113**: 1172–77.
70. Birch EE, Spencer R. Visual outcome in infants with cicatricial retinopathy of prematurity. *Invest Ophthalmol Vis Sci* 1991; **32**: 410–15.
71. Keith CG, Kitchen WH. Ocular morbidity in infants of very low birth weight. *Br J Ophthalmol* 1981; **67**: 302–5.
72. Birch EE, Spencer R. Monocular grating acuity of healthy preterm infants. *Clin Vis Sci* 1991; **6**: 331–4.
73. Dubowitz LMS, Dubowitz V, Morante A, Verghote M. Visual function in the preterm and fullterm newborn infant. *Dev Med Child Neurol* 1980; **22**: 465–75
74. Getz L, Dobson V, Luna B. Grating acuity development in 2-week-old to 3-year-old children born prior to term. *Clin Vis Sci* 1992; **7**: 251–6.
75. Searle C, Horne SM, Bourne KM. Visual acuity development: a study of preterm and full-term infants. *Aust NZ J Ophthalmol* 1989; **17**: 23–6.
76. Van Hof-van Duin J, Mohn G. The development of visual acuity in normal fullterm and preterm infants. *Vis Res* 1986; **26**: 909–16.
77. Sebris SL, Dobson V, Hartmann EE. Assessment and prediction of visual acuity in 3- to 4-year-old children born prior to term. *Human Neurobiol* 1984; **3**: 87–92.
78. Norcia AM, Tyler CW. Spatial frequency sweep VEP: visual acuity during the first year of life. *Vis Res* 1985; **25**: 1399–408.
79. Regan D. *Evoked Potentials in Psychology, Sensory Physiology and Clinical Medicine*. London: Chapman & Hall, 1972.
80. Dobson V, Teller DY. Visual acuity in human infants: a review and comparison of behavioral and electrophysiological studies. *Visi Res* 1978; **18**: 1469–83.
81. Apkarian P. Visual evoked potential assessment of visual function in pediatric neuroophthalmology. In: Albert DM, Jakobiec FA, eds. *Principles and Practice of Ophthalmology*. Philadelphia: WB Saunders, 1994: 622–47.
82. Orel-Bixler DA, Norcia AM. Differential growth of acuity for steady-state pattern reversal and transient pattern onset-offset VEPs. *Clin Vis Sci* 1987; **2**: 1–9.
83. Hamer RD, Norcia AM, Tyler CW, Hsu-Winges C. The development of monocular and binocular VEP acuity. *Vis Res* 1989; **29**: 397–408.
84. Courage ML, Adams RJ. Visual acuity assessment from birth to three years using the acuity card procedure: cross-section and longitudinal samples. *Optom Vis Sci* 1990; **67**: 713–18.
85. Boergen KP, Kau T, Lorenz B, Rosenauer E. Untersuchungen zur normalen und gestorten visuellen entwicklung mit preferential looking (PL). *Klin Mbl Augenheilk* 1991; **199**: 103–9.
86. Suzuki Y, Awaya S. Studies on development of visual acuity in infants measured by the Teller acuity cards. *Jpn J Ophthalmol* 1995; **39**: 166–71.
87. Innis S, Nelson CM, Lwanga D, Rioux FM, Waslen P. Feeding formula without

arachidonic acid and docosahexaenoic acid has no effect on preferential looking acuity or recognition memory in healthy full-term infants at 9 mo of age. *Am J Clin Nutr* 1996; **64**: 40–46.

## Commentary

**Lucas**: The key question in most clinicians' minds is whether the early or transient effects observed in visual acuity or retinal physiology with LCPUFA supplementation *matter* in terms of later visual function. Given now that many hundreds of babies randomly assigned to LCPUFAs in infancy will enter early to mid-childhood in the next 5 years, what tests would you recommend to researchers in this field to explore this issue formally?

AUTHOR'S REPLY: To investigate the nature and severity of visual functional deficits that persist or may become evident in childhood as a function of LCPUFA supplementation in infancy, we suggest the following measures.

(i) Visual (recognition) acuity tests should be done with modern test designs using symbols or letters in a crowded format and with log MAR intervals between acuity levels (1). Children at about age 3 years can accomplish a symbol-matching acuity test and children at age 5–6 years can read letters on a line chart (2). Both distance and near acuity should be tested, the latter because near acuity has more relevance to reading tasks.
(ii) Contrast sensitivity tests using symbols or letters are important for determining whether visual deficits to low spatial frequencies may be present. These tests can be accomplished at the same ages as recognition acuity tests.
(iii) A measure of stereoacuity that assesses the child's threshold for stereo targets will be sensitive to abnormal binocularity, and thus, will assess cortical visual function. Stereoacuity tests can be done in children at the same time as recognition acuity tests.
(iv) Color vision should be measured with tests that are sensitive to acquired color vision deficits. Children at age 5–6 years can be tested using simple color vision tests (3).
(v) A complete eye examination is strongly recommended for follow-up studies of children who were subjects in dietary fatty acid

studies, and in particular for preterm subjects because of the relatively high incidence of eye disorders in preterm infants.

1. Committee on Vision. Recommended standard procedures for clinical measurement and specification of visual acuity. *Adv Ophthalmol* 1980; **41**: 103–48.
2. Cryotherapy for Retinopathy of Prematurity Cooperative Group. Multicenter trial of cryotherapy for retinopathy – Snellen visual acuity and structural outcome at 5½ years after randomization. *Arch Ophthalmol* 1996; **114**: 417–24.
3. Dobson V, Quinn GE, Abramov I et al. Color vision measured with pseudo-isochromatic plates at five-and-a-half years in eyes of children from the CRYO-ROP study. *Invest Ophthalmol Vis Sci* 1996; **37**: 2467–74.

**Carlson**: You commented that the 2-month-olds in the Auestad study had mean scores below the range for 2-month-olds shown in your Table 1. Until I read your paper, I had not been aware of this. The comment that I made elsewhere to Innis could apply here, i.e. that the multicenter trials in an admirable effort to train inexperienced testers for higher reliability have actually altered the accuracy of the test. If this is indeed the case, how are we to consider the validity of some of these large multicenter trials that are finding no effect? Generally most people look to larger trials as better than smaller trials. This comparison could be used to argue that the opposite may in fact be true using the current conservative methods for training.

AUTHOR'S REPLY: Your question seems to be concerned about missing real effects on visual acuity of LCPUFA supplementation in a multicenter trial that might be found in a single site trial due to the larger number of testers conducting acuity tests. Training of testers in any technique should meet certain criteria of accuracy and reliability. For acuity card testing, the training procedures we have conducted make every effort to meet high standards of training and testing to criterion of infants both during training and in the site center before acuities of study infants are measured. We recommend periodic reliability checks throughout the trial. Training methods have become more rigorous with increased experience with the goal of maintaining stability of acuity measures both within and between testers. We believe our current training protocols meet that goal.

CARLSON: My understanding is that the acuity cards were developed primarily to identify visual problems. For this purpose, if one underestimates 'true' acuity by nearly 0.5 octave, it is not critical. However, in research, one would like as accurate an assessment of grating acuity as possible, especially if the effect of diet on acuity is only about 0.5 octave as the acuity card procedure studies suggest. Toward that end, do you think there would be any value in including steps of 0.25 octave rather than 0.5 octave for research protocols? Of course, this might necessitate the use of different sets of spatial frequencies for each age so as not to tire infants.

1. Carlson SE, Ford AJ, Werkman SH, Peeples JM, Koo WWK. Visual acuity and fatty acid status of term infants fed human milk and formulas with and without docosahexaenoate and arachidonate from egg yolk lecithin. *Pediatr Res* 1996; **39**: 882–8.

AUTHOR'S REPLY: Smaller step sizes in the acuity card test theoretically might improve precision in estimating acuity, that is, acuity values intermediate between the 0.5-octave intervals could be measured. However, this is likely to be a small effect. The other theoretical rationale for smaller step sizes would be in terms of variability and test–retest reliability. Two studies in which 0.3-octave steps were used in acuity testing did not show reduced population standard deviations or higher test–retest reliability than studies in which 0.5-octave step sizes were used, as mentioned in our chapter. For acuity card testing, the increased test time required to measure acuity using a larger number of test stimuli would not seem to be a good tradeoff for a theoretical increase in precision.

**Innis**: Does an acuity deficit of a given magnitude have a different impact on the visual world of infants at different ages?

AUTHOR'S REPLY: One way of conceiving of the effect of a given acuity deficit for an infant is to compare the acuity to that of infants at younger ages. For example, based upon the normal (non-linear) growth curve of acuity, a 1.0-octave deficit in the acuity of a 2-month-old represents acuity of a 1-month-old, while the same deficit in a 12-month-old could be obtained by a 4–5-month-old. The implication of this is that an acuity deficit of a given magnitude may have less impact on visual function than for an older infant.

**Singer**: What evidence, if any, is there that the increased visual experience of healthy preterm infants affects visual acuity measurements?

AUTHOR'S REPLY: Studies have shown that acuity values of preterm infants are in better agreement with normative values if the age at test is determined based upon the infant's age relative to term (1–6). There is some evidence that acuity may be slightly higher in the preterm infant than in the full-term infant when acuities are compared based upon post-term ages (5–7).

1. Birch EE, Spencer R. Monocular grating acuity of healthy preterm infants. *Clin Vis Sci* 1991; **6**: 331–4.
2. Brown AM, Yamamoto M. Visual acuity in newborn and preterm infants measured with grating acuity cards. *Am J Ophthalmol* 1986; **102**: 245–53.
3. Dubowitz LMS, Dubowitz V, Morante V. Visual function in the newborn: a study of preterm and full-term infants. *Brain Dev* 1980; **2**: 15–29.
4. Getz L, Dobson V, Luna B. Grating acuity development in 2-week-old to 3-year-old children born prior to term. *Clin Vis Sci* 1992; **7**: 252–6.
5. Searle C, Horne SM, Bourne KM. Visual acuity development: a study of preterm and full-term infants. *Aust NZ J Ophthalmol* 1989; **17**: 23–6.
6. Van Hof-van Duin J, Mohn G. The development of visual acuity in normal fullterm and preterm infants. *Vis Res* 1986; **26**: 909–16.
7. Norcia AM, Tyler CW, Piecuch R, Clyman R, Grobstein J. Visual acuity development in normal and abnormal preterm human infants. *J Pediatr Ophthalmol Strabismus* 1987; **24**: 70–4.

**Colombo**: The chapter on the use of the visual acuity cards is informative and helpful in evaluating the results of many of the studies in this area. At the same time, it does not address the simple question of whether the acuity cards are the best means of evaluating the effects of such nutritional supplements on infant visual function. Obviously, the cards yield a behaviorally mediated measure of the infant's acuity. However, some of the research I am acquainted with in the area of nutritional supplementation suggests that, for example, DHA is a particularly important structural component specifically for the rod photoreceptors. If, in fact, nutritional supplements differentially affected scotopic versus photopic function, would the acuity cards be sensitive to this? The acuity cards provide an excellent test, and they are relatively easy and fast to administer (we have, in fact, used them for rapid visual screening in our own laboratory studies of normal infants). Given, however, that they yield an admittedly small part of the picture of visual function, are there other core visual functions which might be worth investigating beyond high-end acuity thresholds?

AUTHOR'S REPLY: Visual measures of the photopic visual system that might yield functional effects of LCPUFA apart from visual acuity include contrast sensitivity, temporal sensitivity (e.g. flicker), vernier or hyperacuity, stereoacuity, and color vision. A measure of the scotopic visual system that might be useful is dark-adapted light sensitivity.

# Behavioural Science Considerations

# Long-Chain Polyunsaturated Fatty Acids and the Measurement and Prediction of Intelligence (IQ)

ROBERT B. McCALL AND CLAY W. MASH

*Office of Child Development, UCSUR/121 University Place, University of Pittsburgh, Pittsburgh, PA 15260, USA*

## Introduction

A major issue in the decision of whether to add to infant formulas certain long-chain polyunsaturated fatty acids (LCPUFAs), especially docosahexaenoic acid (DHA) and arachidonic acid (AA), is the possibility that such substances are important for normal mental development. DHA and AA, among other LCPUFAs, are natural constituents of human breast milk and may be essential for normal brain functioning. Typically, fetuses obtain these substances prenatally from their mothers and postnatally from breast milk. Infant formulas in the USA, however, do not contain LCPUFAs. Consequently, formula-fed infants, especially preterm infants who are exposed to their mothers' nutritional system for a shorter period of time, have relative deficits in these substances, and it is possible that these deficits play a causal role in the lower IQs of some preterm infants relative to full-term infants, and of formula-fed relative to breast-fed infants (1).

Makers of the potential DHA additive are eager to promote the idea that DHA in particular might contribute to later intelligence. Martek Biosciences Corporation of Columbia, Maryland, for example, describes DHA as 'food for better brains', and some media headlines have read 'bottle-fed babies need brain boosting formula' (2). Although LCPUFAs allegedly contribute to improved visual acuity and provide other neurological and nutritional benefits (see other chapters in this volume), it is the supposed potential of LCPUFAs to promote intelligence that tends to make headlines and stir the emotions of parents.

To evaluate these claims, it is helpful to have an understanding of the nature and measurement of intelligence, principally IQ. In addition, many of the studies in which LCPUFAs have been experimentally added to infant formulas have followed these infants only through 12–18 months of age. These infants were evaluated with standardized tests of general infant development (3) or assessments of the nature of the infants' deployment of attention to novel and familiar stimuli. The assessment of the implications of these studies for mature intelligence requires knowledge of the ability of these infant measures to predict childhood and adult IQs. Consequently, this chapter begins with a discussion of these issues, followed by a review of the portion of the LCPUFA literature that pertains directly to mental performance.

## The Nature and Measurement of Intelligence in Children and Adults

General mental performance has been studied extensively for nearly a century. Much is known, but a great deal remains unknown. Intelligence is one of the most controversial concepts among scientists and the general public alike, especially when claims are made regarding the relative contributions of heredity and environment to differences between individuals and between racial and ethnic groups in intelligence. Consequently, the American Psychological Association, partly in response to the furor caused by the publication of *The Bell Curve* (4), convened a consensus committee to prepare a statement declaring what is known and what is not known about intelligence. This committee published its review in a remarkably readable and succinct paper (5), and what follows in this section is primarily a selective summary of its conclusions.

### Definition

No agreed definition exists for the concept of intelligence. When two dozen prominent theorists were asked to define intelligence, they provided two dozen somewhat different definitions (6). Common themes among these definitions pertain to the ability to understand complex ideas, to adapt effectively to the environment, to learn from experience, to engage in various forms of symbolic or abstract reasoning, and to overcome obstacles through thought.

Intelligence, whatever it is, is probably not an entity or thing we carry around with us. Rather, the term refers to a class of behaviors that we choose to describe as 'intelligent'. In this sense, it is akin to the notion of 'athleticism', which also lacks a tidy definition and also refers to a rather diverse collection of skills and behaviors. While it may seem strange that such an important and common concept as intelligence lacks a consensus definition, the APA committee observed that 'scientific research rarely begins with fully-agreed definitions, though it may eventually lead to them' (5).

## Measuring Intelligence

While it may seem anomalous to try to measure a concept that lacks a precise, consensus definition, measurement provides a tangible quantified entity that can be related to various behaviors of interest, and the pattern of those associations will contribute to the creation of 'construct validity' for the concept underlying the measurement.

## IQ tests

The most widely used and researched measures of intelligence are the so-called IQ (intelligence quotient) tests. They stem from an instrument originally created by Alfred Binet in France in the early 1900s to select school children for specialized educational experiences in a more objective and (ironically) less prejudicial way than simply relying on teacher or administrator judgments. Current versions of such tests are the Stanford–Binet and the Wechsler Scales (e.g. 7,8) among others. Such tests include vocabulary items, analogies, memory for numerical digits, spatial–perceptual tests, psychomotor skills, and other types of items. Except as noted, the focus of this chapter is on intelligence as measured predominantly by these IQ tests.

Since these tests are highly verbal in nature, performance depends on the language environment and culture of the individual. This obvious fact has led to claims that such tests are biased in favor of the upper-middle class, so other tests have been created that do not involve language (e.g. Raven's Progressive Matrices, which requires inductive reasoning about two-dimensional spatial patterns). These so-called 'culture-free' tests correlate highly with the verbal IQ tests, and it is debatable whether they are any more independent of one's cultural experience than the verbal tests.

*How many 'intelligences'?*

Debate has raged for decades over whether intelligence is unitary or a collection of different types of mental abilities. Athleticism can refer to widely different skills, such as a gymnast performing on the nonparallel bars or a shot putter heaving an iron ball. Neither of these two athletes is likely to perform the other's feats very well, yet both can be said to have high degrees of athleticism. Similarly, the global concept of intelligence may embrace both the kind of mental skills typically required in schools (e.g. 'academic intelligence') as well as 'practical intelligence' (e.g. street smarts, social intelligence), for example. Several theorists have proposed multiple intelligences, from Gardner's (9) musical, bodily-kinesthetic, personal, linguistic, logical-mathematical, and spatial intelligences to Sternberg's (10) analytical, creative, and practical intelligences. Further, the ability to solve certain problems can be highly specific to a particular context, or quite general across contents. For example, women shoppers had no difficulty comparing product values at a supermarket but were unable to carry out the same mathematical operations in a paper and pencil test (11).

No conceptual agreement exists on this issue, and the data help only a little. Specifically, individuals who perform well on the assessment of one mental ability also tend to perform well on the assessment of a different mental ability, which communality presumably reflects some rather general mental ability, which is often called 'general intelligence' or '$g$'. This is the unitary notion of intelligence. But tests for over 70 different abilities are available (12), and individuals are not uniformly good or poor in all skill areas. They are good at one general type of ability and less good at another general type. One way to think about all these different abilities is to form a hierarchy or pyramid of abilities, with $g$ at the top, followed by a small set of general categories of abilities (e.g. verbal, spatial-perceptual), followed by yet more specific abilities within each of those categories (e.g. vocabulary, verbal analysis), and so on. So while there appears to be one general intelligence $g$, there are also more and more specific abilities within it that are progressively more independent of each other.

## Stability and change in IQ

Most tests of general intelligence yield a score that in the population averages 100, with a standard deviation of approximately 15 for each age. This means that the test is designed so that it does not reveal general age differences in average IQ performance. Furthermore, after the age of approximately 6 years, the stability of intelligence is very high. That is, the correlation within a single group of individuals of their performance on an intelligence test assessed at one age with their performance assessed a year later is 0.90 or higher, and the correlation between tests given at age 6 and age 18 is nearly 0.80 (13). Therefore, children who score high on an intelligence test taken at one age are likely to score high the next year and even several years later. Year-to-year stability in IQ prior to age 6 years, however, is much less impressive (see below).

Nevertheless, while many children do not change in IQ very dramatically during childhood and adolescence, some do. Despite the high year-to-year correlations, the average change in IQ between the highest and the lowest score for a sample of children who were assessed 17 times between 2½ and 17 years of age was 28.5 IQ points, one-third of the children changed at least 30 points, 1 out of 7 changed 40 or more points, and one individual improved 74 IQ points during this period (14). Therefore, while IQ in childhood and adulthood is perhaps the most stable behavioral measurement we know, substantial change is nevertheless possible for large subsets of individuals.

## The Validity of Intelligence Tests

If IQ tests measure 'general mental ability', then individual differences in IQ scores should be related to the performances of those individuals on activities that presumably require 'general mental ability', such as school achievement and occupational status. In fact, IQ is the best single predictor of school performance, years in school, and job status, and this is true for essentially all racial and ethnic groups in the USA (5). Specifically, IQ scores correlate approximately 0.50 with school grades, 0.55 with attained years of education, 0.50 with social status, 0.30–0.50 (0.54 when corrected for unreliability) with various measures of job performance, and somewhat less than 0.20 with juvenile delinquency (5). However, the squares of these correlation coefficients reflect the proportion of the variability between individuals in the outcome measure that is associated with the predictor (i.e. IQ). Notice, therefore, that in no case does IQ account for more than 30% of the differences between individuals on these important outcomes.

Consequently, while IQ may be the best single predictor of these outcomes, it certainly is not the only predictor, nor does it account for the majority of differences between individuals on these measures.

## Factors Correlating with IQ

### Neural efficiency

A variety of recent studies have shown that people who apprehend, scan, retrieve, and respond to simple perceptual and cognitive tasks more quickly than others tend to score more highly on intelligence tests (e.g. 15–17). This has led to the general hypothesis that the speed or neural efficiency with which the neurological system acts or processes information is related to general IQ. Measures of such speed might include the rapidity with which individuals make same/different judgments or other speeded responses to visual displays, the speed with which individuals react to the presence or absence of a stimulus, the amount of time that people spend inspecting a visual display before making a judgment or response, and certain measures of visual evoked potentials in response to the introduction of stimuli. However, such correlations are rarely more than 0.30, which is less than 10% of the variability in IQ. Nevertheless, these data suggest that at least a component of IQ may consist of speed of information processing, speed of neural conduction, or neural efficiency, attributes that may be influenced by nutrition and other biological agents and may be present and potentially measurable in infancy (see below).

### Heritability

Estimates of the proportion of differences between individuals in IQ that are associated with differences in the genetic backgrounds of those individuals range from approximately 45% during childhood to approximately 75% during adulthood. This fact, plus the neural efficiency results, seem to suggest that a substantial contributor to intelligence is biological in nature, and this factor apparently takes on greater importance with age.

However, environmental factors contribute to differences in IQ as well. For one thing, differences between families (e.g. in terms of the general intellectual climate of the home) seem to influence IQ in early childhood, but the unique (called 'nonshared') experiences of children within a family (who share the same general intellectual home climate) seem to contribute more substantially in adolescence and adulthood. Further, while occupation and schooling are consequences of intelligence, they also contribute to it.

That is, while intelligence predicts school performance and the number of years of attained education, attending school and devoting oneself to the academic task also improve IQ. Heredity and environment may both correlate highly with IQ, because they correlate highly with each other.

*Perinatal factors*

Low birth weight and low birth weight relative to gestational age both contribute to the tendency of low birth weight infants to have lower IQ scores in later childhood (1). These correlations tend to be quite small (e.g. $r = 0.05$ to $0.13$) because only a small portion of infants in the population are of low birth weight. Moreover, the effects of low birth weight are substantial only when the birth weight is very low, less than 1500 g (5). Most low birth weight infants who do not have a known biological insult and who are raised in intellectually stimulating homes are not different in IQ from the population.

*Nutrition*

The role of early nutrition in behavioral development has been studied in a variety of ways. For example, a review of 165 experimental studies using animals does not reveal a consistent disadvantage in terms of learning and memory for animals that are relatively undernourished early in life (18). Work with humans, primarily in developing countries where both poor diet and medical and social problems are common, shows that dietary supplementation following poor diets in infancy produces some behavioral or intellectual benefits (19,20), but some scholars have suggested that at least some of these benefits are associated with stimulation rather than diet (21). Slightly more persuasive, perhaps, is other evidence indicating that undernutrition associated with certain disorders (e.g. infantile hypertrophic pyloric stenosis (22)) or specific nutrient deficiencies (e.g. 23) can be associated with general or specific deficits in behavior and mental performance. Similarly, prolonged malnutrition during childhood is associated with long term IQ differences, but again it is often not clear whether the malnutrition has a direct effect on IQ or it influences the responsiveness, motivation, and activity of the children when tested (5).

The issue, however, is whether naturalistic variations in nutrition, especially the feeding of human infants with human milk versus formula, in developed, not just underdeveloped, countries is associated with developmental and intellectual differences. Reviews of this literature have also declared the results to be inconsistent (24), and even prospective studies of nutritional interventions have not produced clear findings partly because of poor design and procedural problems (25). More recently, however, a large-scale study in the UK, described in more detail below and in Lucas's chapter in this volume, randomly assigned low-birth-weight infants whose mothers chose not to provide breast milk to be fed a standard formula or a special nutritionally enriched preterm formula for an average of one month after birth. Infants fed the enriched diet scored substantially higher on the Bayley Scales of Infant Development at 18 months (26), and Lucas reports at this conference that differences between these groups also are found in verbal IQ at 7.5–8 years of age. This study, more persuasively than most, demonstrates that, even in developed countries, perinatal nutrition can influence infant milestone development and childhood verbal IQ, at least for low-birth-weight infants. It should be noted, however, that infants fed human milk, which was lower in nutritional content than even the standard formula, produced developmental scores (especially on the psychomotor scale) at 18 months that were higher than those associated with the standard formula, suggesting either a maternal selection advantage (see below) for human milk or that general nutrition may not be the only mechanism to prevent slower developmental progress.

Therefore, the role of prenatal and postnatal nutrition on IQ per se is ambiguous. Clearly, some unknown minimum nutrition would seem to be necessary, but more than that minimum may not make much difference.

## The Prediction of Childhood IQ from Infant Assessments

A large portion of the literature on LCPUFAs and intelligence rests on the prediction of later IQ from one of two types of measurements made during the first two years of life – standardized infant tests and assessments of the nature of infants' deployment of attention to novel and familiar stimuli.

### Predictions from Standardized Infant Tests

Originally, standardized tests of infant development, such as the Bayley Scales of Infant Development, the Gesell, and other tests, were not designed to be tests of 'intelligence' or to predict later IQ. Instead, they were simply quantitative normative descriptions of the ages at which an infant attained typical motor, behavioral, and eventually verbal and mental milestones. Their primary use was to determine if an infant was 'on schedule' or seriously 'behind schedule'.

However, these tests were assessments of 'general developmental performance', some even had 'mental' and 'psychomotor' subtests analogous to some IQ tests' 'verbal' and 'performance' scales, and many were scored similarly to an IQ test by having an age-adjusted mean of 100 and a standard deviation of approximately 15–16. As a result, these tests became used as if they were assessments of 'infant intelligence'. Furthermore, the early concept of 'intelligence' declared that it was a unitary, pervasive, and stable characteristic of individuals, so it was assumed that the infant intelligence tests would measure this characteristic and consequently predict later IQ. However, decades of research have consistently demonstrated that scores from the standardized infant tests obtained during the first 2 years of life do not predict childhood IQ to any useful extent (13).

## Correlations within infancy

Table 1 presents the median correlations of scores on standardized infant tests across ages within the infancy period (13). These correlations are modest even across very short time periods. Specifically, correlations across a few months during the first 2 years of life are 0.40–0.52; in contrast, correlations for IQ across 1-year intervals between 9 and 12 years of age are 0.90–0.93 (13). These low correlations during infancy are not due to the unreliability of the infant test, which attains reliabilities of approximately 0.85 after 3 months of age. Notice that these correlations follow the typical developmental pattern in which correlations generally increase as the subject becomes older and decrease the longer the interval separating the two assessments.

Table 1. Median correlations across studies between infant test scores at various ages during the first 2 years of life (13)

| Age at later test (months) | Age at infant test (months) | | | |
| --- | --- | --- | --- | --- |
| | 1–3 | 4–6 | 7–12 | 13–18 |
| 4–6 | 0.52 (8/6) | – | – | – |
| 7–12 | 0.29 (14/6) | 0.40 (18/10) | – | – |
| 13–18 | 0.08 (3/3) | 0.39 (6/6) | 0.46 (9/6) | – |
| 19–24 | −0.04 (3/3) | 0.32 (6/6) | 0.31 (9/6) | 0.47 (7/6) |

* Decimal entries indicate median correlations, the numbers in parentheses give the number of different $r$ values and the number of independent studies used to calculate the median. In the case of more than one $r$ per study, the median $r$ for that study was entered into the calculation of the cell median.

*Predictive correlations to childhood IQ*

Table 2 presents the median longitudinal correlations between scores on standardized infant tests obtained during the first 2 years of life to IQ obtained during childhood for unselected samples of presumably 'normal' infants (13), and Table 3 presents the analogous correlations for samples of at-risk and frankly disordered infants (27).

Notice several themes. First, the correlations to later IQ are essentially zero from assessments made in the first year or two of life. Thereafter, they rise slowly and steadily as the age of the infant assessment increases. But even then, predictions from the first 2 years of life are so small that they are of little scientific interest and essentially no practical importance.

Second, the two principles of longitudinal correlations – they get higher as the age of the subjects increases and lower as the time interval between assessments increases – are found here, but with one important modification. Studies in which the infant test items are factored and cohesive subsets of items are used to predict later IQ (not shown here) demonstrate that once the predictions to later IQ appear (that is, from assessments made at 21 months or older) the size of the correlations rises immediately to asymptotic levels regardless of when the IQ test was given (28); that is, the prediction from 24-month to 10-year IQ is nearly as high as to 4-year IQ. This seems to suggest that once symbolic relations are present in the

Table 2. Median correlations across studies between infant test scores and childhood IQ (13)

| Age at childhood test (years) | Age at infant test (months) | | | | |
| --- | --- | --- | --- | --- | --- |
| | 1–6 | 7–12 | 13–18 | 19–30 | Average $r^\dagger$ |
| 8–18 | 0.06 (6/4) | 0.25 (3/3) | 0.32 (4/3) | 0.49 (34/6) | 0.28 |
| 5–7 | 0.09 (6/4) | 0.20 (5/4) | 0.34 (5/4) | 0.39 (13.5) | 0.25 |
| 3–4 | 0.21 (16/11) | 0.32 (14/12) | 0.50 (9/7) | 0.59 (15/6) | 0.40 |
| Average $r^\dagger$ | 0.12 | 0.26 | 0.39 | 0.49 | |

* Decimal entries indicate median correlation, the numbers in parentheses give the number of different $r$ values and the number of independent studies used to calculate the median. In the case of more than one $r$ per study, the median $r$ for that study was entered into the calculation of the cell median.
† The average of the median $r$ values presented in that row/column.

Table 3. Median correlations across studies between standardized infant test scores and childhood IQ for at-risk and frankly disordered samples* (27)

| Age at childhood test (years) | Age at infant test (months) | | | | |
|---|---|---|---|---|---|
| | 1–6 | 7–12 | 13–18 | 19–30 | Average $r$† |
| 8–18 | – | – | – | 0.46 (5/2) | 0.46 |
| 5–7 | 0.40 (2/2) | 0.51 (5/2) | – | 0.70 (4/2) | 0.54 |
| 3–4 | 0.20 (1/1) | 0.42 (4/3) | 0.68 (2/) | 0.83 (1/1) | 0.53 |
| 2 | 0.39 (4/4) | 0.42 (8/6) | 0.70 (2/2) | 0.86 (1/1) | 0.59 |
| | 0.33 | 0.45 | 0.69 | 0.71 | |

\* Decimal entries indicate median correlation, the numbers in parentheses give the number of different $r$ values and the number of independent studies used to calculate the median.
† The average of the median $r$ values for that row or column.

toddler, which skills are needed for two-word utterances at approximately 2 years and symbolic thought later, correlations to later IQ rise precipitously and they seem to predict far into the developmental future at essentially a constant level.

Third, the correlations are slightly higher for at-risk and frankly disordered infants (Table 3) than for unselected normal infants (Table 2). For example, when predictions are made from 7–12 months to IQ at 3–7 years of age, the median correlation for at-risk and disordered samples is approximately 0.50 versus 0.30 for nonrisk samples. This is likely due to the presence of infants who have biological disorders that are less subject to remediation and improvement by positive home environments. Other data reveal that predictions are also higher for lower scoring infants than for higher scoring infants (13), perhaps for the same reason.

Finally, these correlations are not higher that those based simply on parental education or socio-economic status (SES), which predict childhood or adult IQ at the level of approximately 0.50.

*Prediction of Later IQ from Infant Attentional Measures*

Developmentalists, being obsessed with the concepts of developmental continuity and stability, never accepted the null hypothesis provoked by the data reviewed above that mature intelligence was not detectable in early infancy (29). Instead, they assumed the standardized infant tests were assessing the wrong characteristics, that intelligence was detectable during infancy if we only knew how to measure it, and that once we identified the relevant characteristics, infant assessments would predict later IQ at levels better than the standardized infant tests. Such faith in developmental stability has been partly rewarded, because a variety of measures of the deployment of attention to novel and familiar stimuli obtained during the first year of life have been shown to predict childhood IQ at levels somewhat above those of the standardized infant tests, but again generally no higher than those obtained by predicting with parental education or SES (30,31).

The following summary of this literature derives largely from McCall and Carriger (31), McCall and Mash (32), and McCall (33), and a detailed review by Colombo can be found in this volume.

*The infant assessment contexts*

The measures of attention during infancy that predict later IQ predominantly come from two (but there are more) assessment contexts.

In the habituation context, a simple 'standard' stimulus is illuminated in front of the infant either for a fixed duration of time regardless of how long the infant looks at it, or for a period of time that lasts until the infant looks at and then looks away from the stimulus ('infant control procedure'). The number of stimulus presentations may be fixed ('fixed trial procedure'), or the stimulus may be presented for as many presentations as is necessary for the infant to display some specifically defined decrement in looking relative to the infant's length of looking on the first one-to-three trials ('habituation to criterion procedure'). Once this familiarization phase has been completed, a different or 'novel' stimulus is presented to observe the infant's resurgence of looking to the novel relative to the last familiar (standard) stimulus, which also serves as a check against a decline in looking during familiarization that might be associated with fatigue rather than the acquisition of a memory for the standard stimulus. The two most common measures from this paradigm that predict later IQ are the rate of habituation (e.g. the number of trials to reach a habituation criterion or the percentage decrement in looking over a fixed number of trials) and the response to novelty (e.g. the ratio of looking time to the novel stimulus relative to the last presentation of the familiar standard stimulus).

In the recognition memory context, a single standard stimulus or a pair of identical standard stimuli are presented for one or more exposure periods of fixed duration (i.e. regardless of how long the infant looks at this standard stimulus). Once this familiarization phase is completed, the standard stimulus is presented simultaneously with a novel stimulus for at least one test trial presentation and possibly two (with the left–right position of stimuli reversed on the second trial). The most commonly used measure that predicts later IQ is the ratio of the length of looking to the novel stimulus relative to the sum of looking to the novel plus familiar stimuli during the test trial(s).

Other measures have been used less frequently to predict later IQ from both the habituation and recognition memory paradigms. These include the total length of looking during the familiarization phase, the average length of a fixation, and the number of different fixations. Generally, the more rapid the habituation, the shorter the total looking during familiarization, the more looks and the shorter the average look, and the greater looking to the novel relative to the familiar stimulus during the test phase all have been found to predict higher childhood IQs.

*The prediction phenomenon*

In a review of 31 sample correlations from 10 habituation and 13 recognition memory studies, McCall and Carriger (31) came to the following conclusions:

(i) Measures from habituation and recognition memory assessments made on a variety of risk and non-risk samples in the first year of life predicted later IQ assessed between 1 and 8 years of age with a weighted ($N$) average normalized correlation of 0.36 or a raw median correlation of 0.45.

(ii) There was essentially no difference in the size of the predictive correlation based upon measures from the habituation versus the recognition memory paradigms.

(iii) The prediction phenomenon is not obviously associated solely with studies coming from one laboratory, one particular infant response measure, or a few extremely disordered infants within a sample.

(iv) Relative to the reliabilities of the infant measures (which are quite poor, $r = 0.30$ to $0.40$), the level of prediction to childhood IQ is substantial, with correlations to later IQ frequently higher than the

reliability of the infant measure alone (which is statistically possible because of the high reliability of childhood IQ).
(v) Predictions are somewhat higher for risk than for nonrisk samples.
(vi) Predictions from these attentional measures are somewhat higher than from standardized infant tests of general development for nonrisk samples, but this relative advantage does not exist for risk samples; and the level of prediction in either case is not consistently higher than predicting from parental education or socioeconomic status.
(vii) Prediction coefficients seem to be somewhat higher when the infant assessments are made at or between 2 and 7 months of age than when they are made earlier or later.

The data for the last conclusion are contained in Fig. 1, which presents the predictive correlations as a function of the age of infant and childhood assessments, together with median $r$ values. Notice two themes in this figure. First, the median correlations on the ordinate are remarkably consistent from age to age, suggesting (in parallel with the findings for the standardized infant tests) that the predictions from infancy are almost equal, regardless of the age at which the childhood IQ assessment is made (at least between ages 1 and 7 years, and other data suggest this is likely to be true at least through age 12 years). In contrast, the median correlations on the abscissa are also remarkably consistent, but only between 2 and 7 months inclusive, falling off before 2 months and after 7 months, although the number of data points for these extreme ages is few. It is possible that the lower correlations during the first 2 months of life are associated with the lower reliability of the infant assessments during this age period due to state fluctuations and other extraneous factors. Alternatively, infants this young do not systematically scan stimuli in the environment and may not look at or compare two stimuli presented simultaneously. But no obvious explanation exists for the possible reduction in predictions after 7 months, a phenomenon that runs counter to the usual developmental trend of increasing predictive correlations the later in development the predicting assessment is made. Perhaps new mental capabilities arise at this age that mask the predictive process (see below), or the task is less interesting to older infants.

*Predictive mechanism or process*
Because longitudinal studies are needed to demonstrate this prediction phenomenon, it has taken many years to accumulate enough data reflecting different paradigms and measures to establish that the prediction phenom-

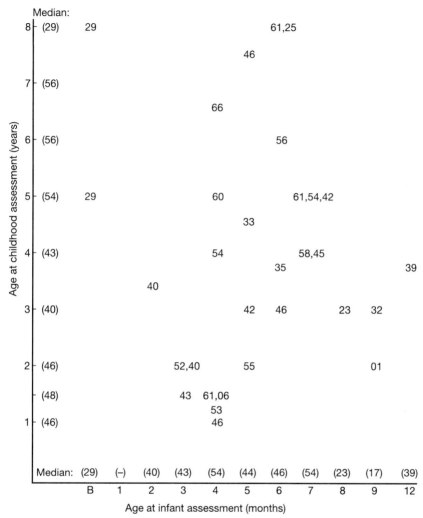

Fig. 1. Predictive correlations (r) as a function of the age of infants and childhood assessments. Marginal median r values are given in parentheses along each axis (31).

enon actually exists. It is only recently, then, that scholars have turned their attention to speculating what the underlying neurological, physiological, or psychological mechanism or process might be that mediates this prediction. The strategy has been to make the most parsimonious assumptions; namely, that the different measures made in different contexts during infancy all tap the same underlying mechanism and that the same mechanism underlies both the infant and childhood IQ assessments. These assumptions may not be true (see below), but they are the simplest starting point. Tentatively proceeding under these assumptions then, the question becomes: What is the common process that underlies all of these assessments at widely different ages?

Unfortunately, almost no data exist that address this question directly. Three primary categories of explanations have been offered by Colombo (34) and McCall (33), and are reviewed by Colombo (35) and McCall and Mash (32). First, scholars in the field have traditionally interpreted the infant measures to reflect speed of information processing. Such assertions have been made primarily on the basis of face validity. Thus it is assumed in the habituation paradigm that fast habituation reflects the more rapid acquisition of a memory for the standard stimulus, and that attention to the novel relative to the familiar stimulus in both habituation and recognition memory paradigms reflects the infant's ability to rapidly encode the standard in the short periods of familiarization typically allowed as a basis for attending longer to the novel relative to the familiar stimulus during the test phase. Similarly, short looking during familiarization and short average looks (or a greater number of looks) also presumably reflect more rapid information processing and encoding of the standard stimulus into memory.

No one questions the basic assumption that the infant must encode some information from the standard stimulus as necessary for behavioral habituation to occur or to respond discriminatively with more looking to the novel than to the familiar stimuli. The debate, forged primarily by McCall (32,33), challenges the requirement that individual differences in the measures that predict later IQ reflect individual differences in the speed of information processing. The argument, too lengthy to present in detail here, is that (i) little information from the standard stimulus needs to be encoded to be able to detect that a completely different novel stimulus is not the same as the familiar standard, and (ii) whatever encoding takes place during familiarization takes very little time, much less than even the short familiarization periods typically offered in these research contexts. Therefore, it is argued, individual differences in the actual speed of encoding the familiar standard are so small a component of the measures of habituation rate and novelty preference that other mechanisms must be invoked to explain the ability of these measures to predict later IQ.

Nevertheless, the rate of information processing hypothesis has great face validity and seems to gel with the emerging evidence that various measures of fundamental processing speed and neural conductivity seem to correlate with IQ performance in adults (e.g. 5,15–17,36–43). It is interesting to note that the correlations of such measures with adult IQ are approximately the same size as the correlations that predict later IQ from the attentional measures taken during infancy. Further, if individuals differ permanently on such a fundamental neurophysiological disposition as neural conduction speed, then the consistency of correlations from infancy to IQ assessed at almost any childhood (and adulthood?) age is reasonable. Why predictions would fall off after 7 months from the infant assessments, however, is not clear, although it is possible that more advanced mental capabilities (e.g. the disposition and ability of the infant to try to attach meaning to a stimulus: 'What is that?' or 'Who is that?') may interfere with the relatively pure assessment of neural conduction speed earlier in infancy.

*Memory skills*

Colombo (34) presents the argument that memory capacity may underlie the longitudinal predictions. McCall and Mash (32) agree that some memory is required for infants to habituate and respond differentially to novel relative to familiar stimuli. But, they argue, the amount and duration of memory capacity that is required in these assessments is minimal, and the contemporaneous correlations of pure memory assessments with child and adult IQ are modest and inconsistent. The case for individual differences in memory capacity, they assert (without evidence), is weak and not persuasive.

*Inhibition or disengagement of attention*

McCall (33) and McCall and Mash (32) argued that the mediating process is the ability of infants to inhibit attention to or disengage from irrelevant and familiar stimuli and of children to inhibit attention to irrelevant information given in the mental problems on IQ tests.

Little direct evidence is available to evaluate this hypothesis either. However, some data suggest that the ability to inhibit attention to extra stimuli in the infant context may be related to the measures that predict later IQ (44). Also, predictions from infancy to assessments at age 12 years are stronger for childhood measures that require the individual to eliminate irrelevant information from problem solving tasks than similar tasks that do not involve such inhibition, and the cognitive task requiring such inhibition correlates more strongly with contemporaneous childhood IQ than does the task not involving such inhibition (45). This general line of reasoning lays interpretative emphasis on brain inhibitory mechanisms and their development during infancy. Such inhibitory mechanisms tend to develop more slowly than excitory mechanisms (e.g. 46), and their development attains reasonable levels by approximately 8 months of age, perhaps explaining the decline after this age in the ability of infant attentional measures to predict later IQ. In this context, it is interesting to note that average looking duration to the familiar stimulus declines more rapidly over age than average looking duration to the novel stimulus in paired comparison tests with monkeys (47,48), suggesting that the ability to inhibit or disengage attention from the familiar improves with age more rapidly than does the speed of processing the information in the novel stimulus.

## Evidence that LCPUFAs Influence IQ

We now turn to reviewing and critically assessing the evidence that levels of LCPUFAs in the prenatal and postnatal diets of term and preterm infants influence intelligence, either as assessed directly in children or as assessed indirectly by noting differences in attentional behavior during the first year or so of life that the previously reviewed literature suggests are related to later IQ. These studies can be grouped into two categories: naturalistic studies of term and preterm infants fed human milk (typically breast fed) versus formula, and studies in which infant formula was or was not experimentally supplemented with LCPUFAs and fed to randomly assigned infants whose mothers chose not to breast-feed.

### Studies of Infants Fed Human Milk versus Formula

Naturalistic studies of infants fed human milk versus formula, typically breast-fed versus bottle-fed infants, tend to show developmental and mental (IQ) advantages for breast-fed infants. Because human milk contains LCPUFAs (among many other unique ingredients) but formulas do not, developmental and IQ advantages of breast- or human-milk-fed infants and children are sometimes interpreted as indirect evidence that LCPUFAs may contribute to such benefits. However, nearly every study conducted in several countries over the last seven decades shows that mothers who choose to breast-feed or provide milk have more education and/or higher social status and/or have infants who are born with fewer perinatal problems than mothers who choose to bottle-feed (47–58). The scientific issue is whether correlates of mothers' choice to provide milk or to breast-feed can account for such observed differences or whether feeding infants human milk versus formula per se is the cause.

Older studies (e.g. 47,48,55,57) made little or no attempt to control for socio-economic, nutritional, or medical factors that might be associated with mothers' choice, and consequently some scholars (e.g. 56) appropriately questioned whether breast-feeding (i.e. human milk) per se produced these benefits. However, after matching or controlling for social and demographic factors in a large, nationally representative sample, Rodgers (54) found a small advantage at 8 and 15 years of age for children who had been entirely breast fed versus those entirely formula-fed. Taylor and Wadsworth (58) found a similar advantage for vocabulary and visual–motor coordination, but Silva et al. (56) found no developmental advantage for breast-feeding. At this point, the preponderance of evidence seems to favor an advantage to breast-feeding over and above differences associated with maternal education, SES, and birth circumstances.

*The Great Britain study*

The most extensive, prospective, and recent longitudinal study of infants fed human milk versus infant formula has been conducted by Lucas, Morley, and co-workers in Great Britain (see the chapter by Lucas in this volume). This five-site study of low-birth-weight infants born less than 1850 g between January 1982 and March 1985 is unique because it represents a combination of nonrandomized and randomized methodological approaches. Mothers chose whether to provide breast milk for their infants within 48–72 hours of delivery. Breast milk was expressed and then fed by tube to all infants, so that any advantages to using human milk were not associated with the breast-feeding process itself. Within each of these two naturalistic groups, mother–infant pairs were randomly assigned to be fed solely (in the case of no mother's milk) or supplemented (in the case of mother's milk) either with human milk, an enriched formula for preterm infants (Osterprem, Farley Health Products, Ltd.) at three of the five centers, or a standard infant formula (formerly called Osterfeed and now Ostermilk, Farley Health Products, Ltd.) at two of the five centers.

In the mother's milk group, supplements varied from 0% to 100%, with a median intake of mother's milk of 39% of total infant needs. This means that, even in the group labeled 'mother's milk', an average of 61% of infant nutritional needs were provided by supplementation, which was either non-mother's human milk, standard formula, or preterm formula. The feeding regimen was implemented until the infant left the neonatal unit or reached 2000 g, whichever came sooner. This meant that the randomized feeding regimens lasted on average only approximately 1 month, but this was highly variable. No control or knowledge of how the infant was fed after termination of the special feeding regimen has been reported. Follow-up assessments were conducted at age from expected date of delivery by individuals blind to the early nutritional status of the infants.

*Factors in mother's choice to provide milk*

Factors that were independently related to mother's decision to provide breast milk were higher educational level, older age, having a male first-born infant, vaginal delivery, and greater likelihood of living with a spouse (52). Other factors that also related to this decision but were not statistically independent of these variables were higher social class, longer gestation, and higher birth weight. However, no differences were observed on these factors or on 5-minute Apgar scores between the randomized human versus nonhuman milk experimental groups within the two mother's choice groups.

*Assessments of developmental progress and IQ*

The infants in this study were followed up at 9 months, 18 months, and 7.5–8.0 years of age. At 9 months, the developmental status was assessed with the Knobloch Developmental Screening Inventory (59). At 18 months, the mental scale of the Bayley Scales of Infant Development (3) and the academic scale of the Developmental Profile II (60) which produced an IQ-equivalent score were administered. At 7.5–8.0 years of age, participants were assessed with the anglicized abbreviated version of the Weschler Intelligence Scale for Children (7). Data pertaining to the possible developmental and IQ benefits of human milk versus formula feeding during the first month of postnatal life from this comprehensive study can be integrated and summarized around the two themes that follow.

*Results for mothers' choice to provide milk*

The first theme pertains to differences in infant and childhood outcomes associated with mother's choice to provide milk, which is related to the relative amount of human milk (and, presumably, LCPUFAs) received by her infant. At each of the three ages, higher infant developmental scores or IQs were associated with the infants and children of mothers who chose to provide milk relative to those of mothers who chose not to provide milk

(51,52,61–62). For example, in a comparison made across all five sites at 18 months of age, mothers who chose to provide breast milk had infants who scored substantially higher on both the Bayley Mental Development Index (MDI) (103.3 vs 95.4) and the IQ-equivalent score from the Developmental Profile II (108.3 vs 101.5) than infants of mothers who chose not to provide breast milk, ignoring the actual type of feeding or supplementation received by the infant (52). These differences were reduced by approximately half but were still significant after the educational and social status of the mother and the infant's birth weight, gestational age, and days on ventilation were covaried.

Similarly and more impressively because of the more advanced age of the participants, there was a 10.1 IQ point advantage on both the verbal and the performance scales for children of 7.5–8.0 years of age of mothers who chose to provide milk for both the verbal and the performance scales, a difference that was reduced to 7.7 and 7.9 IQ points, respectively, after covarying educational, occupational, and birth status factors. Among infants whose mothers chose to provide breast milk, IQ scores varied directly with mothers' success at doing so, which the authors interpreted as a human-milk dose–response effect. It should be pointed out, however, that in this study both mothers' choice groups contained infants who were fed solely (in the case of the infants where mothers chose not to provide milk) or who were supplemented (in the case of infants of mothers who did choose to provide breast milk) with either human milk, standard formula, or specially enriched preterm formula, which makes this comparison one of mothers' choice more than one of whether the infants received human milk or formula.

These results plus the earlier literature clearly indicate that the infant developmental and IQ scores of children whose mothers chose to breastfeed or provide milk are higher than those for infants and children of mothers who chose not to provide milk, and such differences persist even after covarying mothers' education and/or social status and the perinatal condition of the infant at birth. However, these data are much less persuasive of the proposition that it is the feeding of human milk versus formula, not mothers' choice and its behavioral and other correlates, that produces these benefits. This is so because it is necessary to covary all independent factors associated with mothers' choice to focus on human milk versus formula logically as the causal agent. This is nearly impossible to do. For example, none of these studies has attempted to covary factors more proximal to the ability of mothers to provide a stimulating environment for their children. Specifically, Jacobson and Jacobson (50), on a sample of full-term infants assessed at 4 years of age, replicated the mothers' choice advantage after covarying maternal social class and education. But when maternal IQ (a more direct measure of mothers' intellectual capability) and the HOME Scale (an index of the stimulating quality of the home) were added to the regression equation, the benefits of maternal choice/human milk disappeared. Gale and Martin (63) also eliminated breast-versus-bottle differences in adult IQ by covarying maternal age at birth, number of older siblings, and father's occupation.

While it is possible that human milk versus formula differences in IQ might exist in low-birth-weight samples even after measures of the home atmosphere and the parents' ability to stimulate the child intellectually are covaried (the Jacobson and Jacobson study was on full-term infants), one must conclude that these naturalistic observations of the difference between breast-fed and formula-fed infants do not convincingly support the proposition that the feeding of human milk (and thereby LCPUFAs) conveys any later advantage in IQ.

*Random assignment to human milk versus formula*

The second type of result from the Great Britain Study is based on those infants whose mothers chose not to provide human milk who were then randomly assigned to be fed entirely on human donor milk, standard formula, or a special nutritionally enriched preterm formula. Although such results are confined to infants whose mothers chose not to provide human milk, they have the distinct advantage that random assignment presumably balances out any subject selection bias (such as that discussed above for mothers' choice) and that these infants were fed the assigned milk as their sole early diet.

The results across the outcome ages were rather consistent in showing no particular advantage in developmental score or IQ for infants who were randomly assigned to donor human milk versus those assigned to a formula. Further, infants of mothers who chose to provide milk but whose infants required substantial supplementation with formula did better, not worse, than those supplemented with human milk.

The special nutritionally enriched preterm formula produced better outcomes at 18 months than the standard term formula, but this difference cannot be attributed to LCPUFAs because neither formula contained any. The fact that human milk produced better results than the standard infant formula (but not the enriched formula), especially on the psychomotor rather than mental scale, despite the fact that the standard formula had more nutrients than human milk, would seem to suggest that the LCPUFAs (among other unique characteristics) of human milk might compensate for relatively low nutrition. But the lack of difference between human milk and the enriched preterm formula would imply that nutrition alone, without LCPUFAs, can produce developmental outcomes equivalent to human milk. Why the effects of the enriched preterm formula should influence the psychomotor more than the mental skill is not clear. The size of the infant may play a role, but animal work has also shown that early undernutrition affects the cerebellum, an area associated with motor skills and balance, to a greater extent than other parts of the brain (64–66), although this result may depend upon the timing of the undernutrition.

Unfortunately, the report of IQs at 7.5–8.0 years of age did not include comparisons within the group whose mothers chose not to provide human milk between infants receiving human milk, standard formula, or enriched preterm formula.

*Conclusion*

At the risk of glossing over minor results that may have important implications for these conclusions, these studies provide little firm evidence that human milk contains ingredients, including LCPUFAs, that promote developmental and mental performance later in life or even prevent deficits in these outcomes among low-birth-weight and full-term infants. The naturalistic studies of mothers' choice to provide or not to provide milk do show advantages for infants of mothers who choose to provide milk, and while such differences persist even after maternal education or SES and infant perinatal status are covaried, at least one study suggests that such differences do disappear if mother's IQ and an assessment of the stimulating character of the home are added to the covariance regression equation. Moreover, among infants in these same studies whose mothers choose not to provide human milk and who were randomly assigned to receive human milk versus formulas, no advantage to human milk is reported.

Speculatively, however, human milk and LCPUFAs could survive the random assignment results of the Great Britain Study if a developmental advantage is conferred when an infant received his or her own mother's milk as opposed to human milk donated by women other than the infant's mother. Such a nutritional hypothesis requires that human breast milk varies from mother to mother, such variation tends to match certain biochemical characteristics of the mother's own infant, this match produces developmentally beneficial results, and the naturalistic data on mothers' choice holds up for low birth weight infants after all relevant covariates are statistically removed. Alternatively, and less interestingly, the results may be explained by the facts that banked human milk may be processed and stored, producing nutritional changes, while milk from mothers fed to their infants may not be.

*Experimental Studies of Adding LCPUFAs to Infant Formulas*

The most compelling cause-and-effect evidence for the potential benefits of LCPUFAs for cognitive functioning could come from experimental studies in which substances containing or producing certain LCPUFAs (specifically DHA and AA) are experimentally added to formulas. The major limitation of these studies is that the participants have not been assessed at ages older than 18 months, leaving the long-term significance of these results to rest on the relatively modest predictions to childhood IQ of standardized tests of infant development and measures of the nature of attentional behavior in the first year of life (see above).

Much of this literature has been produced by Carlson and her colleagues, who have changed the nature of the DHA supplement they have used as a result of their initial studies. Specifically, in study I, Carlson and colleagues studied the addition of fish oil I to the diets of preterm infants from 3 weeks through 9 months after expected term. Both experimental and control infants were fed an enriched preterm formula, but only until they reached 1.8 kg, after which they were given regular term formula. This procedure left the general nutritional status of infants in both groups at marginal levels. In addition, nutritionally and biochemically, this version of fish oil was not totally satisfactory. While it did raise DHA levels, fish oil I contained eicosapentaenoic acid (EPA) as 0.3% of total fatty acids, which apparently produced a decrease in AA (67). AA levels were highly negatively correlated with individual differences in growth (68) and were generally associated in the first year of life (69) with lower mean Bayley Psychomotor Development Index (PDI) scores in those infants fed the supplemented diet (68). No differences were observed for Bayley MDI scores.

In study II, the biochemical problems of the fish oil I additive were corrected by lowering the amount of EPA to 0.03% from 0.3%; all infants were fed the enriched preterm formula for a longer period of time (i.e. 2 months past expected term), which produced better nutritional status for both supplemented and nonsupplemented groups; and supplementation was given for a shorter period of time.

*Effects on standardized tests of infant development*

Among low-birth-weight infants, supplementing infant formula with DHA may improve general developmental indices, but it is not clear under exactly what circumstances this will occur or on what general developmental index the effects will appear. In full-term infants, however, supplementation may produce no differences or even negative effects.

Specifically, Carlson *et al.* (70) found supplemented low-birth-weight infants to have slightly worse Bayley MDI values at 12 months of age (97.4 vs 102.1; nonsignificant) in study I, but superior MDIs in study II (109.7 vs 98.0, $p < 0.03$). No mention was made of results for the PDI. This rather substantial reversal of effects could be associated with a better balance of DHA relative to AA produced by fish oil II and/or some interaction of the supplement with the length of supplementation and better general nutrition produced by prolonged feeding of infants with enriched formula that occurred in study II. Also, some selective attrition occurred in study II that might have favored the supplemented group.

In a naturalistic study of preterm versus term infants, Rioux *et al.* (71) reported that DHA and AA levels were lower in preterm than in term infants early in the first year of life, and preterm infants had lower PDIs at 4, 8, and 18 months. No MDI values were reported.

In a study of full-term infants randomly assigned to no supplement, AA plus DHA supplement from egg yolk lecithin, only DHA supplement from marine oil (fish oil), and a breast-fed comparison group, Janowsky *et al.* (72) report no differences on Bayley 12-month MDI or PDI. However, the DHA-only supplement group had lower MacArthur Communicative Developmental Inventory scores at 14 months than did breast-fed infants, and the plasma levels of DHA correlated negatively ($-0.20$, $-0.37$) with vocabulary production and comprehension within both formula- and breast-fed groups. There was no such correlation with AA. This result suggests that supplementing term infants may not produce general developmental benefits, or it can produce too much DHA or a disproportionate DHA/AA ratio that leads to poorer vocabulary performance.

*Effects on infant attentional measures*

Several studies have used the recognition memory and habituation paradigms to assess the nature of the deployment of attention in infants. These results suggest that supplementation with DHA can produce 'better' performance for preterm infants on some but not most measures that are otherwise known to predict later IQ, but again there is no evidence that such supplementation produces similar results for term infants.

Specifically, among *low-birth-weight infants*, Carlson *et al.* (70,73,74) report the results of their two supplementation studies on the performance of infants on the Fagan Test of Infant Intelligence, a measure of recognition memory performance, at 12 months of age for study II infants and at 6.5, 9, and 12 months for study I infants. Collectively these results point to three major findings: (i) There were no differences between infants randomly assigned to either marine oil (fish oil) I or II versus no such supplementation on the number of looks or the average length of looks during the familiarization phase of the Fagan Test; (ii) DHA supplemented infants displayed a greater number of looks and a shorter average duration of look (especially to the novel stimulus) at all ages during the test phase of the Fagan; and (iii) there were no differences between supplemented and nonsupplemented infants in the novelty preference percentage score at 6.5 and at 9 months (study I), but DHA-supplemented infants displayed less preference for novelty at 12 months in both studies (significant only in study I). Notice that, whatever biochemistry produced the reversal in Bayley MDI results between studies I and II (e.g. the balance between DHA and AA or general nutritional status), it did not influence attentional behavior.

How likely is it that such results would forecast higher childhood IQs for infants supplemented with DHA? First, one must remember that, although the number of relevant studies is few, predictions to later IQ would be expected to be higher for the 6.5 month relative to the 9- and 12-month data (see above) (31). Second, although a greater number of looks and shorter average looks have been reported to be correlated modestly with later IQ, such predictions might also have been made for such measures obtained from the familiarization, not just the test phase. Further, at least some theoretical perspectives (e.g. 32,33) would suggest that the most cognitively advantageous result would have been shorter average fixations to the familiar, not to the novel, stimuli (see below). In addition, DHA effects also would be expected for the novelty preference ratio and total looking. But the DHA supplement did not have an effect on the novelty preference ratio at 6.5 months and the comparison group had an unexpectedly higher novely preference ratio at 12 months. Why did the supplement not affect all the predictors of later IQ (more predictors were unaffected than affected), and which of these results, the 'positive' or the 'negative', should be accepted as valid?

Actually, there may be more order than chaos in these data than it first appears. Evidence from a variety of sources indicates that different mechanisms may influence average look duration and novelty preference scores. For example: (i) among human infants look duration correlates better with later measures of speed of processing while novelty preference correlates better with later memory and some verbal measures (75–77); (ii) prenatal exposure of infants to alcohol (78), which may lower DHA levels in the brains of adult cats (79), affects looking duration but not novelty preference while prenatal exposure to polychlorobiphenyls (PCBs) affects novelty preference but not look duration (80); (iii) infant monkeys experimentally fed diets deficient in LCPUFAs later showed longer look durations but not lower novelty preference scores than controls (47); (iv) manipulation of dietary taurine (a conditionally essential amino acid) affected novelty preference but not looking durations in monkeys (81); and (v) look duration and novelty preferences show different patterns of change over age for monkeys (81) and human infants (82). Therefore, it is possible that look duration is governed by a different mechanism than novelty preference, and that LCPUFAs are more likely to influence the look duration than the novelty preference mechanism. Further, Carlson's findings in infants that LCPUFAs influence look duration during the test phase more strongly than during the familiarization phase of the paired comparison method was also observed in the monkey research (47,83). This may be a result of the fact that the familiarization allowed is quite long, perhaps long enough to mask important individual differences that are seen more clearly in the shorter test phase.

While this set of studies reflects a rather notable consistency in an otherwise inconsistent literature, one must nevertheless proceed cautiously in arguing from these data that LCPUFAs likely affect later IQ via their influence on look duration, not novelty preference. The size of the correlations and effects are sufficiently small that it is possible that LCPUFAs affect a portion of the variance in look duration that look duration does not share either with novelty preference or, more importantly, with later IQ. However, understanding the LCPUFA/look duration phenomenon may be valuable even if it does not relate to later IQ.

Switching now to research on *full-term infants*, Willatts et al. (84) randomly assigned 48 infants to standard formula and formula supplemented with DHA and observed a breast-fed comparison group ($n = 27$). At 3 months of age, there were no differences overall between these groups in terms of the total fixation, average fixation, or length of peak fixation in an habituation paradigm. Similarly, in a nonrandomized comparison between breast- and formula-fed full-term infants, Innis et al. (85) found no difference in novelty preference at 10 months and no correlation of novelty preference with the duration of breast-feeding.

However, in post hoc analyses, subjects in the Willatts et al. (84) randomized study who did not give their peak fixation on the first trial, were found to be somewhat smaller at birth, and for these infants LCPUFA supplementation was associated with shorter total, average, and peak fixation times. This is consistent with the hypothesis that such supplementation is more likely to alter the attentional behavior of preterm or more immature infants than full-term infants, but even then the implications for later IQ are not clear.

## Conclusions

The conclusions of this chapter fall into several categories.

### Mature Intelligence (IQ) as an Outcome

While no uniform definition of intelligence exists, by age 6 years the standard measures of general intelligence (i.e. IQ tests) represent the most developmentally stable behavior we know (although substantial change in IQ is still possible for many children and adolescents), and IQ is the best single predictor of academic and occupational success for essentially all groups in American society. Therefore, it is a worthy outcome measure for the LCPUFA literature, but certainly not the only one. For example, LCPUFAs could have effects on specific abilities, school performance, personality, and behavioral problems with or without having effects on IQ.

Preterm and low-birth-weight infants have lower IQs later in life, raising the possibility that lower levels of LCPUFAs during the perinatal period produce these consequences. However, the evidence that general nutrition in the perinatal period produces long-term consequences in IQ is inconclusive at best. However, in one major study enriched formula for low-birth-weight infants did produce better developmental outcomes and childhood IQs than did standard formula, but neither of these formulas contained LCPUFAs.

### Predictions to Childhood IQ from Infancy

Neither standardized assessments of developmental status nor assessments of attentional behavior in the first year or two of life have much short-term stability, meaning that individual or group differences (such as those produced by the addition of LCPUFAs) observed at one age ordinarily would not be expected to persist even across a few months during the infancy period.

The prediction to later IQ from standardized infant tests (e.g. the Bayley, Gesell) given in the first 18 months of life is so modest that it has little scientific or practical significance. Therefore, group differences during the first 18 months of life on such instruments produced by adding LCPUFAs to infant formula would not ordinarily be expected to lead to IQ differences years later.

The level of predictions to later IQ from various measures of the nature of attentional deployment observed during the first year of life is somewhat higher than for standardized infant tests for nonrisk samples but about the same for at-risk samples, but the level of such predictions is still rather modest (e.g. median raw observed correlation 0.45). While these predictive correlations are higher for at-risk samples, they may be lower when the infant attentional assessments are made after 7 months of age, when many of the assessments in the LCPUFA literature are made.

Speculations about the process that mediates these predictions focus on the speed of information processing or neural conductivity and on the ability of infants and children to inhibit attention to or disengage from familiar stimuli, low intensity stimuli, and irrelevant information. Such hypotheses, while lacking firm direct evidence, could be the basis for speculations regarding the mechanism by which LCPUFAs influence these behaviors, if such influence is shown to exist.

## Human Milk versus Formula

Naturalistic studies show that infants whose mothers choose to breast feed have higher later IQs than do infants of mothers who choose not to breast feed. While some studies have attempted to control characteristics statistically that distinguish mothers in these two groups and have still found IQ differences in their children, one other study that has controlled such factors plus measures of maternal IQ and the stimulating nature of the home did not find breast-versus-formula differences in IQ. Moreover, human milk contains a whole variety of substances, not just LCPUFAs, that are not contained in formula, and formula contains substances not found in breast milk, any of which could potentially contribute to any differences between breast-fed and formula-fed infants.

Comparisons among infants, all of whose mothers chose not to provide breast milk, who were randomly assigned to be fed human milk from donors versus formula, show no advantage to receiving human milk. Moreover, human milk-fed infants who needed supplemental feeding with formula did better, not worse, than those supplemented with human milk at 9 and 18 months. Therefore, while mother's choice to provide breast milk is associated with better infant and childhood mental test scores, it is not at all clear that such differences are associated with the infant receiving human milk (and thus LCPUFAs).

## Direct LCPUFA Supplementation on Standardized Infant Test Scores

Experimental studies in which infants are randomly assigned to be fed with a formula containing LCPUFAs versus a formula without LCPUFAs have the scientific advantage of possibly revealing directly cause-and-effect evidence on the role of LCPUFAs in IQ, but they currently have the disadvantage that participants have only been tested up to 18 months of age. Studies of full-term infants reveal no developmental advantage of adding LCPUFAs, and at least one study found that LCPUFA supplemented infants had less advanced language at 14 months. It is possible that full-term infants receive sufficient amounts of LCPUFAs, and more is not necessarily better and may even be undesirable.

Low-birth-weight infants given LCPUFA supplements may do better or worse on general tests of developmental milestones at 12 months, perhaps depending on the particular balance among specific LCPUFAs in the supplementation, the length of supplementation, or the general nutritional status of the infants.

## Direct LCPUFA Supplementation on Attentional Behavior

LCPUFA-supplemented low birth weight infants look more times for shorter average durations at faces and other stimuli at 6.5–12 months of age. Analogous results have been observed in infant monkeys. Supplementation does not affect novelty preferences, which a variey of studies suggest may be governed by a different mechanism than looking duration. While other research shows that looking duration (and novelty preference) during infancy predicts childhood IQ at modest levels, it is unknown whether differences in looking duration produced by LCPUFA supplementation will be reflected in differences in later IQ. While waiting for such infants to develop into children, it would be useful to search for and understand the mechanism underlying the effects on looking duration.

LCPUFA supplements given to full-term infants have not produced differences in infant attentional behavior, consistent with results for the standardized developmental exams.

## General Conclusion

The total evidence for any effect of LCPUFAs on early behavioral developmental status and childhood IQ is provocative but not at all persuasive as yet, and any possible benefits are likely to be confined to low-birth-weight infants. Future research should investigate more and more specific mental, social, and emotional outcomes than IQ.

## References

1. Lubchenko LO. *The High-risk Infant*. Philadelphia: WB Saunders, 1976.
2. Coghlan A. Bottle-fed babies need brain boosting formula. *New Sci* 1996; May 18: 12.
3. Bayley N. *Bayley Scales of Infant Development*. New York: Psychological Corporation, 1969.
4. Herrnstein R J, Murray C. *The Bell Curve: Intelligence and Class Structure in American Life*. New York: Free Press, 1994.
5. Neisser U, Boodoo G, Bouchard Jr TJ et al. Intelligence: knowns and unknowns. *Am Psychol* 1996; **51**: 77–101.
6. Sternberg RJ, Detterman DK, eds. *What is Intelligence? Contemporary Viewpoints on its Nature and Definition*. Norwood, NJ: Ablex, 1986.
7. Wechsler D. *Weschler Intelligence Scale for Children*, anglicized revised edition. Sidcup, Kent: The Psychological Corporation, 1974.
8. Wechsler D. *Wechsler's Measurement and Appraisal of Adult Intelligence*. Baltimore, OH: Williams & Wilkins, 1972.
9. Gardner H. *Frames of Mind: The Theory of Multiple Intelligences*. New York: Basic Books, 1983.
10. Sternberg RJ. *Beyond IQ: A Triarchic Theory of Human Intelligence*. New York: Cambridge University Press, 1985.
11. Lave J. *Cognition in Practice*. New York: Cambridge University Press, 1988.
12. Carroll JB. *Human Cognitive Abilities: A Survey of Factor-analytic Studies*. Cambridge: University of Cambridge Press, 1993.
13. McCall RB. The development of intellectual functioning in infancy and the prediction of later IQ: Osofsky JD, ed. *Handbook of Infant Development*. New York: Wiley, 1979: 707–40.
14. McCall RB. Developmental changes in mental performance: the effect of the birth of a sibling. *Child Dev* 1984; **55**: 1317–21.
15. Ceci SJ. *On Intelligence . . . More or Less: A Bioecological Treatise on Intellectual Development*. Englewood Cliffs, NJ: Prentice Hall, 1990.
16. Deary IJ. Auditory inspection time and intelligence: what is the causal direction? *Dev Psychol* 1995; **31**: 237–50.
17. Vernon PA. *Speed of Information Processing and Intelligence*. Norwood, NJ: Ablex, 1987.
18. Smart JL. Undernutrition, learning and memory: review of experimental studies. In: Taylor TG, Jenkins NK, eds. *Proceedings of XIII International Congress of Nutrition*. London: Libbey, 1986.
19. Chavez A, Martinez C. Neurological maturation and performance on mental tests. Quoted in: *Early Nutrition and Later Achievement*. London: Academic Press, 1987.
20. Freeman HE, Klein RE, Townsend JW et al. Nutrition and cognitive development among rural Guatemalan children. *Am J Public Health* 1980; **70**: 1277–85.

21. Sinisterra L. Studies on poverty, human growth and development: the Cali experience. In: Dobbing J, ed. *Early Nutrition and Later Development*. London: Academic Press, 1987.
22. Klein PS, Forbes GD, Nadar, PR. Effects of starvation in infancy (pyloric stenosis) on subsequent learning abilities. *J Pediatr* 1975; **87**: 8–15.
23. Ankett MA, Parks YA, Scott PH, Wharton BA. Treatment with iron increases weight gain and psychomotor development. *Arch Dis Child* 1986; **61**: 849–57.
24. Lloyd-Still JD. Clinical studies on the effects of malnutrition during infancy and subsequent physical and intellectual development. In: Lloyd-Still JD, ed. *Malnutrition and Intellectual Development*. Lancaster: MTP Press, 1976.
25. Grantham-McGregor S. Field studies in early nutrition and later achievement. In: Dobbing J, ed. *Early Nutrition and Later Achievement*. London: Academic Press, 1987.
26. Stein Z, Susser M, Saenger G, Marolla F. *Famine and Human Development: The Dutch Hunger Winter of 1944–45*. New York: Oxford University Press, 1975.
27. Kopp CB, McCall RB. Predicting later mental performance for normal, at-risk, and handicapped infants. In: Baltes PB, Brim Jr OG, eds. *Life-span Development and Behavior*, Vol. 4. New York: Academic Press, 1982: 33–61.
28. McCall RB, Eichorn DH, Hogarty PS. Transitions in early mental development. *Monographs of the Society for Research in Child Development* 1977: **42**: Serial No. 171.
29. McCall RB. Early predictors of later IQ: the search continues. *Intelligence* 1981; **5**: 141–7.
30. Bornstein MH, Sigman MD. Continuity in mental development from infancy. *Child Dev* 1986; **57**: 251–74.
31. McCall RB, Carriger MS. A meta-analysis of infant habituation and recognition memory performance as predictors of later IQ. *Child Dev* 1993; **64**: 57–79.
32. McCall RB, Mash CW. Infant cognition and its relation to mature intelligence. In: Vasta R, ed. *Annuals of Child Development* London: Jessica Kingsley, 1995: 27–56.
33. McCall RB. What process mediates predictions of childhood IQ from infant habituation and recognition memory? Speculations on the roles of inhibition and rate of information processing. *Intelligence* 1994; **18**: 107–25.
34. Colombo J. *Infant Cognition: Predicting Later Intellectual Functioning*. Newbury Park, CA: Sage, 1993.
35. Colombo J. On the neural mechanisms underlying long and short looking: two hypotheses concerning individual differences in infant fixation duration. *Dev Rev* 1995; **15**: 97–135.
36. Eysenck JJ. Speed of information processing, reaction time, and the theory of intelligence. In: Vernon PA, ed. *Speed of Information Processing and Intelligence*. Norwood, NJ: Ablex, 1987: 21–67.
37. Jensen AR. Reaction time and psychometric g. In: Eysenck HJ, ed. *A Model for Intelligence*. Berlin: Springer-Verlag, 1982: 93–132.
38. Lehrl S, Fischer B. A basic information psychological parameter (BIP) for the reconstruction of concepts of intelligence. *Eur J Personality* 1990; **4**: 259–86.
39. McGarry-Roberts PA, Steklmack RM, Campbell KB. Intelligence, reaction time, and event-related potentials. *Intelligence* 1992; **16**: 289–313.

40. O'Donnell BF, Friedman S, Swearer JM, Drachman DA. Active and passive P3 latency and psychometric performance: influence of age and individual differences. *Int J Psychophysiol* 1992; **12**: 187–95.
41. Reed TE, Jensen AR. Arm nerve conduction velocity (NCV), brain NCV, reaction time, and intelligence. *Intelligence* 1991; **15**: 33–47.
42. Reed TE, Jensen AR. Conduction velocity in a brain nerve pathway of normal adults correlates with intelligence level. *Intelligence* 1992; **16**: 259–72.
43. Vernon PA, Mori M. Intelligence, reaction times, and peripheral nerve conduction velocity. *Intelligence* 1992; **16**: 273–88.
44. Frick JE, Colombo J, Saxon TF. Long looking infants are slower to disengage fixation. Presented at the *International Conference on Infant Studies*, Providence, RI, April 1996.
45. Sigman M, Cohen SE, Beckwith L, Asarnow R, Parmelee AH. Continuity in cognitive abilities from infancy to 12 years of age. *Cognitive Dev* 1991; **6**: 47–57.
46. Diamond A. Abilities and neural mechanisms underlying AB performance. *Child Dev* 1988; **59**: 523–7.
47. Broad B. The effects of infant feeding on speech quality. *NZ Med J* 1972; **76**: 28–31.
48. Hoefer A, Hardy MC. Later development of breast fed and artificially fed infants. *JAMA* 1929; **92**: 615–19.
49. Jacobson SW, Jacobson JL. Breastfeeding and intelligence. *Lancet* 1992; **339**: 926.
50. Jacobson SW, Jacobson JL, Frye KF. Incidence and correlates of breastfeeding in socioeconomically disadvantaged women. *Pediatrics* 1992; **88**: 728–36.
51. Lucas A, Morley R, Cole TJ, Lister G, Leeson-Payne C. Breast milk and subsequent intelligence quotient in children born preterm. *Lancet* 1992; **339**: 261–4.
52. Morley R, Cole TJ, Powell R, Lucas A. Mother's choice to provide breast milk and developmental outcome. *Arch Dis Child* 1988; **63**: 1382–5.
53. Morrow-Tlucak M, Haude EH, Ernhart CB. Breastfeeding and cognitive development in the first 2 years of life. *Soc Sci Med* 1988; **26**: 635–9.
54. Rodgers B. Feeding in infancy and later ability and attainment: a longitudinal study. *Dev Med Child Neurol* 1978; **78**: 421–6.
55. Rogerson BFC, Rogerson CH. Feeding in infancy and subsequent psychological difficulties. *J Mental Sci* 1939; **85**: 1163–82.
56. Silva PA, Buckfield P, Spears GF. Some maternal and child developmental characteristics associated with breast feeding: a report from the Dunedin multidisciplinary child development study. *Austr Pediatr J* 1978; **14**: 265–8.
57. Taylor B. Breast versus bottle feeding. *NZ Med J*; 1977 **85**: 235–8.
58. Taylor B, Wadsworth J. Breast feeding and child development at 5 years. *Dev Med Child Neurol* 1984; **26**: 73–80.
59. Knobloch H, Pasamanick B, Sherard ES. A developmental screening inventory for infants. *Pediatrics* 1966; **38**: 1095–108.
60. Alpern GC, Boll TJ, Shearer MS. *Developmental Profile II*. Aspen, CO: Psychological Development Publications, 1980.
61. Lucas A, Morley R, Cole TJ et al. Early diet in preterm babies and developmental status in infancy. *Arch Dis Child* 1989; **64**: 1570–8.

62. Lucas A, Morley R, Cole TJ, Gore SM. A randomised muticentre study of human milk versus formula and later development in preterm infants. *Arch Dis Child* 1994; **70**: F141–6.
63. Gale CR, Martin CN. Breastfeeding, dummy use, and adult intelligence. *Lancet* 1996; **347**: 1072–5.
64. Dobbing J. Undernutrition and the developing brain. In: Himwich WA, ed. *Developmental Neurobiology.* Springfield, IL: CC Thomas, 1970.
65. Dobbing J. Nutritional growth restriction and the nervous system. In: Davison AN, Thompson RHS, eds. *The Molecular Basis of Neuropathology.* London: Edward Arnold, 1981.
66. Lynch A, Smart JL, Dobbing J. Motor coordination and cerebellar size in adult rats undernourished in early life. *Brain Res* 1975; **83**: 249–59.
67. Carlson SE, Cooke RJ, Rhodes PG et al. Long-term feeding of formulas high in linolenic acid and marine oil (fish oil) to very low birth weight infants: phospholipid fatty acids. *Pediatr Res* 1991; **30**: 404–12.
68. Carlson SE, Werkman SH, Peeples JM et al. Growth and development of very low birth weight infants in relation to ω-3 and ω-6 fatty acid status. In: Sinclair S, Gibson R, eds. *Essential Fatty Acids and Eicosanoids.* Champaign, IL: American Oil Chemists' Society, 1993.
69. Carlson SE, Cooke RJ, Werkman SH et al. First year growth of preterm infants fed standard compared to marine oil (fish oil) ω-3-supplemented formula. *Lipids* 1992; **2**: 901–7.
70. Carlson SE, Werkman SH, Peeples JM, Wilson WM III. Growth and development of premature infants in relation to ω-3 and ω-6 fatty acid status. Fatty Acids Lipids: Biol Aspects 1994; **75**: 63–9.
71. Rioux FM, Innis SM, Lupton B, Nelson CM. Waslen P, Whitfield M. Cognitive and visual development, psychomotor performance and growth in term compared to preterm infants. Presented at the *ASP–SPR Meeting*, May 1995.
72. Janowsky JS, Scott DT, Wheeler RE, Auestad N. Fatty acids affect early language development. Presented at the *ASP–SPR Meeting*, May 1995.
73. Carlson SE, Werkman SH. A randomized trial of visual attention of preterm infants fed docosahexaenoic acid until two months. *Lipids* 1996; **31**: 85–90.
74. Werkman SH, Carlson SE. A randomized trial of visual attention of preterm infants fed docosahexaenoic acid until nine months. *Lipids* 1996; **31**: 91–7.
75. Jacobson SW, Jacobson JL, O'Neill JM, Padgett RJ, Frankowski JJ, Bihun JT. Visual expectation and dimensions of infant information processing. *Child Development* 1992; **63**: 711–24.
76. Rose SA, Feldman JF, Wallace IF, Cohen P. Language: a partial link between infant attention and later intelligence. *Developmental Psychology* 1991; **27**: 798–805.
77. Thompson LE, Fagan JF, Fulker DW. Longitudinal predictions of specific cognitive abilities from infant novelty preference. *Child Development* 1991; **62**: 530–8.
78. Jacobson SW, Jacobson JL, Sokol RJ, Martier SS, Ager JW. Prenatal alcohol exposure and infant information processing ability. *Child Development* 1993; **64**: 1706–21.
79. Pawlosky RJ, Salem Jr. N. Ethanol exposure causes a decrease in docosahexaenoic acid and an increase in docosahexaenoic acid in feline brains and retinas. *American Journal of Clinical Nutrition*, **61**: 1284–9.

80. Jacobson SW, Fine CG, Jacobson JL, Schwartz PM, Dowler JK. The effect of intrauterine PCB exposure on visual recognition memory. *Child Development* 1985; **56**: 853–60.
81. Reisbick S, Neuringer M, Graham M, Jacqmotte N, Karbo W, Sturman J. Visual recognition memory in infant rhesus monkeys: effects of dietary taurine. *Infant Behavior and Development* 1995; **18**: 309–18.
82. Rose SA, Feldman JF, McCarton CM, Wolfson J. Information processing in seven-month-old infants as a function of risk status. *Child Development* 1988; **59**: 589–603.
83. Reisbick S, Neuringer M, Connor WE. Effects of n-3 fatty acid deficiency in nonhuman primates: implications for human infant formulas. In: JG Bindels, AC Goedhart, HKA Visser eds, *Recent developments in infant nutrition*. UK: Kluwer Academic Publishers, 1996, pp 157–72.
84. Willatts P, Forsyth J. S, DiModugno MK, Varma S, Colvin M. A randomised study of the effects of long-chain polyunsaturated fatty acids on infant habituation. Poster presented at *10th Biennial International Conference on Infant Studies*, Providence, RI, 18–21 April 1996.
85. Innis SM, Nelson CM, Rioux FM, Waslan P, Lwanga D. Visual acuity, cognitive development and nutrition in term infants. Presented at the *ASP–SPR Meeting*, May 1995.

## Commentary

**Carlson**: You point out that the correlations between infant measures of performance and childhood IQ assessments are best when the infant assessment is made between 2 and 7 months. Perhaps it is the measure of infant performance used after 7 months that is the problem? Colombo has referred to Jacobson (1996) who has correlated look duration at 12 months during the Fagan Test with IQ at 4 years (perhaps the 12-month point in your Fig. 1?). I am unaware of any other comparisons of look duration late in infancy with assessments in childhood. Can you provide the references for the four studies in Fig. 1 in which the infant assessments were done after 7 months or describe the infant attention measures used? I would like to know if these are based on recognition memory (novelty preference) or measures obtained during habituation, such as look duration.

Unless there are studies that have not been included, one must conclude from Fig. 1 that: (i) few investigators have measured attention late in infancy and tried to correlate them with childhood performance; and (ii) there is no real difference in apparent validity between 2 and 12 months.

1. Jacobson SW, Chiodo LM, Jacobson JL. Predictive validity of infant recognition memory and processing speed to 7-year IQ in an inner city sample. *International Conference on Infant Studies*, Providence, RI, 1996; **19**: 524.

AUTHOR'S REPLY: At the time Fig. 1 was created, only four correlations existed between attentional measures assessed after 7 months and childhood IQ. As a result, all types of infant attentional measures cannot be represented in these four correlations. McCall and Carrigan (1) compared habituation versus paired comparison (i.e. recognition memory) measures and found no difference in the level of prediction to later IQ. A related problem is that some data that do not reveal significant predictions are not published. I am aware of a few such cases at older ages using paired comparison measures, but I have no idea of what the total set of unpublished studies would show. In sum, there do not appear to be obvious differences between infant paradigms and measures at all ages considered; and the evidence that predictions decline after 7 months is suggestive, but the number of data points available to confirm it is few and may be unpublished.

1. McCall RB, Carrigan MS. A meta-analysis of infant habituation and recognition memory performance as predictors of later IQ. *Child Dev* 1993; **64**: 57–79.

CARLSON: You raised the issue that shorter look duration occurred during the test but not during the familiarization phase of the Fagan Test in our docosahexaenoic acid supplementation trials. Our data are quite analogous to those of Reisbick *et al.* (1) in that shorter look duration occurs during test but not during familiarization with higher docosahexaenoic acid status. Both the Fagan Test and the method used for the infant monkey assessments required rather long periods of familiarization (perhaps past habituation), which presumably could decrease the sensitivity of the familiarization phase for showing differences in look duration between groups.

Jacobson has published evidence that high exposure to alcohol in utero increases look duration during the Fagan Test familiarization (2), but she has also analysed her paired comparison data and indicated to me that this same group also has longer looks during the test. In infants fed DHA, we find a nonsignificant decrease in look duration during the familiarization ($p < 0.12$) and a significant decrease during the paired comparison phase. Moreover, we have tested infants with bronchopulmonary dysplasia (BPD), who are at risk for long-term cognitive effects. In this group, we find significantly longer look duration during both familiarization and test. Finally, Colombo mentions a correlation between look durations in familiarization and in test phase. All of these findings suggest that look duration during the familiarization and test phases of the Fagan Test are measuring the same aspect of development, but the effects on look duration are more easily observed during the test phase than during the familiarization phase. (3).

1. Reisbick S, Neuringer M, Connor WE. Effects of n-3 fatty acid deficiency in nonhuman primates. In: Bindels JG, Goedhart AC, Visser HKA, eds. *Recent Developments in Infant Nutrition*, Kluwer Academic Publisher, 1996: 157–72.
2. Jacobson SW, Jacobson JL, Sokol RJ, Martier SS, Ager JW. Prenatal alcohol exposure and infant information processing ability. *Child Dev* 1993; **64**: 1706–21.
3. Werkman SH, Carlson SE. A randomized trial of visual attention of preterm infants fed docosahexaenoic acid until 9 months. *Lipids* 1996; **31**: 91–7.

**Lucas**: McCall's critique identifies general IQ as the most developmentally stable human behavior – and the one most predictive of academic and occupational performance. 'Intelligence' is a poorly defined, somewhat abstract term that is used in different ways even by experts in cognitive function. However, intelligence should not be confused with 'intelligence quotient' (IQ) which, for each standard test (e.g. Weschler and Stanford–Binet) reflects the formal and specific measurement of a defined range of human verbal and nonverbal skills. Thus, whilst there may remain sufficient debate over what intelligence is to render it too nebulous a concept for scientific use, the full scale Weschler Intelligence Scale for Children (WISC-R IQ), say, is a well defined entity with useful predictive power, particularly when used in populations rather than individuals.

The demonstration of early nutritional effects on later IQ would establish clear importance for nutritional management in a clinical and public health context. McCall is pessimistic, however, about whether early nutrition has indeed been convincingly related to later IQ. In our own large prospective outcome studies in preterm infants randomly assigned formulas differing in overall nutrient content (term versus preterm formula for, on average, 1 month in the neonatal period), there were major differences between groups at 18 months on the Bayley Scales (1). These differences are now seen, notably in verbal IQ, at 7.5–8 years of age (to be published). We are not aware of other studies of this nature. Indeed there has, until recently, been a lack of investment in studies that follow up cohorts randomized to early nutrition into the age group where formal IQ testing can be done. In full-term infants, most data come from developing countries and are often highly confounded by poverty and poor social circumstances. The validity of many earlier studies apparently showing long-term effects of early malnutrition on later cognitive performance was challenged in a previous

Dobbing workshop (2). Even now, as McCall states, the hard evidence in full term infants that infant nutrition affects later IQ is scanty. Indeed even some of the data McCall cites as 'more persuasive', could be regarded as inconclusive. For instance, he cites Klein's contention that malnutrition in infantile pyloric stenosis (3) could impair later cognition. What Klein showed in a nonrandomized, uncontrolled observational study was that percentage weight loss in pyloric stenosis was related to later cognitive function. However, the percentage weight loss that a baby was allowed to sustain before presentation to a pediatrician could easily have been influenced by the cognitive and parenting ability of the parents, which could itself bear on the child's later cognitive ability. Moreover, children with feeding difficulties of this nature cannot necessarily be treated as a normal population. Clearly, the increasing recognition of the importance of randomized prospective studies has provided the best opportunity to explore the effects of early nutrition, and more convincing data based on this type of study are beginning to emerge (4–7). However, the relative lack of data on nutritional effects on later cognition may, again, reflect the lack of appropriate studies rather than the absence of the phenomenon.

Whether studies comparing breast- and formula-fed infants add further weight to the idea that early nutrition is important for later development is debated. In general I agree with McCall and Mash that the data are inconclusive. However, I would interpret the findings that they cite from our own studies in preterm infants slightly differently (1,8,9). These studies, described by McCall and Mash and also in my own chapter in this volume, provide evidence that: (i) general nutrition affects later neurodevelopmental performance (independently of LCPUFA intake); (ii) when infants whose mothers have chosen not to provide breast milk are assigned to diets of similar nutrient content (donated breast milk versus term formula), the human-milk-fed group have higher psychomotor development scores at follow-up; and (iii) babies fed adequately and on human milk (mother's milk plus preterm formula) have the best scores, suggesting that the effects of human milk and of good general nutrition are to some extent additive. However, I would accept that step (ii) above is based on a nonrandomized comparison (albeit unconfounded by mother's choice to breast feed). But in any case, I agree with McCall that, even if there was an advantage of human milk for cognitive development, at least in the vulnerable preterm infant, it could not necessarily be attributed to LCPUFA intake.

Thus, two basic premises for the LCPUFA field, i.e. (i) that early nutrition is important for later cognitive function, and (ii) that breast-milk-fed babies are neurodevelopmentally advantaged, are challenged by McCall.

My own view is that the data support a slightly less pessimistic stance on these basic premises, though this still leaves us a long way from the establishment of efficacy and safety for LCPUFA supplementation of formula-fed infants.

1. Lucas A, Morley R, Cole TJ et al. Early diet in preterm babies and developmental status at 18 months. *Lancet* 1990; **335**: 1477–81.
2. Dobbing J, ed. *Early Nutrition and Later Achievement*. London: Academic Press.
3. Klein PS, Forbes GB, Nader PR. Effects of starvation in infancy (pyloric stenosis) on subsequent learning abilities. *J. Pediatr* 1975; **87**: 8–15.
4. Joos S, Pollitt E, Meuller W, Albright D. The Bacon Chow study; maternal nutritional supplementation and infant behavioural development. *Child Dev* 1983; **54**: 669–76.
5. Waber DP, Vuori-Christiansen L, Oritz N et al. Nutritional supplementation, maternal education and cognitive development of infants at risk of malnutrition. *Am J Clin Nutr* 1981; **34**: 807–13.
6. Grantham-McGregor SM, Powell CA, Walker SP, Himes JH. Nutritional supplementation, psycho-social stimulation and mental development of stunted children: the Jamaican study. *Lancet* 1991; **338**: 1–5.
7. Husaini MA, Karyadi L, Husaini YK, Karyadi D, Pollitt E. Developmental effects of short-term supplementary feeding in nutritionally at risk Indonesian infants. *Am J Clin Nutr* 1991; **54**: 799–804.
8. Lucas A, Morley RM, Cole TJ, Gore SM. A randomised multicentre study of human milk versus formula and later development in preterm infants. *Arch Dis Child* 1994; **70**: F141–6.
9. Lucas A, Morley R, Cole TJ et al. Early diet in preterm babies and developmental status in infancy. *Arch Dis Child* 1989; **11**: 1042–5.


**Colombo**. Although the evidence is admittedly preliminary, there seems to be some probability that different measures of infant cognition tap different and dissociable underlying cognitive constructs. Jacobson's work suggests a dissociability between look duration and novelty preference, and, more recently (1), between novelty preference and ocular RTs. Susan Rose's data suggest very complex interrelations between infant processing components and later cognitive outcome. Indeed, our own recent work and theorizing (2) suggest that different central nervous system substrates may mediate the processes that underlie novelty preference and look duration. Given the possibility that the infant measures of habituation and novelty preference may reflect different and dissociable cognitive components, isn't it problematic to tether the evaluation of all effects of dietary supplementation to performance on a global test like IQ?

1. Jacobson SW, Jacobson JL, Sokol RJ, Martier SS, Chiodo LM. New evidence for neurobehavioural effects of in-utero cocaine exposure. *Journal of Pediatrics* 1996; **129**: 581–90.
2. Colombo J, Janowsky JS. A cognitive neuroscience approach to individual differences in infant attention and recognition memory. In: Richards JE, ed. *The cognitive neuroscience of attention: developmental perspectives.* Hillsdale, NJ: Erlbaum.

**Singer**: McCall and Mash's review of the nature and definition of the constructs of intelligence and IQ is an important contribution to understanding the debate regarding the value of adding LCPUFAs to formulas, especially for those outside the field of psychology who may overly deify these constructs. One important point is the great overlap of SES factors (such as years of schooling, extra educational activities, exposure and connections to higher paying occupations) related to the IQ measures used as outcomes, since IQ reflects the combination of general ability level ($g$) and educational achievement. Some subtests of the WISC (e.g. vocabulary and arithmetic) are highly dependent on cultural and SES factors, such as language exposure and quality of education, while others may be less so. Our own data on over 300 preterm and term infants followed longitudinally to 3 years of age indicate that SES factors strongly predicted Bayley MDIs at 3 years, but did not predict motor outcomes (1). The data on LCPUFAs would be more compelling if they reflected an impact on a cognitive and behavioral process which was more independent of SES than IQ, such as motor functioning (although, arguably, such measures may be less societally important). This is one reason that early measures of information processing, such as visual recognition memory and look duration, are so promising, as they may potentially detect the effects of an intervention on early learning processes which are relatively independent of SES factors.

1. Singer LT, Yamashita TS, Lilien L, Collin M, Baley J. Three year outcome of infants with BPD and VLBW. *Pediatr Res* 1996: **39**: 280A.
2. Weisglas-Kuperus, Baerts W, Smrkovsky M, Sauer PJJ. Effects of biological and social factors on the cognitive development of very low birthweight children. *Pediatrics* 1993; **92**: 658–65

**Wainwright and Ward**: To date, there are no data available from long-term longitudinal studies on humans on the putative relationship between LCPUFAs during early development and later cognitive function. Thus we are left in the position of having to make inferences based on tests of infant development conducted during the first 2 years. This chapter makes an important contribution to this by presenting the current understanding of the extent of the predictive validity of standardized infant tests and measures of visual attention to later performance on IQ tests, and integrating this with the evidence available from dietary studies conducted in infants. After making the case that selective measures of visual attention during the first 2 years of life have better predictive validity than the standardized tests, such as the Bayley Scales (particularly for samples at risk, and when the measures are taken between 2 and 7 months of age), the authors go on to point out that, although studies of DHA supplementation have shown effects on some of these measures, these effects are not consistent with those that correlate with later performance.

You discuss the possible mechanisms underlying this 'window' of predictive validity, including rate of information processing, memory, and the development of inhibitory mechanisms. It is argued that because inhibitory mechanisms develop slowly, they may only attain reasonable levels at about 8 months, thereby accounting for the decline in predictive capacity after that. It is not clear to me why this should be so, unless performance on the task after 8 months is accomplished using different strategies. Is there any way of assessing this by looking at the pattern of responses in older and younger infants?

Following from this, it would seem that, when comparing two groups of infants, it would be important to be able to compare their 'developmental profiles' on these tasks. Assessments made at only one point in time may be misleading, particularly when they are negative.

AUTHOR'S REPLY: Yes, predictions are likely to be higher if early profiles are used than if simple correlations are computed from single ages during infancy. While repeated testing during infancy using standardized infant tests has been done and could be used to determine early profiles, very few longitudinal data exist for the first year of life using the attentional measures. Therefore we currently have little opportunity to make predictions from profiles for attentional behavior.

**Heird**: Your chapter and those by Columbo and Bornstein discuss the predictability of childhood IQ from tests performed in infancy. From these, I gather that none of the tests during infancy are great predictors of childhood IQ. This being the case, it is unlikely that most practical individuals will be particularly impressed by small differences at various times during the first year of life between infants who receive two different nutritional regimens. On the other hand, a difference in scores at an age when these scores are predictive of later intelligence is likely to impress even the practical skeptic. Thus, if I really want to know whether LCPUFAs (or some other nutrient) affects the brain development of infants, would I not obtain less ambiguous data from a study designed to test intelligence of infants at an age when this test is predictive of later IQ rather earlier? I realize that this strategy may prevent detection of effects of the nutrient being studied on discrete functions at a specific time in development. However, if these effects are not manifest by a poor performance score later in life, are they of practical importance?

AUTHOR'S REPLY: Waiting for children to become at least 6 years old before assessing IQ is the preferred approach; but differences produced in early attentional behavior may be interesting in themselves or help to interpret later IQ differences, if they exist.


**Mayer and Dobson**: McCall and Mash show that correlations are higher between habituation/recognition memory performances in infancy and IQ in childhood when the infant measures are obtained between 2 and 7 months than at younger or older ages. They speculate that the infant's state and 'other extraneous' factors may account for low correlations under age 2 months. They suggest, however, that the low correlations after age 7 months are not obviously explained, and go against the trend in developmental research for predictive correlations to increase with increasing age at which the infant tests are done.

One could argue that the habituation/recognition memory paradigms used in younger and older infants are at fault. We believe it is possible that the same visual search strategies are required in paired-comparison paradigms as in preferential looking acuity tests (see also our comments on Colombo's chapter), and we speculate on explanations for low predictability of visual recognition memory tests based on observations testing a wide age range of infants with the same preferential looking tests of acuity. Notably, with the acuity card procedure (ACP), infants can be tested over the entire developmental period, from as young as 36 weeks gestation (at least) up to school age. The stimulus arrangements in the ACP and other preferential looking acuity tests are worth reviewing. The rectangular Teller Acuity Cards (used in many of the nutrition studies) contain a

12.5 × 12.5 cm square patch of grating with the inner edge of the grating displaced 7.5 cm to one side of a central peephole through which the tester observes the infant's visual behaviors. The other half of the card is blank and the whole card is the same space-averaged luminance as the grating. Other preferential looking stimulus arrangements use two spatially defined positions with one position containing the grating, and in some versions, the positions are back-illuminated in an otherwise black screen.

We observe marked changes in the visual behaviors of infants presented with the same stimulus arrangement throughout development. Young infants (below about age 2 months, but including many 2-month-olds as well) do not voluntarily scan the acuity card or make fixation shifts from a central gaze direction toward the eccentrically located stimulus. By age 3 months, most infants spontaneously shift fixation from central gaze to the grating position, if they can see the grating. Indeed, this shift in fixation from center to the right or left is the single most important cue to grating detection that the tester uses. At about 4–6 months, infants begin to look back and forth between both stimulus positions, especially as the gratings become finer, indicating, we suppose, memory for the spatial positions of the gratings.

Although this has not been formally investigated, we find that testing young infants is more efficient when they are 'rotated' in front of the acuity display so that they have the opportunity to see both the right and left stimulus positions. We speculate that low acuities in infants at term to 2 months, reported in some previous normative studies, were due to lack of adequate exposure of eccentrically located grating stimuli due to seating the infant passively in the parent's lap, in combination with stimulus arrangement factors.

Visual search strategies and other behaviors in preferential looking acuity tests show another important change in the last half of the first year. In our cross-sectional normative acuity card study (1), we found that many 12-month-olds and some 9-month-olds appeared more interested in the social and mechanical aspects of the test than in looking at the gratings. Their subjective responses (e.g. laughing when the tester disappeared behind the acuity card, staring at the peephole when the card was in place) often dominated the simple fixation changes we rely upon in testing nonverbal subjects. Contingent reinforcement for looking at the gratings can be effective in testing acuity of older infants and has been used in other variants of preferential looking tests (operant preferential looking (e.g. 2,3)). However, contingent reinforcement is not easily taught to acuity card testers unless they have a developmental background or are particularly talented with children.

We note that the testing situation in the Fagan Test provides similar social distractions for the older infant as in acuity card testing, as the tester disappears behind a screen when the stimuli are presented. Thus, we suggest that, for the younger infant, the stimulus arrangement may not be optimal for assessing visual discriminations in this developmental stage when infants do not voluntarily scan their visual environment. Further, for older infants, visual discrimination paradigms are probably too passive, and do not tap into the emerging social, emotional, and cognitive capabilities that are the hallmark of this stage in development. Nutrition studies in progress have included Piagetian types of tasks (e.g. A-not-B paradigm) apparently to address the lack of predictive validity of visual discrimination tasks for IQ in childhood.

We also wonder, somewhat more speculatively, whether discrepancies among the results of nutrition studies in the younger infants (i.e. 2 months) might be explained by immature oculomotor development (poor scanning) in those with lower acuities. The 'recovery' of normal acuity at later ages might be due to improved oculomotor behavior and not to sensorineural recovery. Of course, this cannot be argued in studies in which visual evoked potential (VEP) acuity was also tested and was also low, in that oculomotor responses are not required in VEP tests of acuity.

McCall and Mash's view that inhibition is the mediating process in infant cognitive measures seems cogent in light of some of our experiences testing infants using the Fagan Test novelty preference test. Maintaining infants' interest in the visual stimuli in the Fagan Test is an unstated requirement of the test, as it is in acuity card testing. In our testing of preterm infants, the familiarization and trial phases were characterized by many interruptions in data recording by the tester because of the infant's distractibility. Some Fagan Tests in our preterm infants had to be invalidated because enough time could not be accumulated on some problems.

1. Mayer DL, Beiser AS, Warner AF et al. Monocular acuity norms for the Teller Acuity Cards between ages one month and four years. *Invest Ophthalmol Vis Sci* 1995; **36**: 671–85.
2. Mayer DL, Dobson V. Visual acuity development in infants and young children, as assessed by operant preferential looking. *Vision Res* 1982; **22**: 1141–51.
3. Birch EE, Gwiazda J, Bauer JA Jr, Naegele J, Held R. Visual acuity and its meridional variations in children aged 7 yr 60 months. *Vision Res* 1983; **18**: 1019–24.

# Individual Differences in Infant Cognition: Methods, Measures, and Models

JOHN COLOMBO

*Department of Human Development, University of Kansas, 4001 Dole Center, Lawrence, KS 66045-2133, USA*

Introduction

Although the study of individual differences in infant cognition dates back more than two decades, this topic has received much more attention in recent years. Such interest can be traced to the surprising conclusion that measures of infant cognition (visual attention, learning, and memory) are modestly but significantly predictive of performance on standardized assessments of cognitive, language, and intellectual ability in childhood and adolescence. This conclusion was first drawn in a review by Bornstein and Sigman (1). Several other reviews of this literature have since appeared (2–6).

The primary focus of this volume is the evaluation and interpretation of the results of studies on the effects of early fatty acid nutrition on developmental outcome. Recent work in this area has employed outcome measures based on infant sensory, perceptual, and cognitive development. The purpose of this chapter is to describe the basic procedures and measures that are used in assessing these domains of behavioral development during infancy. The chapter will also review information regarding the psychometric properties of the infant measures. It will include an update and recapitulation of research on the predictive validity of the infant measures to cognitive outcome in childhood, and a discussion of the various theoretical constructs thought to underlie the infant measures. Because the research on infant feeding and nutrition has for the most part focused on measures from the habituation and paired-comparison techniques, this review will emphasize research on these two procedures. However, reference will be made to other procedures, as appropriate, during the course of the review.

Presented first is a review of the basic features of the habituation and paired-comparison paradigms. For each paradigm, a brief summary of the theoretical rationale for the paradigm is outlined, followed by a description of methodological variations, and a catalog of the measures that each paradigm yields.

## The Visual Habituation Procedure

### Rationale

Although a large body of literature exists on the phenomenon of habituation in the comparative, anatomical, and physiological realms (e.g. 7), the use of the habituation procedure with human infants can be traced historically to Fantz (8) and Berlyne (9). In the typical habituation procedure, a stimulus is repeatedly presented to an organism, during which one or more of the organism's orienting responses is monitored. Across such repeated presentations, the magnitude of the orienting response usually declines (Fig. 1).

The most widely accepted theoretical framework for interpreting this change in response is comparator theory (e.g. 10), which posits that the decline of the orienting response reflects the acquisition of an accurate internal representation ('engram') of the stimulus presented. Thus, habituation represents a crude 'learning curve', with the magnitude of the orienting response reflecting the degree to which some mismatch exists between the external stimulus and the subject's internal representation of it. Thus, habituation has been used widely as a method for assessing visual learning in preverbal infants since the 1960s.

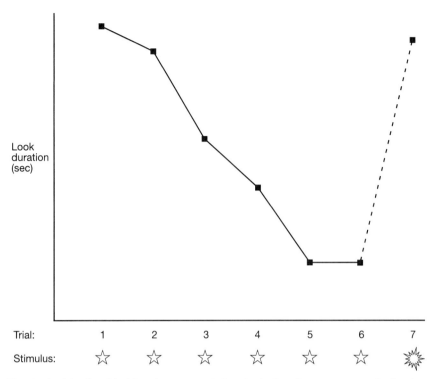

Fig. 1. An idealized habituation curve. Infant look durations to repeated stimulus presentations decline monotonically to some criterion, and then recover to a new stimulus. Recovery would need to be validated by comparing such performance to conditions where presentations of the habituation stimulus continued after attainment of the criterion.

## Methodology

As noted above, the habituation procedure involves a sequence of repetitive stimulus presentations. The methodology for such presentations varies widely, but falls into two broad categories.

## Fixed-trial procedures

The simplest habituation paradigm is the 'fixed-trial procedure', which was developed and used primarily during the 1960s and early 1970s. In most forms of this procedure, a stimulus is provided to the infant for some fixed amount of time, and these intervals are repeated for some fixed number of trials. Such trials are typically conducted without regard to the infant's visual behavior (e.g. the trials begin and end, whether the infant is looking or not; an infant may not look at all during a trial).

The advantage of the fixed-trial methodology rests with its simplicity. Sessions can be conducted easily, quickly, and inexpensively, they are uniform in length, and they yield the same number of data entries for each infant. The most serious disadvantage, however, is that the procedure is less sensitive to individual differences in looking than others in which the infant is 'forced' to habituate. For example, it is a common problem that not all infants will actually habituate to the stimulus presented under fixed-trial procedures (11,12).

*Infant-control procedures*

A second form of habituation paradigm emerged during the early 1970s and has generally been adopted as the preferred method. The strategy of this method was to adapt the number and length of stimulus exposures to the infant's own behavior toward the stimulus. These advanced techniques culminated in a class of 'infant-control paradigms' (13), in which a stimulus is presented in discrete trials until the infant's looking actually reaches some asymptotic decline (i.e. habituates). Each of the discrete trials is defined by the infant's ad libitum duration of looking to a stimulus, usually with some parametric constraint (e.g. the infant must look for at least 1 second to begin a trial, and must look away from the stimulus for at least 1 second to terminate one (14)). The criterion for the attainment of the asymptote (i.e. habituation) is typically calculated as some percentage of decline from levels of looking exhibited by the infant during prior trials of the sequence (e.g. two consecutive looks at 50% of the longest previous look in the sequence (15)).

The advantage of these procedures is that it ensures that each infant tested actually habituates to the stimulus in question. Additionally, because the criterion is set to some constant percentage of the infant's initial looking, the degree of habituation is theoretically equated across infants. The three main disadvantages are that: (i) the procedure is considerably more difficult and expensive to set up and run, (ii) the sessions may vary quite widely in length from infant to infant, and (iii) the data are not typically uniform for each subject (e.g. different numbers of trials).

*Measures*

Visual habituation is necessarily expressed in terms of a curve in which the orienting variable is plotted across presentations. This curve can be quantified in a number of different ways, but the variables it furnishes can be incorporated into a taxonomy of five distinct classes.

*Decrement*

One such class is primarily a measure of the rate of decline of the infant's looking across the repeated stimulus presentations. Given the assumptions of comparator theory, measures of decrement should provide the best index of individual differences in infants' formation of the engram of the habituation stimulus.

If, as in the case of fixed-trial paradigms, the number of trials is constant across infants, then a simple calculation of the magnitude of change across those trials is sufficient; such a measure is usually expressed in terms of a percentage or proportion of decline from the levels of the infants' looking during the initial trial(s) of the habituation session.

If the number of habituation trials varies across infants, however, as in the infant-controlled procedure, then the calculation is slightly more complicated. Some researchers have simply calculated the magnitude of the decline from either the initial look(s) or longest look(s) in the sequence to the last set of looks in the sequence (16,17). It is also possible to calculate a regression line for each infant to quantify the linear trend of decline (18). In light of the fact that not all habituation curves may be linear in nature (see the section on habituation pattern, below), however, it might seem reasonable to derive best-fitting lines to the habituation curve as is typically done in hierarchical linear (i.e. growth curve) modeling; but, to my knowledge, this has not yet been done.

*Attainment of the criterion*

A second set of measures involves the rapidity with which the infant habituates to the presented stimulus. This is also derived from the predictions of comparator theory, and is conceptually related to the decrement measures. However, it is typically calculated in terms of the number of trials or presentations necessary for the infant to reach some sort of asymptotic performance in his or her attention to the visual stimulus in question.

In cases where attainment of such an asymptotic criterion is not a critical feature of the paradigm (i.e. in most fixed-trial applications), this variable is obviously not applicable. In the infant control procedure, however, this variable simply translates to the number of looks to criterion (16,17).

*Look duration*

Another parameter is the relative height of the habituation curve, or of some portions of the curve. This, then, is reflected by measures of look duration. Comparator theory holds that the magnitude of the orienting response corresponds inversely to the status of the match between external stimulus and internal engram. If that is true, then the duration of looks should reflect the speed or efficiency of the processes by which the external stimulus is compared with the internal representation, and the processes by which the internal engram is updated and revised.

The habituation curve yields several possible indices of look duration. If one assumes that comparator theory is correct in holding that learning the stimulus takes place across the entire habituation session, then one might take the average look duration to reflect the average rate with which these processes take place. On the other hand, if one assumes that stimulus processing takes place more quickly, then perhaps one may take the duration of one or two critical looks during the habituation session (e.g. during the longest or 'peak' look), as a characteristic index of those processes. In this latter form, duration may be valid even if no criterion is attained.

*Habituation pattern*

Ideally, the pattern of decline of infants' attention in a visual habituation procedure proceeds in a monotonic pattern, with the longest look first, followed by looks of decreasing length until the criterion is reached. However, in a substantial proportion of cases, the pattern of decline is nonmonotonic (19). Two main patterns of nonmonotonic responding have been described (20). One is an 'increase–decrease' pattern in which look durations show a monotonic increase before a monotonic decrease to criterion; the other is a 'fluctuating,' or 'chaotic' pattern, in which look durations show no obvious trend before attaining a peak (typically late in the session), and then declining rapidly to criterion (Fig. 2).

The meaning of these patterns is generally unclear. From the point of view of comparator theory, one might interpret nonmonotonic patterns as reflecting a tendency not to focus attention on the presented stimulus in a sustained fashion (but see 21).

Alternatively, increase–decrease patterns resemble in part the response profile predicted to occur by dual-process theory (22). Dual-process theory posits that the curves yielded by common visual habituation paradigms are actually a product of both habituation (essentially, a comparator-like process) and sensitization (a transient arousal response elicited by the unexpected appearance of the stimulus per se). The net result of these dual processes is an increase (rather than a decrease) in the magnitude of orienting early in the sequence (23,24).

Finally, there is some evidence that, at least in the infant-control procedure, the pattern of looking may be influenced by initial levels of look durations (16). Recall that the criterion for habituation in this procedure is derived from look durations that occur earlier in the sequence. If such prior looks are very brief in duration (e.g. 5 seconds), the infant will have difficulty attaining the habituation criterion (here, the criterion would be consecutive looks of 2 seconds or less). As a result, infants who tend to exhibit such brief looks (e.g. older infants) may show a greater tendency to exhibit fluctuating patterns of habituation (25).

*Recovery*

In initial uses of the habituation procedure with human infants, investigators were concerned that the decline in attention may have been due to noncognitive factors such as sensory adaptation or fatigue. As a result, a novel stimulus was typically presented at the end of the habituation sequence; if looking increased ('recovered') to this presentation, one could rule out these alternative and uninteresting explanations for the decline in looking. It was quickly recognized, however, that such 'recovery' provided a means to test for detection of stimulus novelty and recognition memory for the habituation stimulus. As a result, this measure has been used in many studies of the normative development of infant perception and cognition (26,27), and in some studies of individual differences (e.g. 28).

The recovery measure is problematic for several reasons. First of all, because the habituation procedure coaxes infants to what is presumably an asymptotic level of looking, there is some threat that what is apparently recovery of looking may actually be attributable to regression to the mean. The degree to which the recovery measure truly reflects detection of the novel stimulus, then, can be assessed by comparing a group receiving a posthabituation novel presentation against a group that does not receive such a presentation. Only the former group should show recovery; the latter group should not. Such nomothetic comparisons are of course difficult to extend to reasoning about the idiographic performance of individual infants.

Second, there are problems with how to calculate recovery. Difference scores between the last habituation trial and the first posthabituation trial can be computed, but these are statistically problematic (29). Calculation of some magnitude of recovery (e.g. percentage increase from criterion looks) is possible, although it is not clear that the magnitude of recovery is meaningful, since the constructs it presumes to assess (discrimination and recognition memory) may well be considered to be dichotomous (i.e. yes/no) products of information acquisition.

*Summary*

As should be evident from the above discussion, the investigator has many choices in terms of how to quantify the habituation curve. The choices may be justified in terms of implied theoretical import (e.g. 30), or may be based on descriptive empirical analyses of data across ages (20,25). Other investigators have instead determined that no one measure is adequate to capture individual differences in performance, and that the construct of habituation may be best represented in terms of some summary factor extracted from, for example, structural equation measurement models (18,31).

A central issue here is whether the habituation measures are considered to reflect a single construct, or whether different measures might reflect different components of infant information processing. For example, Colombo and Mitchell (2,25) have argued that, in infant-control procedures, duration and decrement reflect one important underlying cognitive process, recovery another, with trials to criterion and habituation pattern each reflecting separate processes that may or may not be cognitive in nature. It is also important to note that each of these variables is likely to be influenced by both the cognitive constructs of most interest to the investigators and various noncognitive factors (infant state, hunger, sleepiness, etc.) that simply add error variance to the overall equation.

*Consistency of Individual Differences in Visual Habituation*

Beginning with Bornstein and Benasich's (19) examination of the individual differences in habituation, a number of studies of the psychometric properties of the measures yielded by the habituation paradigm have been reported. Simply put, the consistency of habituation measures does not attain the standards set for adult psychometric work. They are, however, high enough to suggest that they reflect the operation of some consistent behavioral processes of interest.

*Test–retest reliabilities*

Although there are some reports that include within-age test–retest reliabilities in the +0.60 to +0.70 range (32–34), the magnitudes of most

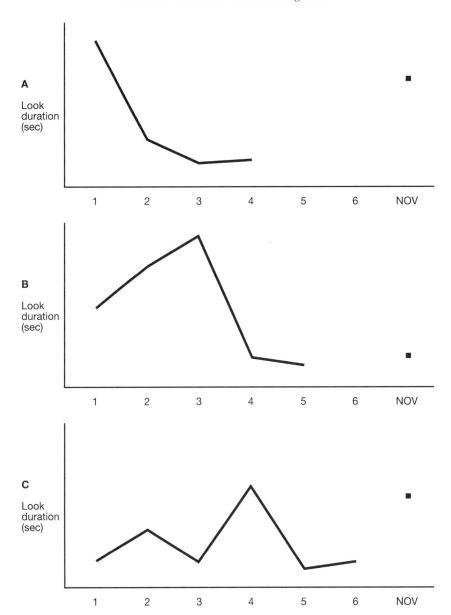

Fig. 2. Three different habituation curves, as obtained by infant control methods. (A) An infant whose look durations decline monotonically and quickly to one stimulus, and then recover to a novel stimulus (NOV). (B) An increase–decrease pattern, with little recovery. (C) The longer 'fluctuating' pattern, and recovery to the novel presentation.

short-term test–retest reliabilities of habituation variables fall below this level, with median values in the +0.30 to +0.40 range (2). As with most behavioral measures, test–retest values drop off as the interval between tests increases (25); those variables that show short-term test–retest reliabilities of about +0.40 generally show longer-term test–retest relationships (e.g. across 5 months) in the range from +0.20 to +0.40 (16).

A logical question at this point is whether some habituation variables show better test–retest reliabilities than others. This is a difficult question to answer, however, because many of the variables are arithmetically or logically interrelated. Derived variables such as total looking time or decrement often involve calculations that employ other 'raw' variables, such as peak look, first look, and number of looks to criterion (25,35). Thus, it becomes difficult to parse the reliability of the derived variables from the nonderived ones. The infant control procedure compounds this problem by adjusting the criterion for habituation separately for each infant during the course of the test session on the basis of what has happened earlier in the session; as a result, even some of the nonderived variables will be procedurally interrelated. Nevertheless, some general patterns have emerged. Although ordinal measures such as looks to criterion and habituation pattern are occasionally reported to show some significant test–retest reliability (16,19), this has not been consistent across studies (20). Similarly, although the test–retest reliability of recovery has been assessed in a number of reports (16,20,34) it has been shown to have significant test–retest reliability in only two reports (19,36).

Most reports suggest that the level of test–retest reliability for look duration within ages is in the region of +0.40 to +0.50 (see (2) for a summary of literature from various laboratories; recent additions to literature that have been previously reviewed or summarized include Colombo et al. (37), Frick and Colombo (15) Saxon et al. (38), although consistency in look duration is not universally reported across different stimuli (e.g. faces and geometric designs) during the second half of the first year (39,40).

The extant literature also generally suggests that decrement shows test–retest reliabilities that are comparable to look duration (16,33). Such findings should be considered, however, in light of the fact that look duration variables are components of most calculations of decrement. Colombo et al. (16) presented data suggesting that the significant test–retest reliability of decrement variables was largely attributable to the reliability of the duration variables that were used in those calculations.

A second logical question is whether these variables are more reliable at some ages than at others. The answer seems to be yes, with two qualifications. The first qualification is that the answer is based only on a handful of data; and the second is that, if indeed the relationship between age and the level of consistency exists, it does not appear to be a simple one. For example, Colombo *et al.* (16) found higher test–retest reliabilities for peak look durations at 4 and 9 months than at 3 and 7 months. More recently, large samples of infants were tested at 3, 4, 6, and 8 months of age across a series of six different stimuli in a look duration assessment 'battery' (see 38,41); the internal consistency for look duration was reasonable across these ages, but showed a distinct drop at 6 months. There is some evidence that the consistency of younger infants' responses may be more affected by the physical properties of the stimuli; for example, although Colombo *et al.* (16) reported relatively low test–retest reliability for look durations in 3-month-olds, subsequent tests under conditions featuring a limited range of stimuli (17) yielded much higher test–retest correlations. A consistent trend in the literature also points to a 'dip' in consistency at 6–7 months (16,38,39). This may reflect a change in what the look duration variable represents at some point during the second half of the first year.

*Predictive validity*

The predictive validity of look duration (in various forms: peak, total, average) has been indicated in a handful of studies (18,34,42–45). Several other studies have reported measures of decrement from fixed-trial procedures, which have been argued to be determined at least in part by, or related to, measures of look duration (see 2). Colombo and Mitchell (25) also show predictive validity (e.g. 46,47) to later measures of cognition and behavior. The predictive coefficients are generally modest, with typical values in the range 0.30–0.40, although the entire range is 0.20–0.70. McCall and Carriger (4) have suggested that infant measures are most predictive when administered in the age range 2–8 months. This seems to be true in general, but it is worth noting that look duration assessed during the newborn period in Sigman's (44,45) sample predicted various cognitive measures in late adolescence, and look duration assessed at 6.5 and 12 months of age in Jacobson *et al.*'s (42) sample predicted to 4 year IQ outcomes.

## The Paired-comparison Paradigm

*Rationale*

Like habituation, the paired-comparison procedure has roots in comparative psychology and ethology (48,49). Its development, however, followed more on logistical or empirical considerations than on a particular theoretical tradition. The modern paired-comparison procedure was originally based on simple 'choice' paradigms in which two stimuli differing in a critical visual property (e.g. color, contrast, or brightness (4,50,51)) were simultaneously presented to an infant. Infants' preference for one property over another could, by logical extension, be used to infer discrimination of and sensitivity to such properties.

Following Fantz's (8) demonstration that infants over 2 months of age prefer a novel stimulus over a familiar one, the procedure was modified to include a period during which the infant was exposed ('familiarized') to one stimulus for some amount of time and then probed for such a preference in choice trials where the familiar stimulus was simultaneously paired with a novel one. As such, the procedure yields a measure that indicates visual discrimination, recognition memory, and/or motivation to attend to the novel stimulus (Fig. 3).

The procedure has been extensively developed and refined in studies of early short- and long-term memory (52,53), and for the study of individual differences (e.g. 54–56). Paired-comparison protocols (stimuli, length of familiarization, and choice trial durations) were eventually standardized for the Fagan Test of Infant Intelligence (FTII) (57).

*Methodology*

*Familiarization*

In the paired-comparison procedure, an infant is familiarized with a stimulus for some predetermined amount of time. Such exposure functions much like a form of habituation session, since it is assumed that the infant will 'acquire' the familiarized stimulus during this period. Indeed, some studies have reported habituation of looking during the familiarization phase, and that subsequent performance on choice trials is dependent on the occurrence of habituation during familiarization (58).

In older studies, the stimulus was simply made available for some interval of time before initiation of the choice trials, and this sequence was conducted without regard to the infant's amount of looking at the stimulus. More modern versions of the procedure are better controlled, in that all infants accumulate some constant amount of looking to a stimulus before conducting the choice trials.

Familiarization may be conducted with a single stimulus presented in the center of the visual field, or to the left/right of midline in one of the lateral positions that will be used in the subsequent choice trials. Alternatively, two identical familiarization stimuli may be presented to the left and right of midline. The effect of these different methods of familiarization has not been widely studied, although it is likely that look durations will be depressed under conditions where presentations are paired, rather than single.

*Choice trials*

The paired-comparison choice trials represent the 'test' of the acquisition of the stimulus that presumably occurs during familiarization. Typically, these trials are administered by simultaneously presenting the familiarized and novel stimulus to the left and right of the infant's midline. In most experimental uses of the technique, the lateral positions of the stimuli are reversed halfway through the trial, because many infants under 5–6 months of age will have a preference for looking either left or right (59–61). In this way, the lateral positions of the familiar and novel stimuli are counterbalanced within subject.

As with familiarization, the choice trials may be conducted without regard to the infant's looking. In most contemporary uses of the paradigm, however, the amount of looking during the choice trials is accumulated until reaching some predetermined length, and thus well controlled.

Presumably, the separation and total visual angle of the paired stimuli during the choice trials might affect the infants' responses. Although this has not been a direct topic of study, it is probably safe to say that most uses of the procedure maintain a separation that keeps both targets within infants' peripheral visual-field thresholds (62).

Additionally, the effect of the length of the choice trials has not been studied parametrically, although in practice it is generally assumed that briefer choice trials are more appropriate as infants get older.

*Familiarization–choice trial interactions*

Success with the technique is virtually dependent on the length of the familiarization that precedes the choice-trial preference tests. Familiarization that is too brief will yield either random responding on the part of the infant, or even a preference for the familiarized stimulus (e.g. 63), and familiarization that is too long will result in attrition and/or variable performance on the part of the infant. As might be expected, the length of familiarization necessary to yield a significant novelty preference varies with both the stimuli employed and the age of the infant. If familiarization is calibrated properly, the novelty preference (which is expressed as either

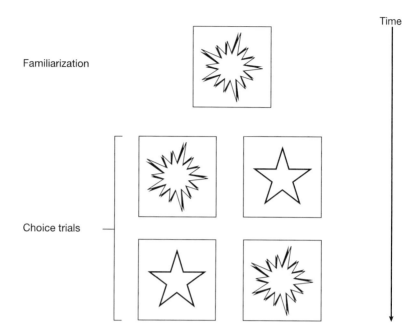

Fig. 3. Schematic drawing of the paired-comparison paradigm. The infant is familiarized to the first stimulus (presentation is portrayed here as singular stimulus, but it could be paired as well). Tests for discrimination/recognition occur in the following choice trials, where the familiarized stimulus is paired with a novel one. Two choice trials are run so that both stimuli are tested in both lateral positions. The percentage time looking to the novel target is calculated, collapsed across the two choice trials.

a percentage or proportion of looking to the novel stimulus) will usually range from 55% to 70%.

## Measures

The major focus of the paired-comparison paradigm is on the visual preference expressed by the infant on the choice trials. However, a number of other measures can be gleaned from both the familiarization and choice trial components of the procedure.

### Look duration/frequency

First, it is possible to derive or record look duration during familiarization. The interpretation of look duration here would presumably be similar, if not identical to, those interpretations delineated for look duration in the previous section on the habituation paradigm.

The monitoring of look duration can be done partly by using the number of looks during familiarization. If familiarization has been accumulated (i.e. if every infant looks at the familiarized stimulus for the same amount of time) it is possible to derive an 'average look' by dividing the number of looks by the length of familiarization. However, this ignores the fact that, under such conditions, the infant's final look is always cut off. It is also possible to keep track of the length of the infant's longest (peak) look during familiarization. Although this also may result in a similar 'ceiling' effect for infants who accumulate all their familiarization in a single look, and whose peak looks are thus cut off at the end of the predetermined length of familiarization.

Generally, look durations have been examined only from familiarization periods. Of recent interest to the readership of this volume are studies of nutritional enhancement in preterm infants (64,65) and nutritional deprivation in infant primates (66) that have analyzed look durations during the choice trials. The duration of choice trial looks has not been a topic of empirical or theoretical scrutiny in the infant literature. The interpretation of look durations, however, might be conceptualized similarly to look duration during familiarization, since the recognition and acquisition processes taking place during familiarization should also be operative during choice trials (see, however, the section on shifts below). Colombo (67) re-analyzed the Colombo *et al.* (60) database and found that look durations during choice trials were positively correlated with look durations during familiarization at both 4 and 7 months of age ($r = +0.56$), but this obviously needs to be replicated with various ages and stimulus combinations.

*Visual preferences*

The main focus of the paired-comparison paradigm is the visual preference yielded in the choice-trial phase. As noted above, if familiarization is sufficient, a 5–20% advantage in selective looking is typically observed for the novel stimulus. However, if familiarization is less than sufficient (63,68,69) or if the stimuli are highly complex (e.g. three-dimensional) (70–73), systematic preferences for the familiarized target have also been observed. A popular 'model' for visual preferences in the paired-comparison paradigm posits that selective looking shifts from random to familiarity references to random and then to novelty preferences as a function of increasing familiarization (74,75). A similar model was proposed for habituation, specifying an increase in attention to the habituation stimulus before the onset of the decline in looking (see the literature on 'backward habituation curves' (36,76)).

Any discriminative response (i.e. familiarity or novelty preference) in the choice trial phase of the paradigm, however, is subject to at least three basic interpretations. First, such a response cannot occur unless the infant is capable of perceiving the differences between the two stimuli presented at some level of processing (i.e. discrimination); thus, higher order cognitive interpretations may be brought to bear in cases of poor performance only if visual problems are ruled out. Second, such a response will not occur unless the infant has been sufficiently exposed to the familiarization stimulus (i.e. has learned, or acquired some 'engram' of the stimulus). Finally, showing such a preference may also rely on processes specific to memory itself; the infant must be able to recognize which of the two stimuli was familiarized.

*Shifting*

A final measure that can be taken from the choice-trial phase of this paradigm is the number of alternating looks ('shifts') made from stimulus to stimulus (60,77). Such shifts indicate an active comparison of the stimuli presented, but it is not clear whether this reflects alertness and motivation on the part of the infant, or whether it reflects confusion in discrimination or recognition of the stimuli in question. In theory, there is probably an inverted curvilinear relationship between shifts and novelty preferences, such that extremely low or high rates of shifting are associated with poor performance, while moderate shift rates are associated with better performance.

It is worth noting that some infants will fixate only one lateral position (i.e. right or left, irrespective of which stimulus occupies the position) during both choice trials. Such complete lateral bias will yield visual preferences at exactly chance-level responding if lateral position of the novel and familiar stimulus is counterbalanced within both stimuli and subjects; if lateral position is counterbalanced between stimuli or subjects, such a complete bias will simply increase the amount of variability in responding. Some researchers exclude infants who do not shift between stimuli within choice trials (15,68,78–80), but to my knowledge such exclusion has not been recommended for psychometric applications of the paradigm such as the FTII.

## Consistency of Individual Differences in Paired-comparison Performance

Once again, the individual consistency paired-comparison performance will not satisfy the psychometrician who works with adults. Generally, test–retest reliabilities are moderate to low for most variables. It is worth noting that the predictive validity of these measures is as high as, or higher than, the reliabilities of these assessments. However, unlike most habituation paradigms, the format of the paradigm allows for repeated or multiple measurement of the same individual at the same test session.

### Test–retest reliabilities

Reliabilities for look duration during the familiarization phase of this paradigm have been reported in a few studies, and are similar to those reported for look duration measured within the habituation paradigm. Colombo et al. (60) report respective week-to-week reliabilities of + 0.54 and + 0.48 for 4- and 7-month-olds, and a correlation of + 0.33 across the ages of 4 to 7 months. Colombo and Frick (41) reported internal consistencies (Cronbach's $a$) for look duration assessed during familiarization-like accumulation periods with six stimuli at 3 (+ 0.73), 4 (+ 0.51), 6 (+ 0.37), and 8 (+ 0.65) months of age; Saxon et al. (38) report a significant correlation of + 0.39 in a longitudinal study spanning the ages of 6 to 8 months.

The test–retest reliability for any single novelty preference task or item is fairly low. Rose et al. (70) report internal consistencies (measured with Cronbach's $a$) in the mid-20s, and Fagan and Detterman (81) report a value of 0.28. Colombo et al. (60) report an average correlation of + 0.24 across up to five tasks administered per visit, although internal consistencies as calculated on this database yielded a much lower value, with some negative correlations in the item matrix. Anderson (82) has also recently reported as for the FTII as being 0.02 at one test site and 0.10 at another. Based on the low internal consistency of the measures, Benasich and Bejar (83) have strongly criticized the use of the FTII/paired-comparison paradigm for diagnostic and screening purposes. The low internal consistency may justifiably concern those who use the FTII, and may have theoretical implications for what the scale actually measures (41). However, significant effects cannot logically be attributed to low reliability (84), and that assessments need not be internally consistent to be predictive (81). This suggestion is borne out to some degree by Anderson's (82) recent publication showing negligible internal consistency yet modest prediction of verbal IQ ($r$ about + 0.20) and auditory reception ($r$ about + 0.25) using the FTII.

While the reliability of single items is less than impressive, test–retest reliabilities for multiple assessments of novelty preferences (e.g. novelty preference batteries; similar to the manner in which the FTII is administered) are generally higher. Such batteries show a range of reliability from + 0.40 to + 0.50, although lower values have been reported (84).

Finally, the psychometric properties of shifts has only been reported on in one study (60). Shifts had poor test–retest reliability at 4 months, and good reliability at 7 months. As might be expected from this result, the frequency of shifts was not correlated across those ages.

Because novelty preferences are typically assessed and aggregated across many tasks involving various stimuli, there are no really good estimates of how the test–retest reliabilities or stabilities vary as a function of different stimuli or stimulus pairings. With respect to age differences, the novelty preference appears to be as (or slightly more) reliable at younger ages as at older ones, although the coefficients are not substantially different, and, as with most behavioral measures, the size of the reliability is inversely related to the length of the test–retest interval (2).

*Predictive validity*

The predictive validity of visual preferences (both novelty and familiarity preferences) has been established in a number of longitudinal studies. Summaries of the existing literature suggested that the predictive coefficients generally fell into the range from 0.40–0.50, with a range of + 0.25 to + 0.66 (2); data on additional samples published since then have yielded prediction at the lower end of this range (85). There is no apparent trend from that review for the preferences to be more highly correlated at some ages than at others, although the bulk of the longitudinal work has been conducted with infants from 3 to 9 months of age, and so such assessments may thus be most useful during that period (4). The range to which the preferences have been shown to predict now extends to 11 years of age (86,87).

## Underlying Constructs and Mechanisms

Given the existence of modest reliability of infant measures, and moderate predictive validity of those measures to childhood cognition and behavior, one of the more important issues currently under consideration involves the delineation of those mechanisms or constructs that are responsible for such continuity of intellect from infancy. We have discussed these issues from various points of view in a number of different recent reviews (41,88,89), but we will recapitulate them here for this audience. Some of these mechanisms will be more relevant to performance in one paradigm than in the other.

## General Intelligence (g)

Perhaps the most historically imbued construct for individual differences in adult cognitive performance is the general factor model of intelligence (e.g. 90). In fact, the concept of $g$ has been raised in the discussion of such continuity of infant cognition to childhood (89,91,92). Initial considerations of the continuity of infant measures to childhood speculated that the infant measures might tap some 'general' construct that also contributed to more mature forms of intelligence (92). Still, simply attributing early individual differences in cognition cannot explain those individual differences; it is necessary to hypothesize an actual process for such theorizing to be useful (89).

## Speed or Rapidity of Stimulus Processing

Toward that end, it has been suggested that both habituation and novelty preferences may be related to speed of processing, which is a construct that has been theoretically linked to the principal component of IQ tests that gives rise to $g$ (93–95). Indeed, some measures of attention and cognition in infancy have been linked with simple measures of reaction time (96,97), and infants' ocular reaction time (98) has been directly correlated with later intelligence itself (84).

The phrase 'speed of processing', however, may also be interpreted as something to explain rather than as an explanation itself. Colombo and Mitchell (25) noted that long look duration was a characteristic of younger (i.e. 'immature') organisms, and it was eventually hypothesized (99) that the types of visual inspection and scanning in which young infants engage while looking at a visual stimulus (100–102) might in fact be a characteristic of those older infants who show prolonged look durations. Bronson (103,104) has recently confirmed this in studies of visual scanning within and across ages. Furthermore, the results of a series of recent experiments (15,37,68,78) suggest that the types of visual encoding that might be predicted to occur from these less mature scanning patterns are also seen in infants who show longer look durations. This has led to speculations

(81,99) that individual differences in look duration might also be related to: (i) subpathways of the geniculostriate visual system that might mediate certain types of analyses of the visual world; (ii) extrastriate areas that mediate focused attention on different visual properties; or (iii) the parietal cortex, which mediates switching of attention from one visual property to another. With some exceptions (see the description of Frick *et al.* (105) below), however, these hypotheses have not yet been the topic of inquiry .

*Memory Processes*

Since recognition memory is a necessary component of performance in both the habituation and paired-comparison paradigms, it is obviously a plausible candidate in accounting for the predictive validity of these measures (see 106). Indeed, individual differences in infant memory in a different paradigm have been found to predict to preschool intelligence at approximately the same magnitude of effect as the habituation and novelty preference measures discussed here (107).

*Inhibition*

Also proposed to account for individual differences in early cognition is the construct of inhibition (3,4). Essentially, this hypothesis holds that both paradigms tap the infant's ability to inhibit looking to irrelevant aspects of the environment (e.g. a familiar stimulus). In the habituation paradigm, such inhibition results in a decline in looking to a repetitively presented stimulus. In the paired-comparison paradigm, such inhibition is expressed by less looking to the familiarized stimulus, and thus (by default) greater looking to a novel one.

Colombo (99) has suggested that a mechanism conceptually related to inhibition might be reflected in individual differences in look duration. Based on Posner's (108) work on the neuroscience of attention, it has been proposed that long durations may be related to difficulties in the disengagement of attention from a visual stimulus. That is, long look durations may not necessarily reflect extended information processing, but rather an inability to 'let go' of a stimulus, even after stimulus processing is ended. In the adult, disengagement of attention function appears to be mediated by the parietal lobe (109–112); it is worth noting that many of the behaviors associated with an attentional subsystem involved in the shifting and disengagement of attention in the adult appear to emerge somewhere between 2 and 6 months of age (113–117). Frick *et al.* (105) found look duration to be correlated with disengagement of attention in both 3- and 4-month-olds.

*Positive Response to Novelty*

It has been argued that a generalized positive response to novelty (118) may represent the major component of infant cognitive performance that is related to adult intelligence. This model was posited mostly in consideration of the predictive nature of novelty preferences, but the general concepts here might well be applied to habituation as well. For example, it could be argued that in progressively looking less and less at a stimulus during habituation, infants are actually looking more and more at some other stimulus elsewhere in the environment. Indeed, the overall session lengths of both habituation and fixed-accumulation familiarization periods for infants who look for briefer look durations is longer than for infants who engage in more prolonged looks (25,90).

*Variation in Reinforcement Value of Stimuli*

A position somewhat related to the positive response to novelty hypothesis is one that attempts to explain the distribution infant attention to stimuli in terms of learning theory (119). Many of these begin by positing that any stimulus possesses some reinforcement value, and that such value declines with repeated presentation. This particular position is problematic for several reasons (25), although, to be fair, the behavioral approach has not been completely exhausted in terms of its relevance to infants' performance in these paradigms.

# Do Measures of Infant Cognition Reflect One Process, or Many Processes?

To this point, I have compiled a catalog of possible processes and functions that may contribute to individual differences in early cognition. A common issue under consideration, however, is whether the continuity of individual differences in both habituation and paired-comparison performance reflect either a single process or multiple, dissociable processes. In the sections that follow, we examine the arguments for each.

## Single-process Models

The first set of models to be discussed holds that the continuity of infant measures to childhood intelligence/cognition is mediated by a single construct. Indeed, a 'general factor' solution would be more parsimonious than one which invoked multiple factors or cognitive processes (2,120). The best argument for such a general factor rests on three points. First, there are claims that the infant measures predict better to aggregate measures of performance in childhood than they do to specific tasks (106,121). Second, it appears that many of the predictive infant measures can be shown to be intercorrelated to some degree (2; see also 122,123), although this seems certain to vary with methodological parameters (e.g. 15). Third, individual differences on some measures of visual habituation and/or recognition memory have been reported to be associated with relatively generic developmental indices, such as motor development (17), reaction time (96,97), autonomic function (124,125), and some manifestations of arousal (126).

The general factor model remains viable in the consideration of the continuity of cognitive function from infancy to childhood. Such a model, however, can be put to a particularly critical test generated by the following prediction. If indeed these measures tap a general construct, then each measure should correlate with the other, and be predictive under any and all conditions where one should expect to encounter cognitive risk.

## Multiple Processes

A second broad class of models posits that habituation and paired-comparison measures may reflect different and dissociable components of information processing in infancy. If these paradigms tap independent and dissociable cognitive processes, then hypotheses that are specific to one paradigm may be considered without regard to their relevance to the other paradigm. As noted above, this might include applying any of the factors discussed above as applying to performance in both paradigms only to performance in one of them; for example, Colombo (2) suggested that look duration during habituation might tap a mechanism involved with processing speed, while novelty preferences might tap a process related to recognition memory (see also 86). Included in this category are models holding that several different such components may be reflected within a single measure of infant cognition. Thus, it is possible that one infant's failure to show a novelty preference on a paired-comparison task may be attributable to slowed encoding, while another infant's suboptimal performance may be traced to difficulty with processes mediating recognition memory.

There are two important corollaries of such multiple-process models. First, if the multiple-process position is correct, one should be able to find situations where the various measures do not correlate with one another. Second, if this position is correct, it should be possible to find some situations in which there is variability in the pattern of prediction from different infant measures to different outcome measures. Both these corollaries are to some degree addressed by Jacobson's work on prenatal exposure to toxins.

With respect to the issue of intercorrelation, Jacobson *et al.* (96) collected data on novelty preference, reaction time, and fixation duration from over 100 alcohol-exposed infants, and subjected these data to a factor analysis. Two factors emerged, with look duration loading positively on the one (a factor shared with reaction time), and novelty preference loading significantly on the other.

With respect to the second corollary, Jacobson has not provided data on differential prediction, but has shown evidence for differential sensitivity of habituation and paired-comparison measures to conditions that might well yield different deficits. During the mid-1980s, Jacobson *et al.* (127) reported infants prenatally exposed to polychlorinated biphenyls (PCBs) showed lower novelty preferences than cohorts who were not so exposed. Subsequently, Jacobson *et al.* (128) found that look duration was significantly longer in infants whose mothers drank significant amounts of alcohol during pregnancy. Recently, Jacobson (129) noted that novelty preference was not sensitive to the degree of prenatal exposure to alcohol, and look duration was not sensitive to the level of PCB exposure. That is, PCBs affected novelty preference, but not fixation duration, and alcohol affected fixation duration, but not novelty preference. This, of course, is a pattern implied by a multiple-process model of infant cognition and, perhaps, of continuity from infancy.

One last study of interest is Bornstein and Tamis-LeMonda's (31) longitudinal investigation of the antecedents of habituation and novelty preference. Habituation performance at 5 months of age was predicted by two variables (visual discrimination performance at 2 months and a concurrent measure of the quality of maternal interaction) that were unrelated to novelty preference performance at 5 months. This further suggests that the two measures have different origins, and may thus tap different constructs.

## Other Issues

The following sections include a discussion of several additional issues concerning the measurement and interpretation of infants' performance in habituation and paired-comparison paradigms. Some of the fundamental issues concerning these measures have been previously reviewed (e.g. 4) and have also been very recently re-summarized and updated (41). Here, however, I will address two additional issues that may be relevant to the audience of the current volume.

### Generalizability Across Paradigms

In the first part of this chapter, I briefly enumerated some of the procedural variations within both the habituation and paired-comparison paradigms. Given such diversity in measurement, it is logical to wonder whether results obtained under specific conditions are applicable or generalizable to results obtained under other conditions.

Several studies have conducted comparisons of various procedures in evaluating fundamental phenomena of infant perception and cognition. For example, Greenberg and Weizmann (134) found that similar stimulus preferences emerged in 8- and 12-week-old infants, whether they were retested in a single-presentation fixed-trial paradigm or in a simple paired-comparison selective-looking procedure. Haaf et al. (131) also compared fixed-trial procedures with infant-controlled procedures; they found that looking time was increased and much more variable under infant-controlled procedures than with fixed-trial procedures. However, the conclusions concerning infant discrimination abilities yielded by these two methods in this study were the same. Millar and Weir (12) recently presented data on fixed-trial versus infant-control procedures that essentially replicate Haaf et al. (131), except that apparent recovery of looking to a novel stimulus was observed only in the infant-control procedure.

The important issue with respect to individual differences, however, has to do with the concordance of findings across different paradigms. Despite the fact that predictive validity has been reported for both recovery in the habituation paradigm (e.g. 28,46,132) and novelty preferences from the paired-comparison procedures (e.g. 55), there are (to my knowledge, at least) no studies that have evaluated the concordance of these measures with one another. Look duration from familiarization and habituation procedures have also been reported as being significantly concordant (37). As one might expect, however, the degree of concordance is apparently dependent upon the length of the familiarization period employed (133). For example, look durations can be cut off under conditions of brief familiarization, and average look durations may be artificially depressed if familiarization is too lengthy. Habituation decrement has been assessed within subjects in fixed trial and infant control in only one study (134), but no measures of individual consistency across these tasks were reported.

## Standardization of Measurement

All of the issues discussed above lead us to a final matter for consideration. A common concern among those who would like to design, use, and interpret these measures of infant cognition for follow-up studies of infants at risk, evaluation of early intervention, or effects of experiential manipulation is that, aside perhaps from the FTII, there is no one widely accepted or standardized format for assaying these #ognitive processes. That is, the applied demand for standardized versions of these measures generally has not been met by those researchers who use them.

I think the reason for this is that these measures were not initially developed for the purposes of prediction per se. Instead, the impetus for the emergence of these procedures was interest in measuring basic cognitive processes in normative and nomothetic laboratory studies of information processing with preverbal human subjects. The fact that they were in any way predictive of childhood cognition was, I think, as much a surprise to those who used them as to the general developmental audience who had long believed that such prediction was not possible. As such, however, the paradigms (again, with the exception of the FTII to some degree) have largely remained in their experimental form, and are not readily accessible for those individuals who would like to use them for standard clinical practice.

## Summary

After all this, what can be said about the relationship of early cognitive performance to later cognitive status? In some ways, we have not gone very far beyond the fundamental conclusions offered by Bornstein and Sigman (1); the conclusions they drew and the issues they raised still ring true even now a decade later.

(i) *Prediction*. First, there appears to be significant prediction of later cognition from measures of infant attention and memory. The persistence of the prediction phenomenon across new samples since the publication of that review (18,34,42–45) does raise confidence in the utility of these measures, even if they account for a relatively modest proportion of the variance in later cognitive status.

(ii) *Reliability.* Second, the modest reliability of the infant measures most likely serves as a significant constraint on the magnitude of this prediction. Colombo (2) intimated that raising the reliability of these measures by conventional means (e.g. more measures per child, standardization of protocols) was all that was needed to quell doubts about the degree of prediction they offered. However, raising the reliability of these measures has been difficult (41), and this leaves the question open of whether these measures will provide some level of useful clinical significance (see, however, 135).

(iii) *Mechanisms.* Third, while the candidates for the processes underlying such prediction may be limited to a finite set, that set is large in number and it is still not clear if one or many of that set apply to the problem in question. It is important to remember that the measurement of infant attention, encoding, and memory in isolation ignores the obvious and powerful contributions that the environment makes to any child's cognitive status (see 18,43).

However, despite these problems, it would appear that many of the measures discussed above are well suited to studies that involve nomothetic comparisons of infants whose experiences early in life may bear on later cognitive skills. Such comparisons include those who have been exposed prenatally to alcohol (128) or environmental toxins (127), infants from varying socio-economic backgrounds (130), and, the topic of the current volume, nutritional manipulations. It should also be obvious that such comparisons conducted to resolve issues with an applied emphasis have great potential for informing basic research on infant cognitive processes.

## Future Directions

A final question concerns how research using measures of infant cognition should be carried out in the future. Some consideration of this question has been offered recently in light of the extant literature of the developmental cognitive neuroscience of visual attention (88), but I will offer two specific suggestions here.

## Design Strategies

First, given the possibility that the infant tasks tap different underlying constructs, investigators interested in using such measures in their longitudinal studies may need to either: (i) use or construct infant tasks that specifically tap the processes or functions that are likely to be affected by their early interventions, exposures, or manipulations; or (b) include more than one of the extant infant measures in their studies.

## Measurement Strategies

Second, because looking can be interpreted or considered in so many ways, we are now well past the time when it should not be employed as the sole index of attention. Although an infant must look to attend to a visual stimulus, it does not logically follow that an infant always attends to the stimulus while looking. Richards (e.g. 125,136–139) has employed convergent psychophysiological measures of attention (which are easily implemented, relatively inexpensive, and now commercially available) in tandem with looking, and his data strongly suggest that the quality and quantity of information processing varies within looks. We have adopted this strategy in our laboratory over the past year, with both illuminating and productive results. Progress in both the applied and basic arenas of study may well be furthered by a multivariate and multilevel approach to cognitive performance.

## Acknowledgement

Preparation of this chapter was supported in part by NIH Grant HD29960.

## References

1. Bornstein MH, Sigman MD. Continuity in mental development from infancy. *Child Dev* 1986; **57**: 251–74.
2. Colombo J. *Infant Cognition: Predicting Childhood Intelligence*. Newbury Park, CA: Sage, 1993.
3. McCall RB. What process mediates prediction of childhood IQ from infant habituation and recognition memory? Speculations on the roles of inhibition and rate of information processing. *Intelligence* 1994; **18**: 107–24.
4. McCall RB, Carriger M. A meta-analysis of infant habituation and recognition memory performance as predictors of later IQ. *Child Dev* 1993; **64**: 57–79.

5. McCall RB, Mash C. Infant cognition and its relation to mature intelligence. *Ann Child Dev* 1995; **11**: 27–56.
6. Slater A. Individual differences in infancy and later IQ. *Child Psychol Psychiatr Allied Disciplines* 1995; **36**: 69–112.
7. Tighe T, Leaton R. *Habituation.* Hillsdale, NJ: Erlbaum, 1976.
8. Fantz RL. Visual experience in infants: decreased attention to familiar patterns relative to novel ones. *Science* 1964; **146**: 668–70.
9. Berlyne DE. The influence of complexity and albedo of stimuli on visual fixation in the human infant. *Br J Psychol* 1958; **49**: 315–18.
10. Sokolov E. *Perception and the Conditioned Reflex.* Oxford: Pergamon, 1963.
11. Lewis ME. A developmental study of information processing within the first three years of life: response decrement to a redundant signal. *Monographs of the Society for Research in Child Development* 1969; **34** (9, No. 133).
12. Millar WS, and Weir CG. A comparison of fixed-trial and infant-controlled habituation procedures with 6- to 13-month-old at-risk infants. *Infant Behav Dev* 1995; **18**: 263–72.
13. Horowitz FD, Paden LY, Bhana K, Self PA. An infant control procedure for studying infant visual fixations. *Dev Psychol* 1972; **7**: 90.
14. Colombo J, Horowitz FD. A parametric study of the infant control procedure. *Infant Behav Dev* 1985; **8**: 117–21.
15. Frick JE, Colombo J. Individual differences in infant visual attention: recognition of degraded visual forms by 4-month-olds. *Child Dev* 1996; **67**: 188–204.
16. Colombo J, Mitchell DW, O'Brien M, Horowitz FD. Stability of infant visual habituation during the first year. *Child Dev* 1987; **58**: 474–89.
17. Colombo J, Mitchell DW, O'Brien M, Horowitz FD. Stimulus and motoric influences on visual habituation at three months. *Infant Behav Dev* 1987; **10**: 173–81.
18. Tamis-LeMonda CS, Bornstein MH. Habituation and maternal encouragement of attention in infancy as predictors of toddler language, play, and representational competence. *Child Dev* 1989; **60**: 738–51.
19. Bornstein MH, Benasich AA. Infant habituation: Assessments of short-term reliability and individual differences at five months. *Child Dev* 1986; **47**: 87–99.
20. McCall RB. Individual differences in the pattern of habituation at 5 and 10 months of age. *Dev Psychol* 1979; **15**: 559–569.
21. McCall RB. *Discussant's Comments: What Underlying Mechanisms Explain the Predictive Validity of Infant Visual Attention?* Tamis-LeMonda CS, Chair). Indianapolis, IN: Society for Research in Child Development.
22. Thompson RF, Glanzman DL. Neural and behavioral mechanisms of habituation and sensitization. In: Tighe T, Leaton R, eds. *Habituation.* Hillsdale, NJ: Erlbaum, 1976: 49–94.
23. Bashinski H, Werner JS, Rudy JW. Determinants of infant fixation: evidence for a two-process theory. *J Exp Child Psychol* 1988; **39**: 580–98.
24. Kaplan PS, Werner J, Rudy J. Habituation, sensitization, and infant visual attention. In: Rovee-Collier C, ed. *Adv Infancy Res* 1988; **4**: 61–110.
25. Colombo J, Mitchell DW. Individual and developmental differences in infant visual attention. In: Colombo J, Fagen JW, eds. *Individual Differences in Infancy.* Hillsdale, NJ: Erlbaum, 1990: 193–227.

26. Pancratz C, Cohen, LB. Recovery of habituation in infants. *J Exp Child Psychol* 1970; **9**: 208–16.
27. Saayman G, Ames EW, Moffitt A. Response to novelty as an indicator of visual discrimination in the human infant. *J Exp Child Psychol* 1964; **1**: 189–98.
28. O'Connor M, Cohen S, Parmelee A. Infant auditory discrimination in preterm and fullterm infants as a predictor of 5 year intelligence. *Dev Psychol* 1984; **20**: 159–65.
29. Appelbaum MA, McCall RB. Design and analysis in developmental psychology. In: Kessen W, volume ed.; Mussen P, series ed. *Handbook of Child Psychology*, Vol. 1. New York: Wiley, 1983: 415–76.
30. DeLoache JS. Rate of habituation and visual memory in infants. *Child Dev* 1976; **47**: 145–54.
31. Bornstein MH, Tamis-LeMonda CS. Antecedents of information-processing skills in infants: habituation, novelty responsiveness, and cross-modal transfer. *Infant Behav Dev* 1994; **17**: 371–80.
32. Barten S, Ronch J. Continuity in the development of visual behavior in young infants. *Child Dev* 1971; **42**; 1566–71.
33. Fenson L, Sapper V, Minner DG. Attention and manipulative play in the one-year-old child. *Child Dev* 1974; **45**: 757–64.
34. Rose D, Slater A, Perry H. Prediction of childhood intelligence from habituation in early infancy. *Intelligence* 1986; **10**: 251–63.
35. Cohen LB. Habituation of infant visual attention. In Tighe T, Leaton R, eds. *Habituation*. Hillsdale, NJ: Erlbaum, 1976: 207–38.
36. Byrne JM, Clark-Tousenard ME, Hondas BJ, Smith IM. *Stability of Individual Differences in Infant Attention*. Toronto, Ontario: Society for Research in Child Development, 1985.
37. Colombo J, Frick JE, Ryther JS, Gifford JJ. Individual differences in infant visual attention: four-month-olds' recognition of forms connoted by complementary contour. *Infant Behav and Dev* 1996; **19**: 113–19.
38. Saxon TF, Frick JE, Colombo J. Individual differences in infant visual fixation and maternal interactional styles. *Merrill–Palmer Q* 1997; **43**: 48–66.
39. Jankowski JJ, Rose SA. The distribution of visual attention in infants. *J Exp Child Psychol* 1997; **65**: 127–35.
40. Vervloed MPJ. *Learning in Preterm Infants: Habituation, Operant Conditioning, and their Associations with Motor Development*. Groningen, Netherlands: Proefschrift Rijksuniversiteit Groningen, 1995.
41. Colombo J, Frick JE. Recent advances and issues in the study of preverbal intelligence. In: Anderson M, ed. *The Development of Intelligence*. London: University College Press (in press).
42. Jacobson SW, Chiodo LM, Jacobson JL. Predictive validity of infant recognition memory and processing speed to 7-year IQ in an inner-city sample. Presented at the *International Conference on Infant Studies*, Providence, RI, 1996.
43. Mitchell DW, McCollam K, Horowitz FD, Embretson SE, O'Brien M. *The Interacting Contribution of Constitutional, Environmental, and Information Processing Factors to Early Developmental Outcome*. Seattle, WA: Society for Research in Child Development, 1991.
44. Sigman MD, Cohen SE, Beckwith L, Parmelee AH. Infant attention in relation to intellectual abilities in childhood. *Dev Psychol* 1986; **22**: 788–92.

45. Sigman MD, Cohen SE, Beckwith L, Asarnow R, Parmelee AH. Continuity in cognitive abilities from infancy to 12 years of age. *Cognitive Dev* 1991; **6**: 47–57.
46. Lewis ME, Brooks-Gunn J. Visual attention at three months as a predictor of cognitive functioning at two years of age. *Intelligence* 1981; **5**: 131–40.
47. Miller D, Spiridigliozzi G, Ryan E, Callan M, McLaughlin J. Habituation and cognitive performance: relationships between measures at four years of age and earlier assessments. *Int J Behav Dev* 1980; **3**: 131–46.
48. Fantz RL. A method for studying early visual development. *Perceptual Motor Skills* 1956; **6**: 13–15.
49. Fantz RL. The origin of form perception. *Sci Am* 1961; **204**: 66–72.
50. Stirnimann, F. Ubr das forbenempfinden neugeborner. *Ann Paediatr* 1944; **163**: 1–25.
51. Valentine CW. The colour perception and colour preferences of an infant during its fourth and eighth months. *Br J Psychol* 1913; **6**: 363–86.
52. Fagan JF. Infant recognition memory for a series of visual stimuli. *J Exp Child Psychol* 1971; **27**: 27–34.
53. Fagan JF. Infant recognition memory: the effects of length of familiarization and type of discrimination task. *Child Dev* 1974; **45**: 351–6.
54. Fagan JF. The paired-comparison paradigm and infant intelligence. *Ann NY Acad Sci* 1990; **608**: 337–64.
55. Fagan JF, McGrath SK. Infant recognition memory and later intelligence. *Intelligence* 1981; **5**: 121–30.
56. Fagan JF, Singer LT. Infant recognition memory as a measure of intelligence. *Adv Infancy Res* 1983; **2**: 31–79.
57. Fagan JF, Shepard PA. *The Fagan Test of Infant Intelligence*. Cleveland, OH: Infantest Corporation, 1986/1987.
58. Caron AJ, Caron RF, Minichiello MD, Weiss SJ, Friedman SL (1977). Constraints on the use of the familiarization-novelty method in the assessment of infant discrimination. *Child Dev* 1977; **48**: 747–62.
59. Cohen LB, Gelber EC. Infant visual memory. In: Cohen L, Salapatek P, eds. *Infant Perception: From Sensation to Cognition*, Vol. 1. New York: Academic Press, 1975: 347–404.
60. Colombo J, Mitchell DW, Horowitz FD. Infant visual behavior in the paired-comparison paradigm: test–retest and attention–performance relations. *Child Dev* 1988; **59**: 1198–210.
61. Haaf RA, Diehl RE. Position bias and the paired comparison procedure in studies of infant attention. *Dev Psychol* **12**: 548–9.
62. Maurer D, Lewis T. Peripheral discrimination by three-month-old infants. *Child Dev* **50**: 276–9.
63. Wetherford MJ, Cohen LB. Developmental changes in infant visual preferences for novelty and familiarity. *Child Dev* 1973; **44**: 416–24.
64. Carlson SA, Werkman SH. A randomized trial of visual attention of preterm infants fed docosahexaenoic acid until two months. *Lipids* 1996; **31**: 85–90.
65. Werkman SH, Carlson SA. A randomized trial of visual attention of preterm infants fed docosahexaenoic acid until nine months. *Lipids* 1996; **31**: 91–7.
66. Reisbick S, Neuringer M, Gohl E. Visual attention in monkeys: effects of dietary fatty acids and age. *Dev Psychol* (in press).
67. Colombo J. Discussant's comments. In: Hartmann EE, chair. *Nutritional*

*effects on Infant Development: Visual, Motor, and Cognitive Effects.* Providence, RI: International Society for Infant Studies, 1996.
68. Freeseman LJ, Colombo J, Coldren JT. Individual differences in infant visual attention: four-month-olds' discrimination and generalization of global and local stimulus properties. *Child Dev* 1993; **64**: 1191–203.
69. Uzgiris IC, Hunt J McV. Attentional preference and experience: II. An exploratory longitudinal study of the effect of visual familiarity and responsiveness. *J Genet Psychol* 1970; **117**: 109–21.
70. Rose SA, Wallace IF, Feldman JF. Individual differences in infant information processing: reliability, stability, and prediction. *Child Dev* 1988; **59**: 1177–97.
71. Rose SA, Gottfried A, Melloy-Carminar P, Bridger W. Familiarity and novelty preferences in infant recognition memory: implications for information processing. *Dev Psychol* 1982; **18**: 704–13.
72. Rose SA, Gottfried A, Bridger W. Cross-modal transfer in infants: relationship to prematurity and socioeconomic background. *Dev Psychol* 1978; **17**: 661–9.
73. Rose SA, Wallace IF. Visual recognition memory: a predictor of later cognitive functioning in preterms. *Child Dev* 1985; **56**: 885–91.
74. Hunt J McV. Attentional preference and experience: II. An exploratory longitudinal study of the effect of visual familiarity and responsiveness. *J Genetic Psychol* 1970; **117**: 109–21.
75. Wagner SH, Sakovits LJ. A process analysis of infant visual and cross-modal recognition memory: implications for an amodal code. *Adv Infancy Res* 1986; **4**: 195–217.
76. Cohen LB, Menten TG. The rise and fall of infant habituation. *Infant Behav Dev* 1981; **4**: 269–80.
77. Ruff HA. The function of shifting fixations in the visual perception of infants. *Child Dev* 1975; **46**: 857–65.
78. Colombo J, Freeseman LJ, Coldren JT, Frick JE. Individual differences in infant visual fixation: dominance of global and local stimulus properties. *Cognitive Dev* **10**: 271–85.
79. Johnson MH, Dziurawiec S, Ellis H, Morton J. Newborns' preferential tracking and its subsequent decline. *Cognition* 1991; **40**: 1–19.
80. Slater A, Mattock A, Brown E, Bremner JG. Form perception at birth: Cohen and Younger (1984) revisited. *J Exp Child Psychol* 1991; **51**: 395–406.
81. Fagan JF, Detterman DK. The Fagan Test of Infant Intelligence: a technical summary. *J Appl Dev Psychol* 1992; **13**: 173–93.
82. Anderson H. The Fagan Test of Infant Intelligence: predictive validity in a random sample. *Psychol Rep* 1996; **78**: 1015–26.
83. Benasich AA, Bejar II. The Fagan Test of Infant Intelligence: a critical review. *J Appl Dev Psychol* 1992; **13**: 153–71.
84. DiLalla LF, Thompson LA, Plomin R et al. Infant predictors of preschool and adult IQ: a study of infant twins and their parents. *Dev Psychol* 1990; **26**: 759–69.
85. Cohen J, Cohen P. *Applied Multiple Regression/Correlation for the Behavioral Sciences.* Hillsdale, NJ: Erlbaum, 1983.
86. Rose SA, Feldman JF. Prediction of IQ and specific cognitive abilities at 11 years from infancy measures. *Dev Psychol* 1995; **31**: 685–96.
87. Rose SA, Feldman JF. Cognitive continuity from infancy: a single thread or a

twisted skein? Presented at the *Meeting of the Society for Research in Child Development*, Indianapolis, IN, March 1995.
88. Colombo J, Janowsky JS. A cognitive neuroscience approach to individual differences in infant attention and recognition memory. In: Richards JE, ed. *The Cognitive Neuroscience of Attention: Developmental Perspectives*. Hillsdale, NJ: Erlbaum (in press).
89. Mitchell DW, Colombo, J. Infant cognition and general intelligence. In: Tomic W, Kingma J, eds. *Advances in Cognition and Education: Reflections on the Concept of Intelligence*. Greenwich, CT: JAI Press (in press).
90. Spearman, C. General 'intelligence' objectively determined and measured. *Am J Psychol* 1904; **15**: 201–93.
91. Fagan JF. Infant intelligence. *Intelligence* 1981; **5**: 239–43.
92. Fagan JF. The intelligent infant: implications. *Intelligence* 1984; **8**: 1–9.
93. Jensen AR. Why is reaction time correlated with psychometric *g*? *Current Directions Psychol Sci* 1993; **2**: 53–5.
94. Reed TE, Jensen AR. Conduction velocity in a brain nerve pathway of normal adults correlates with intelligence. *Intelligence* 1992; **16**: 259–72.
95. Vernon PA, Mori M. Intelligence, reaction times, and peripheral nerve conduction velocity. *Intelligence* 1992; **16**: 273–88.
96. Jacobson SW, Jacobson JJ, O'Neill JM, Padgett RJ, Frankowski JJ, Bihun JT. Visual expectation and dimensions of infant information processing. *Child Dev* 1992; **63**: 711–24.
97. Lamarre G, Pomerleau A. *The Meaning of Individual Differences in Early Habituation*. Tours, France: International Society for the Study of Behavioral Development, 1985.
98. Haith MM, Hazan C, Goodman G. Expectation and anticipation of dynamic visual events by 3.5-month-old babies. *Child Dev* 1988; **59**: 467–79.
99. Colombo J. On the neural mechanisms underlying individual differences in infant fixation duration: two hypotheses. *Dev Rev* 1995; **15**: 97–135.
100. Bronson GW. *The Scanning Patterns of Human Infants: Implications for Visual Learning*. Norwood, NJ: Ablex, 1982.
101. Salapatek PS. Pattern perception in early infancy. In: Salapatek P, Cohen L, eds. *Infant Perception: From Sensation to Cognition*, Vol. 1. New York: Academic Press, 1975: 133–248.
102. Bronson GW. The growth of visual capacity: evidence from infant scanning patterns. *Adv Infancy Res* **11**: (in press).
103. Bronson GW. Infant differences in rate of visual encoding. *Child Dev* **62**: 44–54.
104. Bronson GW. Infants' transitions toward adult-like scanning. *Child Dev* **65**: 1243–61.
105. Frick JE, Colombo J, Saxon TF. Long-looking infants are slower to disengage fixation. Presented at the *International Conference on Infant Studies*, Providence, RI, April 1996.
106. Rose SA, Feldman JF, Wallace IF, Cohen P. Language: a partial link between infant attention and later intelligence. *Dev Psychol* 1991; **27**: 798–805.
107. Fagan JW, Ohr PS. Individual differences in infant conditioning and memory. In: Colombo J, Fagan JW, eds. *Individual Differences in Infancy*. Hillsdale, NJ: Erlbaum, 1990: 155–92.
108. Posner MI. Orienting of attention. *Q J Experimental Psychol* **32**: 3–25.

109. Posner MI, Walker JA, Friedrich FA, Rafal RD. Effects of parietal lobe injury on covert orienting of visual attention *J Neurosci* 1984; **4**: 1863–74.
110. Posner MI, Walker JA, Friedrich FA, Rafal RD. How do the parietal lobes direct covert attention? *Neuropsychologia*, **25**: 135–145.
111. Pierrot-Deseilligny Ch, Gray F, Brunet P. Infarcts of both inferior parietal lobules with impairment of visually guided eye movements, peripheral visual inattention and optic ataxia. *Brain* 1986; **109**: 81–97.
112. Rafal RD, Robertson L. The neurology of visual attention. In: Gazzaniga MS, ed. *The Cognitive Neurosciences*. Cambridge, MA: MIT Press, 1995: 625–48.
113. Clohessy AB, Posner MI, Rothbart MK, Vecera SP. The development of inhibition of return in early infancy. *J Cognitive Neurosci* 1991; **3**: 345–50.
114. Hood BM. Inhibition of return produced by covert shifts of visual attention in 6-month-old infants. *Infant Behav Dev* 1993; **16**: 245–54.
115. Johnson MH, Posner MI, Rothbart MK. Components of visual orienting in early infancy: contingency learning, anticipatory looking, and disengaging. *J Cognitive Neurosci* 1991; **3**: 335–44.
116. Johnson MH, Tucker LA. The ontogeny of covert visual attention: facilitatory and inhibitory effects. Presented at the *Society for Research in Child Development*, New Orleans, LA, April 1993.
117. Vecera SP, Rothbart MK, Posner MI. Development of spontaneous alternation in infancy. *J Cognitive Neurosci* **3**: 351–4.
118. Berg C, Sternberg RJ. Response to novelty: continuity versus discontinuity in the developmental course of intelligence. *Adv Child Dev Behav* 1985; **15**: 1–47.
119. Malcuit G, Pomerleau A, Lamarre G. Habituation, visual fixation, and cognitive activity in infants: a critical analysis and attempt at a new formulation. *Eur Bull Cognitive Psychol* 1988; **8**: 415–40.
120. Fagen JW. Predicting IQ from infancy: we're getting closer. *Contemp Psychol* 1995; **40**: 19–20.
121. Fagan JF. The relationship of novelty preferences during infancy to later intelligence and recognition memory. *Intelligence* 1984; **8**: 339–46.
122. Benasich AA. Tallal P. Auditory temporal processing thresholds, habituation, and recognition memory over the first year. *Infant Behav Dev* 1996; **19**: 339–57.
123. Tamis-LeMonda CS. McClure J. Infant visual expectation in relation to feature learning. *Infant Behav Dev* 1995; **18**: 427–34.
124. Linnemeyer SA. Porges S. Recognition memory and cardiac vagal tone in 6-month-old infants. *Infant Behav Dev* **9**: 43–56.
125. Richards JE. Respiratory sinus arrhythmia predicts heart rate and visual responses during visual attention in 14 and 20 week old infants. *Psychophysiology* 1985; **22**: 101–9.
126. Colombo J, Frick JE, Gorman SA. *Procedural and Individual Differences in Infant Sensitization during Fixed-trial Habituation Sequences*. Washington, DC: Society for Research in Child Development, 1997.
127. Jacobson SW, Fein GG, Jacobson JL, Schwartz PM, Dowler JK. The effects of intrauterine PCB exposure on visual recognition memory. *Child Dev* 1985; **56**: 853–60.
128. Jacobson SW, Jacobson JL, Sokol RJ, Martier SS, Ager JW. Prenatal alcohol

exposure and infant information processing ability. *Child Dev* 1993; **64**: 1706–21.
129. Jacobson SW. Evidence for speed of processing and recognition memory components of infant information processing. Presented at the *Meeting of the Society for Research in Child Development*, Indianapolis, IN, 1995.
130. Mayes LC, Bornstein MH. Infant information-processing performance and maternal education. *Early Human Development* 1995; **4**: 891–6.
131. Haaf RA, Smith PH, Smitely S. Infant response to facelike patterns under fixed-trial and infant-control procedures. *Child Dev* 1983; **54**: 172–7.
132. Bornstein MH. Infant attention and caregiver stimulation. Presented at *Early Cognitive Development: International Conference on Infant Studies*: New York, NY, April 1984 (two papers).
133. Colombo J, Mitchell DW, Coldren JT, Freeseman LJ. Individual differences in infant attention: are short lookers faster processors or feature processors? *Child Dev* 1991; **62**: 1247–57.
134. Greenberg DJ, Weizmann F. The measurement of visual attention in infants: a comparison of two methodologies. *J Exp Child Psychol* 1971; **11**: 234–43.
135. Fagan JF, Singer J, Montie J, Shepard PA. Selective screening device for the early detection of normal or delayed cognitive development in infants at risk for later mental retardation. *Pediatrics* 1986; **78**: 1021–6.
136. Richards JE. Infant visual sustained attention and respiratory sinus arrhythmia. *Child Dev* 1987; **58**: 488–96.
137. Richards JE. Development and stability of HR-defined visual sustained attention in 14, 20, and 26 week old infants. *Psychophysiology* 1989; **26**: 422–30.
138. Richards JE. Baseline respiratory sinus arrhythmia and heart-rate responses during sustained visual attention in preterm infants from 3 to 6 months of age. *Psychophysiology* 1994; **30**: 235–43.
139. Richards JE, Casey BJ. Development of sustained visual attention in the human infant. In: Campbell BA, Hayne H, Richardson R, eds. *Attention and Information Processing in Infants and Adults*. Hillsdale, NJ: Erlbaum, 1990: 30–60.

# Commentary

**Shaw and McEachern**: We raise the following concerns regarding the look habituation and paired-comparison tests. First, the habituation of looking response is taken as a measure of a 'learning curve'. Is it not possible that other variables may negate this conclusion, e.g. looking may not always mean seeing and encoding? Second, different patterns of looking seem to be described: 'declining' versus 'increase-decrease' versus 'fluctuating'/'chaotic'. Various theories are invoked to rationalize these patterns (comparator versus dual process theory), but what this variable response pattern may suggest is that the test measures are labile and poorly understood. This interpretation is supported by Colombo's statements that: (i) many different cognitive and noncognitive processes or 'constructs' (e.g. infant state, hunger, sleepiness) may be measured; (ii) test–retest reliability is not very good – the outcome of the look duration test may depend on different processes in the infant at different times; and (iii) 'systematic preferences for the familiarized target have also been observed'. Do these tests indeed provide accurate measures of infant cognition?

AUTHOR'S REPLY: It is my position that, although there is certainly room for improvement in the area of measurement, these measures do in fact provide accurate measures of infant cognition. Shaw's question is based on the fact that performance is affected by presumably noncognitive influences, that the reliabilities are low and may vary with age, and that the infant's response may vary with what are essentially variations of task parameters. I think that all these factors are also true of well known standardized assessments (such as the IQ tests) in the adult. First, if one were to test an adult under conditions of hunger or sleepiness, one would not get a very accurate picture of that person's performance. Second, the performance of an adult on specific subtests of a test may be arrived at through different strategic or cognitive mechanisms; while this provides important fodder for certain fields of study, it is often not taken as a criticism of the test itself. The issue of reliability is a thorny one and may be subject to certain expectancies. The adult tests are most certainly constructed to provide a highly reliable product, but it is not clear to me that adult cognition, when measured under everyday (i.e. ecologically valid?) conditions, does in fact show such a high uniformity of rank among individuals. Finally, it seems logical that manipulations of IQ task parameters might lead to qualitatively different modes of responding, even in the adult.

SHAW AND MCEACHERN: With respect to these last points, we believe that the variability of response would be less in a test designed along the lines of traditional operant conditioning tasks. An example is a test in which the child is trained to point to a novel stimulus in order to receive a food reward (as cited in Innis, this volume). This type of design has the advantage that the stimulus is imbued with a definite significance to the child, and will motivate an unambiguous behavioural response. A potential disadvantage is that such a design is precluded in very young infants. This may not prove to be a serious limitation, however, given that (i) the view that infant tests of IQ predict later performance is not unanimously accepted, and (ii) a lasting developmental deficit will still be measurable at older ages, and a transient effect may in some senses be unimportant.

AUTHOR'S REPLY: I agree wholeheartedly that investigators interested in examining the effects of LCPUFAs during infancy should explore different paradigms. A number of those paradigms are enumerated in my chapter. Furthermore, others await development, pending the specification by those in the LCPUFA field of exactly those domains in which the effects of LCPUFA would be proposed to be manifest. It would appear that the choice of the current infant tasks for use in this area was governed more by expediency and the perception of standardization, than by a careful analysis of how variation in LCPUFAs might be expressed in the behavior of the infant and young child (see similar sentiments expressed by Bornstein, this volume). I would recommend that the field re-think its commitment to the general concept of IQ as an ultimate 'cognitive' outcome. The particular domains that would theoretically be affected by LCPUFAs should be described, that developmentalists be commissioned to design and implement tasks that reflect performance in these domains, and that the effect of variation in LCPUFAs on performance in those domains be empirically determined.

**Mayer and Dobson**: For those of us interested in visual system development, it is disappointing that the stimulus properties of the stimuli used in visual habituation and novelty preference paradigms appear not to have been investigated despite years of research in this area. For example, we are not aware of any reports on the spatial frequency and spatial phase properties of the stimuli they use. Are investigators in this area confident that infants are able to discriminate between the features of familiar versus novel stimuli? Have studies been done on the strength of novelty preference, e.g. in relation to discriminability or salience of stimulus features? Visual discrimination, after all, is the essential requirement for novelty preference and visual habituation. Models of the transmission of visual information in the infant based upon retinal and cortical anatomy and optics (e.g. 1,2) might be used to good effect in predicting performance on visual discrimination in these paradigms.

1. Banks MS, Crowell JA. Front-end limitations to infant spatial vision: Examination of two analyses. In: Simons K, ed. *Early Visual Development. Normal and Abnormal.* New York: Oxford University Press, 1993: 91–116.

2. Wilson HR. Theories of infant visual development. In: Simons K, ed. *Early Visual Development. Normal and Abnormal.* New York: Oxford University Press, 1993: 560–72.

AUTHOR'S REPLY: As I understand it, stimuli in the FTII are generally pretested for discriminability and 'salience' before being integrated into a standardized paired-comparison problem. In our own research, we pilot tasks for such properties, or rely on tasks that have been previously used in designing or implementing a study. With respect to the issue of the effect of discriminability on novelty preference, Fagan published a series of papers (1,2) in the mid-1970s in which discriminability dimensions were varied in an attempt to predict the strength of novelty preferences. Our own work on individual differences in infants' abilities to recognize and discriminate global (i.e. low spatial frequency) and local (i.e. high spatial frequency) visual properties (e.g. 3) are also relevant to this point as well.

1. Fagan JF. Infant recognition memory: the effects of length of familiarization and type of discrimination task. *Child Dev* 1974; **45**: 351–6.
2. Fagan JF. An attention model of infant recognition. *Child Dev* **48**: 345–59.
3. Colombo J, Freeseman LJ, Coldren JT, Frick JE. Individual differences in infant visual fixation: dominance of global and local stimulus properties. *Cognitive Dev* 1995; **10**: 271–85.

MAYER AND DOBSON: A related concern regarding use of the FTII in studies of at risk infant populations is whether the effect of disorders of the primary visual system (e.g. refractive error) on performance have been investigated. For example, if the infant must discriminate between small features of two faces, which necessarily requires high spatial frequency discrimination, the infant with a high refractive error may not be able to resolve the critical features of the two faces. This is not an imaginary concern; in one of the familiar–novel pairs used in the version of FTII with which we are familiar, the pair consists of photographs of the same woman's face differing only in the mouth area.

AUTHOR'S REPLY: This is a valid concern, since sensory deficits should always be first ruled out before constructing a theory that invokes higher order cognitive constructs. Fagan presumably recommends that poor performance on the FTII be followed up first with a visual screening, although I have no idea as to how often this is actually done in the field. We have also prescreened infants in several studies in our own laboratory to determine whether infants' performance on our 'cognitive' tasks might be attributable to deficits in visual acuity.

MAYER AND DOBSON: It seems from various chapters in this volume that the long-term predictive validity of measures of infants' visual habituation/ novelty preference with IQ is moderate at best. Since these infant measures assess visual discrimination specifically, it might be of interest to assess predictive validity for specific visual–perceptual and visual–constructional abilities in childhood.

AUTHOR'S REPLY: Rose (e.g. 1,2) has conducted some post hoc analyses of the paths from infant cognition measures to mature intellectual function, and has in fact shown some relationships to more specific function. Indeed, some of the variance in infant performance on novelty preferences is related to later performance on visual spatial skills and tests of pure perceptual speed. This is a current focus of research in the area, and represents a welcome departure from the reliance on the global IQ measure as a sole 'meaningful' measure of mature cognitive outcome.

1. Rose SA, Feldman JF. Prediction of IQ and specific cognitive abilities at 11 years from infancy measures. *Dev Psychol* 1995; **31**: 685–96.
2. Rose SA, Feldman JF. *Cognitive Continuity from Infancy: A Single Thread or a Twisted Skein?* Indianapolis, IN: Society for Research in Child Development, 1995.


**Carlson**: Fagan has raised the expectation that a lower novelty preference in one group compared with another is negative, regardless of the test age; but, as far as I have been able to determine, his own published data deal always with infants younger than 12 months of age. I would be interested in your comments on what McCall and Mash have described as a 'negative' effect of docosahexaenoic acid on the FTII novelty preference in our study. There is no overall effect of diet on novelty preference, nor any effect of age on novelty preference in the experimental group (1). On the other hand, the novelty preference increased in the control group at 12 months, so that at that age it is significantly higher than the controls. Perhaps relevant to this discussion, Jacobson finds with the FTII that look duration at 12 months is inversely related to 4 year performance in children (2), but now tells me she finds no relationship between 12 months novelty preference and 4 year performance. In her studies, it is the visual recognition memory at 6.5 months (or sometimes an average of all three measures) that correlates with later outcomes (Jacobson, personal communication).

1. Werkman SH, Carlson SE. A randomized trial of visual attention of preterm infants fed docosahexaenoic acid until 9 months. *Lipids* 1996; **31**: 91–7.
2. Jacobson SW, Jacobson SL, Sokol RJ, Martier SS, Ager JW. Prenatal alcohol exposure and infant information processing ability. *Child Dev* 1993; **64**: 1706–21.

AUTHOR'S REPLY: As they are reported, I do not believe that your mean differences in novelty preferences as a function of LCPUFA supplementation are of much consequence. As I understand your findings, both groups are significantly above chance, which means that both groups significantly discriminated the stimuli in question. It is a focus of current debate as to whether the magnitude of novelty preferences in ranges above or below chance are in fact meaningful with respect to later cognitive outcome (1), and it should be noted that the opinion expressed above represents my bias on this issue. Certainly, the magnitude of the discriminative response has traditionally been held to be irrelevant in interpreting the outcome of normative studies of infant cognition in which the novelty preference has been used, and so I am puzzled as to why it might be considered important in studies of individual differences. The only way in which one might be tempted to examine the magnitude of novelty preferences is if it reflects some greater number of tasks in which the infant performed above, or at, chance (i.e. for infants who have been administered ten problems, one who discriminated on eight of those problems would have a higher novelty preference than one who discriminated only six of these). If you were to take this possibility seriously, a logical follow-up on this issue would be to examine whether the LCPUFA supplemented group has lower numbers of infants performing below chance.

1. Colombo J, Frick JE. Recent advances and issues in the study of preverbal intelligence. In: Anderson M, ed. *Development of Intelligence*. London: University College of London Press (in press).

CARLSON: You have commented on the low internal consistency of measuring novelty preference in the paired-comparison paradigm, but indicated that reliabilities for multiple assessments of novelty preferences (such as the FTII) are generally higher. I wonder if you might also comment on what is known about consistency among ages with the FTII. In our first trial, the 6.5 months FTII score did not correlate at all with 9 or 12 months score (6.5 and 9 months, $r = 0.055$; 6.5 and 12 months, $r = 0.055$), although the 9 and 12 months FTII scores were correlated ($r = 0.361$). On the other hand, total looks were highly correlated among all of these ages (6.5 vs 9 months, $r = 0.545$; 6.5 vs 12 months, $r = 0.48$; 9 vs 12 months, $r = 0.685$) (Carlson, unpublished data). Thus it appears to me that (i) looks (attentional shifts?) were the most reliable assessment of infant attention among these three ages, and (b) that novelty preference and attentional shifts diverged between 6.5 and 9 months.

AUTHOR'S REPLY: The issue of consistency on the FTII is addressed in terms of the sections on cross-age reliability of novelty preferences in the chapter, but does not deal with this specific issue. The drop-off in reliability may reflect any number of problems with the test in maintaining attention at these ages, or may reflect some developmental trend in infants' responses to the stimulus tasks provided on novelty problems. For example, if infants who were intent on 'discriminating' stimuli (which produced novelty preferences when FTII stimuli were only subtly different) at one age began suddenly to behave in terms of 'generalizing' or 'categorizing' stimuli (where subtle differences between FTII stimuli were ignored or glossed over), then it might well be manifest as discontinuity in individual differences across those ages. This is meant, of course, to be a speculative example of the types of explanations that might be tested empirically, and not as a direct explanation of your specific findings.

**Lucas**: Colombo provides a detailed consideration of the validity, meaning and predictive value of infant cognitive testing, focusing principally on habituation and paired comparison techniques used in a number of studies that rely on looking behavior as a measure of cognitive performance in the pre-verbal infant. Many using tests of habituation will have been unaware of the uncertainly over how best to quantify the habituation curve, the problem of test–retest reliability and its possible relation to age (with an apparent dip in reliability at 6–7 months) and the relatively modest predictability for later cognitive function in a number of studies (with correlation coefficients for later function generally in the 0.3 to 0.4 range). Likewise, paired comparison procedures, developed for instance in the FTII, also pose significant methodological issues, often modest test–retest reliability and modest predictive value for later outcome.

The predictability of early testing has now become an important issue in the light of the recent focus on prospective studies that examine the long-term consequences of early nutrition on cognitive function. If early cognitive testing were to prove usefully predictive, then clinical efficacy of nutritional interventions could be established quickly, and considerable research resources required for long-term follow-up could be potentially saved.

In my view, however, we are nowhere near this point. The predictive value of tests of looking behavior for later cognitive function is fairly small. The fact that there is some degree of significant correlation between infant and later cognitive scores could imply that there is at least some common mechanistic basis; but only a small part of the variance in later scores is accounted for by earlier ones. Moreover the strength of the association between earlier and later test scores could have been overestimated. Often the association has been established on the basis of small sample size, which could yield nonsignificant results when the correlation is low; one could speculate that such studies may have led to negative publication bias (1). Our own data suggest correlations might be highest when the subjects come from selected motivated populations, as volunteer populations often tend to be, or when the subjects are from high-risk groups such as preterm infants. Data on this issue are cited in the paper in this volume by McCall and Mash.

Whilst data are presented here and by McCall and Mash (in their chapter) suggesting that looking tests have the best predictive value of the infant cognitive tests, this has been challenged in some aspects. In a recent large study by Laucht (2), visual attention at 3 months was found to show low correlation coefficients with cognitive function up to 54 months, with only 4% of variance of later tests accounted for. In this study visual attention measures were not superior to conventional developmental tests (e.g. the Bayley) in predicting later cognitive function.

At this early stage in our knowledge of the long-term impact of infant nutrition on later function, we cannot generalize in any case from previous predictive indices. At present we are committed to long-term follow-up (whether we do infant tests or not), if we wish to test the hypothesis that early nutrition has lasting consequences.

However, infant testing – whether based on visual attention or more conventional developmental assessment – does have two potential values. First, it tests for the immediate effect of the intervention of the infants' cognitive performance, which has contemporary validity and biological interest in its own right. Second, if data are recorded appropriately, nutritional intervention trials with early and later cognitive measure should continue to tell us more about the predictability of early testing and whether or not it is generalizable for different types of intervention.

1. Malcuit G, Pomerlau A, Lamarre G. Habituation, visual fixation and cognitive activity in infants: a critical analysis and attempt at a new formulation. *Eur Bull Cognitive Psychol* 1988; **8**: 415–40.
2. Laucht M, Esser G, Schmidt MH. Contrasting infant predictors of later cognitive functioning. *J Child Psychol Psychiatr* 1994; **35**: 649–62.

AUTHOR'S REPLY: The logical implication of Lucas' commentary is that infant cognition assessments are of limited (if any) utility and should probably not be used in research on evaluating the effects of early dietary interventions or manipulations. As support for his argument, Lucas makes explicit his own perception of the predictive coefficients between infant cognition and IQ as inadequately low, and also makes indirect claims based on concerning the possible influence of sampling bias, publication bias, and uncertainty of mechanisms.

I should first point out that I fully agree that issues of sampling and publication bias are potential sources of difficulty and should be explored with respect to this phenomenon. I would, however, make two notes for clarification here. First, although Malcuit *et al.* (1) raise the issue of sampling bias, they do not provide any direct evidence for it in their paper. My own experience is that cognitive outcome with self-selected volunteer populations is so homogeneous that one would actually expect it to depress (rather than raise) the magnitude of any predictive correlation. Second, if I understand the Laucht *et al.* (2) study correctly, it features a single assessment of habituation at one age, and yielded predictive correlations in the neighborhood of 0.20. It should be noted that, although this represents a small proportion of the absolute variance involved, this zero-order predictive correlation, when corrected for reliability (the reliability of habituation measures ranges from 0.40 to 0.50 (3)), may actually reflect as much as one-sixth of the accountable-for variance (a correlation of 0.40). This correlation does not, of course, reflect the actual level of prediction, but it does reflect prediction of the variance that is in fact available for statistical 'explanation'. There will be more on this below.

The core of my response, however, is focused on the logical distillation of Lucas' points, which seems to be as follows: the infant measures (i) do not represent adequate surrogates of IQ tests administered at maturity, and (ii) are of no use in this context unless they are in fact surrogates of IQ. Let us examine the assumptions of each of these points in turn.

*On the magnitude of the predictive correlations.* First to be considered is the magnitude of the predictive correlations. The judgment that such correlations are 'low' is based on the comparison of what those correlations are (e.g. on average, 0.30–0.40), compared to what they might be maximally (i.e. 1.00). The expectation that the correlation could attain its maximum presumes a number of statistical conditions (linearity of relationship, full range of values, reliability of assessments) that are probably not fully met in the reported studies of prediction. The expectation of maximal predictive validity rests on two additional presumptions that are in need of explication.

One is that there exists a single, general construct of 'intelligence' that is reflected by and accessible through the IQ test. Furthermore, it is also assumed that the infant measures also tap that same general construct. The point of the inclusion of multiple-process models in my chapter is that this assumption may not be entirely tenable. Individual differences in processing components (e.g. subtests of the overall IQ protocol, or different domains that the IQ protocol is designed to tap) do not correlate perfectly with IQ, simply because they are only part of the overall test. For example, although speed of processing is considered to be an integral component of the cognitive functions that contribute to IQ, it only correlates 0.3 to 0.4 with the total IQ score. If the multiple-process model of infant cognition I've discussed here is correct, the different infant measures may only tap these parts of the puzzle, and thus the degree of prediction to the total score will be necessarily constrained.

The second assumption concerns the role of development in the construction of mature intellectual function. The expectation that the infant tests could correlate + 1.00 with childhood IQ rests on the assumption that all of the variance in IQ at maturity is in fact available to be explained during infancy. That is, it presumes that no condition or event that occurs after the infant assessment has been done will influence performance on an IQ test some years later. This, of course, is patently unreasonable; only some of the variance in mature IQ can be accounted for by measurements taken during infancy. Given this limitation, the size of the predictive correlations may be viewed in a different light; small effects may take on more meaning. Furthermore, the question of infant nutrition may need to be asked not in terms of how much of the 'total' variance in IQ is affected by early dietary conditions, but rather how much of the 'affectable' variance in IQ is influenced by such conditions.

*On IQ as the outcome measure.* Next to be considered is the implicit reliance on the IQ test as the penultimate measure of cognitive outcome.

Perhaps my objections to this position simply echo my previous comments (to Shaw, see above) concerning the conceptualization of what will be meaningful domains of measurement in assessing the effects of LCPUFAs on central nervous system function, and my criticism that researchers in this area have not actively sought alternate measures of cognitive performance in human or infrahuman populations. However, there is the additional point that reliance on a single cognitive 'product' such as the IQ score relegates the developmental processes that give rise to whatever is attained on the IQ test to little or no consequence. Different developmental processes may well predispose two different children to use different cognitive mechanisms in solving the problems or tasks that compose an IQ test. Such mechanisms may give rise to two identical scores on the IQ test, but may prove to be more or less advantageous under ecologically valid conditions of cognitive challenge or stress.

1. Malcuit G, Pomerlau A, Lamarre G. Habituation, visual fixation and cognitive activity in infants: a critical analysis and attempt at a new formulation. *Eur Bull Cognitive Psychol* 1988; **8**: 415–40.
2. Laucht M, Esser G, Schmidt MH. Contrasting infant predictors of later cognitive functioning. *J Child Psychol Psychiatr* 1994; **35**: 649–62.
3. Colombo J. *Infant Cognition: Predicting Childhood Intelligence*. Newbury Park, CA: Sage, 1993.


**McCall and Mash**: The pattern of habituation over repeated presentation of the familiar stimulus has not been examined as a predictor of later IQ. However, certain features of the pattern of habituation have potential implications for interpreting the meaning of behavioral habituation (i.e. the decline in attentional behavior with repeated presentations of the familiar stimulus). Specifically, in a study that examined the patterns of habituation displayed by 5- and 10-month infants to visual stimuli, McCall (1) noted that the primary difference between one response pattern and another during familiarization was not the rate of decline in responding with repeated presentations but the number of trials that preceded the trial of longest look to the familiar standard.

Following that trial, looking decreased rapidly and at the same rate for all groups of infants. These data suggest that, at least in this case, habituation pattern says more about what the infant is doing in the experimental context prior to paying maximum attention to the experimenter's stimulus than it does about how long it takes the infant to encode the standard into memory. Presumably, once the infant invests maximum attention in the standard stimulus, behavioral habituation is roughly comparable and very rapid thereafter, possibly encouraging the belief that whatever the infant encodes about the standard is done primarily on a single trial and with a relatively short amount of looking.

The same data raise another question. Specifically, does habituation rate reflect the rate of information processing by the infant in encoding the familiar stimulus into memory? Given the above interpretation, behavioral habituation rate is not a very accurate index of information processing. In the discussion of underlying constructs and mechanisms, Colombo takes up the topic of speed or rapidity of stimulus processing. It may be helpful in this section to distinguish between 'speed of information processing' on the one hand and 'neural conduction speed' on the other. Traditionally, and in the absence of direct empirical support, behavior in these several infant cognition paradigms has been interpreted to reflect speed of information processing. That is, habituation rate, the ability to differentiate novel from familiar stimuli, duration of looking, and other measures found to predict later IQ have all been interpreted to represent the relative rapidity with which infants encode, retrieve, and use the information (i.e. knowledge of the configural properties of the familiar and novel stimuli) in encoding the familiar and responding to new stimuli. The above data on habituation pattern call into question that interpretation (see 2,3). Similarly, the amount of time the infant is actually allowed to study the familiar stimulus and the novel–familiar stimulus pairing in the paired comparison recognition memory tasks is relatively short, further suggesting that individual differences in stimulus information processing may not be at the heart of the matter. On the other hand, neural conduction speed may nevertheless be an underlying feature of all the infant measures, and has been demonstrated to be a correlate of adult IQ.

The evidence available to distinguish the single-process versus multiple-processes models is ambiguous, I feel. First, it should be pointed out that the correlations to later IQ are not obviously different for the several different measures of infant cognition that have been used to predict later IQ (4). Further, the factor analyses of infant measures that produce different factors may be reflecting different procedural circumstances surrounding the measures loading on one versus another factor. For example, looking duration measures are assessed during familiarization, whereas novelty preferences are assessed during testing. This raises a more general point pertaining to the evidence that some factors (e.g. prenatal toxins) influence some measures of infant cognition more than others or that some infant measures load on one factor versus another. The important issue is whether the variance shared or not shared by different parameters or infant measures with one another is the same variance in those measures that predict later IQ. More specifically, the intercorrelations and predictive correlations are so modest for these measures that one must assume numerous parameters and influences on these measures, not just one. Therefore, it is quite possible for parameters and influences quite unrelated to the process that mediates the prediction to later IQ to produce different factors or to have different influences on some measures versus others without bearing any relationship to the process that both underlies these infant measures and predicts later IQ.

1. McCall RB. The development of intellectual functioning in infancy and the prediction of later IQ. In Osofsky JD, ed. *Handbook of Infant Development*. New York: Wiley, 1979: 707–40.
2. McCall RB. What process mediates predictions of childhood IQ from infant habituation and recognition memory? Speculations on the roles of inhibition and rate of information processing. *Intelligence* 1994; **18**: 107–25.
3. McCall RB, Mash CW. Infant cognition and its relation to mature intelligence. In: Vasta R, ed. *Annuals of Child Development*. London: Jessica Kingsley: 1995: 27–56.
4. McCall RB, Carriger MS. A meta-analysis of infant habituation and recognition memory performance as predictors of later IQ. *Child Dev* 1993; **64**: 57–79.


**Singer**: This chapter focuses on the methodological differences within and between the visual habituation and paired-comparison procedures for assessment of infant cognitive processing as alternatives to traditional assessments, such as the Bayley. Colombo's frank explication of the problematic aspects of the habituation paradigm, particularly the lack of clarity regarding the meaning of the constructs and the measures derived, underscores the need to accumulate a much more extensive database on all aspects of the measurement properties of these procedures before using them as outcome measures in clinical trials, except for heuristic purposes.

A critical problem of using look duration as an assessment measure with the infant control procedure is the inability to equate the stimulus presentation and trials across infants, a basic psychometric prerequisite. Assessment items on IQ tests at older ages incorporate speed of information processing and speed of task performance into item scoring, but only after successful performance. Thus, lower look duration should only have some positive relationship to IQ in the context of successful habituation.

The problem of level of useful clinical significance raised in your paper is of some import. To my knowledge, the Fagan *et al.* (1) paper is the only one which has addressed the issue of individual prediction. This was a small sample ($n = 62$), largely of convenience and included only 28 preterm children. Predictive validity for delay averaged only 0.55 (0.60 for term children, and 0.50 for preterms), indicating that only 6 out of 11 children delayed based on the visual recognition memory items, were actually delayed at 3 years. It was pointed out in the paper that prediction of delay in a subset of children who had received both visual recognition memory (VRM) tasks and Bayley scores at 8 months was significantly better on VRM tasks. Predictive correlation coefficients for infant test scores in this same group was equivalent for VRM and Bayley scores to 3 year Binet scores, however (2). In that study, both 8 month percentage novelty score and 8 month Bayley score were reliable predictors of 3 year Binet IQ ($v = 0.47$, $p < 0.02$, and $r = 0.52$, $p < 0.01$, respectively), but did not correlate with each other ($r = 0.11$, not significant). When sensitivity and specificity were examined, however, there were differences (1), suggesting that predictive group correlations do not necessarily extrapolate to individual predictions.

1. Fagan JF, Singer LT, Montie J. Shepherd PA. A selective screening device for the early detection of normal or retarded cognitive development in infants at risk for later mental retardation. *Pediatrics* 1986; **78**: 1021–6.
2. Singer LT, Fagan JF. The cognitive development of the failure-to-thrive infant: a three-year longitudinal study. *J Pediatr Psychol* 1984; **9**: 363–84.

# Early Nutrition and Behavior: a Conceptual Framework for Critical Analysis of Research

PATRICIA E. WAINWRIGHT and GLENN R. WARD

*Department of Health Studies & Gerontology, University of Waterloo, Waterloo, Ontario, Canada, N21 3G1*

Introduction

Recognition of a deficiency syndrome associated with the feeding of diets devoid of the n-6 and n-3 essential fatty acids (EFA) (1) launched decades of research into the relative roles of these nutrients in health and disease. The fact that the central nervous system (CNS) is particularly rich in docosahexaenoic acid (DHA; 22:6n-3), the long-chain n-3 metabolite, together with the realization that formula-fed human infants exhibit reduced levels of DHA in the CNS (2) intensified the effort to establish the specific functional role of the n-3 series in the developing mammalian CNS. However, even after numerous clinical studies in human infants, as well as experiments using animal models, there is little consensus on the specific functional essentiality of the n-3 series in this regard. This skepticism is often in sharp contradiction to the impression gained from surveys of the literature, where numerous studies report significant alterations in

CNS function and behavior in human infants and laboratory mammals whose brains contain reduced concentrations of n-3 fatty acids.

What is the reason for this contradiction? In our view, it is due at least partly to the multidisciplinary nature of this field, which often presents researchers with the daunting task of having to make critical judgments in areas far removed from their immediate area of expertise. This is true of people with behavioral training, like ourselves, when we attempt to evaluate the biochemical literature, as well as of nutritionists and biochemists who must attempt to reconcile the often conflicting findings of the behavioral studies. As in other fields, all behavioral research designs are not created equal, and critical analysis in behavioral research involves, above all else, the recognition of just what constitutes an appropriate research design to address the question being asked. While this recognition is partly the result of highly specialized training and experience, it is based on a few clearly defined general concepts. Therefore our intention in writing this chapter is first to provide an understanding of this conceptual framework and how it can be used to facilitate the critical analysis of studies which ultimately address the following question: Is brain function compromised by reduced CNS levels of DHA during early development? Then we will ask to what degree the studies being discussed are designed appropriately to answer such a question.

## The Concept of Validity

Critical analysis in behavioral research is informed by the concept of validity, which indicates the range of inferences that can be supported by an experimental outcome. Validity is not a unitary concept, but is based on consideration of several different types of evidence, each looking at different aspects of the experimental procedure and designated by a specific term. Unfortunately, one of the problems is that different researchers will use different terms for the same concept, or the same term in different ways, and this can sometimes be confusing. In this chapter we will use the conceptual framework for assessing validity outlined by Cook and Campbell (3) but, where appropriate, will indicate the ways in which the terms used differ from those used in other disciplines. In this scheme a broad distinction is made between two aspects of the experimental design: those contributing to internal and external validity, respectively.

*Internal validity* refers to the ability to infer that a relationship between two variables is causal. The criteria for designating a variable as a causal factor involve the nature of the relationship between the presumed cause and the putative effect and include: (i) association, i.e. the causal factor must show statistical association with the effect; (ii) time order, i.e. the causal factor must be shown to precede the effect; and (iii) direction, i.e. it must be shown that change in the outcome is a consequence of change in the antecedent factor, rather than an association produced by a common cause or 'third variable' (4). As defined by Cook and Campbell, internal validity is concerned with the causal relationship in terms of the forms in which the variables were manipulated or measured, i.e. irrespective of what they theoretically represent.

*Construct validity*, on the other hand, is concerned with the ability to generalize from the research operations used to measure cause and effect to the theoretical constructs they putatively represent, e.g. to what extent does performance on a scale of infant development (operational definition) actually measure intelligence (theoretical construct). Thus construct validity falls under the broader rubric of *external validity*, which is the ability to infer that the presumed causal relationship will generalize beyond the experimental setting, not only in terms of alternate measures of cause and effect, but also in terms of subjects, settings, and times. In the case of both internal and external validity, issues can be identified that lead the researcher to ask a series of questions with respect to study design and analysis, and then, on the basis of the answers to these questions, to make an informed decision on the validity of the findings.

## Internal Validity

### True Experimental Designs

Research designs differ in terms of the degree of confidence they promote with respect to cause, and this relates directly to the degree of control achieved in manipulation and isolation of the putative causal factor. Questions related to internal validity address sources of *bias*, these being factors other than the treatment that may systematically affect the mean value of the differences between experimental groups. Many potential sources of bias can be eliminated in what are termed true experimental designs, with the random assignment of subjects at the beginning of the experiment to either treatment or control group. However, even with the luxury of controlled experimentation, one cannot afford to relax one's vigilance with respect to internal validity. Randomization as a technique for equating the characteristics of treatment and control groups at the outset of a study is based on probability theory, and may not always be successful. This is especially true in studies with small sample sizes. Again, one can check on this to some extent by doing baseline comparisons on other variables known to be related to both the exposure and the outcome. But even if the groups do start out being equivalent with respect to these variables, there is always a concern about the loss of subjects during the course of the experiment that may not be random across the groups in terms of subject characteristics. Thus, at the conclusion of the study the subjects remaining in the groups will differ for reasons other than the treatment. For this reason it is imperative that records be kept of the distribution of subject attrition across groups, as well as of the characteristics of the dropouts, so that, if necessary, this can be considered in the interpretation of the outcome (5).

### Quasi-experimental Designs

The question of whether an extraneous 'third variable' may be the true cause of the relationship under study becomes of immediate concern in quasi-experimental designs, where there is assignment of subjects to a treatment, but where randomization to groups is not possible because of ethical or other considerations. It is also an important issue in observational research, which includes most epidemiological studies. Although *cross-sectional studies*, where measurements are taken at one point in time, may be useful in showing an association, they are limited with respect to establishing time order or direction. Similar concerns apply to *case-control studies*, which identify persons with and without a condition, and then seek to establish retrospectively whether or not they had prior exposure to the putative cause. While observation of events in sequence is a positive feature of *prospective longitudinal studies*, because subjects assign themselves to a treatment, the evidence they provide with respect to causality still remains equivocal. Jacobson and Jacobson (6) provide an excellent discussion of the importance of identifying and including these possible third variables in the study design so that they can be controlled statistically through the use of multivariate analyses. They also discuss the importance of distinguishing in the analysis between these third (control) *variables* and *mediating variables*, the latter representing intervening variables that mediate the effect of exposure on outcome. It is important to note here that, while the epidemiological literature often refers to control variables as potential '*confounders*', Cook and Campbell reserve use of the term 'confound' for such intervening variables, and include these as an aspect of construct validity (discussed below).

A recent example of such a prospective study is one in which the cognitive development of a cohort of infants fed breast milk was compared with that of those who were fed infant formula (7). Here subjects assigned themselves to groups on the basis of pre-existing maternal characteristics. In this case, concerns that mothers who chose to provide breast milk for their infants may also have differed in other respects important to infant development, such as education or socio-economic status (SES), represent possible sources of bias. Being aware of these issues, the researchers collected relevant information as part of the design, and corrected for this in the analysis. Nevertheless, there is always the concern that the variables identified may not be the best control variables. For example, although SES may account for a large proportion of the variance in cognitive outcome, other factors, such as the quality of the intellectual stimulation provided by the parent, might make an additional independent contribution.

In such circumstances Jacobson and Jacobson advocate the use of more direct measures of the quality of the stimulation provided by the home, as would be provided by the HOME inventory. They also suggest the use of alternative measures of cognitive capacity which are less affected by socio-cultural factors. While further controlled experimentation in humans in terms of randomized clinical trials might be considered ideal in such a case to establish causality, clearly in the case of breast-feeding this is precluded by ethical constraints. This is where animal models become an essential tool, both in establishing plausible biological mechanisms, as well as in addressing related issues of safety and efficacy before large-scale dietary interventions are implemented in human infants.

*Statistical Conclusion Validity*

Because covariation is necessary for inferring causation, internal validity also deals with issues relating to appropriate statistical inference. A prerequisite for all statistical analyses is that the distribution of the data conforms to the assumptions of the test. This is particularly important with respect to behavioral data, as many of these variables are often not normally distributed, and very often the scales used are nominal or ordinal rather than ratio. Moreover, in any experiment, there is always a certain probability of false conclusions based on statistical analysis, referred to as type 1 and type 2 errors, respectively. The probability of *type 1 error* is based on the $\alpha$ level, and is the probability of claiming an effect when none exists. The probability of type 1 error increases in an experiment that includes multiple measures or one that makes multiple comparisons. In such cases it is important to take into consideration the fact that a certain proportion of these outcomes will be significantly different by chance and to use procedures, such as Tukey's or Bonferroni, that adjust the significance level for the number of tests carried out. However, in a well-designed study it is seldom the case that one is interested in all possible comparisons among the groups, but rather in a discrete set of questions dictated by the hypothesis under consideration. As long as the number of questions does not exceed the number of degrees of freedom among the groups, the most powerful analysis is that of preplanned comparisons, using the error term derived from the overall analysis (8).

*Type 2 error* refers to the probability that the study will fail to detect a true effect. Conversely, the *power* of an experiment is the probability of identifying correctly an effect of a specified size (9). Thus, if the probability of type 2 error were 0.2, then the power of the experiment would be 0.8. The probability of type 2 error decreases (and hence power increases) as the variability in the sample decreases, and as the sample size increases. Thus questions relating to power address whether the study is sufficiently sensitive to detect an effect of a size which is meaningful in terms of the question being asked, with effect size commonly being expressed in 'standard-deviation units'. This requires the experimenter to exercise judgment in terms of the effect size being sought, and then to ensure that the sample size is large enough to allow this to be detected with at least an 80% probability. To claim no effects based on small samples, where the power of the study may be very low, can be dangerous, particularly when issues of the safety of an intervention are at stake. It is also important to realize that outcome measures differ in terms of their variability, and that behavioral measures tend to be much more variable than biochemical or anthropometric measures. Thus in studies including different types of measures, power calculations should be based on those that are the most variable, most often the behavioral, thereby requiring much larger sample sizes than dictated by biochemical measures. When one is faced with having to make decisions based on studies of limited power, meta-analytic techniques may prove useful (10). This entails the statistical analysis of the combined results of these individual studies for the purpose of integrating the findings. However, this approach is somewhat controversial, particularly in terms of the advisability of combining outcomes based on different study designs. For example, in trying to reach consensus on the literature relating supplementation of infant formula with long-chain n-3 fatty acids to effects on vision we have studies involving different populations (e.g. preterm versus term infants), different interventions (e.g. fish oil supplementation versus egg phospholipid), and different outcomes (e.g. visual evoked potentials versus forced preferential looking). Whether it would be meaningful to combine these studies in one meta-analysis is debatable, although a case might be made for separating by population (i.e. preterm versus term infants) or by outcome (i.e. electrophysiological versus behavioral measures).

Another problem restricting the value of meta-analysis is the reluctance of some researchers to identify which of their studies involve independent samples. Many researchers, especially those involved in studies of human infants, where the collection of the data can take many years, publish preliminary findings followed by further analyses on the same original sample, or on a larger sample consisting partly of the subjects studied in the original report. If this procedure is not clearly explained in the paper, the reader may be under the impression that the findings are more robust that they really are (i.e. that they occur among many independent samples, whereas, in reality, they have been seen only in one population at one study site).

Problems of low power are also of concern in animal studies. In studies of larger animals, particularly those requiring intensive use of resources such as infant monkeys (11) and piglets (12), sample sizes are often as low as four subjects per group. There are also studies of laboratory rodents in which the statistical power is likely to be quite small due to insufficient sample size if the analysis were based, as it should be, on litters and not on individual pups (13). This is because, in the case of developmental nutritional studies, when treatment is applied to the pregnant and lactating dam, and the outcome measured in the pups, it is the dam (and, consequently, the litter) that is the randomly assigned unit, not the individual pup. Therefore, as power calculations are based on the variation observed among the randomly assigned units, the appropriate degrees of freedom for such an analysis are the number of litters, not the number of pups. Unfortunately, many studies in the published literature on the developmental effects of dietary fatty acids in rats and mice erroneously consider the pup as the unit of analysis. This leads to a situation where a small number of mothers (the true sample size) give rise to a large number of pups (the sample size used in the statistical analysis), thereby spuriously inflating the power of the study, and thus the potential of falsely rejecting the null hypothesis, i.e. claiming effects that may not in fact exist.

## Reliability

A concept that is integral to that of internal validity is that of reliability of measurement: measures with low reliability will be unable to detect change, and are therefore meaningless. Reliability indicates to what extent measures can be depended on to register 'true' differences, i.e. the extent to which individual differences in test scores can be attributed to differences in the characteristic being studied and the extent to which they are attributable to chance error (14). The reliability coefficient is interpreted directly as the percentage of variance in the score that is attributable to the true variance in the trait being measured, with the remainder due to error variance. Reduction of error variance makes scores more reliable, and this can be achieved by ensuring consistency of experimental conditions, both in implementation of the treatment and measurement of the outcome, as well as with respect to subject characteristics, such as age and gender.

Reliability is measured by correlation coefficients that can be computed in various ways. *Test–retest* reliability, which is a measure of temporal stability, describes the consistency of scores obtained by the same persons when re-examined on the same test on a different occasion. In this case it is obviously important to specify the time interval, particularly with developmental measures where a plateau in performance may be reached. *Alternate-form* reliability is a measure of both temporal stability and of consistency of response to a sample of items, different from the first, but designed to be of similar content to the first. *Interobserver* reliability is particularly important where the judgment of the scorer plays a role, and in such cases it is important that observers be trained to a high criterion of agreement ($> 85\%$) before data collection begins.

It is important to realize that high interobserver reliability does not always mean equivalency of scoring. What it denotes is that the ranking of the observations is consistent between the observers, but one observer may nevertheless be assigning much higher (or lower) scores than the other. This could become a problem with respect to internal validity if, for example, the scores were developmental indices in children, such as the Bayley Scales and the higher scoring observer always tested the treated group, a clear bias is introduced. This can be avoided by ensuring that both observers collect an equal number of scores in both treated and control groups: then any difference between the observers is no longer a bias, but does contribute to a larger error term, and hence a less powerful study.

## Construct Validity

### Convergent and Divergent Validity

Broadly defined, construct validity is concerned with what the treatment and outcome are held to represent and how well they do this; in other words, 'What is the relationship between the operations employed in implementation of the treatment and measurement of the outcome, and the ultimate interpretation of the research findings?' Behavioral research is characterized by a high degree of concern about construct validity. Whether gas chromatograms truly represent the relative percentage of DHA in brain fatty acids is probably far less controversial than whether performance on a maze can truly be said to measure learning. Specificity of measurement is an integral aspect of construct validity, and this presupposes that there be consistency across measures that one would expect to covary, i.e. convergent validity, and no change in measures that are related to, but different from the variable of interest, i.e. divergent validity. For example, if the effect of nutrient supplementation on behavior were to be interpreted specifically as a reduction in anxiety, rather than as an antidepressant, scores should be reduced on various measures of anxiety, but not on accompanying measures of depression. In other words, confidence in construct validity is strengthened by looking at the *pattern* of results obtained over the treatment and outcome operationalized in several different ways, either in the same experiment, or over a series of experiments.

## Mediating Variables

Cook and Campbell recognize the identification of mediating variables as an important aspect of the construct validity of the treatment. Mediating variables are a consequence of the treatment and represent the agent(s) actually responsible for the observed outcome. An example of this comes from our own work on the effects of prenatal stress in rodents on offspring development (15). Previous literature had shown that the pups of stressed dams were developmentally delayed relative to those of unstressed controls. Prenatally stressed dams, however, also reduced their food intake. Thus, in repeating the experiment in mice, we included an additional control group that was pair-fed to the stressed group. The outcome of this study was that the pups of the prenatally stressed and pair-fed control group did not differ in their development, but both groups were retarded relative to the unstressed control fed ad libitum. Therefore, while the two-group study might have attributed the results to direct effects of the physiological changes induced by stress on brain development, our three-group study identified malnutrition as the mediating factor, which leads to quite a different interpretation of the findings.

Subject and/or experimenter expectations can also be the source of mediating variables. For example, if, in a randomly controlled trial of nutrient supplementation, mothers of supplemented infants are made aware of the hypothesis that the infants are expected to perform better on developmental tests, this might influence them to engage in other types of environmental stimulation with their offspring that would have a similar outcome. It is therefore important that subjects be *blind* to the experimental hypothesis and their group assignment. This should also be true of the experimenters collecting the data, where their expectations might influence the outcome. For similar reasons, the use of a *placebo* control group is essential in any such study, so that any of the associated effects of being part of a study protocol are similar across groups. Hence the term *randomized double-blind placebo controlled* study. In nutrient supplementation studies such designs may also include a *cross-over* component where subjects formerly on the treatment are switched to placebo, and vice versa. Although this latter is probably not appropriate in developmental nutrition studies, it can make an important contribution in later intervention studies; for example, in administering essential fatty acids therapeutically to children with attention-deficit disorders.

*Sensitivity of Measurement*

Other concerns with respect to construct validity are questions pertaining to level of treatment (What is the shape of the dose–response curve?) as well as timing and duration (Are there sensitive periods during development?). Only by using research designs which vary these parameters systematically can one arrive at an understanding of the true nature of the relationship between treatment and outcome. Another important aspect of this process is the sensitivity of the measurement being used. It is sometimes the case that the measurement scales used in behavioral research may not encompass equal intervals, these being, for example, narrower at the ends of the scale than at the midpoint. Therefore any treatment will be more likely to show an effect in midrange than at the extremes, leading to what are known as *floor* or *ceiling* effects. Floor effects can be a problem with behavioral measures like response latency, where there is a minimum amount of time necessary to perform a particular response, for example, crossing from one arm to another in a maze. If a group were to start out performing at this minimum level on day 1 of testing, it would be impossible for their performance on day 2 to register change, compared to a group that started out slower on the first day.

The sensitivity of the measurement is a particularly important consideration in developmental work when one is faced with choosing the time of testing for group differences using scales based on changes in normal development. As all subjects will eventually reach a plateau on these measures, it is important that the tests be conducted during the time that encompasses the steep rise of the developmental curve. If tested when both groups have reached the plateau, insensitivity of the measure at this point might lead one to a false conclusion of no differences. In fact a better strategy in such instances might be to measure at various time points, enabling comparison of the developmental curves between the two groups.

## Concurrent and Predictive Validity

When considering outcome, some researchers restrict the term 'construct validity' to broader theoretical inferences and use the term *criterion validity* to indicate the effectiveness of a test in predicting behavior in a specified situation (14). Here a distinction is made between concurrent and predictive validity, based on the objectives of the test. Concurrent validation is relevant to tests used in the diagnoses of existing states, whereas predictive validation is concerned with the prediction of future outcomes. An example of the latter is the relationship of various tests of infant development to cognitive capacity in early childhood or adolescence. Here it is important to realize that, because such relationships are strictly correlational, a test may have good predictive validity in one population, e.g. healthy term infants, but not in a different population, e.g. very low birth weight preterm infants. This is because the factors contributing to the shared variability of the measures may differ between the two populations.

## Maternal Factors as Mediating Variables

Issues of construct validity are particularly salient when it comes to evaluating nutritional effects on behavioral development in mammals, since development occurs in the context of complex maternal–infant interactions, in animals as well as humans. For example, because n-6 fatty acid deficiency in rats has been shown to affect the function of the mammary glands (16), any demonstration of behavioral retardation in the offspring of dams fed an n-6 deficient diet would have to include controls for the secondary effects on lactational performance, and hence caloric availability, before this could be attributed directly to the effects of diet on the pup development. Furthermore, postnatal differences in maternal care due to n-3 deficiency may also conceivably lead to behavioral alterations in the offspring. This is especially relevant in this field given the evidence that n-3 deficient lactating mice discriminate less between pups of their own dietary condition and pups born to n-3 adequate mothers (17).

While cross-fostering techniques have been proposed as a way of addressing such problems, they will not work in the case where the treatment alters characteristics of the offspring, which in turn may result in their eliciting different responses from their environment. These then function as mediating variables in influencing behavioral outcomes. For example, if a dietary treatment were to make infants more irritable, this might elicit more hostile behavior from their caretakers, which could then result in differences in behavioral development. Such mechanisms have in fact been invoked in supporting a contribution by genetic factors to the development of individual differences in human behavior (18). One solution to this problem in experimental animals is to raise the offspring from early life on artificial milk formulas differing only in the fatty acid composition, and this approach is one commonly used for the monkey (19), piglet (20), and, recently, the rat (21,22). However, since mammals, by definition, evolved to receive their early postnatal nutrition as part of a co-ordinated maternal system of care, the artificial aspect of this approach could lead to problems of interpretation (discussed under external validity below).

This leads to the problem of defining the nutritional manipulation. It is commonly the case that nutritional manipulations are implemented using available sources such as, for example, various dietary oils that differ in their n-6/n-3 fatty acid ratio, such as safflower or soybean oil, or different dietary sources of DHA, such as single-cell oils, fish oil, or egg phospholipid. However, these sources may differ in terms of their bioavailability, as well as in the presence of other naturally occurring compounds. The provision of nutritional treatments through the maternal diet also presents some problems. This is mainly due to the lack of control over individual fatty acids in the nutrient supply to the developing brain because the maternal system has a highly active fatty acid metabolism and provides, through its milk, a wide spectrum of individual fatty acids. Furthermore, the precise form in which these fatty acids are provided is often vastly different between milk and infant formulas. For example, while most fatty acids in both milk and formula (where they are usually provided by vegetable oils) are in the form of triglycerides, the exact nature of triglycerides present often differs between the two, which again may affect the bioavailability of the fatty acids. It is therefore important in any experiment that the effects of dietary treatment be validated by measurement of tissue levels. This leads then to questions of which levels to measure, what are the appropriate indices for assessing brain levels in humans, and in animal studies, what areas of the brain and which fractions might be relevant to behavioral outcomes? Again, convergent validity would require that the outcome of studies based on different sources of the same fatty acid be similar, particularly when they lead to equivalent tissue levels. If this is not the case, as has been reported with work comparing the effects of soybean and perilla oil (23), then one might legitimately suspect that factors other than the differences in fatty acids are responsible. One way to avoid this is to use pure sources, but this is often not practical when considering that the findings may eventually be generalized to nutritional interventions in large populations.

## External Validity

Construct validity as described above is an important aspect of external validity, which also embraces concerns related to the generalizability of the results to or across subjects, experimental settings and time periods. Basically the question being addressed by external validity is whether the effects of the treatment differ as a result of any of these factors, i.e. whether there is a population by treatment, setting by treatment, or time by treatment interaction. These are important considerations. Researchers who study the vulnerability of children to developmental insult are often faced with considerable variability in the outcome, with some children proving remarkably resilient (6). A model of such differences in vulnerability has been suggested by Rutter (24), based on the premise that there is no main effect of treatment, but rather that the consequences of exposure depend on other factors that interact with the treatment. Such factors might include individual differences in biology due to genetic or other factors, as well as behavioral and socio-cultural influences. For example, long-chain dietary n-3 deficiency may make no difference to developmental outcome in healthy term babies, but may be of considerable importance to a low birth weight premature infant with other health problems. Moreover, such effects may never be manifest within the environmental context afforded by homes of high SES, but may be yet another factor contributing to the risk of the low SES infant. When faced with discrepant findings from different studies the possibility of such interactions becomes a salient consideration.

### Generalizeability of Animal Models

As stated previously, in view of the many difficulties encountered in isolating causal factors in human studies, an important contribution to understanding can be made by animal models. With this approach both amount and timing of exposure can be manipulated by the experimenter under controlled conditions, and valuable insight can be gained in terms of identifying the biological mechanisms involved. However, while animal models can allow considerable control with respect to internal validity, their possible lack of external validity is generally the most problematic issue for many researchers. Not only is the applicability of results from animal studies to human conditions not always clear, but the relative applicability of one animal model over another is also not always easily determined. Furthermore, the issue is compounded by the fact that the most appropriate animal model for one parameter, such as similarity to humans in fatty acid metabolism, may not be appropriate in terms of some other pattern, such as reliance upon visual function during particular behavioral tasks. Thus, the researcher should ask not whether a particular animal model is the best one in general, but whether it is the most appropriate one in the particular instance.

The first question the researcher may ask about a particular animal model is whether or not the metabolism, in terms of n-3 deposition into nervous tissue, is similar to that in the human. While there may be species differences in the activities of the various desaturase enzymes, most animal models currently in use appear similar to humans in at least one important respect: that n-3 fatty acids are taken up rapidly by the developing brain and preferentially retained by the adult during periods of nutritional stress (25). Thus we would expect the period of greatest sensitivity of the brain to dietary essential fatty acid manipulation to be the period of rapid brain development regardless of species. While this generally appears to be true, the issue is complicated by the fact that temporal patterns of brain growth among species are also related to the timing of birth, and their relationship is rarely similar in any two species (26).

In human infants, the period of concern with regard to polyunsaturated fatty acids uptake and deposition is the period of rapid brain development known as the brain growth spurt, generally encompassing the third prenatal trimester and the first few months of postnatal life in the term infant. Thus, in the term infant, much of this period has occurred in the womb, when the maternal blood supply is the primary nutrient source. Changes in maternal dietary fatty acid intake can have a profound impact on milk fatty acid composition (27) and, while this fact makes maternal nutrition a field of interest for many researchers, the focus of most research into the role of n-3 polyunsaturated fatty acids has been the early postnatal period in formula-fed infants. In the premature infant, on the other hand, much of the period of rapid brain development can take place postnatally, when the primary nutrient supply is provided by artificial formulas. These formulas are often deficient in the appropriate n-3 fatty acids and are fed during the time when DHA is usually deposited rapidly in the brain, coinciding with the formation and maturation of axons and dendrites and their resulting synaptic membranes (28). Therefore, the early postnatal period in the preterm infant may be the time of greatest susceptibility to n-3 deficiency, and reduced tissue levels of n-3 fatty acids in formula-fed preterm infants have been noted in several studies (to be discussed below). Consequently, if an animal model is to have external validity, it must ensure that the dietary manipulation is applied during a period of time relevant to the clinical observations of human infants. This is complicated further by the fact that the brain growth spurt can occur at various times relative to birth in different animal species. For example, in the infant monkey, much of the development of the brain is complete at birth, and dietary deficiency beginning at birth may not mimic, in its effects on the brain, the effects of deficiency during the brain growth spurt in human infants. In the case of the piglet, the stage of development at birth is somewhat similar to that of the human term infant, but not of the premature infant (29). Finally, the various rodent species run the full range of possibilities, from the rat and mouse, in which the postnatal period is the most important, to the guinea pig, in which brain development is almost fully completed by birth. Complicating the issue further still is the fact that direct feeding of infant animals is possible only in certain cases, usually when the infant is born somewhat precocious and, therefore, at a relatively advanced stage of brain development. In the case of altricial species, i.e. the rat and mouse, technical constraints often prevent the use of artificial formulas to provide nutrients directly during the period of interest, although artificial rearing of rat pups is possible (30) and has recently been used to study formula polyunsaturated fatty acid composition and development (22). Therefore, each of these models can be expected to have characteristic methodologies appropriate to the timing of birth relative to the timing of the period of most rapid brain development. In the rodent, the nutritional manipulation is generally provided to the maternal diet throughout both pregnancy and lactation. In the monkey, the newborn infant is often manipulated directly through formula feeding while the maternal diet may or may not have been manipulated during pregnancy. In the piglet, formula feeding is generally used, although maternal manipulations during pregnancy have been used in a number of studies.

Unfortunately, timing of the experimental treatment is not the only important factor to be considered from the point of view of external validity. The appropriateness of the species as a model of human infant behavioral development is also of primary importance. In this respect, it is interesting to note that the piglet, a species most similar to the human infant in terms of the temporal relationships between brain development and birth, has not been studied extensively as a behavioral model. The nonhuman primate, on the other hand, has been used often to evaluate functional effects of n-3 FA deficiency during early development. However, because its brain is more developed at birth relative to that of the human infant, its use as a model of brain development in newborn term and, especially, preterm human infants may have some limitations. Furthermore, it is also particularly expensive and labor intensive, often requiring resources beyond the reach of many investigators. The rat, on the other hand, is the most common of all animal models, being both relatively inexpensive and easy to maintain. Furthermore, its behavioral repertoire is the most well known of any nonhuman species. Unfortunately, the applicability of rat behavioral development to human behavior is a contentious issue, particularly with regard to species-typical behavior. However, it is a model employed extensively in behavioral neuroscience to elucidate the basic mechanisms of brain–behavior relationships.

A particular problem for external validity in animal models is the nature of the behavioral outcome assessed by the researcher. In particular, the primary sensory modality targeted by an experimental procedure must be appropriate for the species under study. For instance, given the known sensitivity of the developing retina to n-3 deficiency, and the documented functional differences between supplemented and deficient primates as assessed on the electroretinogram and visual tasks, it is not surprising that many researchers have used visually oriented tasks in rodent studies. However, the rat and mouse are nocturnal species, and therefore rely much more on their sense of smell when navigating their environments, and this fact is apparent to anyone who has noted the relative size of the rodent olfactory bulbs compared with those of the human. Therefore, forcing the rodent to use visual cues to navigate may create artificial demands upon the subjects, limiting again the applicability of the results to humans. Moreover, the nocturnal nature of these species should be taken into consideration when determining the laboratory setting under which they are tested.

## Functional Effects of n-3 Deficiency

Now to return to the initial question: Is brain function compromised by reduced CNS levels of DHA during early development? The main ways in which this question has been approached have been by looking at effects on visual function across a range of species, and by addressing effects on learning and memory, primarily in rodents. How do criteria of validity outlined above apply with respect to this work? In the following discussion, we will not attempt to provide an exhaustive review of the literature, but rather we raise specific issues which are often not discussed in other reviews. There is a great deal of overlap between the issues relevant to the literature on n-3 fatty acids and vision and that on cognitive function. Therefore, we wish to advise the reader that many of the points raised in each section will also apply to the other.

## Vision

The vertebrate retina is especially rich in DHA (31), and much of the evidence for a functional role of DHA in nervous tissue comes from research into n-3 deficiency and retinal function. Therefore researchers who wish to learn about the role of DHA in the CNS have chosen retinal function as a potentially informative model, and there are numerous published reports of controlled experiments involving vision in human infants fed various dietary fats in formula.

Some studies report that premature infants fed formulas devoid of long-chain polyunsaturated fatty acids exhibit higher rod (but not cone) electroretinogram (ERG) b-wave thresholds and lower amplitudes, and that those receiving formulas supplemented with DHA show responses similar to those of breast-fed controls. Moreover, these differences, while present at 36 weeks postconception (32) are no longer apparent at 57 weeks (33). The ERG is a measure of physiological responding of the retina to flashes of light, and, as such, does not measure changes in other parts of the visual system. Visual evoked potentials (VEPs), on the other hand, offer a measure of the physiological responding by the cortex to signals transmitted by the retina via other structures, such as the optic nerve. Finally, behavioral measures of visual acuity, such as the forced choice preferential looking (FPL) task and the Teller Acuity Card offer a measure of behavioral experience as influenced not only by the structures and responses described above, but also by oculomotor behavior. Therefore, many researchers have subsequently expanded their measures to include other aspects of vision.

Both preterm and full-term formula-fed infants have been reported to exhibit reduced acuity as assessed by VEPs (34,35), the FPL task (34), and the Teller Acuity Card (36–38), although differences were not found in some studies (39–41) or were no longer apparent after several months of age (36,39,42) or on a number of ERG measures at 4 months of age (43) in formula-fed term infants.

Different measures of visual function may exhibit different sensitivities to alterations in polyunsaturated fatty acid status in the brain and retina. For example, in a study by the same group which found that ERG measures recovered by 57 weeks postconception in n-3 deficient preterm infants, both VEP and FPL acuity were reported to remain impaired at this age (34). Furthermore, measures of behavioral acuity may be less sensitive than VEP to altered polyunsaturated fatty acid status, since in the latter study the FPL scores were only slightly lower in deficient infants at 57 weeks.

The overall impression given by these studies is that dietary n-3 deficiency during early development can have detrimental but transient effects on visual development. The question to be asked, however, is to what degree the studies described meet the criteria for internal and external validity. The studies of human infants meet some of the criteria for well-controlled quasi-experimental designs, in that the groups of infants were similar in terms of gestational age, weight, length, absence of significant neonatal morbidity, etc. However, they are still not without their problems. With regard to internal validity, for example, most of these studies randomly assigned subjects to the various formula-fed groups, but not to the breast-milk-fed groups (32–35,38). This is understandable from a practical and ethical viewpoint, and is often an inherent property of research on human infants. Still, it leaves open the possibility that breast-fed infants were different from formula-fed infants due to a third variable relating to maternal care or even some other variable. Ameliorating this concern somewhat is the fact that many of these studies reported significant differences between different formula-fed groups (i.e. between infants fed formula deficient in n-3 fatty acids versus those fed formula supplemented with short- or long-chain n-3 fatty acids). However, this leaves open the possibility that formula-fed infants comprise a subgroup on some unforeseen variable, and that the effects of the experimental treatment, in fact, interract with that variable. In such a case, the effects on n-3 deficiency seen in formula-fed infants may not be seen were those infants made equivalent to breast-milk-fed infants on all variables other that n-3 status. Another reason for caution with respect to external validity is the fact that infants whose mothers have agreed to participate in the research may not be representative of all infants in general. The degree to which they may comprise a self-selected subset of infants is not well understood, and should be kept in mind when attempting to interpret the findings.

A related problem is the extent to which these results can be generalized to the population at large. One reason for this is that the actual number of independent studies conducted on different populations appears to be quite small. In many of these cases, often several reports appear to have been conducted on the same sample, with later studies using slightly different measures and/or larger sample sizes. This in itself is not necessarily a problem, but, if not made explicitly clear to the reader, can lead to false assumptions about the robustness of the findings. Sometimes it has not been made explicit that preliminary findings have already been published, while at other times it is clear that the same group of subjects has been included in two separate reports. For example, in the reports of ERG function, which are sometimes cited by others in the literature as two studies, it appears that the premature breast-milk-fed group is composed of the same infants in both studies (32,33). While this does not detract from the reality of the specific findings of these studies, it may restrict the confidence with which the findings can be generalized.

Finally, as in all studies of human infants where measures are taken at various times throughout the first months or years of life, the degree to which subject attrition occurs randomly across groups is always a matter of concern. For example, in one study which compared visual function in infants at both 36 and 57 weeks postconception, there was a substantial loss of subjects at the later age, due mainly to difficulty in locating the parents (32). Given that the loss was heavily concentrated in some groups, and that the findings varied with age, the fact that the groups may no longer consist of the same infants warrants caution in the interpretation of the results. The reduction in the number of subjects will also reduce the power of the study at these ages.

While research into n-3 fatty acids and visual function relies heavily on studies done in humans, many of these findings were preceded by similar observations in other animal species. It should be noted, however, that the animal research focused more on the effects of n-3 deficiency per se, rather than the effects of supplementation of formula during early development. As such, the direct applicability of these animal studies to human infants may be somewhat problematic. This is due to the fact that the dietary treatment was long term, and included the prenatal period, thereby implicating a potential role for maternal variables.

When female rhesus monkeys were fed n-3-deficient diets from at least 2 months before conception and throughout pregnancy, their offspring were reported to exhibit reduced visual acuity, as measured on a preferential looking task (44), and reduced recovery of the dark adapted ERG response to saturating flashes (45). Furthermore, when infant monkeys were switched to formulas supplemented with fish oil beginning at from 10 to 24 months and maintained on the diet for 28 weeks, retinal DHA levels were no longer different from n-3 adequate controls, but delays in peak latency responding by both rods and cones were still present (46).

Experimental essential fatty acid deficiency in rats reduces both a- and b-wave amplitudes (47), and normal ERG function is regained only after n-3, rather than n-6, replacement (48). Both a- and b-wave amplitudes are reported to be reduced in 4-week-old n-3 deficient rats and, although adult rats will show improvements in both functions, only b-wave amplitude is completely restored to control levels (49). Furthermore, a recent study has reported reduced a-wave amplitudes in response to bright light exposure in 6- to 9-week-old n-3 deficient guinea pigs (50).

*Behavior*

Compared to the research on visual function, which is relatively unified in terms of its methods and, to some extent, its findings, research on n-3 fatty acid status and behavior is still in a very early stage of organization. Very little work has been conducted in human infants and, of the studies using animal models, there has been very little unity in the approach taken. The animal literature consists of various reports describing effects of n-3 deficiency on such phenomena as shock avoidance (49), exploratory activity (51), Y-maze performance (52), brightness discrimination (53,54), and habituation (55). This literature has been reviewed previously (56) and, although there has been more work done since then, the issues and criticisms raised remain relevant. Therefore, the following is not intended to be another exhaustive review, but rather a discussion of what we consider to be some of the major critical concerns, illustrated by particular studies only where relevant. Finally, because most of this work has been carried out in rats and mice, we will address behavioral effects of n-3 fatty acids by beginning with the studies of learning conducted in rodents.

Based on the present literature, it is our opinion that claims that tissue levels of n-3 fatty acids can influence learning remain to be substantiated. There is clearly no strong evidence of convergent validity over many different studies conducted in different laboratories. In our own laboratory, using a variety of different learning paradigms in mice, including social learning and both acquisition and reversal learning on the Morris water maze (57,58), we have not been able to show effects of dietary n-3 deficiency on learning ability. Although this may conceivably be due to the possibility that our dietary manipulations did not vary tissue levels of n-3 fatty acid sufficiently, these levels ranged over studies from a low of 20% of control values in a saturated-fat-fed group in one study (57) to 120% in a supplemented group in another (58). The many published studies that do report significant behavioral findings, on the other hand, reported deficiencies within the same range (i.e. about 50% of control values). One problem with the majority of these studies reporting positive findings is in terms of internal validity: specifically that the individual pup and not the litter has been used as the unit of analysis with the attendant concern that the significance of the findings may be spurious because of artificially inflated power. Another bone of contention with respect to this work lies in terms of construct validity, i.e. do the operations used in these experiments support an interpretation of effects on learning and memory, or is there an alternative explanation?

First let us consider what is meant by 'learning'. This has been defined as an adaptive change by the animal in response to an environmental stimulus, such that the animal responds differently upon re-exposure to the stimulus, based upon the animal's memory of what it has experienced. Short-term memory refers to the processes necessary for the acquisition of the response, whereas the mechanisms operating to store that information over longer periods of time are designated as long-term memory. Thus, from a biological perspective, memory is not considered a unitary phenomenon, but one that involves multiple systems with distinct anatomical organization. Learning and memory can be studied at different levels of organization, ranging from electrophysiology (e.g. recordings of long-term potentiation in the hippocampus), to analysis of changes in overt behavior, which is the level under consideration here. What it is most important to realize is that behavioral tests measure *differences in performance*, which in turn lead to *inferences of differences in learning and memory* at the physiological level. However, before such inferences can be supported, there are many intervening factors involved that need to be considered.

The more complex tasks commonly used to assess learning ability involve instrumental conditioning, where the animal has to learn to make a specific response in order to obtain reinforcement. Positive reinforcement refers to access to a pleasurable stimulus, such as food, whereas negative reinforcement refers to the opportunity to escape from an aversive stimulus, such as immersion in cool water in a water maze. Typically the animals are required to perform the task over consecutive days, during which time performance improves. In comparing the performance of two groups of animals fed different diets, differences between the groups over time, i.e. differences in slope, would be indicated, not by a main effect of diet, but by a *group × time interaction*. However, further steps must still be taken before inferring that these differences in performance are due to differences in learning and memory, or whether they might be accounted for by other differences between the groups. For example, if the groups differ on the first day, when there has been no opportunity for learning, this may be due to dietary effects on performance factors unrelated to learning. Possible performance factors include initial test reactivity, inability to meet the motor demands of the task, or differences in motivational levels. If they subsequently perform equivalently on the following days, the group × time interaction will be significant, but this certainly does not support differences in learning. On the other hand, were they to start out equivalently, and then improve their performance at different rates to reach the same level of asymptotic performance, this would then be supportive of differences in learning ability. At the risk of being repetitive, it must be emphasized that only by looking at a *pattern* of results can one come to some understanding of what performance on any particular test means. That is why it is important to include other measures, as well as to do other experiments, in order to assess the various possibilities related to performance factors.

The assessment of performance factors is far from simple, however, and depends very much on the specific task at hand. If, for example, we were to hypothesize that n-3 fatty acid deficiency impaired spatial learning in animals, convergent validity would require that we find differences in performance in both the Morris water maze as well as the radial arm maze, both of which have been shown to be measures of spatial ability in rodents. The presence of an effect on only one of these measurements would raise a concern that there might be differences in the performance factors related to each task, rather than spatial learning ability per se.

This is where the consideration of divergent validity becomes useful, where a study includes control tasks that have the same requirement in terms of performance variables, but differ in the process of interest. Therefore in the Morris maze, for example, spatial learning is measured by learning the location of a hidden platform relative to extra-maze cues, whereas the cued version, where the location of the platform is visible, can be used to determine differences in performance factors, including vision and swimming capability. This is illustrated by some of our recent work which showed that a low dietary n-6/n-3 ratio will impair performance in a Morris maze (58). In this study, however, through the inclusion of a cued version of the task, we were able to ascertain that the differences were not due to reduced learning ability, but to reduced swimming speed in the low ratio group! Similarly, in a study of operant conditioning requiring bar pressing for food reinforcement, the presence of baseline differences in overall rate of responding becomes an important consideration (52,53). Furthermore, with negative reinforcers like shock, are there group differences in shock threshold? The rationale of some recent work has been that one can avoid some of these problems by looking at simpler forms of learning such as habituation (55, 59). But this is clearly not the case. The interpretation of reaction time in a plus maze, for example, is certainly not straightforward, and the findings from this and other measures used in these studies do lend themselves to alternative explanations. But this is not to say that these findings are not interesting, or that effects on performance factors are 'nuisance factors'. In fact, effects on motor performance or behavioral reactivity can be as important as effects on learning in contributing to understanding the functional effects of dietary manipulation.

Concerns with respect to validity also apply to the use of developmental scales to measure cognitive development in human infants, such as the Bayley Mental Development Index (60) or the Brunet–Lezine Scale (61). With these measures it is very important to control for the independent contribution to test scores of social and environmental factors. It has been suggested that more narrowly defined assessment instruments, such as those assessing visual recognition memory and attention (Fagan Test), or expectation (Haith's paradigm), are influenced less by such factors, and therefore may be better indicators of true cognitive capabilities (62). But it is still important to control for temperamental and attentional factors that may affect performance on these tests. Another test has been developed recently, which assesses an infant's search strategy for a hidden toy, and builds into the test independent measures of motor capabilities and motivational levels (63). Interestingly, both human infants (64,65) and rhesus monkey infants (66) with the better DHA status have been shown to have shorter look durations in comparisons of novel with familiar stimuli. Although these visual tests have been shown to have some predictive validity with respect to later intellectual performance (62), in many cases the disadvantaged populations in which they have been used to assess effects of n-3 fatty acids do differ from the healthy populations in which validity has been established.

## Conclusion

At the beginning of this chapter, we stated that there are few very firm conclusions regarding the functional essentiality of n-3 fatty acid in the developing CNS. While the evidence with respect to visual function across different species is fairly convincing, in humans these effects appear to be transient. This is not to say, however, that early transient effects at the level of visual function during sensitive periods of brain development may not affect subsequent brain organization, such that effects will be seen at other levels of function later in life. At present, the data from long-term studies in humans needed to answer this question are not available. We do have evidence from animal models of behavioral effects of long-term dietary n-3 deficiency, but the interpretation of these findings remains controversial and requires further investigation. As stated at the outset, it is our opinion that this is because the proper interpretation of much of the behavioral research in this field may be inaccessible to researchers in other fields. Our intent in writing this chapter was to help nonbehavioral specialists draw their own conclusions by applying the concepts of validity as used by researchers in the behavioral sciences. We believe that the appropriate application of these concepts will aid progress in this field by contributing to the resolution of these controversies.

Acknowedgements

The authors would like to thank Dr Ruth Morley for her helpful comments on this chapter, together with those of the other participants in this workshop.

References

1. Burr GO, Burr MM. On the nature and role of the fatty acids essential in nutrition. *J Biol Chem* 1930; **86**: 587–621.
2. Farquharson J, Cockburn F, Patrick WA, Jamieson EC, Logan RW. Infant cerebral cortex phospholipid fatty acid composition and diet. *Lancet* 1992, **340**: 810–3.
3. Cook TD, Campbell DT. *Quasi-experimentation: Design and Analysis for Field Settings*. Chicago, IL: Rand McNally College, 1979: 37–94.
4. Susser M. What is a cause and how do we know one? A grammar for pragmatic epidemiology. *Am J Epidemiol* 1991; **133**: 635–48.
5. Galtman D. Better reporting of randomized controlled trials: the CONSORT Statement. *Br Med J* 1996; **313**: 570–1.
6. Jacobson JL, Jacobson SW. Prospective longitudinal assessment of developmental neurotoxiciy. *Environ Health Perspect* 1996; **104**(Suppl): 275–83.
7. Lucas A, Morley R, Cole TJ, Lister G, Leeson-Payne C. Breast milk and subsequent intelligence quotient in children born preterm. *Lancet* 1992; **339**: 261–4.
8. Cook RJ, Farewell VT. Multiplicity considerations in the design and analysis of clinical trials. *J R Statist Soc, Ser A* 1996; **159**: 93–110.
9. Cohen J. *Statistical Power Analysis for the Behavioral Sciences*, 2nd edn. Hillsdale, NJ: Erlbaum, 1988.
10. Rosenthal R. Meta-analysis: a review. *Psychosom Med* 1991; **53**: 247–71.
11. Connor WE, Neuringer M, Barstad L, Lin D. Dietary deprivation of linolenic acid in rhesus monkeys: effects on plasma and tissue fatty acid composition and on visual function. *Trans Assoc Am Physicians* 1984; **97**: 1–9.
12. Hrboticky N, MacKinnon MJ, Innis SM. Effect of a vegetable oil formula rich in linoleic acid on tissue fatty acid accretion in the brain liver plasma and erythrocytes of infant piglets. *Am J Clin Nutr* 1990; **51**: 173–82.

13. Abbey H, Howard E. Statistical procedure in developmental studies on species with multiple offspring. *Dev Psychobiol* 1973; **6**: 329–35.
14. Anastasi A. *Psychological Testing*, 5th edn. New York: Macmillan, 1982: 102–20, 131–55.
15. Ward GR, Wainwright PE. Reductions in maternal food and water intake account for prenatal stress effects on neurobehavioural development in B6D2F$_2$ mice. *Physiol Behav* 1988; **44**: 781–6.
16. Ollivier-Bousquet M, Guesnet P, Seddiki T, Durand G. Deficiency of (n-6) but not (n-3) polyunsaturated fatty acids inhibits the secretagogue effect of prolactin in lactating rat mammary epithelial cells. *J Nutr* 1993; **123**: 2090–100.
17. Wainwright PE, Huang YS, Coscina DV, Levesque S, McCutcheon D. Brain and behavioral effects of dietary n-3 deficiency in mice: a three generation study. *Dev Psychobiol* 1994; **27**: 467–87.
18. Plomin R. The genetic basis of complex human behaviors. *Science* 1994; **264**: 1733–9.
19. Reisbick S, Neuringer M, Connor WE, Barstad L. Postnatal deficiency of omega-3 fatty acids in monkeys: fluid intake and urine concentration. *Physiol Behav* 1992; **51**: 473–9.
20. Arbuckle LD, Rioux FM, MacKinnon MJ, Hrboticky N, Innis SM. Response of (n-3) and (n-6) fatty acids in piglet brain, liver and plasma to increasing, but low fish oil supplementation of formula. *J Nutr* 1991; **121**: 1536–47.
21. Winters BL, Yeh SM, Yeh YY. Linolenic acid provides a source of docosahexaenoic acid for artificially reared rat pups. *J Nutr* 1994; **124**: 1654–9.
22. Woods J, Ward G, Salem N Jr. Is docosahexaenoic acid necessary in infant formula? Evaluation of high linolenate diets in the neonatal rat. *Pediatr Res* 1996; **40**: 687–94.
23. Okuyama H. Minimum requirements of n-3 and n-6 essential fatty acids for the function of the nervous system and for the prevention of chronic disease. *Proc Soc Exp Biol Med* 1992; **200**: 174–6.
24. Rutter M. Psychosocial resistance and protective mechanisms. In: Rolf J, Masten AS, Cicchetti D, Nuechterlein KH, Weintraub S, eds. *Risk and Protective Factors in the Development of Psychopathology*. Cambridge: Cambridge University Press, 1990.
25. Crawford MA, Hassam AG, Stevens PA. Essential fatty acid requirements in pregnancy and lactation with special reference to brain development. *Prog Lipid Res* 1981; **20**: 31–40.
26. Dobbing J, Sands J. Comparative aspects of the brain growth spurt. *Early Human Dev* 1979; **3**: 79–83.
27. Innis SM. Human milk and formula fatty acids. *J Pediatr* 1992; **120**(Suppl): 56–61.
28. Wiggins RC. Myelin development and nutritional insufficiency. *Brain Res Rev* 82; **4**: 151–75.
29. Innis SM. The colostrum-deprived piglet as a model for study of infant lipid nutrition. *J Nutr* 1993; **123**(Suppl): 386–90.
30. Patel MS, Vadlamudi S. Artificial-rearing of rat pups: implications for nutrition research. *Ann Rev Nutr* 1994; **14**: 21–40.
31. Fliesler SJ, Anderson RE. Chemistry and metabolism of lipids in the vertebrate retina. *Prog Lipid Res* 1983; **22**: 79–131.
32. Uauy RD, Birch DG, Birch EE, Tyson JE, Hoffman DR. Effect of dietary

omega-3 fatty acids on retinal function of very-low-birth-weight-neonates. *Pediatr Res* 1990; **28**: 485–92.
33. Birch DG, Birch EE, Hoffman DR, Uauy R. Retinal development in very-low-birth-weight infants fed diets differing in omega-3 fatty acids. *Invest Ophthalmol Vis Sci* 1992; **33**: 2365–76.
34. Birch EE, Birch DG, Hoffman DR, Uauy R. Dietary essential fatty acid supply and visual acuity development. *Invest Ophthalmol Vis Sci* 1992; **33**: 3242–53.
35. Makrides M, Simmer K, Goggin M, Gibson RA. Erythrocyte docosahexaenoic acid correlates with the visual response of healthy term infants. *Pediatr Res* 1993; **34**: 425–7.
36. Carlson SE, Werkman SH, Rhodes PG, Tolley EA. Visual acuity development in healthy preterm infants: effect of marine-oil supplementation. *Am J Clin Nutr* 1993; **58**: 35–42.
37. Carlson SE, Ford AJ, Werkman SH, Peeples JM, Koo WWK. Visual acuity and fatty acid status of term infants fed human milk and formulas with and without docosahexaenoate and arachidonate from egg yolk lecithin. *Pediatr Res* 1996; **39**: 882–8.
38. Jorgensen MH, Hernell O, Lund P, Holmer G, Michaelsen KF. Visual acuity and erythrocyte docosahexaenoic acid status in breast-fed and formula-fed term infants during the first four months of life. *Lipids* 1996; **31**: 99–105.
39. Auestad N, Montalto MB et al. Visual acuity, erythrocyte fatty acid composition, and growth in term infants fed formulas with long-chain polyunsaturated fatty acids for one year. *Pediatr Res* 1997; **41**: 1–10.
40. Innis SM, Nelson CM, Rioux MF, King DJ. Development of visual acuity in relation to plasma and erythrocyte omega-6 and omega-3 fatty acids in healthy term gestation infants. *Am J Clin Nutr* 1994; **60**: 347–52.
41. Innis SM, Nelson CM, Lwanga D, Rioux FM, Waslen P. Feeding formula without arachidonic acid and docosahexaenoic acid has no effect on preferential looking acuity or recognition memory in healthy full-term infants at 9 mo of age. *Am J Clin Nutr* 1996; **64**: 40–6.
42. Hartmann EA, Neuringer M. Longitudinal behavioral measures of visual acuity in full-term human infants fed different dietary fatty acids. *Invest Ophthalmol Vis Sci* 1995; **36**: S895.
43. Neuringer M, Fitzgerald KM, Weleber RG *et al*. Electroretinograms in four-month-old full-term human infants fed diets differing in long-chain n-3 and n-6 fatty acids. *Invest Ophthalmol Vis Sci* 1995; **36**: S48.
44. Connor WE, Neuringer M, Barstad L, Lin D. Dietary deprivation of linolenic acid in rhesus monkeys: effects on plasma and tissue fatty acid composition and on visual function. *Trans Assoc Am Physicians* 1984; **97**: 1–9.
45. Neuringer M, Connor WE, Lin D, Barstad L, Luck S. Biochemical and functional effects of prenatal and postnatal ω-3 fatty acid deficiency on retina and brain in rhesus monkeys. *Proc Natl Acad Sci USA* 1986; **83**: 4021–5.
46. Connor WE, Neuringer M. The effects of n-3 fatty acid deficiency and repletion upon the fatty acid composition and function of the brain and retina. In Karnovsky ML, Leaf A, Bolis LC, eds. *Biological Membranes: Aberrations in Membrane Structure and Function*. New York: AR Liss, 1988: 275–94.
47. Benolken RM, Anderson RE, Wheeler TG. Membrane fatty acids associated with the electrical response in visual excitation. *Science* 1973; **182**: 1253–5.
48. Wheeler TG, Benolken RM, Anderson RE. Visual membranes: specificity of

fatty acid precursors for the electrical response to illumination. *Science* 1975; **188**: 1312–4.
49. Bourre JM, Francois M, Youyou A et al. The effects of dietary α-linolenic acid on the composition of nerve membranes enzymatic activity amplitude of electrophysiological parameters resistance to poisons and performance of learning tasks in rats. *J Nutr* 1989; **119**: 1880–92.
50. Weisinger HS, Vingrys AJ, Sinclair AJ. Effect of dietary n-3 deficiency on the electroretinogram in the guinea pig. *Ann Nutr Metab* 1996; **40**: 91–8.
51. Enslen M, Milon H, Malnoë A. Effect of low intake of n-3 fatty acids during development on brain phospholipid fatty acid composition and exploratory behavior in rats. *Lipids* 1991; **26**: 203–8.
52. Lamptey MS, Walker BL. A possible essential role for dietary linolenic acid in the development of the young rat. *J Nutr* 1976; **106**: 86–93.
53. Yamamoto N, Hashimoto A, Takemoto Y et al. Effect of the dietary α-linolenate/linoleate balance on lipid compositions and learning ability of rats. II: Discrimination process extinction process and glycolipid compositions. *J Lipid Res* 1988; **29**: 1013–21.
54. Yamamoto N, Okaniwa Y, Mori S, Nomura M, Okuyama H. Effects of a high-linoleate and a high linolenate diet on the learning ability of aged rats: evidence against an autoxidation-related lipid peroxide theory of aging. *J Gerontol Biol Sci* 1991; **46**: B17–22.
55. Frances H, Monier C, Clement M, Lecorsier A, Debray M, Bourre JM. Effect of dietary α-linolenic acid deficiency on habituation. *Life Sci* 1996; **58**: 1805–16.
56. Wainwright PE. Do essential fatty acids play a role in brain and behavioral development? *Neurosci Biobehav Rev* 1992; **16**: 193–205.
57. Wainwright PE, Huang YS, Bulman-Fleming B, Levesque S, McCutcheon D. The effects of dietary fatty acid composition combined with environmental enrichment on brain and behavior in mice. *Behav Brain Res* 1994; **60**: 125–36.
58. Wainwright PE, Xing H-C, Mutsaers L, McCutcheon D, Kyle D. Arachidonic acid offsets the effects on mouse brain and behavior of a diet with a low (n-6):(n-3) ratio and very high levels of docosahexaenoic acid. *J Nutr* 1997; **127**: 184–93.
59. Frances H. Monier C, Bourre JM. Effects of dietary α-linolenic acid deficiency on neuromuscular function in mice. *Life Sci* 1995; **57**: 19, 35–47.
60. Carlson SE, Werkman SH, Peeples JM, Wilson WM. Long-chain fatty acids and early visual and cognitive development of preterm infants. *Eur J Clin Nutr* 1994; **48**: S27–30.
61. Agostoni C, Trojan S, Bellu R, Rive E, Giovanni M. Neurodevelopmental quotient of healthy term infants at 4 months and feeding practice: the role of long-chain polyunsaturated fatty acids. *Pediatr Res* 1995; **38**: 262–6.
62. Jacobson SW, Jacobson JL, O'Neill JM, Padgett RJ, Frankowski JJ, Bihun JT. Visual expectation and dimensions of infant information processing. *Child Dev* 1992; **63**: 711–24.
63. Willatts P, Forsyth JS, DiModugno MK. The infant planning test: a new method for assessing infant intelligence based on means–end problem solving. In: *PUFA in Infant Nutrition: Consensus and Controversies*. Champaign, IL: American Oil Chemists' Society, 1996 (abstr).

64. Carlson SE, Werkman SH. A randomized trial of visual attention of preterm infants fed docosahexaenoic acid until two months. *Lipids* 1996; **31**: 85–90.
65. Werkman SH, Carlson SE. A randomized trial of visual attention of preterm infants fed docosahexaenoic acid until nine months. *Lipids* 1996; **31**: 91–7.
66. Reisbick S, Neuringer M, Gohl E, Wald R, Anderson GJ. Visual attention in infant monkeys: effects of dietary fatty acids and age. *Dev Psychol* 1997; **33**: 387–95.

## Commentary

**Neuringer and Reisbick**: We would like to respond to comments in this chapter and other chapters about the human relevance of studies in developing nonhuman primates. It is true, as noted in several chapters, that the human infant is born at a somewhat earlier stage of brain development than the rhesus monkey. However, neuronal cell division is nearly complete by midgestation in both monkey and human fetuses, and in both monkeys and humans there is dramatic postnatal brain growth spurt, which includes the process of exuberant cortical synaptogenesis followed by cell death and synaptic reorganization. (See our comments on Shaw.)

A basic principle of cross-species developmental comparisons is that the species must be matched with respect to the relevant stage of development, not with respect to the time of birth (e.g. 1). Thus, the evidence demonstrates that combined prenatal and postnatal dietary manipulation in monkeys is necessary to provide an accurate model of human infants, particularly those born prematurely. In one study, we compared the effects of postnatal deficiency in rhesus monkeys with those of combined prenatal and postnatal deficiency and found that postnatal deficiency alone, while it produced measurable effects on the ERG, did not significantly affect acuity development. As anticipated from the relative stage of development at birth, postnatal deficiency underestimated the effects found in the human infant. Therefore most of our studies have included dietary treatment of both the pregnant female throughout gestation and the infant postnatally.

We are comparing rhesus monkeys deficient in n-3 fatty acids (that is, fed diets low in $\alpha$-linolenic acid) to human infants fed similar formulas but, more importantly for the present discussion, to infants fed formulas with higher $\alpha$-linolenic acid but no source of preformed long-chain polyunsaturated fatty acids (LCPUFAs). One important question, then, is the comparative effect of these dietary manipulations on tissue fatty acid levels. The DHA content of the cerebral cortex is reduced in all three cases. In ethanolamine and serine phosphoglycerides, the magnitude of this reduction is approximately 75% perinatally in deficient monkey infants (your reference 45), 35–40% in term infants fed corn oil formulas (0.4% α-linolenic acid), and 20–25% in term infants fed balanced formulas (> 1.0% α-linolenic acid) compared with those who received human milk (2). In both species, the magnitude of the difference increases with age, up to about 50% in human full term infants by 30–40 weeks of age compared with 83% in monkeys at 2 years. Compositional data are available for very few preterm infants, but, as expected, show greater reductions than in term infants (2). In the one preterm (born at 30 weeks) reported by Makrides *et al.* (3), which received a balanced formula for 10 weeks, DHA was reduced by 60% compared to breast-fed full terms. Thus, effects on cerebral DHA are greater in n-3 fatty acid deficient monkeys than in term infants fed balanced formulas, but may be comparable to effects in preterm infants. Even if the combined maternal and postnatal deficiency in monkeys overestimates effects on human infants, the model is valuable because of the amplification of these effects, making them easier to identify and document with the necessarily small numbers of subjects.

With respect to functional outcomes, however, the magnitude of the effects we found on acuity development in infant monkeys closely resemble those found in human preterm studies (0.4–1 octave; see chapter by Mayer and Dobson, Tables 2 and 3). Furthermore, the primary measures of vision and behavioral development used in human infant studies in this area came directly from our work with rhesus monkeys. Our laboratory identified three major and distinct effects in n-3 fatty acid deficient infant monkeys, each of which was then documented in preterm, and in some cases, term human infants fed standard formulas as compared with those receiving supplemented formulas or human milk: (i) delayed visual acuity development; (ii) abnormalities in several aspects of retinal function, as measured with the electroretinogram; and (iii) increased look durations in visual paired comparison tests. Our findings led directly to increases in the α-linolenic acid content of many US infant formulas, even before human data were available. In our current studies of cognitive function, we have sought to increase further the applicability to human infants by explicitly selecting a battery of tests that can be used with both human and monkey infants, and which have shown similar sensitivity in both species to factors affecting CNS development.

1. Dobbing J, Sands J. Comparative aspects of the brain growth spurt. *Early Human Dev* 1979; **3**: 79–83.
2. Farquharson J, Jamieson EC, Abbasi KA, Patrick WJA, Logan RW, Cockburn

F. Effect of diet on the fatty acid composition of the major phospholipids of infant cerebral cortex. *Arch Dis Child* 1995; **72**: 198–203.
3. Makrides M, Neumann MA, Byard RW, Simmer K, Gibson RA. The fatty acid composition of brain, retina, and erythrocytes in breast and formula fed infants. *Am J Clin Nutr* 1994; **60**: 189–94.

AUTHOR'S REPLY: We agree that the monkey studies have made an important contribution to this field. We wished to make the point that each animal model has specific strengths and weaknesses and that a comprehensive account of the animal contribution will have to recognize that each model contributes to specific aspects of our understanding of the problem, but perhaps not to others. Furthermore, by attempting to understand where the specific strengths and weakness of each model lie, we can make better informed decisions.

NEURINGER AND REISBICK: As described in this chapter, behavioral changes in n-3 fatty acid deficient rodents have almost always been described as differences in 'learning ability', but the results are open to alternative interpretations (also reviewed in (1)). Just to amplify this point, we would argue that n-3 fatty acids may affect attention, reactivity or responsiveness to stimuli or temperament variables which are separable from cognitive function, and that these effects are important in themselves. For example, in three studies using a simple multiple schedule (animals were rewarded in the presence of a bright stimulus, S+, but not in the presence of a dimmer one, S−), Yamamoto *et al.* (e.g. your references 53, 54) consistently found increased response rates in rats and mice deficient in $\alpha$-linolenic acid compared to controls. Because of the higher response rates during the S−, they interpreted this result as a difference in learning. However, acquisition of responding was similar in control and deficient rats, and differences appeared only because the response rates of the two groups leveled out at different levels. Thus, these results suggest a difference between the diet groups in motivation, responsivity to reward or emotional reactivity rather than a difference in learning. The same interpretation can be applied to the finding of slower extinction (your reference 53), and to differences in shock escape and avoidance (your reference 49). Part of the inconsistency in behavioral studies of n-3 fatty acid deficiency might reflect differences in the testing environment and the type and schedule of reinforcement used in the tasks, all of which would interact with these noncognitive effects. Also consistent with this hypothesis are your own studies (your references 17, 57), which found effects on swimming speed but not spatial learning in the Morris maze, as well as differences in open field activity and paw-lick latency to heat stimulation (your reference 17).

In addition, in our tests of n-3 fatty acid deficient monkeys, we have found no effects on several tests of learning, memory, or response inhibition (1). The power to detect such effects was not high, because of the small numbers of monkeys and individual variability on the tasks. However, despite these factors we have found consistent differences on tasks which measure attention and responsivity to stimuli (described in our general commentary, page 517). Finally, differences in attention and responsiveness are consistent with differences in frontal cortex dopamine and dopamine $D_2$ receptors which have been reported in n-3 fatty acid deficient rats (2) (see (3) and our comments on Shaw).

1. Reisbick S, Neuringer M, Connor WE. Effects of n-3 fatty acid deficiency in non-human primates. In: Bindels JG, Goedhart AC, Visser HKA, ed. *Recent Developments in Infant Nutrition*. Dordrecht: Kluwer, 1996: 157–72.
2. Delion SS, Chalon S, Herault J, Guilloteau D, Besnard JC, Durand G. Chronic dietary α-linolenic acid deficiency alters dopaminergic and serotonergic neurotransmission in rats. *J Nutr* 1994; **124**: 2466–76.
3. Reisbick S, Neuringer M. Omega-3 fatty acid deficiency and behavior: a critical review and directions for future research. In: Yehuda S, Mostofsky DI, eds. *Handbook of Essential Fatty Acid Biology: Biochemistry, Physiology, and Behavioral Neurobiology*. Totowa, NJ: Humana Press, 1997: 397–425.

AUTHORS' REPLY: We agree.


**Innis**: In premature infant studies, infants may be enrolled up to a weight limit of 1500 g birth weight with a considerable variation in individual weights. For example, Carlson *et al.* (1) enrolled infants between 748 and 1398 g. Could you comment as to whether you think it an important design aspect that the distribution of weights, rather than the mean group weight, age at birth, etc., is the same among treatment and nontreatment groups. In the paper referenced above, the birth weight for the control group is (mean ± SD) 1074 ± 193 and for the supplemented group is 1133 ± 163. This could reflect a greater proportion of smaller babies in the control group, with the mean and SD increased by a few heavier babies. The issue is theoretical, since of course the distribution of birth weights in Carlson's study was not published. However, discussion is relevant as the view may be developing that smaller babies may be more sensitive to diet than older babies. Given the potential for very wide (almost twofold difference in body weight) differences in degree of maturation of infants within a premature infant study, could differences in weight/gestational age at birth distribution among groups result in apparent treatment differences.

Please comment with regard to your discussions on validity. An attempt is made to control for disparate gestational age at birth for visual function/developmental test results by correcting the infant's age to a common age based on obstetric or pediatric dating, relative to expected term delivery. What happens to the data if light exposure, some aspect of medical/nursery management, etc., influences the processes which are tested?

1. Carlson SE, Werkman SH, Rhodes PG, Tolley EA. Visual acuity development in healthy preterm infants: effect of marine oil (fish oil) supplementation. *Am J Clin Nutr* 1993; **58**: 35–42.

AUTHORS' REPLY: Your question about considering the distribution of characteristics in a group of premature infants in addition to the mean values is an interesting one. I think the concern you are raising is one of whether the data obtained from groups constituted of premature infants are normally distributed, or whether they may in fact represent a bimodal distribution, with very small infants constituting one extreme. I would expect this to be apparent from checking the distribution of these data for normality and homogeneity of variance, which should be standard initial procedure in any data analysis.

The question of whether or not to correct for gestational age and length of postnatal experience in premature babies is a thorny one. The problem addressed by such correction procedures is that, when comparing infants of the same chronological age from birth, differences in postmenstrual age may be a confounding factor. One way in which correction is commonly done is to correct according to postmenstrual age, i.e. measure all infants at the same postmenstrual age, regardless of birth date. An assumption that appears to be presupposed by this approach is that of a genetic 'program' or 'blueprint', where endogenous factors serve as the driving force of change over time. There is an alternative viewpoint, based on developmental systems theory, which views the impetus for developmental change deriving from interactions between the genome and those species-typical environmental events that are essential to allowing the developmental interplay to proceed in an appropriate fashion. Here the environment is defined by the needs and the capacity of the organism to respond, and these may well differ at different stages of development. Thus, what might be considered appropriate stimulation for the newborn could conceivably be entirely inappropriate for a premature infant at an earlier stage of CNS development. From this perspective, the relationship between extrauterine environment and development becomes an empirical question which might be best addressed using a modeling approach. For example, one way to address this might be to correct to postmenstrual age, but then to include gestational age and chronological age as covariates in the analysis, thereby equating the groups statistically for these factors.

**Carlson**: One of your comments brought to mind a question I have had for some time regarding the number of questions that may be asked within a given study. My discussions with Lucas lead me to believe he would support your somewhat conservative position about the number of questions that can be asked in a trial. On the other side, at least two of the chapters mention the importance of getting multiple measures of behavior as a way to sort out the diversity of effects of any variable on behavior (and presumably to plan better future trials). How does one reconcile these views or are they different?

AUTHORS' REPLY: The point that we were making is that if one analyses a trial that (i) asks many questions in terms of mean comparisons among groups, or (ii) uses multiple outcome measures, it is always important to keep in mind the increased probability of a type I error. So, for example, if one decides not to use preplanned comparisons, but rather to compare all groups in the study, then the use of more conservative analytical techniques, such as Tukey's or Bonferroni's, becomes a necessity. Similarly, with multiple outcomes, it is the overall pattern of results that must be considered, not stray effects that happen to be significant. The use of an overall multiple ANOVA prior to individual analyses has been advocated as a possible way of dealing with this. However, this too has been criticised on the grounds that different types of questions are addressed by MANOVA and multiple ANOVA analytic strategies in reference 1. MANOVA is appropriate when one is interested in understanding the nature of the possible correlations among the outcomes, whereas multiple ANOVA addresses the question of independent effects on individual outcomes. In using the latter, the possibility of type 1 error can be controlled by adjusting the significance level for the number of outcomes analyzed. In my view, in the long run, there probably is really no good analytical strategy that will substitute for hypothesis-generated, preplanned questions.

CARLSON: I could not agree with you more about multiple publications of the same outcome from the same center having the potential to elevate falsely the robustness of the findings. By the same token, many independent trials from the same center that measure similar outcomes also confuse casual readers of the literature who somehow come away with the conclusion that a single trial was done and the findings are less than robust. Perhaps these two forces balance each other out in the end?

AUTHORS' REPLY: Both lead to confusion; it would be optimal if it were to be made very clear in describing a study which portions of the data have been presented previously and where.

CARLSON: You say that, based on the present literature, it is your opinion that claims that n-3 fatty acids are necessary for learning remain to be substantiated. I know you have studied the monkey and infant data, and it would be helpful to have your candid views on what the data from these studies suggest. For example, do you not agree that there are effects on behavior related to n-3 fatty acids? Alternatively, do you agree there are effects, but believe they are more likely due to motivational or sensory than cognitive changes?

AUTHORS' REPLY: My interpretation, based on present evidence, is that behavioral effects of n-3 fatty acids in animals are more likely to be manifestations of behavioral 'state', i.e. reactivity and motivation, than to cognitive ability. It seems to me that factors such as these may also contribute to performance on tests such as the forced choice preferential looking paradigms; a 'fussy' baby may have a different profile of looks compared with one who is calmer. But, as Neuringer and Reisbick have repeatedly pointed out, this is not to say that effects of this type are any less interesting or important than those that can be termed purely cognitive.

**Mayer and Dobson**: Wainwright and Ward's elucidation of validity in experimental research designs provides a good framework for evaluating the results of the nutrition studies that are the focus of this Symposium.

The authors point out that in studies in which multiple measures or multiple comparisons are made, corrections for post hoc analyses can be done that minimize the probability of a type I error (erroneously finding a significant effect). In prospective longitudinal studies, this might occur if data from multiple age points are treated as if the design were cross-sectional, ignoring that repeated measures are made on the same individuals.

Is it not possible in longitudinal studies to make the opposite, type II error (not rejecting the null hypothesis when it is false)? I know that repeated measures analysis of variance is the statistically appropriate analysis, but it seems possible that in a study with many age points, a difference between groups in one or two age groups might be swamped by the absence of between-group differences in the other age groups. That is, a single group difference will be masked if there is no overall main effect nor a significant interaction. For example, Auestad *et al.* (1) found no statistically significant differences between mean acuities of dietary groups tested at five age points. In one age group (2 months), however, inspection of mean acuities shows a difference between the breast milk and control formula group that is in the direction and in the range of differences found to be significant in other nutrition studies at the same age, but in which fewer age groups were studied. Can these data be reanalyzed legitimately in any other way?

1. Auestad N, Montalto MB, Hall RT, *et al.* Visual acuity, erythrocyte fatty acid composition, and growth in term infants fed formulas with long-chain polyunsaturated fatty acids for one year. *Pediatr Res* 1997; **41**: 1–10.

AUTHORS' REPLY: Yes, I think that there is a concern when one has many time-points that the overall repeated-measures analysis may be too conservative. The same thing can happen in a standard ANOVA, where the presence of many control groups can dilute the treatment effect such that the overall analysis is not significant. Certainly if one were in the position of being able to predict a priori, based on a theoretical understanding of the basic processes involved, that effects would be apparent at some ages and not at others, this might support doing individual analyses at particular ages. Possibly other members of this group could comment on this as well.

At least a couple of studies have reported that polyunsaturated fatty acid levels in the retina react to dietary n-3 deficiency in a manner similar to that seen in the rest of the central nervous system. In the rhesus monkey, for example, dietary n-3 deficiency during both the prenatal and postnatal period produced reductions in DHA, and reciprocal increases in 22:5n-6, in both the occipital and frontal cortex, as well as in the retina (1), both in newborn infants and at 22 months of age. In the artificially reared infant rat, feeding of formulas containing 18:3n-3 but no long-chain n-3 fatty acids for 2 weeks was reported to lead to reductions in DHA, and reciprocal increases in long-chain n-6 fatty acids, in both retina and the rest of the CNS (2). Interestingly, in the latter study brain DHA levels equal to those of dam-reared controls could be obtained when the amount of formulae 18:3n-3 was increased to 25% of fatty acids, but retinal DHA levels, while increasing somewhat, still remained lower than those of controls. Therefore, while DHA levels in both the retina and CNS can be sensitive to dietary n-3 manipulations, they may not necessarily do so to the same extent.

1. Neuringer M, Connor WE, Lin DS, Barstad L, Luch S. Biochemical and functional effects of prenatal and postnatal ω-3 fatty acid deficiency on retina and brain in rhesus monkeys. *Proc Natl Acad Sci USA* 1986; **83**: 4021–5.
2. Woods J, Ward G, Salem N, Jr. Is docosahexaenoic acid necessary in infant formula? Evaluation of high linolenate diets in the neonatal rat. *Pediatr Res* 1996; **40**: 687–94.


**Singer**: This valuable chapter recapitulates salient design and analysis issues necessary to address in clinical trials to maintain the quality of the intervention studies, and particularly reminds that, while the use of animal models can eliminate many potential sources of bias intrinsic to human studies, they are not without their own sources of error.

The authors make the important point that the measurement scales used in behavioral research may not encompass equal intervals, which might produce floor or ceiling effects if there were many scores in the extreme ranges. The possibility of such effects should be considered in the high-risk, preterm samples studied in LCPUFA trials, since a higher rate of scores > 3 SD below the mean would be expected. Some scales (the WISC and the Bayley) restrict standard score ranges from 50 to 150, whereas the Stanford–Binet scale is sensitive to differences below and above those ranges. We have used extrapolated scores which were published for the old Bayley Scale by Naglieri (1), but these are not available for the new Bayley Scale.

Another issue to consider in measurement is the assessment of children with cerebral palsy and other sensory/motor deficits who are likely to be identified in high risk preterm studies. Morely *et al.* (2) identified 50 infants with cerebral palsy out of 834 infants in a follow-up. The Bayley Scales, requiring age-appropriate manipulative skills, may not provide valid cognitive outcome data for such children. Morely *et al.* handle this problem by removing these children from analysis on mental outcomes and through the use of more appropriate alternative measures which provide an IQ equivalent. Others (3) have assigned the lowest score possible for such children.

Wainwright and Ward conclude that claims that n-3 fatty acids are essential for learning in animal models have yet to be substantiated, making the important distinction between performance factors and ability, per se. Pretest differences noted between deficient and nondeficient samples on motor capabilities, e.g. reduced swimming speed, may be mediators of the effects of n-3 fatty acid deficiencies, however, and be equally important behavioral endpoints to assess in human studies.

1. Naglieri J. Extrapolated developmental indices for the Bayley Scales of Infant Development. *Am J Mental Deficiency* 1981; **85**: 548–50.
2. Morely R, Cole TJ, Powell R, Lucas A. Mother's choice to provide breast milk and developmental outcome. *Arch Dis Child* 1988; **63**: 1382–5.
3. Hack M, Taylor HC, Klein N, Eiben R, Schatschneider C, Mercuri-Minich N. School age outcomes in children with birthweights under 750 gms. *N Engl J Med* 1994; **331**: 753–9.

**Lucas**: I hope that Wainwright's intellectual synthesis on the way we should critically interpret experimental data, establish causation, and consider the validity and generalizability of findings will be more widely read than by those in the LCPUFA field. I imagine that nearly all of us conducting nutritional intervention studies will recognize in her article pits that we have already fallen into. Whilst the jargon used in her article may not readily enter most research scientists' vocabulary, an understanding of the concepts should raise standards of research and appraisal of results.

I agree with Wainwright that much of the problem of conflicting assertions in the nutritional field in general, and specifically in the field of LCPUFA research in humans, relates to its multidisciplinary nature. Nutrition is not an 'organ speciality' like ophthalmology, but it is a diffuse discipline that cuts across virtually every biological, clinical, behavioral, and social field from molecular biology to anthropology. In our view it is no longer possible for lone scientists to mastermind effectively nutritional intervention trials, as still frequently occurs. Trials on the effects of LCPUFAs should involve a multidisciplinary team including biological, clinical, behavioral, biochemical, and statistical inputs together with practical expertise in the technical problems and pitfalls of running clinical trials themselves. Such inputs make intervention trials more expensive, but more likely to succeed in their objectives – and industry should take this into account in its clinical research collaborations.

I would like to highlight one important point in Wainwright's article. In the section on predictive validity, she emphasizes that a test that has good predictive validity in one population may not have this in a different population. The relationships between two tests conducted at different times are correlational and, as Wainwright points out, the factors contributing to the shared variability of the two measures may differ between populations. This is highly relevant to the LCPUFA issue. Much of the focus in this volume is on the predictive value of infant cognitive testing (since that is most of what we have at present in terms of outcome findings), in relation to later function, e.g. IQ. It would be a mistake to infer from previous correlative data on early versus later test scores that early testing in new or existing LCPUFA studies now obviates the need for later testing. We simply cannot make that inference because of the heterogeneity in both the interventions and the study populations used in LCPUFA trials. We are committed to longitudinal studies into childhood if we wish to confirm whether or not LCPUFA interventions have lasting significance for cognitive function.

**Heird**: Your chapter and that of Singer should be required reading for all who are involved in or plan to become involved in clinical studies, particularly clinical studies concerning nutritional issues in low birth weight infants. The discussions of confounding variables, mediating variables, etc., are particularly relevant and important. On the other hand, these variables are inherent in any low birth weight infant population and a broader discussion of how to deal with them in realistic clinical studies is desirable. For example, should study populations be highly selective or inclusive of all members of the population?

With respect to formulas for feeding low birth weight infants, it is unlikely that multiple formulas will ever be available for specific subgroups of this total population. Further, even if multiple formulas were available, most medical practitioners lack the expertise to decide which infants should receive which formula. Thus, as is now the case, availability of multiple formulas for this diverse population is unlikely. Considering this, further discussion of the merits and/or pitfalls of studying well-defined subgroups of the total population versus a group that is more representative of the total population would be desirable. For example, while the variance of outcome variables is likely to be much less if the study population includes only infants with a narrow range of birth weights, gestational ages, etc., and with few, if any, confounding and/or mediating variables, the results obtained in such a study will only be applicable to the type of infant studied. Despite a greater variance of outcome variables and, therefore, the necessity for studying more infants, the results obtained in a population that includes representatives of all infants likely to receive the intervention under study are more likely to be applicable to the entire population. Moreover, this approach should permit detection of confounding and/or mediating factors which contribute to the variance in the results obtained. A broader discussion of the relative merits vs the disadvantages of these two approaches would be helpful.

AUTHORS' REPLY: In terms of design, one way to have your cake (reduced within-group variability) and eat it too (greater generalizability) would be to include the variety of subgroups as factors in the analysis. But this would no doubt make such studies in human infants prohibitively large. The alternative approach of sampling across the entire population and then using regression analyses to identify pertinent variables seems a reasonable option, but the concern would then be whether one would have sufficient power to detect significant interactions.

# The Effects of Early Diet on Synaptic Function and Behavior: Pitfalls and Potentials

CHRISTOPHER A. SHAW AND JILL C. McEACHERN

*Department of Ophthalmology, c/o Department of Anatomy, 2177 Westbrook Mall, University of British Columbia, Vancouver, BC, Canada V6T 1Z3*

## Introduction

The central problem of this Workshop is to examine the hypothesis that the diet in early development, involving long-chain polyunsaturated fatty acids (LCPUFAs) in breast milk versus a lack of the same in infant formula, has an effect on brain development and function, specifically, producing a deficit in cognitive function in formula-fed relative to breast-fed infants. Previous chapters in this volume have described the biochemistry and function of these fatty acids, and the reader is referred to them for detailed information.

In this chapter we examine this hypothesis, which has the following intrinsic assumptions:

(i) LCPUFAs from breast milk enter the brain:
(ii) LCPUFAs are required either directly or indirectly (e.g., via a synthesis or breakdown product or an element/interaction later in

a biochemical cascade) for some aspect of brain development at the time of breast-feeding; conversely, a lack of LCPUFAs negatively impacts upon development.

(iii) Given that (i) and (ii) are correct and LCPUFAs do affect brain development, then changes in brain structure/function must affect intelligence/behavior as measured by appropriate tests.

Of these assumptions, (i) appears to be correct (e.g. 1), (ii) is the heart of this Workshop and will be examined in detail in this and the companion chapters, and (iii) may be the most important assumption to query, since if it is not correct then assumptions (i) and (ii) do not matter very much except in an academic sense.

Whether or not this last assumption is answerable within the context of current concepts of brain function and the techniques used to measure them is the subject of much of this chapter. We believe that the following discussion will build a conceptual framework that will apply equally well to molecules identified in the future as potentially necessary for brain development.

The following terms will be used throughout this chapter and are defined as follows:

- *Cognitive function*: processes of learning, memory, and/or reasoning.
- *Neural function*: the activity of a particular neuron, neural circuit, pathway, or system.
- *Behavior*: an observable act that can be described and/or measured.
- *Neuroplasticity* (*neuronal plasticity, plasticity*): any change in response properties of a neuron, neural network, etc., following alterations in input (see (2) for a more complete definition of this term).
- *Critical period*: the period in life during which neuroplasticity may occur; note that this period is often during early postnatal development, as in the visual system, but may also occur at later stages as well. See also (3) for a more complete definition of critical and sensitive periods.

At first glance, it seems obvious that alterations in the machinery of the brain would have to impact upon its function. Starting at a molecular level, the complement of neurotransmitters, neurotransmitter receptors, regulatory kinases and phosphatases, etc., will determine in large part the nature of the response of any neuron to synaptic stimulation. In turn, the activity of a single neuron will affect the pattern of activity of neural networks and modules, with effects on whole subregions of brain (e.g. the visual cortex). Ultimately, the cumulative functioning of such a system will affect perception, cognition, and behavior.

Viewed in the above context, developmental alterations in putatively essential molecules should have an impact at all levels from synapse to behavior, especially if the lowest elements, the synaptic and nuclear molecules, are both developmentally regulated and subject to activity-dependent modification. Thus, LCPUFAs, which have been demonstrated to alter neuronal membrane composition, affect membrane fluidity, etc., may be expected to affect the development and function of membrane proteins such as neurotransmitter receptors (4). Receptor development and function will in turn affect synaptic activity, neural response properties, etc. If synaptic activity is indeed altered in development, concern for a potential role of molecules like LCPUFAs in diet may well be justified.

In general, such appears to occur. Intense synaptic modification in early development is a hallmark of neural development for many parts of the mammalian central nervous system, notably those thought to be associated with cognitive function (e.g. the cerebral cortex and hippocampus). Synaptic number, type, and morphology may be extensively altered, and age-dependent alterations in synaptic molecules, notably neurotransmitter receptors and their regulatory enzymes, are key features of synaptic development in prenatal and early postnatal life (5). At the level of receptor expression/function, the observed modifications transcend the receptor superfamily: both ionotropic receptor populations and G-protein receptors show similar age-dependent variations which include alterations in subunit composition (6–8), receptor characteristics (e.g. number and/or affinity) (6–17), distribution (9,18), and regulation (see 5). An example of a developmental profile for one receptor population is shown in Fig. 1 for monkey visual cortex. The change in number of the inhibitory $GABA_A$ receptor and overall cortical synaptic density is plotted against developmental age (13). Both receptor and synapse number increase in parallel during late prenatal and early postnatal life. While prenatal data are not always available from other studies, similar increases in receptor number along with increasing synaptogenesis in early postnatal life have frequently been observed (e.g. 19,20). In many such cases, receptor number increases to a peak value at a relatively early stage in postnatal life, declining thereafter into adulthood (9,12,16,21). In Fig. 1, synaptic number shows a parallel peak and decline. The changes in receptor number and synaptogenesis occur in synchrony with major alterations in receptor distribution, the latter representing transient expression and elimination of a variety of receptor populations. At various stages in development, receptor alterations of the types described may underlie some forms of critical period neural plasticity and may be key to the development of the mature nervous system. Transient expression of receptors may also serve as a final constraint to

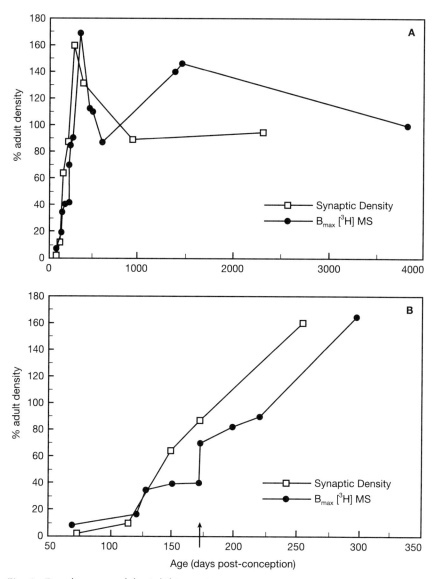

Fig. 1. Development of the inhibitory $GABA_A$ receptor in macaque monkey striate (visual) cortex. (A) Comparison of the development of receptor number ($B_{max}$) to synaptic number across all cortical laminae from prenatal day F72 to postnatal age 11.3 years. Note that this period spans the critical period for ocular dominance plasticity (approximately 6–20 months). $GABA_A$ receptor number was determined by saturation binding experiments using [$^3$H]muscimol. Synaptic density is expressed

modifications of the nervous system because of the dynamic relationship to neural activity and regulation (2).

The control of receptor development occurs at several levels. The first is genomic, specifying the expression of receptors/receptor subunits early in prenatal life. The second is dynamic, dependent on the process of receptor regulation, and influenced by neural activity at various levels. Receptor regulation is thus very likely to be affected by the presence, or absence, of particular molecules during development (22), including those which alter membrane composition and fluidity. Figure 2(A) provides a simplified schematic view of the stages of receptor regulation for one class of neurotransmitter receptors, the ligand-gated or ionotropic receptors (2,23). The proposed cycle of neural activity and receptor regulation is constantly balanced by the opposing actions of the various elements. The net outcome should be a stable system in which cellular response fluctuates around some mean level. It should be clear from the above, however, that the situation in the developing nervous system must be much more complex. Figure 2(B) shows some of these complexities, including the action of input activity, synaptogenesis, and feedback between different levels. A major point of this figure is to illustrate that receptor regulation and development are intertwined, in that the basic genetic instructions forming synapses, receptors, regulatory enzymes, and ion channels are constantly being modified by their own, and each others', actions. Such a system is dynamic, yet usually balanced and constrained by the interactions of the various elements. Perturbations of normal activity, e.g. the absence of *essential* molecules or alterations in sensory input, may lead to a realignment of the various elements (24–26). Thus, alterations in any of the elements forming the receptor regulation cycle will have an impact on the final contents/ configuration of the system. For example, LCPUFAs, some of which may be involved in receptor regulation, and others which affect membrane composition and function (for a review of these points see (4)) are likely to affect aspects of receptor function and regulation in the developing central nervous system (CNS), albeit perhaps only transiently. Such an impact may be measurable at the receptor level, but may not be manifest across all levels of organization. This may be most obvious in the mature CNS where a number of other factors come into play. Gene knockout

---

Fig. 1. Continued
as synapses/100 µm$^2$ neuropil (see (74) for details). Each point represents measurements from one animal. (B) Expanded view of ages from F72 days to P18 weeks to show the close correspondence between receptor development and synaptogenesis. The arrow marks birth. For other details, see (13). (Reproduced with permission of the Society for Neuroscience.)

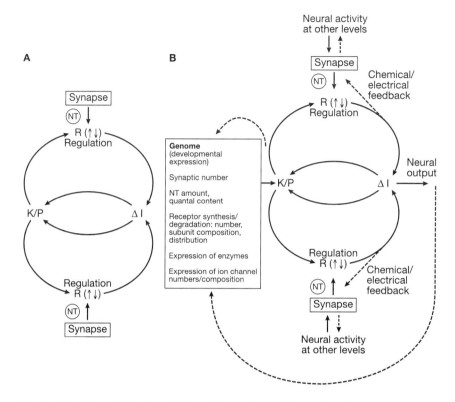

Fig. 2. How diet might affect the interdependence of synaptic elements in receptor regulation. (A) In this simplified scheme, ligand-gated (ionotropic) receptor activation by neurotransmitter leads to increases in membrane ionic currents. The increase in selective ions stimulates receptor specific kinases (K) or phosphatases (P), leading to a phosphorylation or dephosphorylation, respectively, of specific amino acid residues and an alteration in the number of functional receptors. Altered receptor number will subsequently alter the magnitude of the input signal (5). (B) A more complex version of the above with interactions at various levels (5). In this example, input from other neurons is included, as are genomic and developmental influences on receptor regulation. Alterations in LCPUFAs could affect a number of stages in this cycle (e.g. membrane composition, neural morphology, neurotransmitter release) with long-lasting consequences for receptor expression, regulation, and action. (Reproduced with permission of Elsevier Science Publishers B.V.)

studies often show that particular neural functions/behaviors survive the absence of particular molecules during development (e.g. a-calcium dependent calmodulin kinase II (α-CaMKII)) (see below, and 27,28), presumably due to compensation by genes/molecules serving related functions. In this context, the impact of LCPUFAs (e.g. docosahexaenoic acid; DHA) on a particular receptor population may be measurable, yet without overall manifestation on neural function or on performance in a behavioral task if another receptor population is able to provide compensatory action.

Sorting out the potential role of any molecule for affecting higher level function must thus recognize two quite opposing actions. The first may be termed the 'butterfly effect', and refers to the tendency of even small initial events to lead to cascades of progressively more severe consequences in interdependent dynamic systems. Balanced against this is the ability for compensation in systems displaying redundant or overlapping functions. As the developing CNS appears to show both these features, the assumption that alterations in individual neural properties will be expressed at a behavioral level cannot be simply accepted at face value. We believe, however, that there is a solution to be found in experiments designed to traverse the various levels of neural organization from molecules to behavior. This solution will be presented in the following sections.

## Biomechanical Cascades and Neuroplasticity

As stated above, LCPUFAs can affect membrane elements, including the action of neurotransmitter receptors. The development of the latter, in turn, dictates aspects of synaptic activity, and may lead to altered neural activity, particularly if occurring within a critical period of development. How can we test the assumption that a change in neural development/function early in postnatal life will have an impact upon cognition?

### Answering the Question

The following example will help to lay the conceptual framework for answering the above question. Assume that we wish to demonstrate experimentally whether or not a lack of DHA in infant formula has altered neurotransmitter receptor function, thence changing neural activity and cognitive function in formula-fed relative to breast-fed subjects. This type of problem has two facets. First, there are the *component* parts – chemical, neural, and behavioural. How is each defined and measured? Secondly, there are the *causal links* that must be established between components in order to complete a cause–effect chain of events from chemical modification to behavioural change. These various facets of the question will be discussed

## Chemical/Nutritional Component

In the specific example given above, the chemical/nutritional variable is DHA, present in breast milk but not formula. Before the potential for a causal relationship between DHA and behavior can be evaluated, certain groundwork must be laid. First, as mentioned above, DHA would have to be shown to cross the blood–brain barrier to gain access to the brains of breast-fed infants during the period of breast-feeding. Next, some neural or behavioral functional difference between breast- and formula-fed infants would have to be demonstrated. This timing of susceptibility during the period of breast/formula-feeding is crucial, as an effect of DHA later in postnatal life (i.e. post-suckling) is obviously not relevant to the infant formula question. Prenatal effects of DHA are certainly of interest, but involve additional assumptions. Without going into undue detail at this point, these issues illustrate the importance of determining the critical period for DHA requirement (if any) in the developing brain. This then brings us to the question of how to measure a putative effect of DHA on the brain. Function can be assessed at the neural or behavioral level, as detailed in the next two sections.

## Neural Component

This term in itself spans multiple levels of organization, from molecular/cellular changes in neurons to neural activity in networks and systems. Our example above invokes changes in receptors as the effect of DHA deprivation (e.g. 28) and the cause of altered neural activity and behavior in formula-fed subjects. Here again, some foundation must be set before investigating a causal effect of neural modifications on behaviour. For example, differences between breast- and formula-fed groups will be revealed only if the indices of neural change (in this case receptors and neural activity) are measured at an appropriate site and time. Which brain areas or pathways should be tested? Will putative receptor/activity changes be evident in adult animals, or could such events occur transiently during early postnatal development, possibly triggering some separate action before themselves returning to baseline level? A number of examples of transient receptor modifications exist in the experimental literature (5). Behavioral tests may provide some hint whether perturbations at the chemical or neural levels have propagated to affect function at higher levels, or have been 'dampened out' by compensatory mechanisms.

Without the advantage conferred by a prior knowledge of major, and clearly definable, behavioral modifications, observations on molecular alterations in the brain as putative consequences of diet may be hard to interpret. As examples, we cite two relatively recent studies which have examined the role of LCPUFAs on biochemical aspects of brain development and neural activity. These studies are not directly comparable in that the first examined the effects of early depletion of α-linolenic acid in vivo while the second looked at the immediate actions of DHA and arachidonic acid (AA) on neural activity in vitro. Nevertheless, they serve to illustrate the point of the relatively complex actions of fatty acids on neural function/dysfunction. Delion *et al.* (28) examined the effects of α-linolenic deficient diet on several receptors and neurotransmitters in 60-day-old male Wistar rat offspring of females deprived of α-linolenic acid for two generations. Prior to mating, females of the second generation were either switched to an α-linolenic acid replacement diet or maintained on the original diet. Frontal cortex, striatum, and cerebellum were assayed for lipid and monoamine concentration, and the characteristics of $D_2$ dopamine and serotonin 5-hydroxytryptamine ($5-HT_2$) receptors. DHA declined in all areas examined in the deficient rats; AA showed a slight increase in the cerebellum. Of the monoamines dopamine, noradrenaline, and serotonin, only the first was affected, showing a decline in frontal cortex. Also in frontal cortex, $D_2$ receptors declined while $5-HT_2$ receptors increased. It is of interest that, while quite selective neurochemical alterations were found in some areas, the effects across several generations of linolenic acid deficient rats were as modest as they were. This may suggest either that lack of linolenic acid is relatively inconsequential, or that it is compensated for by the presence of other fatty acids. The authors do not report behavioral abnormalities, which, in the absence of testing does not indicate that no behavioral/cognitive deficits were present. An earlier paper by the same group reported some behavioral differences between rats of the two treatment groups (29), but the effects seem neither pronounced nor well characterized. Data such as the above do not lend themselves to a simple interpretation beyond our earlier assertion (see above) that scattered biochemical alterations may be present which may or may not have either a long-lasting or profound behavioral effect. The interpretation of these results is not uncomplicated, however. DHA and AA can have immediate effects as described by Nisikawa *et al.* (30), who showed an opposing and reversible alteration by these LCPUFAs on *N*-methyl-D-aspartate (NMDA) and kainate-induced currents in pyramidal cells of juvenile rat cerebral cortical slices. Possibly related neuromodulatory roles of fatty acid amides have also been described recently (31). Since receptor regulation is under the control of a number of factors (see 2,5,22,23), the α-linolenic acid deficiency in the rats of Delion *et al.* cannot alone be assumed to be causal to any observed alterations in neurotransmitter content or receptor levels.

## Behavioral (Cognitive) Component

Despite the ill-defined nature of cognition, there is general consensus among behavioral scientists that aspects of 'higher function' of the brain are amenable to study using a variety of tests of learning, memory, and reasoning ability. The goal of our working example is to determine whether cognitive ability is affected in formula-fed infants. Several approaches are possible here, depending on species and task preference (choice of human versus animal models is discussed below). Among just a few of the diverse approaches that could be considered are: the Morris water maze and conditioning task performance in animal models; delayed match to sample tests in animals and humans; and visual pathway evaluation and IQ tests in humans. Without entering into detailed discussion of these and other tests of behavior, a set of general caveats applies for each. First, cognitive function may well require the effective performance of the entire brain. In accordance with principles of the 'butterfly effect' in dynamic systems, learning/memory may be disrupted by a perturbation at any level, and, before it is concluded that an observed difference reflects a specific change in cognitive learning or memory ability, a number of alternatives must be ruled out. These include generalized changes in neural excitability and regulation; sensory, motor or motivational changes that would impair performance on the behavioral test in question (e.g. 29); and differential environmental influences. (See (32) for a discussion of the appropriate controls required for modeling dietary influence on brain development and function; included in these is the essential requirement that the period of deficiency should occur within an equivalent developmental stage to that in humans.) There are pros and cons to this situation. One disadvantage is that proving that a performance deficit in one of these tasks is due to a *specific effect on cognition* is not possible. However, if cognitive function *is* affected, from the perspective of the organism it is to some extent irrelevant where along the pathway the problem occurs (i.e. in a sensory input pathway, in more central processing pathways like hippocampus or cortex, or elsewhere). Therefore, it may be beneficial to use sensitive tests of cognitive function as the 'canary in the mine' to flag problems and then better to localize changes with targeted testing. On the other hand, according to principles of redundancy and compensation, also relevant here, certain neural and behavioral functions may be spared, while others are quite abnormal. A case in point is the genetic knock-out mouse that develops without the enzyme α-CaMKII, an enzyme system implicated in receptor regulation (33) and neuroplasticity (34). Various measures of neural excitability and behavior appeared normal in the mutants lacking this enzyme. However, hippocampal long-term potentiation (LTP; a neuroplasticity model widely assumed to be the substrate of learning/memory) was impaired in some mice. LTP was normal in a small proportion of the mutants, however (35). LTP-deficient mice learned a visible-platform version of the Morris water maze as well as did controls, but learned a hidden-platform version of the test (a test of spatial learning) more slowly (36).

It becomes apparent from this example that selective testing, either neural or behavioral, would not have adequately described function in these mice. From certain tests it could have been argued that the mice had compensated for the lack of the enzyme, while other tests revealed them to be impaired. It follows from these ideas that a battery of tests be used to assess brain function following any manipulation.

## Causation

Once chemical/nutritional, neural, and behavioral components of the question are definable and measurable, an attempt can be made to establish causal links between levels of organization. As stated above, the chain of causality that must be satisfied in our working example is as follows:

(i) DHA from breast milk gains access to the brain.
(ii) A lack of DHA during the period of formula-feeding has a developmental consequence in formula-fed individuals (in this example, specifically upon neurotransmitter receptors and neural activity) relative to breast-fed subjects.
(iii) These developmental changes are responsible for a measurable change in cognitive ability compared to breast-fed subjects.

An enormous literature exists that is directed at proving just such a causal relationship between a specific neural change (neuroplasticity) and learning and/or memory. The success achieved in this difficult quest is too date limited (37), for reasons that should become obvious from the ensuing discussion.

## Role of Chemical/Nutritional Elements in Neural and Behavioral Alterations

Returning to our working example, assume that preliminary work has now demonstrated that DHA does indeed enter the brains of infant human and animal subjects. Can an observed difference in neural or behavioral function in formula-fed infants then be attributed to a lack of DHA? Such a conclusion would obviously be unfounded, since DHA is only one of a very large number of variables that differ between formula and breast milk. A functional difference between groups could equally be attributable to factors, including: the absence of another fatty acid, such as AA; a different iron content (a question addressed at a previous Dobbing Workshop (38)); a lack of immune molecules, such as cytokines, which are known to serve an important role in brain development (39,40); or any number of other molecules. The converse situation could also be considered, i.e., rather than something essential being missing from formula, something detrimental may have been added. We note also that some molecules, for example DHA, may have biphasic effects such that high concentrations may increase the propensity for free radical damage to photoreceptors in some circumstances, as noted by Neuringer (41–43). Because of the dynamic interactions of chemical cascades in living systems, changing any element can have unknown ramifications. In such a way, essential nutrients added to formula, which are sometimes derived from sources foreign to humans, and may be 'packaged' differently (44), could have unpredictable and possibly detrimental effects (45). Since certain nutrients in breast milk are not present in formula, and others are present in a different form, the two represent entirely different 'chemical soups'. For this reason, the comparison of formula-fed groups to breast-fed groups cannot be used as a technique to pinpoint the contribution of any single nutrient to changes in neural or behavioral function. Instead, groups fed formula differing in composition by only a single variable, such as DHA content, must be used to address this type of question. In this way, many of the confounding variables and interactions are eliminated and the role of a single nutrient can better be deduced. Again, the role of the *nutrient* in the formula condition cannot be directly extrapolated to the breast milk condition; however, *functional differences* can be compared in a breast-fed versus two formula groups, one containing DHA, and one without. If performance on a behavioral test is equal in the DHA formula group and the breast milk group, but is impaired in the formula group lacking only DHA, a strong case is made for considering this fatty acid an essential nutrient in formula. Conversely, if performance in the DHA formula group is inferior to the breast milk group, it would suggest that something other than, or in addition to, DHA is lacking from the formula in question.

A second option for varying a single nutrient might be to manipulate that nutrient in the mother in an effort to produce maternal milk differing in a single variable, and then to compare breast-fed groups. This would have the advantage of maintaining the ability to make comparisons to normal breast-feeding, the condition of primary interest. A possible disadvantage, however, is that manipulations in the mother might have hidden consequences, introducing variables into the equation other than the intended nutritional variation.

## The Causal Role of Neural Changes in Behavior

Determining a potential causal role of neural modifications in behavioral expression is governed by the same principles as outlined for the chemical/nutritional modifications, and is equally confounded by the large number of molecules and dynamic interactions present at this level of organization. To repeat the reasoning used above, an observed change in neural function, for example receptor regulation, may or may not be the specific agent responsible for a behavioral change; the relationship may be coincidental rather than causal. Equally plausible is a variety of other neural substrates, e.g., altered synaptic number. Without an understanding of basic neural mechanisms of development, plasticity, or behavior in a physiologically *normal* system, it is difficult to know where to search for the effect of a perturbation such as a nutritional deficit; even if an effect is found, it is hard to know what significance it might have for behavior, particularly cognitive function. Models of neuroplasticity currently being studied in the context of learning and memory are described below.

As mentioned above, a neural change may propagate to affect behavior at the global level, or may have no effect, due to redundancy and/or to compensatory mechanisms. Given this uncertainty, the significance of a given neural change is difficult to interpret without complementary behavioral assessment. The advantage of studying this level of organization, though, may be to direct specific behavioral testing to a task dependent on the affected pathway.

## The Significance of a Functional/Behavioral Change for Cognition

Ignoring for the moment the identity of the putative specific causal agent(s) involved, certain functional differences have been reported between animals and human infants fed on diets differing in LCPUFA content. Electroretinogram (ERG) studies in rats (29) and monkeys (41–43) have demonstrated lowered amplitudes and/or increased latencies in n-3 fatty acid deficient animals. Such effects are often transient. Similarly, alterations in behavioral and visual evoked potential (VEP) acuity have been reported for monkeys (41–43) and human infants (46–48); for the latter, visual acuity decreased in the deficient groups (47). Assuming that such functional alterations do occur (for a contrasting view, see (49–51) and below), what they may mean for cognitive abilities, and whether 'different' functions in formula-fed infants is always synonymous with 'inferior' function is, at this stage of our knowledge, more a matter of philosophical than scientific debate. As such, further discussion of this topic will be relegated to the more speculative final section.

## Neuroplasticity Models, Functional Tests, and Learning/Memory

If the relationship of infant formula nutrition to cognitive ability is to be properly evaluated, the tools for assessing neural and cognitive function must be adequate. However, as mentioned above, studies at each of these levels of organization are fraught with ambiguity, and causality often cannot be sifted from coincidence. These are general problems associated with traversing organizational levels, an idea that will be developed below within the context of certain models of neuroplasticity and behavioral function that are currently used to study mechanisms of higher brain function. Examples to be considered include three forms of neuroplasticity (LTP, ocular dominance (OD), plasticity, and kindling) and various tests of behavioral function and intelligence (e.g. VEP and IQ tests).

The neural changes associated with LTP, OD, and kindling are studied as a 'bottom-up' approach to understanding larger scale brain function, in which alterations in neural responsivity are taken to indicate, or even be causal to, processes such as learning. LTP is defined as the sustained increase in postsynaptic response induced in neurons following delivery of a high-frequency presynaptic tetanus (52). Many researchers in the field accord LTP the status of neural substrate of learning/memory, despite a failure to demonstrate an unambiguous correlation with either (37). OD plasticity refers to the developmental change in connectivity and response properties of neurons in the visual cortex of mammals possessing binocular vision. Monocular deprivation within a critical period can have a major impact on the function of the visual system (53). In young, but not adult, cat visual cortex, monocular eyelid closure leads to a shift in functional connectivity to favor the nondeprived eye, and a corresponding impaired responsivity and visual acuity of the deprived eye in response to stimulation. Like LTP, the OD plasticity model has as a strength a measurable change in neural function. The weakness in both models lies in their failure to achieve the crossover from neural to behavioral functional levels, in particular, cognition. LTP induction does not produce an observable change in behavior. Visual deprivation does, however, being most likely relevant only as an adaptive response to a pathological perturbation, and not to normal learning or memory processes. Kindling, an accepted model of temporal lobe epilepsy, is another form of neuroplasticity that occurs in response to pathological neural activation. In the kindling paradigm, daily trains of initially subconvulsive electrical stimuli delivered to brain cortical or limbic sites eventually culminate in seizure expression and altered emotionality (54,55). Again, relevance of this model to cognition is doubtful. It does have as an advantage the clear demonstration of neural (kindling-induced potentiation (56) and neurotransmitter receptor modifications (57–60)) and behavioral (motor seizures and altered emotionality (55,61)) correlates in central brain pathways (e.g. amygdala, hippocampus).

Various forms of IQ and visual function tests, to name just a few, examine brain function at the behavioral level. As was the case with the neuroplasticity models described above, each of these behavioral tests has strengths and weaknesses for linking cognitive and neural changes following a nutritional perturbation. IQ tests may measure certain cognitive abilities (this issue is not unequivocal), but are not reinforced by correlates of neural change. Tests of visual acuity and VEPs provide valuable measures of effects in a peripheral and cortical sensory pathway, but say nothing of cognitive abilities such as learning/memory. Such a list of pros and cons could be prepared for every neural and behavioral index used to measure brain function. However, the above examples should be sufficient to demonstrate two important points. First, causal links between specific neural modifications and altered cognitive ability have not been demonstrated, including for LTP, which is often attributed an established role in learning/memory. Second, no single test will adequately describe alterations in the brain in different nutritional paradigms. Failure to measure a neural change in one pathway or a behavioral deficit on one task does not guarantee normalcy in other areas. Even when a neural change is observed, its relevance to behavior in general, and cognition in particular, cannot be stated with certainty. Further, impairment in one aspect of higher function may go hand in hand with improvement in another, as evidenced in studies of infant language and memory function. One such study found that a group of infants with the best-developed language function at 16 months of age faired worse on long-term memory measures than did the group with the most poorly developed language skills (62). It is apparent that measuring either of these attributes in isolation would have resulted in opposite and incomplete views of function in these infants. It seems important to bear in mind as well that differences in diet may produce differences in behavior that are not necessarily detrimental.

Given these pitfalls, and our current inability to demonstrate causal links across levels from a nutritional to a cognitive change, what is the potential for gaining insight into this matter? It would appear that hope lies in pooling available advantages. While no test is on its own sufficient, each has a strength which can provide a valuable clue to the big picture. Multiple tests within levels can be used as a wide net that will minimize the chance of missing localized alterations. Observed changes in function at one level would suggest particular brain areas and behavioral tasks for more targeted testing. For example, a deficit in a conditioning or problem-solving task involving both sensory and motor components would be followed up by focused electrophysiological analysis of neural activity in the sensory and motor pathways implicated and by examining performance on tasks specifically designed to test sensory or motor function. If no changes are revealed at neural or behavioral levels by a battery of tests encompassing sensory, motor, cognitive, social and motivational function (see Table 2 in (48) for a list of relevant tests), the conclusion that function is equal in different nutritional groups can be considered quite strong. Further, if a neural change is not accompanied by altered behavior in a task requiring function of the same neural pathway, it may be concluded that higher function has been spared.

There are caveats to and requirements for successfully crossing disciplines and organizational levels. First, it is clear that human and animal models will each have selective advantages and disadvantages for testing in any given domain. Measurement of certain standards of neural function is invasive, and is therefore possible only in animal models. The ability to manipulate nutrient content of formula (and possibly breast milk) is more easily achievable in experimental animals; in addition, psychosocial and environmental variability, a major confound in human studies, can be eliminated. However, as for all areas of scientific inquiry, results from animal models are only of practical value insofar as they can be extrapolated to the human condition. Therefore, care must be taken in selecting an appropriate species, ideally one that presents no gross difference in biochemical processing of LCPUFAs. Similarly, behavioral tests with parallels in humans are of greatest benefit for relating results between species, assuming that equivalent periods in development can be studied (32). This is not to say that IQ and language tests should not be used in humans. In areas where it is possible to do so, however, standardization of experimental models and paradigms (e.g. similar nutrient addition/deprivation protocols, one animal species for invasive experiments) will increase the power of comparison and level crossing.

## Effects of Diet on Visual Function in Human Infants: do Measures of Visual Tasks Equate to Cognition and are Comparisons to Visual Deprivation Valid?

Many studies of the effects of early diet have used indices of visual function to examine breast- versus formula-fed infants or animals fed with differing LCPUFAs. Differences between these groups are discussed as examples of the types of CNS function that may be dependent upon certain fatty acids during development. In some cases, comparisons may be made between deficiencies in diet and the well-known detrimental effects of visual deprivation. In this latter context, the concept of the visual critical period becomes highly important. The critical period is that period in early postnatal life during which alterations in a sensory pathway (in this case vision) can influence the final state of the whole system. In the visual system, various forms of visual deprivation produce profound alterations in the morphology, biochemistry, and activity of parts of the visual pathway, notably the visual cortex (63,64). Generally, such visual deprivation paradigms induced later in adulthood are without effect (an exception being for monocular enucleation). Dietary deficiencies occurring during the critical periods for visual development might thus have a significant effect on the performance of the visual system through some form of neuroplasticity. In our view, none of the above assumptions is without serious constraints. In the following, we will attempt a brief outline of the problems that seem most apparent.

First, the attempt to study cognition in infants is a difficult problem both theoretically and behaviorally. It is thus quite understandable that some studies have utilized the visual pathway as a means to assess cognitive abilities. The accessibility of the visual system to measurement, both electrophysiologically and behaviorally, makes this an attractive choice. Two fundamental types of measurements have been used: The first attempts to assess the visual transduction pathway, measuring ERG and VEP representing, respectively, retinal function and the visual pathway from retina to visual cortex. A second set of measurements (e.g. visual habituation, visual recognition) attempts to use this pathway to assess presumed cognitive functions. Clearly, intact function in the first, the transduction pathway, is required for proper assessment of the second, since in a sensory pathway any malfunction at the sensory receptors, or at early stages in the pathway, will affect measurements designed to probe 'higher' levels. Thus, for example, differences in the ERG between breast- versus formula-fed infants (46) or the equivalent in animal models (29,41) preclude an unambiguous assessment of function at more central synapses and make impossible any conclusion beyond a statement that the periphery is affected. This point has been discussed in a recent review by Reisbick and Neuringer (43).

A second crucial point is that comparisons made between indices of visual acuity in normal versus visually deprived infants/children (41,47) and the effects of early diet on visual function are, at least in part, misleading. These comparisons may be made in a sincere attempt to 'inform the nutrition community' about the 'potential' for environment to affect nervous system development. They are, however, very different types of event. Visual deprivations in humans as well as in animal models, particularly those imposing competition between the two eyes, lead to widespread anatomical, biochemical, physiological, and behavioral alterations in vision. In some aspects, particularly in the loss of terminal arborizations in the lateral geniculate nucleus, the neural outcome more resembles a form of pathological plasticity in which synaptic pruning and remodeling occur on a large scale. Nothing of this magnitude has been suggested, to our knowledge, by anything in the literature on the effects of LCPUFAs.

A more appropriate comparison from the visual deprivation literature may be to the effects of dark rearing. In this paradigm, animals, usually cats, are born and raised in the total absence of light. In such cases, many aspects of visual cortex development (e.g., ocular dominance band segregation, ocular dominance plasticity, development of mature receptive field properties, neurotransmitter ontogenesis) (65,66) are abnormal. After removing the animals from the dark, however, many of the above features show a recovery toward the normal status. It is as if dark rearing imposes a delay on the development of normal visual cortex morphology and function, rather than a permanent deficit. Such data are reminiscent of some of the dietary studies comparing various indices of visual function of breast- versus formula-fed infants or animal models at different ages (29,41). Notably, deficits which appear at early testing may disappear at older ages. Third, it is clear that much of the early nutrition literature relies on measures of visual function in infants as indices of the effects of dietary deficiency on cognitive/behavioral competency. Even assuming that visual deprivation studies in humans could be compared to dietary deficiency studies, critical periods for visual development/deprivation effects in humans do not correspond sufficiently well with the periods of normal breast-feeding during which alterations in diet would be expected to exert maximum impact. Daw (67) defines three periods in the development of the human visual system: (i) pre-stereoscopic, in which acuity first develops (postnatal 1–4 months); (ii) onset of stereopsis, during which stereoscopic acuity develops and approaches adult values and ocular dominance columns segregate (4–6 months); and (iii) post-stereoscopic, in which acuity continues to develop (6 months to 2 years). Regarding the effects of visual abnormalities on later visual function, Daw notes that, 'few deficits lead to amblyopia before 6 months of age' (amblyopia being a loss of acuity of one or both eyes) and further states that some visual abnormalities which can induce amblyopia 'need to be persistent for 2 years or more from an early age to lead to a permanent deficit in acuity'. Cataracts have a more severe impact on visual acuity in that shorter periods of affected vision may be more damaging, but the critical period appears to be much the same as the above. The peak critical period for the effects of strabismus (inability to align the two eyes) on acuity appears to lie between 9 months and 2 years. The age at which improvement of visual acuity can no longer be reliably elicited appears to be 7–8 years, although older ages have been reported (67). Breast-feeding may thus coincide with the lower end of the human visual plasticity range, but probably not the peak period of developmental plasticity (approximately 1–2 years) nor the majority of the critical period. The outcome of visual input on visual cortex development and function is dependent on both age of onset and duration of visual deprivation (68). Thus, while dietary deficiencies might affect later vision, especially that dependent on cortical integrity, the effects would likely need to be as severe as those attributed to cataract. As noted previously, some effects of diet on retinal function, often of a transient nature, have been reported in animal models (29,41–43). However, the outcome for visual acuity in humans is not uniform under all conditions. For example, Uauy *et al.* (46) studying term and preterm infants and Carlson (48) studying preterm infants report greater visual acuity for diets containing n-3 fatty acids. In contrast, Auestad *et al.* (51) found no effects on visual acuity in term infants fed breast milk versus those fed formula containing or missing LCPUFAs. Similarly, Innis (50) reported no effects on visual acuity in term breast-fed infants versus those fed formula containing linoleic acid and α-linolenic acid but minus DHA and AA. The lack of comparable experimental outcome may depend on a number of key variables including: the use of term versus preterm infants, socio-economic status, the use of breast milk versus formula containing (or missing) precursor fatty acids or preformed DHA and AA, relative concentrations/ratios of these fatty acids, types of acuity tests performed (VEP versus Teller acuity cards), and others. These points not withstanding, nothing that we have seen in the literature appears to justify comparison of the effects of LCPUFAs in early diet to the effects of early visual deprivation.

## Traversing Levels of Neural Organization to Determine the Effects of Early Diet on Brain Development

Much of the preceding discussion has highlighted caveats to a hypothesis that LCPUFAs play a critical role in brain development and function. The potential impact of these molecules in formula may be considerable; these caveats suggest only that the hypothesis is still very debatable. Given this, and the obvious fact that a very large number of additional molecules present in breast milk are absent from formula, what is the best way to screen for molecules which are essential for normal brain development? A bottom-up approach appears to be precluded, since: (i) virtually any molecule can be expected to alter some aspect of synaptic function (see, for example, (37) on molecules affecting the expression of LTP); (ii) LCPUFA-induced alterations at a molecular, synaptic, or neural level may either be compensated for by the presence of other fatty acids (or other molecules or pathways) or may not occur in neural subsystems obviously linked to behavior; and (iii) they may have secondary, tertiary, etc., effects via biochemical cascades leading to behavioral changes far removed from the LCPUFAs' original site of action, the latter including organ systems other than the CNS.

The above considerations suggest that the most reliable approach is a top-down one, whose goal would be to provide a description of the effects of LCPUFA deficits across all levels of organization starting from a specific behavioral alteration, identified by broad-based testing. This mode of attack would proceed to isolate specific neural systems and subsystems subserving the behavior, leading ultimately to effects at single neurons and synapses, and culminating in alterations in the functions of particular synaptic molecules. Several laboratories have attempted to provide such data in animal models of LCPUFA deficiency (28,29,41–43). Ideally, as a first step, the top-down approach would successfully identify profound changes in behavioral ability which could be unambiguously attributed to neural networks. Such an approach would require selection of an appropriate model system, as most of the levels could not be analyzed in human subjects.

We offer the kindling model as an appropriate system to be used as part of a broad spectrum test paradigm. As described above, kindling is believed to model many of the *important* aspects of temporal lobe epilepsy in humans. How then is this an appropriate model for the effects of early diet, and what advantages are offered? First, unlike many of the tests presently used, the kindling model offers the opportunity to examine the impact of a particular developmental event, e.g. the presence or absence of a particular LCPUFA, across all levels of neural organization in one organism. At the behavioral level, two measurable forms of modification, motor seizures and emotional disturbances, are associated with kindling. Seizures are produced by stimulation of a definable brain locus and spread within predictable circuits. These motor effects are stereotypic and quantifiable. Interictally (between seizures), clear indices of emotional modification are apparent, also linked to specific neural pathways. Evaluation of performance in the elevated plus maze and resistance to capture tests (69,70) indicates increased amygdala-mediated defensiveness in kindled rats (61). At the molecular level, kindling has been demonstrated to induce quantifiable, consistent modification in neural excitability (56), and in the regulation of particular neurotransmitter receptor populations within the affected neural circuits in amygdala, hippocampus, and specific cortical areas (57–60).

To some extent, visual pathway models also have the capacity to traverse levels, providing information about peripheral and cortical function in a sensory pathway. Since retinal function will affect measurements at all later stages in the pathway, however, effects at these sites may not be separable. Accordingly, alterations at the cortical level may simply reflect the fact that retinal function is compromised. In contrast, kindling-related alterations are directly induced in central pathways (e.g. hippocampus, amygdala, and cortex), bypassing peripheral input pathways. Therefore, this model can benefit our knowledge of central neural function, independent of peripheral integrity. Animals given varying LCPUFA diets may differ in various characteristics of kindling, including rate of development and intensity of kindling, expressed at behavioral, neural circuit, and biochemical levels. Many of the underlying substrates remain unknown, and the only relationship to cognition that could be claimed is as a possible misuse of neuroplasticity mechanisms normally subserving learning and memory. Even as a form of pathological neuroplasticity, it nevertheless is a form of neural modification that can be quantified and described across all levels of organization. It may thus serve as a useful diagnostic tool to evaluate various influences in early development on later neural function, including alterations in diet.

## Conclusions

The number of neurons, pathways, synapses, and molecules involved in the activity of the central nervous system is astronomical. Very large numbers of proteins, most likely not yet identified, are developmentally expressed. Of these, most, or all, are certain to be regulated in part by other molecules (e.g., other proteins, lipids, ions), the latter themselves regulated in a like manner. In turn, the dynamical interactions of molecular expression, neural activity, and sensory input combine to make the developing nervous system an almost overwhelmingly complex dynamical system. All of this is to illustrate the point that most molecules which interact with any of these elements will have a *potential* impact on the development of the system.

In ways still not understood, order and stability usually emerge in the mature nervous system. It is perhaps surprising, given the above, that more gross abnormalities in structure and function of the CNS do not occur. Reference is sometimes made to some form of brain 'sparing' in deficit conditions in development (see 71), and to some extent some sparing/protection must occur in the presence of toxic substances. Otherwise, prenatal exposure to substances such as alcohol and cocaine (72,73) would be expected to cause far more severe abnormalities in neural activity and behavior than are in fact documented (see also Commentary by Wainwright and our reponse).

The above highlight our concerns for studies of infant diet and the impact of dietary deficiencies on human brain development. A bottom-up approach will almost certainly find affected circuits and molecules, but perhaps no associated effect at higher levels of neural organization. Only a top-down approach which first identifies detrimental abnormalities in behavior has a reasonable hope of proceeding down through the levels of organization to discover the cause and, perhaps, provide a solution.

## Acknowledgements

The authors thank Drs N. Auestad, J. Weinberg, and D. Giaschi for providing important references. This work was supported by grants from the Natural Science and Engineering Research Council of Canada (CAS) and Epilepsy Canada (JCM).

# References

1. Innis SM. Fatty acid requirements of the newborn. *Can J Physiol Pharmacol* 1994; **72**: 1483–92.
2. Shaw CA, Lanius RA, Van den Doel K. The origin of synaptic neuroplasticity: crucial molecules or a dynamical cascade? *Brain Res Rev* 1994; **19**: 241–63.
3. Bornstein M. Sensitive periods in development: structural characteristics and causal interpretations. *Psychol Bull* 1989; **105**: 179–97.
4. Murphy MG. Dietary fatty acids and membrane protein function. *J Nutr Biochem* 1990; **1**: 68–79.
5. Shaw CA. Age-dependent expression of receptor properties and function in CNS development. In: Shaw CA, ed. *Receptor Dynamics in Neural Development*, Boca Raton, FL: CRC Press, 1996: 3–17.
6. Laurie DJ, Wisden W, Seeburg PH. The distribution of thirteen $GABA_A$ receptor subunit mRNAs in the rat brain. III. Embryonic and postnatal development. *J Neurosci* 1992; **12**: 4151–72.
7. Kim HY, Olsen RW, Tobin AJ. GABA and $GABA_A$ receptors: development and regulation. In: Shaw CA, ed. *Receptor Dynamics in Neural Development*. Boca Raton, FL: CRC Press, 1996: 59–72.
8. Watanabe M. Developmental regulation of ionotropic glutamate receptor gene expression and functional correlations. In: Shaw CA, ed. *Receptor Dynamics in Neural Development*, Boca Raton, FL: CRC Press, 1996: 73–89.
9. Shaw C, Wilkinson M, Cynder M, Needler MC, Aoki C, Hall SE. The laminar distribution and postnatal development of neurotransmitter and neuromodulator receptors in cat visual cortex. *Brain Res Bull* 1986; **16**: 661–71.
10. Monyer H, Seeburg PH, Wisden W. Glutamate-operated channels: developmentally early and mature forms arise by alternative splicing. *Neurone* 1991; **6**: 799–810.
11. Hendrickson A, March D, Richards G, Erickson A, Shaw C. Coincidental appearance of the a1 subunit of the $GABA_A$ receptor and the type 1 benzodiazepine receptor near birth in macaque monkey visual cortex. *Int J Dev Neurosci* 1994; **12**: 299–314.
12. Pelligrini-Giampietro DE, Bennett MVL, Zukin RS. Differential expression of three glutamate receptor genes in developing rat brain: an in situ hybridization study. *Proc Natl Acad Sci USA* 1991; **88**: 4157–61.
13. Shaw C, Cameron L, March D, Cynder M, Zielinski B, Hendrickson A. Pre- and postnatal development of GABA receptors in *Macaca* monkey visual cortex. *J Neurosci* 1991; **11**: 3943–59.
14. Kalb RG, Lidow MS, Halsted MJ, Hockfield S. N-Methyl-D-aspartate receptors are transiently expressed in the developing spinal cord ventral horn. *Proc Natl Acad Sci USA* 1992; **89**: 8502–6.
15. Rossner S, Kumar A, Kues W, Witzemann V, Schliebs R. Differential laminar expresssion of AMPA receptor genes in the developing rat visual cortex using in situ hybridization histochemistry. Effects of visual deprivation. *Int J Dev Neurosci* 1993; **11**: 411–24.
16. Catania MV, Landwehrmeyer GB, Testa CM, Standaert DG, Penney JB, Young AB. Metabotropic glutamate receptors are differentially regulated during development. *Neuroscience* 1994; **61**: 481–95.

17. Laurie DJ, Seeburg PH. Regional and laminar heterogeneity in splicing of rat brain NMDAR1 mRNA. *J Neurosci* 1994; **14**: 3180–94.
18. Bode-Greuel KM, Singer W. The development of *N*-methyl-D-aspartate receptors in cat visual cortex. *Brain Res* 1989; **46**: 197–204.
19. Tremblay E, Roisin MP, Represa A, Charriaut-Marlangue C, Ben-Ari Y. Transient increased density of NMDA binding sites in the developing rat hippocampus. *Brain Res* 1988; **461**: 393–6.
20. Endo SL, Wolff JR. Postnatal development of the excitatory amino acid system in visual cortex the rat. Changes in ligand binding to NMDA, quisqualate, and kainate receptors. *Int J Dev Neurosci* 1990; **8**: 199–208.
21. Shaw C, Needler MC, Wilkinson M, Aoki C, Cynder M. Modification of neurotransmitter receptor sensitivity in cat visual cortex during the critical period. *Dev Brain Res* 1985; **22**: 67–73.
22. Pasqualotto BA, Shaw CA. Regulation of ionotropic receptors by protein phosphorylation. *Biochem Pharmacol* 1996; **51**: 1417–25.
23. Shaw CA, Wilkinson M. Receptor characterization and regulation in intact tissue preparations: pharmacological implications. *Biochem Pharmacol* 1994; **47**: 1109–19.
24. Hendry SHC, Huntsman M-M, Vinuela A, Mohler H, de Blas AL, Jones EG. $GABA_A$ receptor subunit immunoreactivity in primate visual cortex: distribution in macaques and humans and regulation by visual input in adulthood. *J Neurosci* 1994; **14**: 2383–401.
25. Hendry SHC. Regulation of $GABA_A$ receptors and other transmitter-related molecules in adult macaque visual cortex: neurochemical correlates of functional plasticity. In: Shaw CA, ed. *Receptor Dynamics in Neural Development*, Boca Raton, FL: CRC Press, 1996: 205–21.
26. Hendry SHC, Kennedy MB. Immunoreactivity for a calmodulin-dependent protein kinase is selectively increased in macaque striate cortex after monocular deprivation. *Proc Natl Acad Sci USA* 1986; **83**: 1536–40.
27. Routtenberg A. Reverse piedpiperase: is the knockout mouse leading neuroscientists to a watery end? *Trends Neurosci* 1996; **19**: 471–72.
28. Delion S, Chalon S, Herault J, Guilloteau D, Besnard J-C, Durand G. Chronic dietary α-linolenic acid deficiency alters dopaminergic and serotoninergic neurotransmission in rats. *J Nutr* 1994; **124**: 2466–76.
29. Bourre J-M, Francois M, Youyou A et al. The effects of dietary α-linolenic acid on the composition of nerve membranes, enzymatic activity, amplitude of electrophysiological parameters, resistance to poisons and performance of learning task in rats. *J Nutr* 1989; **119**: 1880–92.
30. Nisikawa M, Kimura S, Akaike N. Facilitatory effect of docosahexaenoic acid on *N*-methyl-D-aspartate response in pyramidal neurones of rat cerebral cortex. *J. Physiol* 1994; **475**: 83–93.
31. Cravatt BF, Giang DK, Mayfield SP, Boger DL, Lerner RA, Gilula NB. Molecular characterization of an enzyme that degrades neuromodulatory fatty-acid amides. *Nature* 1996; **384**: 83–7.
32. Wainwright PE. Lipids and behaviour: the evidence from animal models. In: Dobbing J, ed. *Lipids, Learning, and the Brain: Fats in Infant Formula*. Columbus, OH: Ross, 1993: 69–101.
33. Tan S-E, Wenthold RJ, Soderling TR. Phosphorylation of AMPA-type gluta-

mate receptors by calcium/calmodulin-dependent protein kinase II and protein kinase C in cultured hippocampal neurones. *J Neurosci* 1994; **14**: 1123–29.
34. Stevens CF, Tonegawa S, Wang Y. The role of calcium calmodulin kinase II in three forms of synaptic plasticity. *Current Biol* 1994; **4**: 687–93.
35. Silva AJ, Paylor R, Wehner JM, Tonegawa S. Impaired spatial learning in α-calcium calmodulin kinase II mutant mice. *Science* 1992; **257**: 206–11.
36. Silva AJ, Stevens CF, Tonegawa S, Wang Y. Deficient hippocampal long-term potentiation in α-calcium-calmodulin kinase II mutant mice. *Science* 1992; **257**: 201–6.
37. McEachern JC, Shaw CA. An alternative to the LTP orthodoxy: a plasticity–pathology continuum model. *Brain Res Rev* 1996; **22**: 51–92.
38. Dobbing J. *Brain, Behaviour, and Iron in the Infant Diet*. London: Springer-Verlag, 1990.
39. Sawada M, Suzumura A, Marunouchi T. Cytokine network in the central nervous system and its roles in growth and differentiation of glial and neuronel cells. *Int J Dev Neurosci* 1995; **13**: 253–64.
40. Sei Y, Vitkovic L, Yokoyama MM. Cytokines in the central nervous system: regulatory roles in neuronel function, cell death, and repair. *Neuroimmunomodulation* 1995; **2**: 121–33.
41. Neuringer M. The relationship of fatty acid composition to function in the retina and visual system. In: Dobbing J, ed. *Lipids, Learning, and the Brain: Fats in Infant Formulas*. Columbus, OH: Ross, 1996: 134–63.
42. Reisbick S, Neuringer M, Connor WE. Effects of n-3 fatty acid deficiency in nonhuman primates. In: Bindels JG, Goedhart AC, Visser HKA, eds. *Recent Developments in Infant Nutrition*. Dordrecht: Kluwer, 1996: 157–71.
43. Reisbick S, Neuringer M. Omega-3 fatty acid deficiency and behavior. In: Yehuda S, Mostofsky DI, eds. *Handbook of Essential Fatty Acid Biology: Biochemistry, Physiology, and Behavioral Neurobiology*. Totowa, NJ: Humana Press, 1997: in press.
44. Martin J-C, Bougnoux P, Antoine J-M, Lanson M, Couet C. Triacylglycerol structure of human colostrum and mature milk. *Lipids* 1993; **28**: 637–43.
45. Salvati S, Attorri L, Di Felice M et al. Effect of dietary oils on brain enzymatic activities (2′-3′-cyclic nucleotide 3′-phosphodiesterase and acetylcholinesterase) and muscarinic receptor sites in growing rats. *J Nutr Biochem* 1996; **7**: 113–17.
46. Uauy R, Birch DG, Birch EE, Hoffman D, Tyson J. Visual and brain development in infants as a function of essential fatty acid supply provided by the early diet. In: Dobbing J, ed. *Lipids, Learning, and the Brain: Fats in Infant Formulas*. Columbus, OH: Ross, 1993: 215–38.
47. Birch EE, Birch D, Hoffman D, Hale L, Everett M, Uauy R. Breast-feeding and optimal visual development. *J Pediatr Ophthalmol Strabismus* 1993; **30**: 33–8.
48. Carlson SE. Perceptual and cognitive function: methods of assessment and relation to LCPUFA status during infancy. In: *Assessment of Infant Vision and Cognitive Function in Relation to Long Chain Polyunsaturated Fatty Acids*. Basel: Editiones Roche, 1996: 49–69.
49. Jacobson SW, Jacobson JL. Breastfeeding and intelligence. *Lancet* 1992; **339**: 926–7.
50. Innis SM. Feeding formula without arachadonic acid and docosahexaenoic

acid has no effect on preferential looking acuity or recognition memory in healthy full-term infants at 9 mo of age. *Am J Clin Nutr* 1996; **64**: 40–6.
51. Auestad N, Montalto MB, Hall RT et al. Visual acuity, erythrocyte fatty acid composition, and growth in term infants fed formulas with long chain polyunsaturated fatty acids for one year. *Pediatr Res* 1997; **41**: 1–10.
52. Bliss TVP, Collingridge GL. A synaptic model of memory: long-term potentiation in the hippocampus. *Nature* 1993; **361**: 31–9.
53. Hubel DL, Wiesel TN. The period of susceptibility to the physiological effects of unilateral eye closure in kittens. *J Physiol* 1970; **206**: 419–36.
54. Goddard GV. Development of epileptic seizures through brain stimulation at low intensity. *Nature* 1967; **214**: 1020–1.
55. Pinel JPJ, Treit D, Rovner L. Temporal lobe aggression in rats. *Science* 1977; **197**: 1088–9.
56. Racine RJ, Cain DP. Kindling-induced potentiation, In: Morrell F, ed. *Kindling and Synaptic Plasticity*. Boston: Birkhauser, 1991: 38–53.
57. Cincotta M, Young NA. Unilateral up-regulation of glutamate receptors in limbic regions of amygdaloid-kindled rats. *Exp Brain Res* 1991; **85**: 650–8.
58. McEachern JC, Kalynchuk L, Fibiger HC, Pinel JPJ, Shaw CA. Increased emotionality after amygdala kindling in rats may be related to receptor regulation. II. AMPA, kainate and acetylcholine. *Soc Neurosci Abstr* 1995; **21**: 1476.
59. Kalynchuk LE, McEachern JC, Barr KN, Pinel JPJ, Shaw CA. Increased emotionality after amygdala kindling in rats may be related to receptor regulation. I. $GABA_A$ and benzodiazepines. *Soc Neurosci Abstr* 1995; **21**: 1476.
60. McEachern JC, Kalynchuk LE, Burgmann T, Pinel JPJ, Shaw CA. Altered regulation of NMDA receptors and protein kinase C in long-term kindled rats. *Soc Neurosci Abstr* 1996; **22**: 2096.
61. Kalynchuk LE, Pinel JPJ, Treit D, Kippin JR. Changes in emotional behaviour produced by long-term kindling in rats; *Biol Psychiat* 1997; **41**: 438–51.
62. Schwade JA, Dropik P. A little language hurts a lot: productive vocabulary and nonverbal recall in 16- to 20-month olds. *Infant Behav Dev* 1996; **19**: 732–8.
63. Movshon JA, Van Sluyters RC. Visual neural development. *Annu Rev Psychol* 1981; **32**: 477–522.
64. Shaw C, Cynder M. Unilateral eyelid suture increases $GABA_A$ receptors in cat visual cortex. *Dev Brain Res* 1988; **40**: 148–53.
65. Mower GD. The effect of dark rearing on the time course of the critical period in cat visual cortex. *Dev Brain Res* 1991; **58**: 151–8.
66. Shaw C, Aoki C, Wilkinson M, Prusky G, Cynder M. Benzodiazepine receptors in cat visual cortex: ontogenesis of normal characteristics and the effects of dark rearing. *Dev Brain Res* 1987; **37**: 67–76.
67. Daw NW. *Visual Development*. New York: Plenum Press, 1995: 146–52.
68. Olson CR, Freeman RD. Profile of the sensitive period for monocular deprivation in kittens. *Exp Brain Res* 1980; **39**: 17–21.
69. Pellow, S, Chopin, P, File, SE, Briley, M. Validation of open:closed arm entries in an elevated plus-maze as a measure of anxiety in the rat. *J Neurosci Meth* 1985; **14**: 149–67.
70. Albert DJ, Richmond SE. Septal hyperreactivity: a comparison of lesions within and adjacent to the septum. *Physiol Behav* 1975; **17**: 339–47.
71. Dobbing J. *Lipids, Learning, and the Brain: Fats in Infant Formulas*. Columbus, OH: Ross, 1993.

72. West JR. *Alcohol and Brain Development*. New York: Oxford University Press, 1986.
73. Little Z, Teyler TJ. In utero cocaine exposure decreases dopamine $D_1$ receptor modulation of hippocampal long-term potentiation in the rabbit. *Neurosci Lett* 1996; **215**: 157–60.
74. Zielinski B, Hendrickson A. Development of synapses in macaque monkey striate cortex. *Vis Neurosci* 1992; **8**: 491–504.

# Commentary

**Neuringer and Reisbick**: Several possible mechanisms of action for LCPUFAs within the brain are related to the fact that dietary LCPUFAs influence tissue LCPUFA levels, and that these molecules, as part of membrane phospholipids, form a major and integral part of membrane structure. Thus, they provide the physical environment for all membrane-associated proteins and may alter membrane protein function. There is specific evidence for such a mechanism in the case of rhodopsin within synthetic phospholipid membranes of varying fatty acid composition. In vitro studies show that the photochemical activity of rhodopsin is reduced in membranes with low levels of DHA (1). Different biophysical mechanisms have been proposed to explain this phenomenon, including changes in characteristics of the membrane, such as flexibility and expandability, which are necessary to accommodate the shape change associated with rhodopsin activation and its lateral mobility within the membrane. However, the critical comparison between membranes with high DHA versus those with high 22:5n-6 (as seen in n-3 fatty acid deficiency) has not been done, because a source of pure 22:5n-6 has not been available. Functional changes related to changes in membrane DHA content are not likely to be mediated by simple bulk membrane fluidity, which is not a monotonic function of the number of double bonds and is unlikely to change with a shift from DHA to 22:5n-6 (2).

1. Weidmann TS, Pates RD, Beach JM et al. Lipid–protein interactions mediate the photochemical function of rhodopsin. *Biochemistry* 1988; **27**: 6469–74.
2. Dratz EA, Deese AK. The role of docosahexaenoic acid (22:6n-3) in biological membranes: examples from photoreceptors and model membrane bilayers. In: Simopoulos AF, Kifer RR, Martin RE, eds. *Health Effects of Polyunsaturated Fatty Acids in Seafoods*. New York: Academic Press, 1986: 319–51.

AUTHORS' REPLY: We have deliberately left vague the possible mechanisms of LCPUFA action. If the reader comes away with an impression that such fatty acids can act like drugs or neurotoxins, so be it. In fact, such can obviously occur. Cravatt *et al.* (1) have recently reported possible neuromodulatory actions of so-called fatty acid amides; Nisikawa *et al.* (2) show direct AA and DHA actions on NMDA and kainate receptors, the former of which is certainly potentially neurotoxic. We also cite a recent study by Salvati *et al.* (3) on other possibly detrimental effects of n-3 fatty acids in myelination during early development.

1. Cravatt BF, Giang DK, Mayfield SP, Boger DL, Lerner RA, Gilula NB. Molecular characterization of an enzyme that degrades neuromodulatory fatty-acid amides. *Nature* 1996; **384**: 83–7.
2. Nisikawa M, Kimura S, Akaike N. Facilitatory effect of docosahexaenoic acid on $N$-methyl-D-aspartate response in pyramidal neurones of rat cerebral cortex. *J Physiol* 1994; **475**: 83–93.
3. Salvati S, Attori L, Di Felice M *et al*. Effect of dietary oils on brain enzymatic activities (2′-3′-cyclic nucleotide 3′-phosphodiesterase) and muscarinic receptor sites in growing rats. *J Nutr Biochem* 1996; **7**: 113–17.

NEURINGER AND REISBICK: The timing of the component processes in neural development in relation to the timing of formula feeding (reviewed in 1) is an important consideration for estimating the possible impact of dietary LCPUFAs. Developmental processes occurring relatively late in gestation and during the postnatal brain growth spurt are most likely to be influenced by LCPUFA supply. This is especially true for preterm infants, because more of their neural development occurs without the benefit of the selective supply of LCPUFAs from the placenta, so they are dependent on dietary sources at an earlier stage, and for a longer proportion, of the brain growth spurt. It has been estimated that, during the last trimester (or the equivalent postnatal period in preterm infants) and the first three post-term months, the human infant brain requires several milligrams per day of both DHA and AA (from dietary sources or from synthesis) to provide for the normal accumulation of these fatty acids into neural membranes (2), particularly for the massive growth of the cerebral and cerebellar cortices. Unlike cerebral neuronal cell division, which is essentially complete by the end of the second trimester, synaptogenesis is a prominent event in postnatal primate development. The number of synapses in the human visual cortex, for example, doubles between 2 and 8 months from term birth (3). Thus, the relatively late timing of synaptogenesis, and the presence of high levels of DHA in synaptic membranes, suggest that this process might be relatively likely to be influenced by altered DHA supply. Studies in cultured fetal mouse brain tissue suggest that DHA and AA may increase the number, diversity and complexity of synaptic contacts (4). In addition, neuronal migration could be affected, particularly for cells in the superficial layers of the cerebral cortex which are the last to migrate to their mature positions (5). This possibility is particularly interesting in light of Auestad and Innis' recent report that n-3 fatty acid deficiency selectively affects DHA levels in growth cones, but not neuronal cell bodies, in the neonatal rat brain (6).

Many neuroscientists also would disagree with the statement that 'causal links between specific neural modifications and altered cognitive ability have not been demonstrated'. Although this may be true at a molecular level, a number of specific relationships have been shown at the systems level; for example, with regard to specific effects of damage to small, localized cortical regions and neurophysiological correlates of well-defined aspects of cognitive performance. To pick just one example, we could cite studies of the neural basis of spatial working memory in nonhuman primates (e.g. 7) which have related this specific cognitive ability to the response properties of individual neurons in dorsolateral prefrontal cortex and to specific dopamine receptor subtypes and neural circuits within this cortical area.

The statement that 'differences in ERG . . . preclude an assessment of function at more central synapses' is somewhat misleading. Differences in peripheral (retinal) sensory function certainly constrain the interpretation of more central (visual cortical) function. For example, differences in visual acuity, measured by either behavioral or visual evoked potential methods, may be parsimoniously explained by differences in retinal function (or they may not, depending on the specific nature of the retinal changes), and measurement of a change in cortical responses does not mean that the origin of an effect is primarily within the cortex rather than secondary to a retinal effect. But the statement made here implies that it is impossible to evaluate cognitive function in the presence of sensory effects, which is not correct. For example, intelligence can be measured, and is generally entirely normal, in the visually impaired. The use of appropriate tasks, stimuli and controls, together with specific and quantitative measurement of the nature of the sensory deficit, can overcome this problem. In the case of n-3 fatty acid deficiency or lack of dietary LCPUFAs in human and monkey infants, when effects on acuity are found they represent rather mild losses in spatial resolution (0.5–1.0 octave). It is not difficult to design stimuli which are suprathreshold and clearly discriminable to subjects, and to confirm this by measuring acuity thresholds and discrimination. The problem with many of the rodent studies which used visual discriminations as measures of learning is that the vision of n-3 fatty acid-deficient rats has never been adequately assessed, so it is not known to what extent sensory factors may contribute to the effects.

The term sensitive period is more appropriate than the term critical period for primates (monkey and human) where such periods are less tightly defined and there tends to be a gradual and progressive decrease in vulnerability to deprivation and in the potential for reversal. In addition, the specification of critical periods depends on the outcome. There is no one 'critical period' for all aspects of visual development or even all aspects of visual image deprivation. It depends, for example, whether one is talking about the sensitive period for induction or reversal of changes in cell size in the LGN, which is different for the magnocellular and parvocellular systems, or for shifts in ocular dominance columns in primary visual cortex, which are different for the different cortical layers, or for acuity and contrast sensitivity loss, or for failure to develop binocular vision. Shifts in ocular dominance and loss of binocularity apply primarily to monocular deprivation or interocular asymmetries, and, therefore, are less relevant to the possible effects of n-3 fatty acid deficiency, whereas changes in LGN cell size and acuity/contrast sensitivity loss can occur with both monocular and binocular deprivation of pattern vision.

Shaw and McEachern criticize attempts to evaluate the possible ramifications of poorer infant acuity, as produced by n-3 fatty acid deficiency or by lack of dietary LCPUFAs, in light of the literature on early visual deprivation. Such comparisons have been made to inform the nutrition community, which may be unfamiliar with this literature, that transient effects on early vision should not necessarily be dismissed as trivial. The comparisons to early visual deprivation are clearly very speculative, and have been made with appropriate qualifications about the distinctions between monocular and binocular deprivation and competitive and deprivation mechanisms, and the far more serious nature of monocular deprivation/interocular assymetries. However, competition between inputs from the two eyes may not be the only form of competition involved in the development of cortical inputs. The strengthening of synaptic connections by repeated use, and the loss of unused inputs, is a general principle in neural plasticity. Thus, Movshon *et al.* (8) and others have proposed possible competition between magnocellular and parvocellular LGN inputs to visual cortex, and there could also be competition between different spatial frequency channels. It seems important to be aware of these possibilities, even if they have not been demonstrated. We entirely agree that the magnitude of the effects on early visual deprivation is greater than any effects likely to result from early acuity loss associated with low LCPUFA status.

It may be helpful for us to spell out more clearly the hypothesis underlying comparisons of the effects of visual deprivation to acuity loss due to n-3 fatty acid deficiency. The hypothesis is 1) that the primary effect of the deficiency is to alter retinal development in ways (as yet unknown) which result in slower acuity development, and 2) that poorer spatial vision due to this altered retinal development affects cortical development in ways similar to loss of sharp spatial vision induced optically (by cataracts or other ocular media defects, by artificial blur or refractive errors). In both cases, the input of visual information (specifically high spatial frequency information) to the LGN and visual cortex are restricted. There are no primate studies which provide an optimal model for binocular mild acuity loss, so we can only extrapolate from the available data for 1) mild monocular acuity loss induced by optical blur (8), and 2) binocular pattern deprivation (9).

1. Reisbick S. Neural development. In: Carlson SE, Neuringer M, Reisbick S, eds. *Assessment of Infant Visual and Cognitive Function in Relation to Long Chain Polyunsaturated Fatty Acids*. Basel: Editiones Roche, 1996: 10–18.
2. Clandinin MT, Chappell JE, Heim T. Do low weight infants require nutrition with chain elongation–desaturation products of essential fatty acids? *Prog Lipid Res* 1981; **20**: 901–4.
3. Huttenlocher PR, deCourten C, Garey LJ, Van der Loos H. Synaptogenesis in the human visual cortex: evidence for synapse elimination during normal development. *Neurosci Lett* 1982; **33**: 247–52.
4. Tixier-Vidal A, Picart R, Loudes C, Bauman AF. Effects of polyunsaturated fatty acids and hormones on synaptogenesis in serum-free medium cultures of mouse fetal hypothalamic cells. *Neuroscience* 1986; **17**: 115–32.
5. Rakic P. Neurons in the monkey visual cortex: systematic relation between time of origin and eventual disposition. *Science* 1974; **183**: 425–7.
6. Austead N, Innis SM. Dietary deficiency of n-3 fatty acids affects the fatty acid composition of neuronal growth cones. *Proceedings of the International Conference on Highly Unsaturated Fatty Acids in Nutrition and Disease Prevention*, Basel: Hoffmann-La Roche, 1996: 89.
7. Sawaguchi T, Goldman-Rakic PS. The role of DI-dopamine receptor in working memory: local injections of dopamine antagonists into the prefrontal cortex of rhesus monkeys performing an oculomotor delayed-response task. *J Neurophysiol* 1994; **71**: 515–27.
8. Movshon JA, Eggers HM, Gizzi MS, Hendrickson AE, Kiorpes L, Boothe RG. Effects of early unilateral blur on the macaque's visual system: III. Physiological observations. *J Neurosci* 1987; **7**: 1340–51.
9. Hawerth RS, Smith EL III, Paul AD, Crawford MLJ, vonNoorden GK. Functional effects of bilateral form deprivation in monkeys. *Invest Ophthalmol Vis Sci* 1991; **32**: 2311–27.

AUTHORS' REPLY: We have addressed the effects of early visual deprivation on the developing nervous system in comparison to those of early diet since, as you point out, the former is used to 'inform the nutrition community, which may be unfamiliar with this literature, that transient effects on early vision should not necessarily be dismissed as trivial'. We agree that the effects of visual deprivation are not trivial at any level. However, raising the spectre of a truly damaging neural modification in inappropriate context to contentious data on diet is the equivalent of saying, 'Since wolves and dogs are canids, and since wolves are known to kill livestock, don't be too complacent about old Lassie next door'. Yes, dogs do go 'bad' (reference the movie 'Babe', but one needs clearer proof before making such linkages), i.e. these are different levels of problem entirely.

NEURINGER AND REISBICK: We would agree that a top-down approach is necessary as a starting point. However, once some behavioral effects have been identified (as has now been done), then work at the behavioral and mechanistic levels can be complementary and mutually reinforcing. Identified effects on specific neurotransmitter systems or brain loci can help to focus behavioral work as much as the reverse. The fact the n-3 fatty acid deficient rodents have not been examined for basic effects on neurochemistry or brain structure has been an impediment to generating clear hypotheses about underlying mechanisms. As an example, we would like to present a hypothesis (adapted from 1) which builds on the interaction between behavioral and neurochemical studies, and which may serve as a basis for further discussion and hypothesis generation.

Dopamine hypothesis. As noted in this chapter, Delion *et al.* (1) reported decreased levels of dopamine and D2 receptors, as well as increased serotonin receptor density, in rat frontal cortex but not in the striatum or cerebellum. Serotonin and norepinephrine levels were not affected in any of the three brain areas. This result indicates a possible specificity in the neurochemical effects of n-3 fatty acid deficiency. It suggests mechanisms for some of the behavioral differences found in n-3 fatty acid-deficient animals, and these mechanisms might also apply to human infants with low brain DHA. A number of studies of n-3 fatty acid deficiency suggest changes in attention, motivation or reactivity to stimuli and rewards, all of which are consistent with a deficit in the function of the prefrontal dopamine pathways.

There are two ascending dopamine pathways. One sends processes from cell bodies in the ventral tegmental area (VTA) to the nucleus accumbens (nAcc) in the ventral striatum, a key brain area involved in reward.

Presentation of novel or reward-related stimuli produces dopamine release within the nAcc. The second, the mesocortical pathway, sends dopamine fibers to terminate in the cortex, including the medial prefrontal cortex in rodents. In rats, this second dopamine pathway indirectly inhibits the dopamine response of the first (3). For example, dopamine antagonists infused into the medial prefrontal cortex caused a dose-dependent increase in dopamine released by the nAcc, and they also increased lever pressing on DRL and FI operant schedules (4).

This circuit can be related to a number of the reported behavioral effects of n-3 fatty acid deficiency. Deficient rats consistently showed higher rates of lever-pressing for food rewards and slower extinction (decrease in responding after discontinuation of the rewards) (e.g. 5). The effects reported by Delion *et al.* (2), decreased dopamine levels and receptors in the prefrontal cortex in deficient rats, would decrease inhibition of the nAcc, a change consistent with an overall increase in response rates for rewards (6), and with slower extinction from the higher response rates. Disinhibition of the striatal dopamine system would also be consistent with several findings in n-3 fatty acid-deficient mice, including increased activity in a novel environment and faster swimming speeds in a water maze (7) and increased time in the open arms of a plus maze (8), as well as with increased activity and increased behavioral responsivity in our deficient monkeys (1). This hypothesis is also consistent with effects on shock escape and avoidance. Depletion of dopamine in prefrontal cortex decreased escape and had a lesser effect on avoidance responding (9), a pattern similar to that reported in n-3 fatty acid deficient rats (10). The same dopamine circuits also appear to be involved in the response to novelty. Novel stimuli produce a dopamine surge in the nAcc (11,12), and the surge decreases with habituation (13). If there is a decrease in prefrontal dopamine in deficient monkeys, it could be expected to cause disinhibition of the nAcc surge to novelty and slowing of the nAcc dopamine decrease with habituation, changes that could underlie the increased look durations of deficient infant rhesus monkeys to both novel and familiar stimuli. Thus, n-3 fatty acid deficiency and the resulting low frontal cortex DHA levels may dampen the prefrontal dopamine system, disinhibit the striatal dopamine system, and thereby cause an increased tendency to engage with stimuli.

1. Reisbick S, Neuringer M. Omega-3 fatty acid deficiency and behavior: a critical review and directions for future research. In: Yehuda S, Mostofsky DI, eds. *Handbook of Essential Fatty Acid Biology: Biochemistry, Physiology, and Behavioral Neurobiology*. Totowa, NJ: Humana Press, 1997: 397–425

2. Delion SS, Chalon S, Herault J, Guilloteau D, Besnard JC, Durand G. Chronic dietary α-linolenic acid deficiency alters dopaminergic and serotonergic neurotransmission in rats. *J Nutr* 1994; **124**: 2466–76.
3. Banks KE, Gratton A. Possible involvement of medial prefrontal cortex in amphetamine-induced sensitization of mesolimbic dopamine function. *Eur J Pharmacol* 1995; **282**: 157–67.
4. Sokolowski JD, Cousins MS, Nelson J, Salamone JD. Effects of medial prefrontal cortex dopamine antagonism on nucleus accumbens dopamine release and motor activity in the rat. *Soc Neurosci Abstr* 1994; **20**: 823.
5. Yamamoto N, Hashimoto A, Takemoto Y et al. Effects of the dietary alpha-linolenate/linoleate balance on brain lipid compositions and learning ability of rats. II. Discrimination process, extinction process and glycolipid compositions. *J Lipid Res* 1988; **29**: 1013–21.
6. Salamone JD, Cousins MS, McCullough LD, Carriero DL, Berkowitz RJ. Nucleus accumbens dopamine release increases during instrumental lever pressing for food but not free food consumption. *Pharmacol Biochem Behav* 1994; **49**: 25–31.
7. Wainwright PE, Huang YS, Bulman-Fleming B, Levesque S, McCutcheon D. Brain and behavioral effects of dietary n-3 deficiency in mice: a three generation study. *Dev Psychobiol* 1994; **27**: 467–87.
8. Frances H, Monier C, Bourre JM. Effects of dietary alpha-linolenic acid deficiency on neuromuscular and cognitive function in mice. *Life Sci* 1995; **57**: 1935–47.
9. Sokolowski JD, McCullough LD, Salamone JD. Effects of dopamine depletions in the medial prefrontal cortex on active avoidance and escape in the rat. *Brain Res* 1994; **651**: 293–9.
10. Bourre JM, Francois M, Youyou A et al. The effects of dietary alpha-linolenic acid on the composition of nerve membranes, enzymatic activity, amplitude of electrophysiological parameters, resistance to poisons and performance of learning tasks in rats. *J Nutr* 1989; **119**: 1880–92
11. Hooks MS, Kalivas PW. The role of mesoaccumbens-pallidal circuitry in novelty-induced behavioral activation. *Neuroscience* 1995; **64**: 587–97.
12. Williams GV, Rolls ET, Leonard CM, Stern C. Neuronal responses in the ventral striatum of the behaving macaque. *Behav Brain Res* 1993; **55**: 243–52.
13. Broderick PA, Phelan FT, Eng F, Wechsler RT. Ibocaine modulates cocaine responses which are altered due to environmental habituation: in vivo microvotammetric and behavioral studies. *Pharmacol Biochem Behav* 1994; **49**: 711–28.

AUTHORS' REPLY: A dopamine hypothesis is fine. It might even be correct. However, returning to the visual deprivation literature, virtually all neurotransmitter agonists tested were found to alter ocular dominance plasticity (see 1). These data suggested either that a common process was activated or that each agonist had fairly nonspecific actions. In either case, in a dynamic system like the developing nervous system, sorting out cause versus coincidence is a very difficult process, not often successfully accomplished. One of the reasons we all get paid the 'big bucks' is to perform such difficult experiments. For a variety of reasons (strange grouping of data, animals compared as individuals rather than litters, lack of saturation analysis, etc.), the Delion *et al.* (2) paper is a mighty thin reed on which to base a dopamine-receptor-mediated effect on behavior.

1. Shaw CA, Lanius RA, Van den Doel K. The origin of synaptic neuroplasticity: crucial molecules or a dynamical cascade? *Brain Res Rev* 1994; **19**: 241–63.
2. Delion S, Chalon S, Herault J, Guilloteau D, Besnard J-C, Durand G. Chronic dietary α-linolenic acid deficiency alters dopaminergic and serotoninergic neurotransmission in rats. *J Nutr* 1994; **124**: 2466–76.

**Lucas**: Shaw and McEachern depict the immense organizational complexity of the nervous system and hence the diverse ways in which a molecule such as DHA could impact on brain development to influence function. Such an impact theoretically could operate via effects on membrane fluidity that influence membrane proteins, including neurotransmitters, and in turn change receptor development, synaptic properties, and ultimately affect the functioning of neural networks and regions of the brain. The potentially explosive nature of such cascade effects is counterbalanced by compensating systems that help create stability and prevent lower order effects necessarily affecting the overall functioning of the brain in behavioral terms. It becomes, therefore, virtually impossible to predict the impact of a molecule on brain functioning, even if initial components of a cascade effect were to be detected, which would clearly be difficult in humans.

Thus Shaw and McEachern are drawn by their own logical construct to the conclusion that any understanding of the impact of a molecule like DHA on the brain must be through what the authors term a 'top-down' approach, which first identifies abnormalities in cognitive performance or behavior of the dietary intervention or deficiency, so that subsequent 'downward' research through levels of brain organization can be focused. This of course requires animal models and raises inevitable difficulties in finding appropriate models for human brain development, which the authors consider.

It is interesting, in historical terms, to note that mechanistic biologists now increasingly recognize the value of the pragmatic approach of demonstrating the outcome effect early on and then targeting the mechanistic studies in the light of the outcome. Clinical researchers in the therapeutic field have often used this approach. After a certain amount of mechanistic work – enough to generate a therapeutic hypothesis – and after safety studies, therapeutic outcome trials can be initiated. Often drugs have been shown to be efficacious and safe long before their detailed mode of action has been intensively explored. The use of beta-blockers for high blood pressure is a good example. Indeed this must be true for all drugs that were shown to be effective prior to the development of the 'new biology'.

In contrast, in biological sciences, and notably in the field of nutrition, attempted mechanistic studies have often preceded the demonstration of outcome effects. Thus, huge sums of money have been spent on investigating the mechanisms of presumed outcome phenomena that may turn out not to occur. In the field of LPUFAs and brain development, there are enough data to generate the hypothesis that dietary LPUFA supplementation in formula-fed infants could influence neurodevelopment.

I would strongly support the logical implication of the Shaw and McEachern chapter that further major investment in mechanistic research on LPUFAs should be made only as a consequence of clinically relevant, positive, outcome findings, which now require major experimental intervention studies in humans. Fortunately, a number of these studies are now well underway.

AUTHORS' REPLY: It might be useful to put our chapter into a broader perspective concerning molecules that may affect the development of the CNS. First, consider that the possible number of gene products in humans is something like $10^{30\,000}$ (1). These are only the proteins. Even with a conservative assumption that the number of proteins actually made is approximately 105, adding other molecules (e.g. lipids, amino acids), as well as a large number of molecular interactions, makes the potential number of crucial molecules very large. How many of these can be considered essential? This is not known, since (i) we don't yet know how many are expressed in the developing or mature nervous system, and (ii) 'essentiality' may be defined best by behavioral outcome. We note that major enzyme systems can be knocked out in transgenic animals with relatively selective effects. In some cases, surprisingly, there is no obvious effect (2). Thus, at least in such cases, compensation must have been provided, possibly by the presence and action of molecules of related function (e.g. protein kinase A for CaMKII). Just as clearly, this is not always the case: prenatal deficiencies of thyroid hormone (hypothryroidism) and postnatal deficiencies of phenylalanine hydroxylase (phenylketonuria) produce profound retardation in humans (3). The action of LCPUFAs may fall at any stage of a continuum from no effect to disastrous effect. Judged by this standard and the literature we have seen, LCPUFA-deficient diets in term infants would appear to have relatively minor effects, if any.

1. Kauffman SA. Antichaos and adaptation. *Sci Am* 1991; **Aug**: 78–84.
2. Silva AJ, Paylor R, Wehner JM, Tonegawa S. Impaired spatial learning in α-calcium-calmodulin kinase II mutant mice. *Science* 1992; **257**: 206–11.
3. Denhoff E, Feldman SA. *Developmental Disabilities*. New York: Marcel Dekker, 1981.


**Carlson**: Shorter look duration has been used to infer higher cognitive function. Dietary n-3 LCPUFA has been shown to increase early visual acuity in at least five randomized trials of preterm and term infants. It has also been shown to decrease look duration in 6.5- to 12-month-old preterm infants. Finally, look duration at 12 months has been correlated with IQ at 4 years. Colombo has suggested several mechanisms by which look duration might be affected, two of which I read to be sensory (Colombo, this volume). If look duration were mainly the result of the quality of sensory pathways, these examples would appear to illustrate your point that cognitive function requires the effective performance of the entire brain. On the other hand, in our studies look duration is significantly influenced by the years of parental formal education, and it is difficult for me to conceive how parental education could influence pathways for visual function.

Several of the chapters have emphasized cognitive function virtually to the exclusion of other types of CNS function that might also be important. Personally, I think that the effects of n-3 LCPUFAs are interesting, regardless of the neural domain they influence. Even those who imply that the addition of n-3 LCPUFAs to formula cannot be justified unless IQ is higher, should take to heart your message that less than optimal functioning in any neural domain has the potential to impact on the ability to learn or reason.


AUTHORS' REPLY: All you have demonstrated is that multiple processes (e.g. environmental and sensory) contribute to look duration, not that the 'health' of the sensory pathway is not involved. We completely agree that defects in the latter due to diet are certainly interesting and important, regardless of other areas of the CNS affected and can ultimately impact on 'higher' cognitive function. For example, Wainwright (this volume) mentions mice that swim poorly, and Neuringer and colleagues discuss increased reactivity to stimuli, alterations in species-specific behaviors.

CARLSON: The evidence does not seem to support your contention about critical periods for development, at least with regard to the influence of diet. Birch *et al.* (your reference 47) found that breast-fed infants had better stereoacuity at 3 years of age. We found shorter look duration 10 months after LCPUFA was discontinued. Couldn't LCPUFA provided before the peak period of developmental plasticity alter neural composition or impact behavior maturing sometime later? Your comments here seem to imply that the effects of n-3 LCPUFAs on the visual system can only occur concurrently with n-3 LCPUFA feeding. As such they ignore that these compounds increase membrane n-3 LCPUFA content, and are not necessarily readily removed.

AUTHORS' REPLY: There are many forms of neuroplasticity, each with distinct critical/sensitive periods. It is certainly possible that LCPUFA differences on the maturing nervous system will be detectable at some stages and not others. A similar statement can be made for any molecule that has the potential to affect CNS development. Support for your view would seem to necessitate experiments on older individuals. We look forward to seeing these data.

**Wainwright and Ward**: The authors of this paper provide an integrative overview of the various levels of neural and functional organization that interact in mediating putative dietary effects on behavioral outcomes. They advocate as an appropriate research strategy an integrated program that incorporates all these levels and which is based on a 'top-down' approach. Using this approach, inquiry into putative dietary effects on behavior begins with the use of a battery of sensitive behavioral measures to identify dietary effects on specific behavioral domains. The specific nature of the findings then contribute to the further development of mechanistic hypotheses, based on our current understanding of brain–behavior relationships. These hypotheses can then be used to direct inquiry at 'lower' neural and functional levels in the attempt to establish the chain of causal links.

The discussion in this paper provides cogent support for prescribing caution in the interpretation of data based on experiments conducted on unitary outcomes at only one or other level of organization. This is particularly true when the hypothesis of these experiments has not been informed by any theoretical perspective other than 'DHA levels change in the brain, and we should therefore expect to see functional effects'. Some of the points raised are of particular importance, and are therefore seen to be worth emphasizing.

Because of the complexities of neuronal organization, changes in membrane DHA concentration may be merely a correlate and not an immediate cause of any functional effect. For example, diets high in DHA raise membrane DHA levels, but also decrease AA and increase 22:5n-6. We have shown recently in mice that the behavioral effects associated with such changes are more likely attributable to tissue AA than DHA levels (1).

Just because DHA is found in breast milk, it does not automatically follow that this is either necessary or favorable for appropriate developmental outcome. Evolutionary strategies, as we understand them, appear to be opportunistic and based on historical contingencies. Thus, although the overall package may be optimal, not every component need be so. It is therefore important to entertain the possibility that levels of LCPUFA (e.g. DHA) found in breast milk may not in fact contribute to normal developmental outcomes, and that providing these artificially, independent of the other components of the package, may contribute to less than optimal outcomes. I think the information we now have supporting the importance of considering AA as well as DHA is an example of this.

Differences in behavioral function between groups of infants fed different diets need to be interpreted carefully in terms of which is 'better or worse'. Bigger is not always better.

1. Wainwright PE, Xing H-C, Mutsaers L, McCutcheon D, Kyle D. Arachidonic acid offsets the effects on mouse brain and behavior of a diet with a low (n-6):(n-3) ratio and very high levels of docosahexaenoic acid. *J Nutr* 1997; **127**: 184–93.

AUTHORS' REPLY: The context in which a given molecule operates is crucial. We have commented on this at length elsewhere (1). The basic point is that modern molecular biology has given us the ability to place molecules out of context to normal molecular constraints. For example, we might try to induce the expression of a class of receptor in a neuron at a stage when that receptor would not normally be present. However, in the absence of appropriate regulatory enzymes, such a receptor would not function properly. Similarly, nature appears to provide the developmental expression of particular receptor proteins before synaptic contacts have formed (2) or in the absence of other modulatory proteins. In the first example, the unregulated receptor would likely be more harmful than beneficial; in the second case, such receptors would simply be ineffective.

The idea that functional deficits may be apparent or exaggerated only under stressful or challenging conditions is one worth considering. Examples are seen at molecular/neural and behavioral levels. An example from the neuroplasticity literature demonstrates the former where, despite normal responses to instantaneous or baseline stimuli, regulatory function may be impaired (3). Single gene deletions are currently a popular method used to assess the effect of loss of a particular enzyme system on long-term potentiation (LTP), an artificially induced measure of synaptic efficacy. Mice that developed without the *fyn* gene (encoding a tyrosine kinase) had LTP comparable to controls except when a low-threshold stimulus was used, at which point their impairment became significant (4).

A related example is the $\alpha$-calcium calmodulin kinase II ($\alpha$-CaMKII) genetic mutant, which appeared normal under particular electrophysiological tests, but which experienced severe seizures in response to a single, normally subconvulsive kindling stimulus (5). At the behavioral level, there are countless examples of deficits that become obvious under emotionally stressful test conditions, pharmacological challenge (e.g. 6), advancing age, or neural pathology due to disease. Broad-based testing ideally should reveal whether specific manipulations and insults, alone or in combination, have been functionally compensated or rather have brought the organism closer to a threshold for dysfunction.

These points also suggest that we beware summation of stressors. It is well known that variables, including sensory input, nutrition, environment, socio-economic status, etc., all can affect neural development and behavioral outcome. While a deficit in one of these areas may not alone be significant, in combination they may surpass a threshold for tolerance, particularly in the face of additional stressors such as those mentioned above.

Observations from the $\alpha$-CaMKII genetic mutants suggest that kindling might constitute a valuable neural and behavioural level 'challenge test' as part of a wide-spectrum assessment of functional outcomes.

1. Shaw CA, Lanius RA, Van den Doel K. The origin of synaptic neuroplasticity: crucial molecules or a dynamical cascade? *Brain Res Rev* 1994; **19**: 241–63.
2. Shaw C, Cameron L, March D, Cynder M, Zielinski B, Hendrickson A. Pre- and postnatal development of GABA receptors in Macaca monkey visual cortex. *J Neurosci* 1991; **11**: 3943–59.
3. McEachern JC, Shaw CA. An alternative to the LTP orthodoxy: a plasticity–pathology continuum model. *Brain Res Rev* 1996; **22**: 51–92.
4. Grant SCN, O'Dell TJ, Karl KA, Stein PL, Soriano P, Kandel ER. Impaired long-term potentiation, spatial learning and hippocampal development in fyn mutant mice. *Science* 1992; **258**: 1903–5.
5. Silva AJ, Paylor R, Wehner JM, Tonegawa S. Impaired spatial learning in α-calcium calmodulin kinase II mutant mice. *Science* 1992; **257**: 206–11.
6. Saucier D, Cain DP. Spatial learning without NMDA receptor-dependent long-term potentiation. *Nature* 1995; **378**: 186–8.

**Mayer and Dobson**: The authors argue effectively against a simplistic view of the effect of nutritional deficiency and other perturbations on neuronal and behavioral development. They point out that selective effects on neural/behavioral function may occur due to specific environmental modifications, and raise the possibility that enhanced function in one domain may be accompanied by reduced function in another. These points and others argue for investigation of the interrelationship between infants' performance on different functional measures. For example, within the visual sensory domain, one would like to know whether electrophysiological measures of visual function (retinal and cortical) and behavioral measures of acuity in infants are correlated. One might predict no correlation of acuity with flash electroretinograms, but positive correlation of behavioral acuity with visually evoked potential acuity. Regarding the broader domain of visual processing, Carlson reported in her chapter no correlation between behavioral visual acuities at 2 months and later performance on a visual novelty preference task at 12 months, suggesting a dissociation between these functions, or that acuity deficits in the early months are transient. Of course, correlation analysis of these and other data gathered in nutrition studies may be limited by restricted range of values in the variables.

In Section 4, the authors discuss how visual measures relate to cognition and evaluate a 'visual deprivation model for neural/behavioral effects of fatty acid deficiencies', asserting the importance of a visual sensitive period.

See our comment on Innis' chapter, for a somewhat different view of the nature and timing of sensitive periods.

We suggest, contrary to the authors' assertion, that most investigators have not assumed that cognitive function is reflected by visual acuity measures, whether behavioral or electrophysiological, and certainly cognitive function is not reflected by electroretinographic responses to light. The pattern-evoked, cortical VEP was undoubtedly used in nutrition studies because this measure does not require oculomotor responses and is thought to be more sensitive to central retinal abnormalities than are behavioral tests of infants' visual acuity. The function(s) assessed with novelty preference paired-comparison tests is (are) assumed to reflect higher-level, cognitive processes, or at least, more complex attentional processes than are required in tests of behavioral visual acuity. (However, we also think that some of the same visual search behaviors are elicited by both procedures; see our comment on McCall and Mash's chapter.)

We agree with the authors that a model of amblyopiagenesis due to monocular deprivation is inappropriate for nutritional studies, as the effect clearly must be upon both retinas. Dark rearing conditions would seem to be a more appropriate deprivation model, as the authors discuss. However, in both types of visual deprivation, effects on retinal function are not seen. Does not a model of nutritional deficiency have to consider the effects at these multiple levels of the visual system?

Moreover, receptor systems generally may be affected in nutritional deficiencies. This brings up a question for investigators in this area concerning possible effects of fatty acid deficiency on other sensory receptor systems, for example, somatosensory and auditory. Have other sensory receptors been assayed for fatty acids, and have other sensory systems been evaluated in animal models of nutritional deficiencies? Habituation can be measured in primate infants using other orienting responses than visual orienting.

The meaning is unclear of a crucial statement regarding lack of knowledge regarding underlying substrates of kindling, ending in '. . . the only relationship to cognitive learning/memory that could legitimately be claimed is as a misuse of neuroplasticity mechanisms in the brain'.

The authors compare the findings of Birch *et al.* (1) from 3-year-old children and the Innis study of 9-month-olds (2) and conclude that there is no evidence for a permanent effect on visual function, 'especially when the latter is taken as a measure of central cognitive ability'. Outcome at age 9 months on a single behavioral measure of cognitive development is hardly evidence for lack of permanent effects. Moreover, it is a stretch to conclude that the measures on which Birch and colleagues found significant differences between corn oil formula and breast milk groups reflect cognition: stereoacuity is a measure of binocular function and letter matching at age 3 years may be strongly influenced by socio-economic/parental factors. Note that Birch *et al.* also found no differences between groups in visual recognition acuity, just as Innis *et al.* showed no relationship between duration of breast-feeding and grating acuity.

1. Birch EE, Birch D, Hoffman D, Hale L, Everett M, Uauay R. Breast-feeding and optimal visual development. *J Pediatr Ophthalmol Strabismus* 1993; **30**: 33–8.
2. Innis SM, Nelson CM, Lwanga D, Rioux FM, Waslen P. Feeding formula without arachidonic acid and docosahexaenoic acid has no effect on preferential looking acuity or recognition memory in healthy full-term infants at 9 mo of age. *Am J Clin Nutr* 1996; **64**: 40–6.

AUTHORS' REPLY: First, the terms 'critical' and 'sensitive' periods could probably be used more precisely (see (1) for a complete discussion). Nevertheless, critical periods in vision, at least in animal plasticity studies (2), have conventionally referred to a discrete window in early postnatal development during which time visual system morphology and neural response properties are modifiable by sensory input. These plasticity studies appear to be cited in the nutrition literature (reported in this volume and elsewhere) to support the possibility that early events can have long-term effects on the nervous system. It is certainly true that various forms of visual deprivation have severe consequences in the visual system and it is certainly possible that dietary effects may have similar long-term results. It is, perhaps, only our misreading of the intentions of some in the field, but it seems to us that the analogy that is attempted is inappropriate, for reasons discussed below. The only shared aspect is the use of the visual pathway.

Visual deprivation in early postnatal life can have a severe impact on the development of the visual system with dramatic consequences for animal behavior. In particular, major deficits in acuity and stereopsis are often observed. As cited by Mayer and Dobson, unilateral cataract can be extremely damaging to long-term visual capability if left untreated in the first few months of life. However, cataract is an extreme form of visual deprivation, involving, among other things, competition between visual afferents from the two eyes. Other forms of deprivation appear to have much later onsets for their effects (3). In animal models of visual system plasticity (animals with stereoscopic vision such as cats and monkeys), visual deprivation has relatively little effect at or near birth (2,4). Nothing that we have seen from the dietary manipulation literature shows outcomes that are comparable to those seen in most cases of visual deprivation, particularly monocular deprivations. Some investigators in the early diet literature frequently cite a series of studies by Movshon *et al.* using macaque monkey. These studies purportedly demonstrate that visual deprivation effects can begin near birth and be life-long in their impact (5–7). In fact, the Movshon *et al.* studies do not show that unilateral visual deprivation effects are present from birth, merely that the deprivation paradigm began at that time. The timing of these experiments appears to result from a cautious approach designed not to miss the critical period entirely, an approach that precludes precise determination of this period. On the contrary, to cite one example, transient ERG differences were reported in rat following unilateral eyelid suture (8). Yes, indeed a model of nutritional deficiency does have to consider effects at multiple levels of the measured pathway. Animal studies by both Neuringer and colleagues (9) and Bourre and colleagues (10) in primates and rats, respectively, have attempted such cross-level analyses.

This is a good question (and we don't know the answer). Given that individual molecules generally have multiple functions and interactions during development, we would predict the answer to be 'yes'. Further, from the perspective of different types of developmental molecular insults, e.g. Fetal Alcohol Syndrome studies, it would appear likely that other systems would be affected as well (11).

1. Bornstein MH. Sensitive periods in development: structural characteristics and causal interpretations. *Psychol Bull* 1989; **105**: 179–97.
2. Hubel DH, Wiesel TN. The period of susceptibility to the physiological effects of unilateral eye closure in kittens. *J Physiol* 1970; **206**: 419–36.
3. Daw NW. *Visual Development*. New York: Plenum Press, 1995: 146–52.
4. March DF, Goodwin H, Shaw C, Hendrickson A. Visual deprivation alters GABAA, but not NMDA receptor distribution in macaque visual cortex. *Soc Neurosci Abstr* 1993; **19**: 1800.
5. Kiorpes L, Boothe RG, Hendrickson AE, Movshon JA, Eggers HM, Gizzi MS. Effects of early unilateral blur on the Macaque's visual system. I. Behavioral observations. *J Neurosci* 1987; **7**: 1318–26.
6. Hendrickson AE, Movshon JA, Eggers HM, Gizzi MS, Boothe RG, Kiorpes L. Effects of early unilateral blue on the Macaque's visual system. II. Anatomical observations. *J Neurosci* 1987; **7**: 1327–39.
7. Movshon JA, Eggers HM, Gizzi MS, Hendrickson AE, Kiorpes L, Boothe RG. Effects of early unilateral blur on the Macaque's visual system. III. Physiological observations. *J Neurosci* 1987; **7**: 1340–51.
8. Yinon U, Shaw C, Auerbach E. Retinal and cortical changes in the visual system of pattern deprived rats. *Adv Behav Biol* 1975: 42–53.
9. Reisbick S, Neuringer M, Connor WE. Effects of n-3 fatty acid deficiency in

nonhuman primates. In: Bindels JG, Goedhart AC, Visser KHA, eds. *Recent Developments in Infant Nutrition*. Dordrecht: Kluwer, 1996: 157–72.
10. Bourre J-M, Francois M, Youyou A *et al*. The effects of dietary α-linolenic acid on the composition of nerve membranes, enzymatic activity, amplitude of electrophysiological parameters, resistance to poisons and performance of learning tasks in rats. *J Nutr* 1989; **119**: 15–22.
11. West JR. *Alcohol and Brain Development*, New York: Oxford University Press, 1986.

# Appendices: General Commentary on Behavioural Science Implications and Methodology

# Appendix A
# Nutrition and Development – Observations and Implications*

MARC H. BORNSTEIN

*Child and Family Research, NICHD, Building 31 – Room B2B15, 9000 Rockville Pike, Bethesda, MD 20892–2030, USA*

## LCPUFA in Human Infant Studies

Apparently, no studies so far exist that *reliably* show that visual or other aspects of neural or behavioral development in healthy term infants fed human milk, in comparison with formula-fed infants with or without arachidonic acid (AA) or docosahexaenoic acid (DHA), manifest any practically or statistically significant differences in the common dependent criteria of visual acuity or cognition. Any interpretation of this literature that favors a difference between the two groups may be 'overgenerous,' committing a Type I error (supporting a hypothetical relation where none exists).

### Differences in Behavior?

Auestad *et al.* (1) found essentially no differences in looking acuity or visually-evoked potential acuity between 2 and 12 months of age in a

---

* Dr Bornstein wrote this paper after seeing all ten chapters in this book. It was not discussed at the Workshop (see Preface).

multicenter prospective study concerned with the effects of feeding formula (with no AA or DHA or with AA and DHA).

Carlson et al. (2) found differences between long-chain polyunsaturated fatty acid (LCPUFA) deficient and normal groups of infants in acuity at 2 months, but no differences at 4 months or between 4 months and 12 months, suggesting that: (i) no differences ever existed between the groups; (ii) infants in the LCPUFA-deficient group 'caught up' in looking acuity between 2 and 4 months; or (iii) consistent differences exist, but the study was insufficiently powered ($n < 20$ infants per group) to detect them. Carlson et al. also studied infant performance on the Fagan Test of Infant Intelligence (FTII) (3) and, according to McCall and Mash (this volume), found no differences in infants randomly assigned to fish oil 1 or 2 on number of looks or average length per look during familiarization; DHA-supplemented infants displayed a greater number of looks and a shorter average duration look during a test phase following familiarization; but no differences emerged between supplemented and nonsupplemented infants in recovery (novelty preference) scores at 9 months. (DHA-supplemented infants displayed less preference for novelty at 12 months.)

Innis (4) found no differences in breast- versus bottle-fed infants in acuity at 14 days or 90 days, despite lower blood lipid levels of AA and DHA in formula-fed infants. In a follow-up multicenter study, Innis et al. (5) involved a larger sample size, but again found no differences in acuity at 3 months between breast-fed and formula-fed infants, even though plasma and red cell phospholipid levels of AA and DHA were lower in infants fed formula than in those breast-fed. In a larger scale study still ($n = 400$ term infants), Innis et al. (6) found no differences in looking acuity between infants fed formula from birth (never breast-fed) and those who were breast-fed to 39 weeks (never bottle-fed), the extremes of the sample. No relation emerged between the duration of breast-feeding and acuity either. Innis et al. also studied novelty preferences by the FTII in these same babies at 39 weeks postnatal, and again found no differences between breast-fed and bottle-fed infants or among infants who were breast-fed for different durations.

Mayer and Dobson (this volume) report that three prospective studies of the presumed effects of diet in term infants produced no significant differences among several dietary conditions. In one, 'dietary supplementation resulted in decreased acuity in a group of tiny, high-risk infants with bronchopulmonary dysplasia (BPD).' Similarly, Janowsky and Rose (7) reported no differences in Bayley 12-month mental development index (MDI) or psychomotor development index (PDI), but a DHA-supplemented group had lower McCarthy CDI scores at 14 months than did breast-fed infants, and plasma and red blood cell levels of DHA correlated negatively with vocabulary production and comprehension in both formula- and breast-fed groups, suggesting that supplementing term infants may not produce general developmental benefits, or it may produce a disproportionate DHA/AA ratio that could lead to poor cognitive outcomes (e.g. in vocabulary).

McCall and Mash (this volume) conclude optimistically that, given certain controls, 'the preponderance of evidence seems to favor an advantage to breast-feeding over and above differences associated with maternal education, SES, and birth circumstances.' In support of this conclusion, McCall and Mash focus on a study conducted in Great Britain by Lucas, Morley, and co-workers (8) and review carefully its design and conclusions. At each of three outcome ages (9 months, 18 months, and 7.5–8 years), infants and children of mothers who chose to provide milk relative to those of mothers who chose not to provide milk had higher developmental scores or IQs, and differences maintained, but only by one-half, when educational and socio-economic status (SES) variation in mothers and birth weight, gestational age, and days on ventilation in infants, were covaried. Nonetheless, even McCall and Mash conclude, it is logically necessary to covary all independent factors associated with mothers' choice to eliminate social factors as explanatory and to focus on human milk versus formula as the causal agent, and 'this is nearly impossible to do'. For example, Jacobson and Jacobson (9), studying term infants assessed at 4 years of age, showed that mothers' choice to breast-feed exercised an advantage for children, after covarying maternal social class and education; but when maternal IQ (a more proximal measure of mothers' intellectual capability) and the HOME Scale (an index of the stimulating quality of the home) were added to the regression equation, the benefits of maternal choice/human milk attenuated. McCall and Mash also recount that results across outcome ages were consistent in showing no advantage in developmental score or IQ for infants who were randomly assigned to donor human milk versus those assigned to formula, consistent with the previous results that other mediating home factors (proximal or distal) may be carrying any predictive associations.

More recent and perhaps promising studies mentioned in passing, including those by Willatts and Innis *et al.*, require closer examination than possible at this writing.

## Differences at Autopsy?

Although problems are frequently encountered in analyses of autopsy samples, including the normally small number of infants studied, differences in the ages of 'treatment' and 'control' groups, confounding effects of the circumstances surrounding death, time delays due to tissue recovery, and collection of dietary information (Innis, this volume), autopsy studies show no differences in AA in brain or DHA or AA in retinae of infants who had been fed formula compared to infants who had been breast-fed (10-13).

*Differences, but Recovery?*

In the human nervous system, even small differences (presumably) caused by dietary deficiencies may propagate to affect behavior at a global level. However, they may also be accommodated by (i) redundant systems, (ii) the enormous plasticity of compensatory learning, or (iii) later recovery. According to Wainwright and Ward (this volume), dietary n-3 deficiency during early development may have detrimental, but only transient effects on visual development. So, for example, deficits in rod electroretinogram recordings attributed to dietary deficiencies, apparent at 36 weeks, are no longer apparent at 56 weeks (14). Similarly, in terms of catch up, Birch *et al.* (15) tested 43 infants who were fed a controlled diet to 12 months and tested at 36 months. Although at 4 months, breast-fed infants had significantly higher mean visual-evoked potential acuity and forced-choice preferential looking acuity than infants fed formula (the number of infants tested and dispersions of the measures were not given), when tested at 36 months of age (same subjects?) no differences obtained.

In this regard, it could be, as Innis (this volume) points out, that the absence of dietary intake of LCPUFAs affects neural development only in the presence of other environmental deprivations 'which either prevent compensation or compound the insult'.

*Summary*

Shaw and McEachern (this volume) take for granted what many others question in terms of functional differences between breast-fed and formula-fed babies. They argue that electroretinograms, visually-evoked potentials, and visual acuity are affected by diet, as is IQ. By contrast, Wainwright and Ward (this volume) are very cautious in concluding that nutrition and behavior are integrally and causally linked. According to Lucas (this volume), 'A clear case of the superiority of breast milk in promoting short-term development and visual function cannot be made'. The general results in this field, contra Innis, so far fail to 'endorse' any approach to modify infant formula to attain functional differences in children. Indeed, functional differences in humans, as well as potential mechanisms of action, are essentially unproved and unknown. Certainly, there is little reason to believe thus far that, controlling for other pertinent factors, LCPUFAs exert any demonstrably positive effects on developmental outcomes in the normal term population; their effects in the preterm populations are still undetermined (McCall and Mash, this volume). Significantly, this literature (as described) is troubled by methodological problems, notably surrounding specification of the independent and dependent variables.

## Specification of Independent Variables

### Basic Assumptions

Lucas (this volume) makes the case for a biological connection between LCPUFAs and retinal and brain (leading to cognitive) development. However, what are the implications of Innis's assertion that 'the brain and retina, in contrast to other tissues, are highly resistant to changes in fatty acid composition, despite wide variations in circulating fatty acids'? Apparently, the brain and retina maintain fatty acid composition despite varying dietary intakes (partly explained by efficient conservation as well as highly selected pathways of n-6 and n-3 fatty acid uptake). Moreover, 'LCPUFA concentration in the circulation is at least an uncertain and unreliable index of visual or cognitive function' (Lucas, this volume).

Furthermore, linoleic and α-linolenic acids may not be the 'active ingredient' in dietary effects (see Lucas, this volume). As Innis argues, 'dietary AA and DHA may not be necessary directly because dietary linoleic and α-linolenic acids can be converted to AA and DHA and so fulfill tissue needs for n-6 and n-3 fatty acids during growth and development'. And 'there are no studies to show that adding AA or DHA acid to diets containing adequate linoleic or α-linolenic acid confers any benefit to visual or other neurological functions of any mammal, with the exception of felinae'. There may be reason to believe that inadequate supplies of n-6 and n-3 fatty acids during development affect the central nervous system deleteriously, but no studies so far at the biochemical level indicate whether AA or DHA are significant in themselves rather than the precursors linoleic acid or α-linolenic acid (see Innis, this volume).

### Composition of Human Milk

Innis describes general distrust of specification of levels of DHA and other relevant LCPUFAs, as well as plasma or red blood cell level, as effective independent variables. The composition of actual human milk needs to be analyzed and reported in appropriate comparisons. The composition of human milk fatty acids appears to vary from country to country. Indeed, levels of linoleic acid in human milk in North America (alone) appear to have changed over the last 40 years (from 10% to 16% fatty acids). A key issue, as McCall and Mash note, is that human milk contains a variety of substances (not just LCPUFAs) that are not contained in formula, and reciprocally formula contains substances not found in breast milk, any of which could potentially contribute to any differences between breast-fed and formula-fed infants.

Moreover, mild-to-moderate malnutrition, poor quality diet, and sporadic food shortages might affect children's development, but are difficult to study for several reasons. Serious malnutrition can be identified by the physical condition of the child, but mild-to-moderate malnutrition relies on recall of food consumption, judgments about what should be recorded, as well as self-report measures, and information on feeding patterns, typically collected by questionnaire, is, as Innis (this volume) notes, notoriously unreliable. For example, the duration of breast-feeding may be an important variable in the detection of differences between breast-fed and formula-fed infants (breast-feeding is often discontinued or supplemented after 3 months, whereas assessments continue to later ages).

## 'Background Variables'

Most studies do not account for SES differences, ethnic differences, and gender differences among experimental and control groups. 'It appears to be an inherent assumption [in this line of work] that the measures of visual function used will provide information about the development of the infant's visual system, with a little or no effect of potential confounding variables, such as home environment, birth order, or ethnic background . . . [however] the assumption that confounding variables are unimportant has neither been established nor considered in any study reported to date' (Innis, this volume). So, for example, alcohol and maternal smoking have negative effects on child neurodevelopment (16) and exist in an inverse association with DHA (17); thus, correlations of DHA and infant development may simply be a 'marker' of other effective maternal pre- or postnatal variables which may (independently) affect both infant lipids and neurodevelopment.

Where differences between breast- or supplement/formula-fed babies have been detected, they are small, and may be accounted for by multiple other genetic and/or environmental differences between groups. Results are subject to numerous other methodological critiques as well (see the section on procedural and methodological issues below). Twentieth-century women who breast-feed may engage in other behaviors and create other environments which covary with breast-feeding and affect cognitive development in their children. Studies in which expressed mother's milk is delivered via bottle would unconfound some types of mother–infant interactions with mothers' milk content. Lucas (this volume) cites a study in which preterm babies were fed by nasogastric tube, so that breast-feeding per se was ruled out. But it is known that mothers who have children in such dire straits frequently compensate in parenting, providing additional kinds of stimulation (18) that could benefit long-term outcomes; therefore, 'cumulative' and 'contemporary' effects models of development (see below) could still hold. Singer (this volume) underscored the several shortcomings and cautions associated with any 'pure' assessment of infant performance by including a discussion of 'confounders' as well as procedural and ancillary considerations.

Consider studies that pit nutritional supplementation against parenting of various forms. Ramey *et al.* (19) gave one group of failure-to-thrive infants a nutritional supplement and a comparable group a nutritional supplement plus weekly response-contingent stimulation. The latter group performed better later on a vocal conditioning task. Experimental study supports the hypothesis that a postnatal environment which is nourishing and stimulating can remedy some deficits caused by prenatal malnutrition. Zeskind and Ramey (20) adopted the hypothesis that the withdrawn and irritable dispositions of malnourished infants would 'put off' their caregivers, thus leading to a non-supportive environment and a decline in intellectual performance (secondary effects). They assigned the babies to two environments that differed in their intellectually supportive characteristics. Both groups received nutritional supplements and healthcare, but one group also received cognitive and social stimulation at a special daycare program. This investigation found a comparative recovery in children who experienced cognitive and social stimulation in addition to nutritional and medical care, but continuing detrimental effects on intellectual, behavioral, and social development in children who received only nutrition and medical care. As predicted, mothers of children in the enriched environment became increasingly responsive to their children over time, whereas mothers in the nutrition/healthcare only group became less so. The intellectually supportive environment helped to break the cycle of apathy, carrying the clear message that biology alone is not determinative of ultimate development, and some forms of subsequent remediation can compensate for early dietary deprivation.

In a related study (21), low SES infants born to 10th grade education, single, teenage mothers and placed at about 3 months of age into the nurturing and intellectually stimulating North Carolina Abecedarian daycare program differed significantly in IQ at 36 months from control infants reared at home and provided with free formula and pediatric services. Mothers of the control infants were less interactive with their babies as they grew. IQ scores of the program infants were normal at 36 months in comparison with the control infants who score below normal. Notably, in the daycare group, even intrauterine growth-retarded infants scored about average; in contrast, fetally growth-retarded infants in the control group were already performing in the retarded range by 36 months, despite program provision of free nutritional supplements and pediatric care from birth onward. These cognitive disparities were not solely due to the high quality of daycare, but also surely partly attributable to the quality of maternal–infant interaction in single parent, low SES families. Of critical importance were the child care services of the center and the role models that the caregivers provided. Children were directly influenced by the quality care they received. In turn, the quality of parental relationships with the child was influenced by center support for the family and positive adult role models provided. Both processes enhanced child outcomes.

## Models of Experience and 'Critical Periods'

Most authors argue (implicitly or explicitly) that 'early experience' affects functions in the domain of dietary influence of child development where the presence or absence of a particular experience (dietary component) at a particular time in the life cycle (infancy) exerts an extraordinary and dramatic influence over structure or function well beyond that point in development. However, different models of developmental effects of diet could apply, and the attribution of an early experience effect (so-called sensitive or critical period) needs actually to meet several criteria.

*Correspondence* describes consistency in the rank-order status of specific experiences and child outcomes, as indexed by correlation and regression; correspondence may be concurrent or predictive. Concurrent correspondence describes an association between exogenous experience and performance in children at the same time. Predictive correspondence describes an association between an exogenous experience at one time and child performance at a later time. Several additional models of predictive correspondences, indexed by hierarchical regression, define unique effects of experience on the child (22–24). In a simultaneous relations model, an experience affects the child at only one point in development, independent of stability in the child and independent of earlier effects of that experience. In a lagged effects model, an experience uniquely affects the child from an early point in time to a later one, independent of stability in the child and independent of later relations between that experience and the child. Three additional models examine experience effects through time. In an average effects model, early and later experience influences are combined, and their mean exerts a significant effect on the child independent of stability in the child. In a developmental change model, increments (or decrements) in experience from early to later points in time exert a significant effect on the child independent of stability in the child. Finally, in a consistency effects model, the degree to which experiences maintain their level from early to later points in time exerts a significant effect on the child independent of stability in the child. To differentiate among different models of experience effects, it is necessary to measure the child as well as pertinent early and late experiences. Increasing evidence suggests that *specific* experiences (nutrition, parental activities) relate concurrently and, perhaps, predictively to *specific* aspects of child development.

Early experience or lagged effects models are often associated with the concept of 'sensitive' or 'critical periods,' and sensitive periods are thought to be widespread in animal and in human neurobiology and psychology. But sensitive periods have multiple requirements for attribution (25). Attribution of a sensitive period needs minimally to include information about what its structural characteristics are, as well as an interpretation of its causes, including why the sensitive period arises in terms of the natural history of the species and how the sensitive period is regulated in terms of physical, physiological, and psychological processes. In actuality, sensitive periods are comprehensively described by at least 14 distinguishing, essential, and operationally definable structural characteristics. A comprehensive statement about a sensitive period ought ideally to include:

(a) Information about temporal and intensive contours:
 (i) when and how often in the life cycle the sensitive period occurs,
 (ii) the rise of the sensitive period,
 (iii) the decay of the sensitive period,
 (iv) the window of the sensitive period, and
 (v) how sensitivity changes and the stability of sensitivity.
(b) Information about its mechanisms of change:
 (vi) the nature and origins of the effective experience,
 (vii) what structure or function changes, and
 (viii) the channels by which experience affects the structure or function that changes.

(c) Information about its consequences:
  (ix) the outcome in later development,
  (x) how the outcome is effected,
  (xi) when and under what circumstances the outcome occurs, and
  (xii) how long the outcome lasts.

Beyond these defining parameters of the sensitive period and its consequences, information about two evolutionary and ontogenic considerations of sensitive periods is also helpful-to-requisite:

  (xiii) individual and species variation in sensitive-period characteristics, and
  (xiv) the modifiability of the individual parameters of the sensitive period.

Thus, 14 conceptually distinct structural characteristics arrayed in four groups can be identified as inhering in the sensitive period. Researchers view the quantitative delineation of some or all of these parameters to be a primary goal of studying sensitive periods. The descriptive specification of these characteristics constitutes a first-order achievement of sensitive-period study; additionally, it is critical to illuminating causes underlying sensitive-period phenomena.

On an early experience or sensitive period model in nutritional supplementation, therefore, one would want to know, among other things: when and how in the life cycle of the human being a sensitive period to LCPUFAs occurs; its rise and decay times; how long it lasts, how sensitivity changes and whether the level of sensitivity is stable or fluctuating; the specific nature of the nutritional components that are effective in the sensitive period; what structures and/or functions change in relation to those nutritional components; the channels by which nutrition affects those structures or functions; the specific outcomes(s) in development of nutritionally affected changes; how the specific outcomes were affected; when and under what circumstances those effects took hold and how long they endured; individual variation in sensitive periods related to nutrition effects; and the modifiability of the foregoing parameters.

## Specification of Dependent Variables

Two sets of dependent measures have thus far been the principal target functional areas of comparisons of breast- and formula-fed comparison or

LCPUFA-supplementation in infants: visual system function and mental development in infancy and childhood. However, several authors point out: (i) research and logic concerning biochemical or neurological mechanisms connecting LCPUFAs with these two specific target areas are wanting; (ii) each of these target areas is plagued by its own 'problems'; and (iii) few other target outcomes have been delimited, hypothesized, or extensively examined in this literature (see below). Which physiological systems may be sensitive to deficiency and which may benefit from supplementation of LCPUFAs is described by Innis (this volume). The possible importance of AA is thought to lie in biochemical and physiological processes in the central nervous system (CNS) related to intra- and intercellular messengers for the action of a variety of hormones, neurotransmitters, and growth factors on neural and non-neural cells. However, as Innis makes clear, no mechanisms of action have been described at the biochemical level. Without explicit theoretical connections or expectations between diet and acuity or mental growth hypothesized, researchers risk assuming the guise of the proverbial drunk searching for his keys under the street light, because that's where the light is . . . and not where he dropped his keys.

Furthermore, Shaw and McEachern (this volume) assert that 'If cognitive function is affected, from the perspective of the organism it is to some extent irrelevant where along the pathway the problem occurs (i.e. in a sensory input pathway, in more central processing pathways like hippocampus or cortex, or elsewhere).' However, it may be valuable to know where along the pathway the problem occurs if a solution is to be targeted to the problem origin.

Singer and others review the standards and requirements that must be followed to ensure psychometric adequacy of assessments, and they bear repeating here: development and publication of standardized equipment, stimuli, and test items; administration and scoring procedures for test items; training requirements for examiners; development of tasks on a representative, normative population; and provision of normative, reliability (intra-rater or test–retest reliability, chronological age stabilities, inter-rater agreements, and internal consistency), and validity data (derived in part as a function of a number of established relations documented for a particular assessment over time along a variety of parameters). Thus, Jorgensen et al. (26) reported the development of looking acuity in 17 breast-fed differed from 16 formula-fed infants in prospective but not blinded study: experimenter awareness and power factors alone are sufficient to call the results of this study into question.

Published studies have also focused on a limited range of outcomes including, as Lucas notes, neural development, visual function, and 'growth'. However, LCPUFAs presumably have multiple influences on multiple different systems, and the beneficial versus deleterious effects of LCPUFA supplementation on other systems remain to be explored.

### Visual System Function in Infancy: Rationale, Results, and Reservations

DHA is thought to have a specific functional role in visual and perhaps other neural processes which extend beyond contributing to the physical properties of retinal or synaptic terminal membranes (Innis, this volume). The retina is rich in DHA, and much of the evidence for a functional role of DHA in nervous tissue comes from research into n-3 deficiency in retinal function. Therefore, researchers who wish to learn about the role of DHA in the CNS have chosen retinal function as a potentially informative criterion, and experiments that test the vision of human infants fed various dietary fats in formula populate the literature (see Mayer and Dobson, and Wainwright and Ward, this volume).

The acuity card procedure (ACP) has, according to Mayer and Dobson (this volume), 'become a major outcome measure of the effects of dietary fatty acids on infant development'. Insofar as visual acuity has been a dependent measure in nutritional studies, and the ACP is the pre-eminent measurement strategy for visual acuity in infants, it is reasonable to review the history, methodology, and psychometric characteristics (validity and reliability) of the ACP. Study in the nutritional field normally limits itself to the first year of postnatal life. The ACP, as measure of grating acuity (only one kind of visual acuity, grating acuity is the ability to resolve a pattern consisting of repeating black and white stripes), may have many possible sources of variance which the authors try to decompose. The authors summarize the detailed history and the psychometric adequacy (as well as inadequacies) of the ACP, and describe in important detail standard clinical procedures for use of the ACP. Content, construct, and criterion validity measures of the ACP are also described. The ACP has advantages in that it allows rapid assessment of visual acuity in infants. In the ACP, the infants' eye and head movements indicate the infant's ability to resolve the stripes and the grating, and therefore is inferential. The visual evoked potential (VEP) has also been used on a limited basis to study effects of nutrition on grating acuity in infants. Mayer and Dobson make several recommendations for improving the use of ACP in dietary studies, basically boiling down to reducing methodological and other sources of error variance in order better to distinguish signal from noise.

Some specific and general questions that arise concerning reliance on visual acuity as an outcome measure of nutritional manipulation include the following:

(i) Is the conceptual connection between diet and acuity sufficiently strong to promote acuity to such a singularly prominent place in the firmament of possible outcome variables of human diet? What physiological mechanisms would mediate a (presumed) deterioration in acuity as reflective of nutritional status? Is acuity permanently or only transiently affected by variations in nutritional status? of what nutrients? deprivation when in the course of prenatal and postnatal development?

(ii) Is grating acuity the measure of choice? A key consideration (and perhaps limitation) from a developmental point of view is that 'data from older children and adults, who can be tested using both grating acuity and recognition acuity tests, indicate that grating acuity is not necessarily identical to recognition acuity, and that the relation between the two is complex and depends upon the visual pathology.' Recognition acuity is 'the type of acuity that is most frequently measured in adults' (Mayer and Dobson, this volume).

(iii) Is the procedure plagued by too many intrinsic problems? For example, although tester differences are not large, testers themselves do show 'systematic differences' (i.e. holding child and procedure constant, different testers give different patterns of results). Notably, the 'ACP relies on the subjective judgment of the tester', and testers may be aware, for having seen them, of the nutritional status of children: 'One group of investigators noted that it was *not* possible to mask the tester as to whether an infant was breast-fed or bottle-fed' (emphasis added; Mayer and Dobson, this volume).

(iv) Is the measure sufficiently sensitive? Table 1 in the Mayer and Dobson chapter shows 'the large range of mean acuity values published for infants at each age from birth through 1 year'. This means that there is substantial individual variability in infant visual acuity in the first postnatal year, and, as the authors describe, sources of that variability are multivariate themselves. Moreover, infants may have normal acuity, but because of other factors (e.g. their attentional or health state) provide low acuity values. Thus, nutrition affects health, and health may affect visual acuity, and nutritional status may not affect visual acuity directly.

(v) Is the measure sufficiently independent of other influences? Visual acuity deficits are a product of many factors, including, for example, in preterm infants who may have stage III or worse retinopathy of prematurity (ROP), and these other biological causes need to be separated from dietary ones. Studies of very ill infants, such Carlson's nutritional study (this volume), show that dietary effects even in healthy infants can be masked by illness.

## Cognitive Status in Infancy: Rationale, Results, and Reservations

As all authors agree 'cognitive development is a complex process which is influenced individually and interactively by many genetic and environmental factors' (Innis, this volume). Which are the measures of cognitive development most often used as criteria in studies of nutrition? As summarized by Singer, Colombo, and McCall and Mash, of the diverse potential measures in infancy which may have (predictive) validity for cognitive development, habituation and recovery/responsiveness to novelty have shown the most promise. (In childhood, psychometric IQ has been the criterion of choice.)

### Habituation

Habituation is the decrement in attending infants show to a continuously available or repeated stimulus. Bornstein describes basic procedures and measures that are used in assessing its development during infancy, including the history, methodology, and quantification as well as psychometric adequacy and predictive validity of these infant measures, and he discusses theoretical constructs thought to underlie these infant measures (see 27). Habituation has been found to satisfy two prerequisite psychometric criteria reasonably well. Habituation is characterized by adequate individual variation, and it has been shown to be a moderately stable infant behavior, at least over the short term. Successful habituation minimally implies neurological integrity and sensory competence in the infant. However, the decrement in attention that is habituation is also thought to comprise processes that reflect the infant's passive or active development of some 'mental representation' of the stimulus as well as the infant's ongoing comparison of new stimulation with that representation. As a consequence, infant habituation is construed as (at least) the partial analog in adults of encoding, construction, and comparison with some kind of internal representation; that is, as *information processing*. Critically, habituation appears also to possess moderate predictive validity for childhood cognitive development. Infants who habituate efficiently in the first 6 months of life later, between 2 and 12 years of age, perform better on assessments of cognitive

Table 1. Habituation of attention in the first half-year of life in relation to measures of cognitive competence in the second year of childhood and later: Longitudinal studies

| Study* and measure | Age (years) | No. of subjects | Habituation measure, correlation† | Age (months) |
|---|---|---|---|---|
| Bornstein (27,105) WPPSI ($n = 14$) | 4 | 20 | Amount, 0.54 | 4 |
| Bornstein (60) RDLS-R | 2 | 18 | Index,‡ 0.55 | 5 |
| Laucht et al. (106) Total IQ | 4.5 | 221§ | Amount, 0.20 | 3 |
| Lewis and Brooks-Gunn (107) Bayley | 2 | 22 | Amount, 0.61 | 3 |
| Miller et al. (108) Language Comprehension | 3.3 | 29 | Amount, 0.39 | 2–4 |
| Rose et al. (109) WPPSI BAS | 4.5 4.5 | 21 16 16 | Index,$^{\|}$ 0.63 0.77 | 1.5–6.5 |
| Sigman et al. (110)§ WISC-R ($n = 59$) | 12 | 67§ | Amount,$^{\|}$ 0.33 | Newborn |

* Listed alphabetically, by author.
† All correlations are absolute values and significant at $p < 0.05$; direction and nature of the correlation depend on the measures.
‡ Latent variable of baseline, slope, and amount.
§ Sample consists of or includes preterm and/or at-risk infants; testing carried out at corrected age.
$^{\|}$ Mean of total fixation time, duration of first fixation, average fixation duration, and average trial duration.

competence, including standardized psychometric tests of intelligence as well as measures of childhood representational ability, including language and symbolic play. The averaged weighted normalized predictive correlation coefficient across studies of populations of normal babies is about 0.57; for at-risk samples, it is 0.22; and for all samples combined, 0.33 (see Table 1 and 28).

Findings of predictive validity of measures taken so early in life often entice infancy researchers into believing that endogenous processes may be at work. Indeed, the *information-processing* capacities, like the speed, accuracy, and completeness of encoding a stimulus into memory, that may be indexed by habituation appear to relate to mental capacities which are measured by standard tests of intelligence in childhood.

Using habituation as a dependent measure in studies of nutrition may be premature, however. Wainwright and Ward (this volume) assert that more narrowly defined assessment instruments (visual recognition memory, attention, or expectation) are less likely influenced by social and environmental factors, but there is reason to believe that attention measures may be so affected. Experience manifestly influences sensation, perception, and cognition as well as their development at every level of ontogenesis from cells to culture (29). In order to know how and why development proceeds in the individual, it is necessary to understand endogenous processes as well as potential contributions of experience. Indeed, habituation is influenced by external factors like maternal responsiveness. To determine the antecedents of habituation and recovery in 5-month-olds, Bornstein and Tamis-LeMonda (30) obtained several sets of data in the context of a multivariate, short-term prospective, longitudinal design. Naturalistically occurring maternal behaviors were studied at home identically at 2 and at 5 months. At 2 months in the laboratory, infant visual discrimination ability was measured; at 5 months, infant habituation and recovery were evaluated. Laboratory and home sessions were conducted by different observers to eliminate observer knowledge about infant and mother performance at other times. In addition, maternal IQ was assessed. Models of stability and experience on habituation were then evaluated. Habituation at 5 months was uniquely predicted by infant visual discriminative capacity at 2 months, and habituation also shared unique variance with maternal responsiveness in the home at 5 months. So, there is some stability in very early infant functioning, and infant information-processing performance also proved sensitive to concurrent maternal behavior.

*Recovery*

Colombo (this volume) points out that recovery is also problematic for several reasons. First, because the habituation procedure leads the infant to what is presumably an asymptotic level of looking, there is a threat that recovery may actually constitute regression to the mean. Second, calculation of difference scores (posthabituation trial minus last habituation trial) is statistically problematic (31). The most formalized measure of recovery derives from the FTII. However, Anderson (32) reported Cronbach α values of only 0.02–0.10 for the FTII, and Benasich and Bajar (33) reviewed problems with the measure from diagnostic and screening perspectives (such as poor short-term reliability in infants; see 34), although McCall and Carriger (35), among others, point out that reliability is not a bar to adequate prediction. As Wainwright and Ward argue, reliability indicates the degree to which measures register true differences, that is the extent to which individual differences in test scores can be attributed to differences in the characteristic being studied and the extent to which they are attributable to chance error. Bornstein and Tamis-LeMonda (30) found that recovery shared variance only with maternal IQ.

Several questions surround reliance on habituation and recovery/responsiveness in studies of nutritional effects in infancy.

(i) *Are these measures the same?* Length of familiarization necessary to yield a significant novelty preference has been found to vary with both stimuli and child age. Habituation and novelty preference may relate to speed of information processing which is a construct that has been theoretically linked to the principal component of IQ tests that gives rise to $g$. Individual differences in measures of visual habituation and/or recovery tend to covary for different measures and to covary with other measures of reaction time and autonomic function . . . suggesting a single process or general factor model underlies infant performance. Alternatively, as Colombo points out, it could be that habituation and recovery are different processes or that the mechanisms that underlie them are different. Support for a multiple process perspective comes from Jacobson *et al.* (36) who collected novelty preference, reaction time, and fixation duration data from more than 100 alcohol-exposed infants. Factor analysis showed look duration and reaction time as one factor and novelty preference as a second factor. Moreover, Jacobson *et al.* (36) showed that one toxic agent, polychlorinated biphenyls (PCBs), affected novelty preference but not fixation duration, whereas another toxic agent, alcohol, affected fixation duration but not novelty preference. Bornstein and Tamis-LeMonda (30) likewise found that habituation performance and recovery were predicted by different concurrent and antecedent variables.

(ii) *What are habituation and recovery?* For all the progress which has been made in understanding this domain of infant assessment in recent years, the nature and meaning of habituation and recovery are far from settled. Theoretical understanding of the processes of habituation dates back at least to the physiological speculation of Sokolov (37), which in itself is not altogether satisfying. Theory as to how habituation occurs is simple or complex, but still theory (e.g. 38, 39). What are the underlying neurological, physiological, or psychological mechanisms or processes that mediate prediction? Colombo and McCall and Mash address the issue of constructs and mechanisms underlying predictive performance from infant habituation and recovery to child cognitive (IQ) outcomes. The McCall and Mash and Colombo chapters (this volume) are overlapping and mutually supportive in their perspectives, with the exception of McCall's reliance on inhibition of attention as the underlying mechanism and Colombo's pointing to information processing. McCall and Mash argue that information processing is probably not the explanation of habituation and recovery because (a) little information from the standard stimulus needs to be encoded to be able to detect that a completely different novel stimulus is not the same as the familiar standard, and (b) whatever encoding takes place during familiarization takes very little time, much less than even the short familiarization periods typically offered in this research paradigm. McCall and Mash favor an inhibition of attention argument. While McCall and Mash point to the development of brain inhibitory mechanisms (reaching reasonable developmental levels by 8 months of age), socio-emotional and temperamental factors surely also play a role in self-regulation of inhibition.

Alternatively, various measures of fundamental processing speed in neuroconductivity correlate with IQ performance in adults. Speed of processing is a pervasive individual differences factor and underlies various aspects of childhood cognition (see also 40–43). Nettelbeck (44) and Anderson (45), among others (46), have speculated on the association between measures of information processing and IQ test performance, and Nettelbeck and Young (47, 48) confirmed a correlation in 6- and 7-year-olds between inspection time and IQ. Of course, individual variation in processing speed may help to account for variation in childhood IQ (46), but differences in speed may still not sufficiently explain differences in intelligence. That is, processing speed may be 'necessary for the appropriate development of intellectual aptitudes but not sufficient to determine them' (46, p. 279). However, in a developmental study of 7- to 19-year-olds, Fry and Hale (49) showed that individual differences in speed of information processing exerted a direct effect on working memory capacity which in turn was a direct determinant of individual differences in analytical or fluid intelligence. Cognitive development appears to reflect a 'cascade' wherein age-related changes in processing speed effect changes in working memory that enhance analytical intelligence (50–52).

Are LCPUFAs likely to affect either?

(iii) *How are habituation and recovery measured?* As Colombo (this volume) wrote, 'Aside perhaps from the FTII there is no one widely accepted or standardized format for assaying these cognitive processes.' Richards (53) and Ruff (54) both indicate that the quality and quantity of information processing varies even within looks.

(iv) *Do habituation and recovery stand alone in their predictive validity?* Other criteria in infancy might include psycho-physiological indexes, including vagal tone and heart period (see Wainwright and Ward, this volume), and behavioral measures, including visual expectation (55), cross-modal transfer (57), learning (58), or memory (59): learning and memory skills are also integral to infant performance, even if claims that n-3 acids are necessary for learning still remain to be substantiated.

In addition, exogenous factors to the infant may play a significant role. In one long-term prospective study extending from 4 months to 4 years, Bornstein (60) saw mother–child dyads at three points in development. At 4 months, infant habituation was assessed in the laboratory; at 1 year, infant productive vocabulary was ascertained; and at 4 years, children's intelligence was evaluated using the Wechsler series. At 4 months and at 1 year, didactics in mothers' interactions with children were recorded during home observations (didactics included mothers' pointing, labelling, showing, demonstrating, and the like). Path analysis determined direct and unique longitudinal effects of independent variables on dependent variables. Infant habituation showed predictive links both to toddler productive vocabulary size at 1 year and to childhood intelligence test performance at 4 years. However, maternal didactic interactions in infancy contributed to both the 1-year and 4-year child cognitive outcomes *independent* of infant habituation. These findings were later replicated and expanded upon in two short-term longitudinal follow-up studies. Thus, habituation picks up predictive validity, but so do other factors, which may themselves be affected by child. Some children might intentionally or continuously seek more optimal amounts, kinds, or patterns of environmental stimulation and experience, and through such consistent exposure improve their performance. Berg and Sternberg (61) proposed, for example, that infants who experience a greater variety of stimulating experiences may be in a better position to incorporate novel information into their thinking than infants with a more restricted history of experiences and, therefore, experience with novelty leads to more varied learning and improves the child's cognitive capacities. Or, early malnutrition could retard physical growth and psychomotor development, so the child is small in size and delayed in action (a primary effect); then, children who are small and motorically immature could be cared for in inappropriate ways which further limits their cognitive and verbal development (a secondary effect).

## Cognitive Status in Childhood: Rationale, Results, and Reservations

The third main dependent measure in studies of nutrition and development has been general cognition in later childhood, most often operationalized as IQ. Is using IQ the ideal outcome? As McCall and Mash point out, general mental performance has been studied extensively for nearly a century, and much is known, but a great deal remains unknown. They continue: intelligence is one of the most controversial behaviors for scientists and the general public alike, especially when claims are made regarding the relative contributions of biology and heredity versus experience to differences amongst individuals. IQ is the single best predictor of school performance, years in school, and job status, correlating approximately 0.50 with school grades, 0.55 with years of school and education, 0.50 with social status, and 0.54 with job performance (62). But in no case does IQ account for more than 30% of the differences between individuals on these important life outcomes. At minimum, IQ needs to be separated from achievement and other factors with which it shares so much variance. Therefore, IQ may be the single best predictor of these outcomes, but it is certainly not the only factor, nor does it account for the majority of differences between individuals on these measures.

McCall and Mash recount some psychometric issues with respect to intelligence and its measurement. What intelligence is a question in itself: Prominent theories contrast about its structure and nature (63–67) and its development (see below), whatever its value or validity (68–73). Spearman (74–76) asserted a general $g$ factor theory of IQ, positing that $g$ is inherited and universal and enters into all abilities. Thurstone (77,78) theorized that no general factor could account for all mental abilities, but rather he defined several independent, primary mental abilities (see also 51, 52). Some evidence supports a synthesis of these positions: Burt (79) and Vernon (80,81) proposed that $g$ is a higher-order, more important factor than are secondary ability factors. Garrett (82,83) advanced a view that, across childhood, the organization of abilities changes from general and unified to specific and differentiated. Of course, domain general and specific domain abilities could coexist.

## Alternative Outcomes: A Multivariate View

### Digression: A Homily

When children who were exposed to cocaine in utero first reached school-age, the climate of social response, from *Rolling Stone* magazine to the

*New York Times*, was uniformly grave. These were children with 'holes in their heads' and who were 'irremediably damaged'. The problem is not a small one. National estimates suggest that 10–20% of *all* infants in the USA are exposed to cocaine prenatally (84), and in many inner-city populations nearly 50% of women giving birth report or test positive for cocaine use at the time of delivery (85,86).

Moreover, theory and research alike point to a relation between prenatal cocaine or crack exposure and disturbances in orientation, attention, information processing, learning, and memory (87,88). There are at least two reasons for this. First are the effects of cocaine on those areas of both developing and adult brain that are involved in the regulation of attentional states. Cocaine acts at the presynaptic neural junction to block the reuptake of monoaminergic neurotransmitters (dopamine, norepinephrine, and serotonin). Because reuptake is primarily responsible for the inactivation of neurotransmitters, blocking reuptake leaves more neurotransmitters available within the synaptic space and results in enhanced activity of these agents in the CNS, and in specific physiological reactions and behaviors. The exaggerated alertness seen with cocaine use is related to the effect of cocaine on the norepinephrine system, and human infants exposed prenatally to cocaine exhibit increased norepinephrine levels in cerebrospinal fluid. The norepinephrine system plays a role in the regulation of states of arousal, and in turn the maintenance of attention. Second, prenatal cocaine exposure results in vasoconstriction of both placental and fetal blood vessels. Repeated prenatal cocaine exposure results in chronic fetal hypoxemia and decreased nutrient transfer. Diminished placental and fetal blood flow has deleterious consequences, therefore, for fetal growth. Demographics confirm that exposed infants manifest more obstetric complications, lower birth weight, shorter birth length, and smaller head circumference. For these reasons, we prospectively studied habituation in babies exposed to cocaine in utero and in groups of socio-demographic matched, drug-free controls. All babies were tested blind to drug exposure status, and analyses controlled for maternal socio-demographic and systematic perinatal differences.

In our study of newborn performance on the Brazelton Scale, relative to matched controls cocaine-exposed babies performed significantly worse on the habituation cluster (89). These findings bolstered initial hypotheses about the detrimental effects of cocaine on very early information processing. But cocaine-exposed babies showed depressed performance on other clusters as well. In our second study (90), we tested 3-month-old babies to determine the effects of prenatal cocaine exposure on habituation and recovery specifically. Surprisingly, the majority of infants in exposed and control groups reached habituation criterion, and amongst those who did no significant differences emerged between cocaine-exposed and drug-free infants in habituation or in recovery.

However, compared to the drug-free group, infants exposed prenatally to cocaine were significantly more likely to fail to start the habituation procedure, and those who did were significantly more likely to react with irritability early in the procedure. The effects of cocaine may not be solely or exclusively 'cognitive' in nature. Common to the neurotransmitter reuptake and vasoconstriction effects of cocaine in infants is also the dysregulation of state of arousal. Arousal level increases to novel stimulation, and the regulation of arousal serves as a gating mechanism to optimze orientation and attention and thus, information processing, learning, and memory. For example, infants with atypical fetal growth patterns are more likely to have problems with state regulation and with hyperarousal in the face of novel stimulation. We next, therefore, employed a standardized 'novel–repeat' laboratory procedure to evaluate infants' coping with novelty in order to examine arousal regulation (91). Specifically, we studied affective expressions and behavioral states in 3-month-olds who were cocaine-exposed or drug-free. Compared to nonexposed infants, infants exposed prenatally to cocaine were less likely to exhibit positive affect and more likely to display negative affect, and they were less likely to be alert and more likely to exhibit a crying state on novel stimulus presentations.

By way of a homily, then, we have been carried by these data away from a cognitive and information-processing prediction of cocaine effects to one concerned with arousal regulation. With respect to cocaine exposure, the good news is that all drug-exposed babies may not be disadvantaged in terms of cognition and information processing per se. Those babies who completed habituation did not differ in their information-processing characteristics from drug-free babies from the same socio-economic conditions. With respect to cocaine exposure, the bad news is that our testing yielded information on an arousal regulation problem in drug-exposed babies. These impairments in the regulation of arousal need to be considered in terms of their own predictive implications for children's cognitive and social development. Obviously, too, these (unexpected) findings reinforce concerns about specifying sources of influence – beyond experience – and outcome when making deductions about long-term development of performance.

*Nutrition and Development*

The main focus of nutrition–development studies has fallen on visual system function (acuity) and early manifestations of the development of intelligence in infants (habituation and recovery) and cognition in children (IQ), when other sensory systems and other domains of development may well be appropriate target functions for LCPUFAs. With respect to nutrition and development, then, our cocaine-and-development experience suggests that it may be beneficial to look at other target systems for nutrition deprivation effects. As McCall and Mash observed, because of its psychometric characteristics and historical rootedness in the literature, IQ could be a reasonable outcome measure in many nutrition studies. For these reasons, it probably should be included. However, the child is much more than her or his IQ: child growth and development are complex and multidimensional phenomena, to state the obvious, and other perspectives on the child as well as potential outcome measures should be considered. Indeed, a broad-ranging constellation of developmental biological, cognitive, emotional, and social capacities exists – a catalog that goes well beyond the determination of IQ – that could be explored to assess whether and how specific nutritional deficits affect a full range of childhood proclivities and capabilities. Such an investigation should not constitute a 'fishing expedition', although it has seemed ultimately reasonable to a large number of investigators that nutritional composition should connect up with some aspect of child development. The question is '*Which* aspect'? Innis (this volume) argued that the 'nonmyelin membranes of the central nervous system contain high proportions of AA and DHA', and therefore there should be generalized outcomes of AA and DHA deficiency rather than one specific one, say, visual acuity. For example, Innis also pointed out that 'Unlike the brain and eye, tissues such as the heart and kidney . . . depend on uptake of AA and DHA from circulation in order to maintain membrane levels of these fatty acids'. Why not, therefore, measure advantages or disadvantages in pulmonary outcomes in children? What other organ systems are likely or necessary targets for LCPUFAs?

With the relative dearth of results concerning visual system function and mental development, some deep hypothesizing or an expedition of 'hypothesis generation' followed up with confirmatory study may be in order. Motivation and temperament, two parallel and related constructs in the infant, have not been examined adequately in relation to nutritional status. Succeeding in infancy at habituation or recovery and in childhood on mental assessments presumably requires a persistent or vigilant temperament style. Pasilin (92) found that mothers' ratings of their preschoolers' persistence and attention to tasks related positively to children's performance on achievement tests. Motivation is often included in constructs of intelligence (93). Indeed, in discussing both the philosophy and administration of his intelligence scales, Wechsler (94, p. 6) himself defined intelligence as a global multidetermined and multifaceted construct that embraces nonintellectual factors like motivation: 'Intelligent behavior . . . may also call for one or more of a host of aptitudes (factors) which . . . involve not so much skills and know-how as drives and attitudes . . . .. They include such traits as persistence, zest, impulse control, and goal awareness . . .'. In this regard, as Rothbart (95) has argued, in order to attend to the environment, adults and infants alike must be able to modulate their state of arousal. Indeed, Moss *et al.* (96) found that newborn performance on the Range of State cluster of the Neonatal Behavioral Assessment Scale (NBAS) (97) was a factor in infant habituation-recovery performance: infants with better state control (those who were aroused from sleep to an intentive state without becoming distressed) discriminated novel from familiar stimuli better. On these accounts, motivation or self-regulation could be noncognitive constructs in infancy and childhood to explore from the vantage of nutritional status.

McCall and Mash re-raise the valuable point that there are also 'many kinds' of intelligence (66). It may be that assessments of diverse childhood skills, capabilities, and interests, ones that afford a richer opportunity to capture the many theoretical and pragmatic questions that surround biological, mental, emotional, and social development, would constitute a more apt battery of criteria. Included could be measurements of brain function, language skills, problem solving, motor skills, visual–spatial co-ordination, intellectual maturity (broadly conceived: IQ as well as understanding of number concepts, planfulness, and natural science abilities, musical and artistic talents, logical mathematical capacities), emotions and personality, perceived self-confidence, social cognition, and so forth. When investigators have looked to other scales, such as the Brunet–Lazine psychomotor development scale, as Agostoni *et al.* (98) have done, deficits in developmental quotient emerged in infants fed formula without DHA.

## Experimental Design

Wainwright and Ward (this volume) review the basic mechanics and assumptions of experimental design, including psychometric characteristics of assessments and tests as well as modeling. To date, approaches to evaluating nutritional effects have involved largely static between-group comparisons; some longitudinal designs also appear in the literature, but few control for 'background variables'. Certainly, as Lucas and Singer (this volume) observed, a randomized trial may be helpful, but consistent follow-up data of multiple kinds are also necessary, and Singer underscored important sources of error and bias that enter into group comparison studies, even in randomization trials. Once identified, confounders can be managed in several ways (exclusion, matching, stratification, randomization, or data analytic strategies).

Experiments in the field of nutrition and development in human beings are not always possible. For example, if researchers limited some rat pups to an inadequate and specifically LCPUFA-impoverished diet, they might find pups affected vis-à-vis pups offered unlimited quantities of nutritionally balanced LCPUFA foods. Such a deprivation experiment using humans is not permissible, of course, although comparable data might be obtained in natural or quasi-experiments that use supplementation or natural variation in levels of LCPUFAs.

Two alternative approaches might be intervention studies and an ecological, systems framework.

## Intervention Studies

An invervention study is attractive from the causal inference point of view, and manipulation of nutrition relative to outcome may be a low risk-to-benefit topic to explore. Developmental psychology is sorely lacking in proper intervention studies. A short-term longitudinal design is proposed to investigate the potential of nutritional supplementation for children. Independent measures of nutrition supplement and child are used in conjunction with a classic experimental approach that allows clear inferences about the value of the supplement as an 'intervention'. The longitudinal design is experimental in nature and geared to test the power of the intervention for outcome. The study is built on a Solomon–Lessac design (99,100). Four groups participate in three phases:

|         | Pretest | Intervention | Posttest |
|---------|---------|--------------|----------|
| Group 1 | √       | √            | √        |
| Group 2 |         | √            | √        |
| Group 3 | √       |              | √        |
| Group 4 |         |              | √        |

The *pretest* consists of diverse family and child 'background variables' and criteria measures. Criteria measures provide a baseline of child ability. The *intervention* consists of the nutritional supplement(s). Whether there are one, two, or more levels or types of intervention is to be determined, but 'the comparison of formula-fed groups to breast-fed groups cannot be used as a technique to pinpoint the contribution of any single nutrient to changes in neural or behavioral function. Instead, groups fed formula differing in composition by only a single variable, such as DHA content, must be used to address this type of question'. The no-pretest and pretest control groups are observed as often as the intervention groups to control for Hawthorne effects. The *posttest* consists of the same measures as used in the *pretest*.

This design ensures that the experimental groups are equivalent before the intervention (or can be made so), isolates the effects of pretesting on the subsequent intervention (experimental manipulations), and evaluates the effectiveness of nutritional supplementation for child development. Because testing itself may affect future development, the design specifies that one group undergoing intervention is pretested, and one group not pretested; post-intervention comparisons of the pretested and non-pretested groups reveal any effect(s) of pretesting. Participants are assigned randomly to the four groups. Boys and girls are tested, and if no gender differences emerge, their data are pooled within groups. The timing of pretest, intervention, and posttest is to be determined.

## An Ecological, Systems Framework

As several authors point out, transactional processes between the child and the environment, the responsiveness and influence of the environment, both in terms of proximal and distal factors as predictive of cognitive outcome, make for complications in naturalistically measuring developmental outcomes related to early nutritional interventions. Confounding variables can exert influences on both the independent (intervention) and dependent (outcome) variables. As McCall and Mash (this volume) point out, there may be developmental IQ advantages of breast-fed over bottle-fed infants, interpreted as indirect evidence that LCPUFAs contribute to such benefits, but most studies are compromised by the fact that mothers who choose to breast-feed have more education and/or higher social status and/or infants were born with fewer perinatal problems in mothers who choose to bottle feed. Innis claims that 'When the effects of confounding variables such as family and social characteristics are considered, breast-feeding still seems to confer a small, but significant developmental advantage', citing six papers, but do these papers comprehensively control for all relevant 'family and social characteristics' that have been shown to affect cognitive development in children? If so, then nutritional components of breast- versus bottle-feeding may come into play, but if these other 'family and social characteristics' exhaustively account for the developmental advantage, then what is left for nutritional components, growth factors, or other biologically active substances in human milk but not formula? For example, as Innis points out, other variables which extend 'beyond the breast or bottle feeding period' may differentiate families who breast-feed from those who bottle feed (the contemporary or cumulative effects models). Thus social, educational, and demographic factors known to relate to, say, cognitive development discriminate mothers who breast-feed from those who do not. Which specific characteristics of children's experiences influence the expression of individual variation in development?

It may be advantageous, therefore, to measure children and experiences in multiple ways, and simultaneously evaluate the contributions of a comprehensive set of pertinent factors thought to predict individual variation. Most developmental research fails to incorporate multiple variables from different sources of influence simultaneously, and so the unique and combined, the direct and indirect, the independent and interdependent contributions that each source may make to development in childhood are unexplored (101). Systems theorists have, however, emphasized the importance of considering multiple sources and the possible interdependence of different determinants in predicting the emergence, ontogenetic course, and eventual resting level of a developmental ability or activity. Many aspects of development are acknowledged to be embedded in one another, and at a given point in development variables from any and all of these levels may contribute to the status of a structure or function in development. Moreover, these multiple levels do not function independently of one another; rather, they interpenetrate and mutually influence one another.

A framework of research based on the ecological, systems perspective is recommended to identify the role of nutritional variables underlying variation in children's development separate from other experiences and from child ability and activity. The general model of theoretical relations within which this work might be conducted would include family sociodemographic characteristics, maternal and child personological characteristics, and maternal and child behaviors in addition to nutritional status or experience in increasing predictive relation to child outcome(s). It appears, for example, that there are individual differences among women in milk levels of DHA and if so, a naturalistic study could be undertaken which analyzes the effects of levels of DHA in breast-feeding families, all other things being equal. The optimal design should meet several prerequisites for evaluating both specificity and models of stability and experience in development.

Singer reviewed a number of contributing factors which can be conceptualized in the context of such an ecological model. By way of example, consider the following effort to measure direct as well as indirect sources of individual variation in language development in children (102). The relevant factors included maternal sociodemographic characteristics (that is, education and occupation (parental education or SES predict childhood or adult IQ at approximately 0.50)), personological characteristics (including intelligence, personality (interpersonal affect, social participation, and anxiety), locus of control (causal attributions to explain successes/failures in diverse parenting activities), attitude toward parenting (parental competence, satisfaction, role balance), and knowledge of child development (norms and milestones, developmental processes, health and safety)), and mothers' actual language (MLU and vocabulary), as well the children's own gender and social competence (has friends, shares, shows manners). Figure 1 shows the general hypothetical model of theoretical relations that instantiated the design and analyses. Structural equation modeling was used to identify unique predictive relations: we found that child gender and social competence predicted child language (with girls higher), and

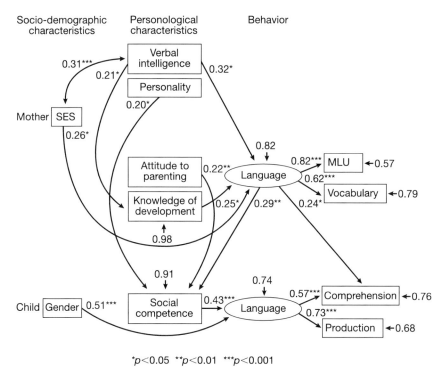

$*p<0.05 \quad **p<0.01 \quad ***p<0.001$

Fig. 1. Predictive relations between child and mother characteristics and child language in an ecological, contextual design.

mothers' language predicted child language comprehension. There were also several significant indirect paths from 'distal' sociodemographic as well as 'proximal' personological characteristics of mother to child language. Not only did maternal intelligence and knowledge about parenting influence maternal language, and therefore child language, but so did SES. This kind of multivariate approach allows parsing of the independent and interdependent roles of different sources of influence on different aspects of child development.

The significance of various 'background variables' should not be underestimated. Within affluent countries, such as the USA, children born into low socio-economic classes grow more slowly than those from high socio-economic classes, and they do not reach equivalent levels of height and weight. These differences presumably reflect the effects of prenatal and postnatal nutrition and care, maternal health and education, and genetic variation. Furthermore, cultural beliefs about the appropriateness of breast-feeding, and about when weaning from the breast should occur, can affect not only the nutritional status of the infant but, because mother's milk may make some children less likely to develop allergies later in life, it can also affect their health status. Diet provides an avenue whereby beliefs about pregnancy influence the health of the baby. Societies that ban certain foods which contain proteins important to prenatal growth place newborn infants at risk (103). Cole (104) observed that the Siriono (South America) do not allow women to eat the meat of animals and birds that are normally part of other people's diets because they fear that characteristics of the animals women eat while pregnant will be transferred to the unborn child. Further validation of this kind of influence is found in the results of studies of nutrition-related motivational strategies of Mexican–American mothers. Serving and helping children with their food is associated with food consumption compliance, whereas threats and bribes are negatively associated with healthy food consumption.

## Safety Issues

LCPUFAs occur naturally, and therefore the argument given is that supplementation 'couldn't hurt'. As several authors point out, however, safety concerns in supplementation of infant formula (e.g. infection, bleeding) are an issue. Studies conducted by Carlson showed no difference in unsupplemented and LCPUFA-supplemented groups of children on acuity by 1 year, but, notably, weight, length, and head circumference measures were significantly *reduced* in supplemented groups of children at all time periods to the 12-month follow-up, as were mental and psychomotor development indices in the Bayley Scales. It appears that through intervention n-3 fatty acids depress AA status and growth. Thus, as noted by Lucas, growth suppressing or possibly deleterious effects have been a concern of nutrition supplementation advocates above and beyond inconsistent findings. Perhaps the relation between essential fatty acids and outcome is not a linear, dose–response association, but a threshold or more complex relation wherein a minimum amount of LCPUFAs suffices; above that amount supplementation is harmful; but deficiency below that amount is meaningful. Safety issues, given possible deleterious effects already observed, should not be minimized. Term infants might receive sufficient amounts of LCPUFAs; more is not necessarily better and may even be undesirable. The conclusions from many papers are that LCPUFAs thus far have limited, if not potentially detrimental, effects on child outcome.

# References

1. Auestad N, Montalto MB, Hall RT et al. Visual acuity, erythrocyte fatty acid composition, and growth in term infants fed formula with long-chain polyunsaturated fatty acids for one year. *Pediatr Res* 1997; **41**: 1–10.
2. Carlson SE, Cooke RJ, Werkman SH, Tolley EA. First year growth of preterm infants fed standard compared to marine oil in n-3 supplemented formula. *Lipids* 1992; **27**: 901–7.
3. Fagan JF, Detterman DK. The Fagan Test of Infant Intelligence: a technical summary. *J Appl Dev Psychol* 1992; **13**: 173–93.
4. Innis SM. Human milk and formula fatty acids. *J Pediatr* 1992; **120**: S56–61.
5. Innis SM, Nelson CM, Rioux FM, King DJ. Development of visual acuity in relation to plasma and erythrocyte ω-6 and ω-3 fatty acids in healthy term gestation infants. *Am J Clin Nutr* 1994; **60**: 347–52.
6. Innis S, Nelson CM, Lwanga D, Rioux FM, Waslen P. Feeding formula without arachidonic acid and docosahexaenoic acid has no effect on preferential looking acuity or recognition memory in healthy full-term infants at 9 months of age. *Am J Clin Nutr* 1996; **64**: 347–52.
7. Jankowski JJ, Rose SA. The distribution of visual attention in infants. *J Exp Child Psychol* 1997; **65**: 127–40.
8. Lucas A, Morley RM, Cole TJ, Gore SM. A randomized multicentre study of human milk versus formula and later development in preterm infants. *Arch Dis Child* 1994; **70**: F141–6.
9. Jacobson SW, Jacobson JL. Breastfeeding and intelligence. *Lancet* 1992; **339**: 926.
10. Farquharson J, Cockburn F, Patrick WA, Jamieson EC, Logan RW. Infant cerebral cortex phospholipid fatty-acid composition and diet. *Lancet* 1995; **340**: 810–13.
11. Farquharson J, Jamieson EC, Abbasi KA, Patrick WJA, Logan RW, Cockburn

F. Effect of diet on the fatty acid composition of the major phospholipids of infant cerebral cortex. *Arch Dis Child* 1995; **72**: 198–203.
12. Farquharson J, Jamieson EC, Logan RW, Patrick WJA, Howatson AG, Cockburn F. Age- and dietary-related distributions of hepatic arachidonic and docosahexaenoic acid in early infancy. *Pediatr Res* 1995; **38**: 361–5.
13. Makrides M, Neumann MA, Byard RW, Simmer K, Gibson RA. Fatty acid composition of brain, retina, and erythrocytes in breast- and formula-fed infants. *Am J Clin Nutr* 1994; **60**: 189–94.
14. Carlson SE, Werkman SH, Tolley EA. Effect of long chain n-3 fatty acid supplementation on visual acuity and growth of preterm infants with and without bronchopulmonary dysplasia. *Am J Clin Nutr* 1996; **63**: 687–97.
15. Birch DG, Birch EE, Hoffman DR, Many RD. Retinal development in very-low-birth-weight infants fed diets differing in omega-3 fatty acids. *Invest Ophthalmol Vis Sci* 1992; **33**: 2365–76.
16. Simon JA, Fong J, Bernet JT, Browner WS. Relation of smoking and alcohol consumption to serum fatty acids. *Am J Epidemiol* 1996; **144**: 325–34.
17. Niemala A, Jarvenpaa A-L. Is breast-feeding beneficial and smoking harmful to the cognitive development of children? *Acta Paediatr* 1996; **85**: 1202–6.
18. Goldberg S, DiVitto B. Parenting children born preterm. In: Bornstein MH, ed. *Handbook of Parenting*, Vol. 1. Mahwah, NJ: Lawrence Erlbaum Associates, 1995: 209–31.
19. Ramey CT, Starr RH, Pallas J, Whitten CF, Reed V. Nutrition, response-contingent stimulation, and the maternal deprivation syndrome: results of an early intervention program. *Merrill–Palmer Q* 1975; **21**: 45–53.
20. Zeskind PS, Ramey CT. Preventing intellectual and interactional sequelae of fetal malnutrition: a longitudinal, transactional, and synergistic approach to development. *Child Dev* 1981; **52**: 213–18.
21. Ramey CT, Bryant DM, Wasik BH, Sparling JJ, Fendt KH, LaVange LM. The Infant Health and Development Program: Program elements, family participation, and child intelligence. *Pediatrics* 1992; **3**: 454–65.
22. Bornstein MH. Between caretakers and their young: two modes of interaction and their consequences for cognitive growth. In: Bornstein MH, Bruner JS, eds. *Interaction in Human Development*. Hillsdale, NJ: Lawrence Erlbaum Associates, 1989: 197–214.
23. Bornstein MH, Tamis-LeMonda CS. Activities and interactions of mothers and their firstborn infants in the first six months of life: covariation, stability, continuity, correspondence, and prediction. *Child Dev* 1990; **61**: 1206–17.
24. Bradley R, Caldwell B, Rock S. Home environment and school performance: a ten year follow-up and examination of three models of environmental action. *Child Dev* 1988; **58**: 852–67.
25. Bornstein MH. Sensitive periods in development: structural characteristics and causal interpretations. *Psychol Bull* 1989; **105**: 179–97.
26. Jorgensen MH, Hernell O, Lund P, Holmer G, Michaelsen KF. Visual acuity and erythrocyte docosahexaenoic acid status in breast-fed and formula-fed term infants during the first four months of life. *Lipids* 1996; **31**: 99–105.
27. Bornstein MH. Habituation of attention as a measure of visual information processing in human infants: summary, systematization, and synthesis. In: Gottlieb G, Krasnegor NA, eds. *Measurement of Audition and Vision in the*

*First Year of Postnatal Life: A Methodological Overview*. Norwood, NJ: Ablex, 1985: 253–300.
28. Bornstein MH. Stability in mental development from early life: Methods, measures, models, meanings, and myths. In: Butterworth GE, Simion F, eds. *The Development of Sensory, Motor Cognitive Capacities in Early Infancy: From Sensation to Cognition*. Hove: Psychology Press (in press).
29. Bornstein MH. Perception across the life span. In: Bornstein MH, Lamb ME, eds. *Developmental Psychology: An Advanced Textbook*. Hillsdale, NJ: Lawrence Erlbaum Associates, 1992: 155–209.
30. Bornstein MH, Tamis-LeMonda C. Antecedents of information-processing skills in infants: habituation, novelty responsiveness, and cross-modal transfer. *Inf Behav Dev* 1994; **17**: 371–80.
31. Appelbaum MA, McCall RB. Design and analysis in developmental psychology. In: Kessen W, vol. ed.; Mussen P, series ed. *Handbook of Child Psychology*, Vol. 1. New York: Wiley, 1983: 415–76.
32. Anderson H. The Fagan Test of Infant Intelligence: predictive validity in a random sample. *Psych Rep* 1996; **78**: 1015–26.
33. Benasich AA, Bejar II. The Fagan Test of Infant Intelligence: a critical review. *J Appl Dev Psychol* 1992; **13**: 153–71.
34. DiLalla LF, Thompson LA, Plomin R et al. Infant predictors of preschool and adult IQ: a study of infant twins and their parents. *Dev Psychol* 1990; **26**: 759–69.
35. McCall RB, Carriger MS. A meta-analysis of infant habituation and recognition memory performance as predictors of later IQ. *Child Dev* 1993; **64**: 57–79.
36. Jacobson SW, Jacobson SL, Sokol RJ, Martier SS, Ager JW. Prenatal alcohol exposure and infant information processing ability. *Child Dev* 1993; **64**: 1706–21.
37. Sokolov E *Perception and the Conditioned Reflex*. Oxford: Pergamon, 1958/1963.
38. Cohen LB. A two process model of infant visual attention. *Merrill-Palmer Q* 1973; **19**: 157–80.
39. Miller GA, Galanter E, Pribram KH. *Plans and the Structure of Behavior*. New York: Holt, Rinehart & Winston, 1960.
40. Hale S. A global development trend in cognitive processing speed. *Child Dev* 1990; **61**: 653–63.
41. Kail R. Sources of age differences in speed of processing. *Child Dev* 1986; **57**: 969–87.
42. Kail R. Developmental functions for speeds of cognitive processes. *J Exp Child Psychol* 1988; **45**: 339–64.
43. Kail R. Developmental change in speed of processing during childhood and adolescence. *Psychol Bull* 1991; **109**: 490–501.
44. Nettelbeck T. Inspection time and intelligence. In: Vernon PA, ed. *Speed of Information Processing and Intelligence*. Norwood, NJ: Ablex, 1987; 295–346.
45. Anderson M. *Intelligence and Development: A Cognitive Theory*. Oxford: Blackwell, 1992.
46. Nettelbeck T, Wilson C. Childhood changes in speed of information processing and mental age: a brief report. *Br J Dev Psychol* 1994; **12**: 277–80.

47. Nettelbeck T, Young R. Inspection time and intelligence in 6-year-old children. *Person Ind Diff* 1989; **10**: 605–14.
48. Nettelbeck T, Young R. Inspection time and intelligence in 7-year-old children. *Person Ind Diff* 1990; **11**: 1283–9.
49. Fry AF, Hale S. Processing speed, working memory, and fluid intelligence: evidence for a developmental cascade. *Psychol Sci* 1996; **7**: 237–41.
50. Kail R. Processing speed, speech rate, and memory. *Dev Psychol* 1992; **28**: 899–904.
51. Kail R, Park YS. Global developmental change in processing time. *Merrill-Palmer Q* 1992; **38**: 525–41.
52. Kail R, Salthouse TA. Processing speed as a mental capacity. *Acta Psychol* 1994; **86**: 199–225.
53. Richards JE. Heart rate responses and heart rate rhythms, and infant visual sustained attention. *Adv Psychophysiol* 1988; **3**: 189–221.
54. Ruff HA. Individual differences in sustained attention during infancy. In: Colombo J, Fagan J, eds. *Individual Differences in Infancy: Reliability, Stability, Prediction*. Hillsdale, NJ: Erlbaum, 1990: 247–70.
55. Dougherty TM, Haith MM. Infant expectations and reaction time as predictors of childhood speed of processing and IQ. *Dev Psychol* 1997; **33**: 146–55.
56. Canfield RL, Smith EG, Brezsnyak MP, Snow KL. Information processing through the first year of life: a longitudinal study using the visual expectation paradigm. *Monog Soc Res Child Dev Serial No. 250*. Chicago, IL: University of Chicago Press, 1997.
57. Rose SA. Measuring infant intelligence: new perspectives. In Bornstein MH, Krasnegor NA, eds. *Stability and Continuity in Mental Development: Behavioral and Biological Perspectives*. Hillsdale, NJ: Lawrence Erlbaum Associates, 1989; 171–88.
58. Fagen JW, Ohr PS. Individual differences in infant conditioning and memory. In: Colombo J, Fagen J. eds. *Individual Differences in Infancy: Reliability, Stability, Prediction*. Hillsdale, NJ: Erlbaum, 1990: 155–92.
59. Rovee-Collier C. Learning and memory in infancy. In: Osofsky JD, ed. *Handbook of Infant Development*, 2nd edn. New York: Wiley, 1987; 98–148.
60. Bornstein MH. How infant and mother jointly contribute to developing cognitive competence in the child. *Proc Natl Acad Sci USA* 1985; **82**: 7470–3.
61. Berg C, Sternberg RJ. Response to novelty: continuity versus discontinuity in the developmental course of intelligence. *Adv Child Dev Behav* 1985: **15**: 1–47.
62. Neisser W, Boodoo G, Bourchard TJ et al. Intelligence: knowns and unknowns. *Am Psychol* 1996; **51**: 77–101.
63. Carroll JB. The measurement of intelligence. In: Sternberg RJ, ed. *Handbook of Human Intelligence*. Cambridge: Cambridge University Press, 1982: 29–119.
64. Carroll JB. *Human Cognitive Abilities: A Survey of Factor-analytic Studies*. Cambridge: Cambridge University Press, 1993.
65. Ceci SJ. *On Intelligence . . . more or less: A Bioecological Treatise on Intellectual Development*. Englewood Cliff, NJ: Prentice Hall, 1990.
66. Gardner H. *Frames of Mind: The Theory of Multiple Intelligences*. New York: Basic Books, 1983.
67. Sternberg RJ. *Beyond IQ: A Triarchic Theory of Intelligence*. Cambridge: Cambridge University Press, 1985.

68. Barrett GV, Depinet RL. A reconsideration of testing for competence rather than for intelligence. *AM Psychol* 1991; **46**: 1012–24.
69. Deese J. Human abilities versus intelligence. *Intelligence* 1993; **17**: 107–16.
70. Howe MJA. Intelligence as an explanation. *Br J Psychol* 1988; **79**: 349–60.
71. Hunter JE. Cognitive ability, cognitive aptitudes, job knowledge, and job performance. *J Voc Behav* 1986; **29**: 340–62.
72. McClelland DC. Testing for competence rather than for 'intelligence'. *Am Psychol* 1973; **28**: 1–14.
73. Rabbit P. Human intelligence: critical review of RJ Sternberg's work. *Q J Exp Psychol* 1988; **40A**: 167–85.
74. Spearman C. 'General intelligence' objectively determined and measured. *Am J Psychol* 1904; **15**: 201–93.
75. Spearman C. *The Nature of 'Intelligence' and the Principles of Cognition.* London: MacMillan, 1923.
76. Spearman C. *The Abilities of Man.* London: Macmillan, 1927.
77. Thurstone LL. *Primary Mental Abilities.* Chicago, IL: University of Chicago Press, 1938.
78. Thurstone LL, Thurstone J. *Test of Primary Mental Abilities*, rev. ed. Chicago, IL: Chicago Science Research Association, 1962.
79. Burt C. *The Factors of the Mind: An Introduction to Factor Analysis in Psychology.* London: University of London Press, 1940.
80. Vernon PE. *The Structure of Human Abilities.* London: Methuen, 1961.
81. Vernon PE. *Intelligence and Cultural Environment.* London: Methuen, 1969.
82. Garrett HE. Differentiable mental traits. *Psychol Rec* 1938; **2**: 259–98.
83. Garrett HE. A developmental theory of intelligence. *Am Psychol* 1946; **1**: 372–8.
84. Chasnoff IJ, Landress HJ, Barrett ME. The prevalence of illicit drug or alcohol abuse during pregnancy and discrepancies in mandatory reporting in Pinellas County, Florida. *N Engl J Med* 1990; **322**: 102–6.
85. Amaro H, Fried LE, Cabral H, Zuckerman B. Violence during pregnancy and substance use. *Am J Publ Health* **80**: 575–9.
86. Osterloh JD, Lee BL. Urine drug screening in mothers and newborns. *Am J Dis Child* 1989; **143**: 791–3.
87. Bornstein MH, Mayes LC, Tamis-LeMonda CS. Habituation information processing mental development and the threat of cocaine exposure in infancy. In: Hepper PG, Kendal-Reed M, eds. *Perinatal Sensory Development: Psychology and Psychobiology.* Cambridge: Cambridge University Press (in press).
88. Struthers JM, Hansen RL. Visual recognition memory in drug exposed infants. *J Dev Behav Pediatr* 1992; **13**: 108–11.
89. Mayes LC, Granger RH, Frank MA, Schottenfeld R, Bornstein MH. Neurobehavioral profiles of infants exposed to cocaine prenatally. *Pediatrics* 1993; **91**: 778–83.
90. Mayes LC, Bornstein MH, Chawarska K, Granger RH. Information processing and developmental assessments in three-month-old infants exposed prenatally to cocaine. *Pediatrics* 1995; **95**: 539–45.
91. Mayes LC, Bornstein MH, Chawarska K, Haynes OM, Granger RH. Impaired regulation of arousal in 3-month-old infants exposed prenatally to cocaine and other drugs. *Dev Psychopathol* 1996; **8**: 29–42.
92. Pasilin H. Preschool temperament and performance on achievement tests. *Dev Psychol* 1986; **22**: 766–70.

93. Scarr S. Testing for children: assessment and the many determinants of intellectual competence. *Am Psychol* 1981; **36**: 1159–66.
94. Wechsler D. *Wechsler Intelligence Scale for Children – Revised.* New York: The Psychological Corporation, 1974.
95. Rothbart MK. Longitudinal observation of infant temperament. *Dev Psychol* 1986; **22**: 356–66.
96. Moss M, Colombo J, Mitchell JW, Horowitz FD. Neonatal behavioral organization and visual processing at three months. *Child Dev* 1988; **59**: 1211–20.
97. Brazelton TB. Neonatal behavioral assessment scale. *Clinics in Developmental Medicine*, No. 50. Philadelphia PA: Lippincott, 1973.
98. Agostoni C, Riva E, Bellu R, Trojan S, Luotti, D, Giovannini M. Effects of diet on the lipid and fatty acid status of full-term infants at 4 months. *J Am Coll Nutr* 1994; **13**: 658–64.
99. Solomon RL, Lessac MS. A control group design for experimental studies of developmental processes. *Psychol Bull* 1968; **70**: 145–50.
100. Lessac MS, Solomon RL. Effects of early isolation on the later adaptive behavior of beagles: a methodological demonstration. *Dev Psychol* 1969; **1**: 14–25.
101. Bronfenbrenner U, Crouter AC. The evolution of environmental models in developmental research. In: Kessen W, vol. ed.; Mussen PH, series ed. *Handbook of Child Psychology*: Vol. 1. *History Theory and Methods.* New York: Wiley, 1983: 357–14.
102. Bornstein MH, Painter KM, Haynes OM, Rose AR. *Child Language Competence: Sources of a Multivariate Model.* National Institute of Child Health and Human Development. Unpublished manuscript. Bethesda, MA.
103. Mead M, Newton N. Cultural patterning of perinatal behavior. In Richardson S, Guttenmacher A, eds. *Childbearing: Its Social and Psychological Aspects.* Baltimore MD: Williams Wilkins, 1967.
104. Cole M. Culture in development. In: Bornstein MH, Lamb ME, eds. *Developmental Psychology: An Advanced Textbook*, 3rd ed. Hillsdale NJ: Lawrence Erlbaum Associates, 1992.
105. Bornstein MH. *Infant attention and caregiver stimulation: Two contributions to early cognitive development.* Paper presented at the International Conference on Infant Studies, New York City, 1984.
106. Laucht M, Esser G, Schmidt M. Contrasting infant predictors of later cognitive functioning. *Journal of Child Psychology and Psychiatry* 1994; **35**: 649–62.
107. Lewis M, Brooks-Gunn J. Visual attention at three months as a predictor of cognitive functioning at two years of age. *Intelligence* 1981; **5**: 131–40.
108. Miller DJ, Ryan EB, Arbeger E, McGuire MD, Short EJ, Kenny DA. Relationships between assessments of habituation and cognitive performance in the early years of life. *International Journal of Behavioral Development* 1979; **2**: 159–70.
109. Rose DH, Slater A, Perry H. Prediction of childhood intelligence from habituation in early infancy. *Intelligence* 1986; **10**: 251–63.
110. Sigman M, Cohen SE, Beckwith L, Asarnow R, Parmelee AH. Continuity in cognitive abilities from infancy to 12 years of age. *Cognitive Development* 1991; **6**: 47–57.

# Appendix B
# Design, Measurement, and Statistical Approaches

MARK APPELBAUM

*Department of Psychology–0109, University of California, San Diego, CA, USA*

---

The emphasis of the following commentary is upon those chapters which deal with measurement and design issues, as opposed to the predominantly nutritional and biochemical ones. As a person new to this field, I find it remarkable that much of what I have read is so very similar in form to the early childhood intervention literature. The problems of theory, design, measurement, analysis, and interpretation are virtually identical. As a science we are faced with a problem in which there is no well-established theory (at least not one strong enough to suggest either the form of a dose–response function or even to make a strong statement concerning the domains of behavior/function in which we necessarily would see such a relationship). While there is an empirical knowledge base (mainly from the animal literature) dealing with relationships associated with long-chain polyunsaturated fatty acid (LCPUFA) dietary variation, the bulk of that literature seems to be based upon extremely low levels of these agents and the impact of deficiency upon more anatomically or neurally defined events. The bridge between that literature and experimentation with complex human information processing under less extreme variation in levels of the dietary agents is a strained (or at least a tentative) one. Given this lack of theoretical guidance (and in the absence of a large and unambiguous empirical literature) it is difficult to know exactly what to make of new, weak, or inconsistent empirical results, particularly when trying to assess the likely 'payoffs' to be expected from further empirical research.

The design issues seen in the LCPUFA research literature also have an uncanny similarity to many of those of the early childhood intervention literature. Difficulties in conducting the empirical research as truly randomized clinical trials stand at the center of these design issues. The choice of whether to breast-feed is a family decision not subject to randomization. Other factors, such as the health and birth status of the child, complicate simple attempts at randomization even among those who are willing participants in a randomized clinical trial. While it is true that randomized clinical trials are the preferred method of understanding the impact of a treatment or intervention, the simple, uncritical use of a randomized trial does not solve all of the problems inherent in research of this type. While a randomized clinical trial can serve to 'balance out' sources of confounding, the adoption of this form of research does not automatically solve problems such as precision, power, differential take of treatment, differential dropout, and multi-collinearity, to name but a few of the thorny problems seen in the empirical literature. The adoption of the uncritical assumption that simply conducting large (or even very large) clinical trials will solve all of the problems of our current lack of knowledge in the field is ill-advised and likely to produce very costly errors.

The profound impact of other non-nutritional influences upon human information processing usually mandates the use of some form of covariate adjustment in all but the largest and most complex of randomized clinical trials in order to achieve sufficient precision. Here our lack of knowledge of the factors which may influence cognitive development make the selection of appropriate covariates extremely difficult, and the analysis of covariance is a very delicate tool. Differential dropout, which includes not only discontinuation of participation in the trial but also movement among the treatment arms of the study, is another problem shared by both forms of empirical study. The appropriate use of both 'intention to treat' and 'efficacy' analyses in the face of differential dropout is an essential component of the analytic treatment of data. Without both forms of analysis there is a high probability that pattern of results will be substantially biased and usually of the form that the intervention will show greater treatment effect than is actually present in nature. Recent advances in the analysis of situations in which the pattern of missing data is not 'ignorably random' (e.g. 1) have provided increasingly sophisticated analytic approaches to this problem. A related issue which deserves mention is that of replacement of subjects lost to randomization. In our irrational desire to keep the size of the treatment arms equal, researchers will sometimes replace 'dropouts' with new cases in a way that is inconsistent with the initial randomization plan and will, further, exclude the 'dropout' case from the 'intent to treat' analysis. Attempts to correct the naturally occurring 'accidents' which happen during the conduct of a clinical trial are usually ill-advised and should be avoided.

In addition to these issues, other problems occur which make unambiguous interpretation of results difficult from the design perspective. One of the most difficult of these issues, particularly from the point of view of trying to establish a dose–response function, is the 'effective dosage' level. Even in a randomized clinical trial, there may be a substantial difference between the amount of agent specified and the amount of agent effectively delivered. Differences in feeding patterns, ingestion efficiency, metabolism, and biochemical processing may well add substantially to the 'with cell variance'. (In behavioral studies this phenomenon is sometimes referred to as the 'differential take problem'.) When these forms of uncontrolled (and uncontrollable) variance are added to the already sizable variation due to individual differences in human information processing, plasticity of the organism, and the host of environmental, psychological, and parenting effects with influence development, it may be virtually impossible to detect effects which are small and/or subtle unless the size of the trial is very large and extremely carefully controlled, particularly with the range of experimental manipulation which is ethically or pragmatically possible. It is incumbent upon the research team to strive to lower or eliminate all extraneous sources of variation (e.g. unreliability of observation) in the face of so many uncontrollable sources of variation.

Perhaps the most profound problem, however, in research such as this is the measurement issue. Much has been said in the submitted chapters about measurement, particularly the measurement of infant information processing and its later relationship to more adult-like cognitive functioning, particularly verbal functioning. As was noted in the McCall and Marsh, Colombo, and many other chapters, the problem of measurement of early intellectual functioning is a very old and very difficult one, particularly when the criterion rests on prediction to later intellectual functioning. While some degree of progress has been made in that domain, the problem has certainly not been solved, and there are those who argue that it is a misguided task. While the view of continuity in intellectual functioning is a widely held belief, it is certainly not a logically necessary one – particularly where continuity is defined as the maintenance of rank position within a population, as is implied by the use of the correlation coefficient as the index of continuity. It is certainly a logical possibility that there are some aspects of childhood or adult intellectual functioning which simply do not exhibit themselves in the first years of life, and that individual differences in those aspects of cognitive functioning simply cannot be predicted from behavior observable in the earliest years of life. Further, the human organism is blessed with a fairly high degree of plasticity, and early deficits (or gifts) are not, of necessity, with the organism for life. The selection of the fundamental outcome measures of a study is perhaps the most critical decision that one makes in the planning of a study, although far more is written about the design and analysis issues. The degree to which the measurement issue is overlooked, or at least only superficially considered, is troubling, and the organizers of this Workshop should be commended for their careful attention to the issue.

Most of the analytic issues that one detects in studies such as those discussed in these chapters are fairly standard ones, and arise for all of the reasons outlined above. While the wise, and very cautious, use of quantitative tools can be of considerable utility, they cannot fix what is fundamentally broken. The analysis of covariance, one of the basic tools of the trade, is a very delicate instrument and must be used with great care. For those researchers who regularly use analysis of covariance, it would be a weekend well spent to get a basic text such as Brad Huitema's (2); or, if they can only afford a lunch break, a paper like Janet Elashoff's classic (3), and to read carefully about the analysis of covariance. The analysis of covariance can be a powerful tool, but it can also be horribly misused. In fact, I can think of no statistical technique which has been more misapplied than this one.

Two other techniques are of great importance and should be mentioned, for there seems to be little evidence of the use of either of them in the chapters of this volume. First is the utilization of power analysis for the purpose of determination of sample size. Studies are neither large nor small in any absolute sense. A study with 300 cases may be very large (even excessively large) for some purposes and not nearly large enough for others. The wise and, again, cautious use of power analysis can help in making the expenditure of funds and subject time far more rational than is often the case. A further benefit which results from the use of power analysis is the need to specify the effect sizes which may be expected or desired. Facing the effect size issue can be a sobering but important exercise. A nontrivial determination of effect size (i.e. not just opting out for a 'moderate' effect size) forces one to explore issues such as within cell variation to be expected and can force one to deal with a number of control issues which are far too often forgotten after having passed the 'experimental design' course required in graduate school. In the process of sample size determination, power is not the only factor to be considered, however. Many of the analytic techniques upon which the study will depend have their own minimum sample size requirements. One should also be aware of these needs as one considers the power issue. Finally, the designer of clinical trials in the area of developmental nutrition should be aware of a class of techniques known as optimal design (4) when first planning the design of a trial.

The second class of techniques which I would like to mention are those grouped under the term 'exploratory data analysis and graphical display of data' (5). Following the usual custom of scientific writing, the chapters in this volume are filled with $p$ values, group means, and standard deviations, but there is not a single graphical display of data. It is almost impossible to get any idea of what the data look like: of the distributional forms, of the presence or absence of outliers, of clustering or clumping of data, etc., from the summary statistics provided. The lack of use of graphical/exploratory techniques may simply reflect the 'FDA model' which stresses the fairly atheoretical approach of demanding that an effect be shown to be statistically present, but given the lack of theoretical (and even empirical) bases for our understanding of the phenomena, as scientific discourse much could be added by the use of these techniques.

Finally, there seems to be a degree of conceptual confusion that runs throughout the literature as presented in these chapters. There really seems to be two, albeit related, issues. First, is there evidence that LCPUFAs are a factor in the development of neural, brain, visual, and/or intellectual functioning? Second, will supplementation of 'formula' with these substances have a measurable impact on these domains? While it is clearly the case that these two issues are related, they are not the same. What serves as evidence for the confirmation/disconfirmation of the one does not necessarily serve as evidence for the other. As I was reading the material (and please be clear that I am quite new to this domain), I was somewhat surprised at a few of the things that I did not see. Perhaps the one that I found myself returning to over and over was the question of what the natural variability among nursing mothers is in their production of LCPUFAs (Innis's paper gives some indication that there is considerable variation among nursing mothers); and what the relationship between that natural variation and outcome is among those children who are breast-fed; i.e. the natural variations study. The second issue (and it really is again the issue mentioned under the Exploratory Data Analysis issue) was one dealing with the degree to which the results of studies are carried by a relatively small number of outlying cases, i.e. cases with very high influence.

## References

1. Little R, Yau L. Statistical techniques for analyzing data from prevention trials. *Psychol Methods* (in press).
2. Huitema BE. *The Analysis of Covariance and Alternatives*. New York: Wiley, 1980.
3. Elashoff JD. Analysis of covariance: a delicate instrument. *Am Educ Resource J* 1969; **6**: 383–401.
4. McClelland G. *Optimal Design in Psychological Research. Psych Methods* 1997; **2**: 3–19.
5. Behrens JT. Principles and procedures of exploratory data analysis. *Psych Methods* 1997; **2**: 131–60.

# Appendix C
# General Commentary

## MARTHA NEURINGER and SYDNEY REISBICK

*Section of Clinical Nutrition & Lipid Metabolism, Department of Medicine, Oregon Health Sciences University and Oregon Regional Primate Research Center, 505 NW 185th Avenue, Beaverton, OR 97006, USA*

---

Multiple Process Models of Infant Attention and the Effects of LCPUFAs

Both Colombo and Bornstein cite the studies of Jacobson *et al.* (1) as evidence for the dissociation of effects on novelty preference and look duration outcomes within the paired comparison paradigm. Evidence for this dissociation also comes from our longitudinal studies of infant rhesus monkeys, in which paired comparisons were administered at four ages within the first 3 postnatal months (roughly equivalent to the first year in human infants). We found that dietary taurine (a conditionally essential amino acid) affected the proportion of time looking at novel stimuli but did not influence look durations (2). Conversely, infants fed diets low in n-3 fatty acids, and therefore having poor long-chain polyunsaturated fatty acid (LCPUFA) status, showed no effect on novelty preference, but had longer look durations in the choice phase (3,4). The difference in look durations was consistent across novel and familiar stimuli, across stimulus sets (photographs of faces and abstract drawings), and across different delays between familiarization and the novelty test (immediate and 24 hours). Furthermore, we have now confirmed this difference in look duration in a second study (described in more detail below). The pattern of results seen in these monkey infants was similar to that found by Carlson and co-workers in their two randomized studies of n-3 fatty acid supplementation in preterm human infants (5,6): higher LCPUFA status was associated with shorter look durations but no consistent difference in novelty preference. Furthermore, Forsyth and Willatts (7) reported that, among a subset of infants with nonmonotonic patterns of habituation in an infant control paradigm, look durations again were shorter in those receiving LCPUFA supplementation. Therefore, the data now suggest (i) that look duration and novelty preference measures are dissociable by at least four different perinatal variables (alcohol, polychlorinated biphenyls (PCBs), taurine and LCPUFA status), and (ii) that look duration, but not novelty preference, is sensitive to both LCPUFA status and prenatal alcohol exposure (8). We do not know whether the differences found in our monkeys reflect differences in speed of information processing, ability to disengage attention, or responsiveness to visual stimuli. We have some evidence for the latter hypothesis, in that responsiveness to a variety of visual, tactile and social stimuli was increased in older monkeys fed diets low in n-3 fatty acids (9). It will require a number of additional studies to evaluate these alternative hypotheses.

It has been noted that look duration in the familiarization phase of paired comparisons is most comparable to look duration in habituation paradigms, but that the effects on look duration in Carlson's studies and in our first study are found in the test rather than the familiarization phase. We concur with Carlson that a lack of effect in the familiarization phase is likely to be related to its length. In our first study, described above, the criterion for familiarization was a very long 30 seconds of stimulus fixation, which we chose based on the previous paired-comparison studies of Gunderson *et al.* in infant pigtailed macaques (e.g. 10). When the infant is required to accumulate this duration of looking, typically a few early long looks are followed by a large number of very short ones (we had no minimum duration requirement for counting a look), so that any effect on look duration early in the sequence is not detectable. That the monkey infants had habituated to the stimulus during this 30 seconds of exposure was confirmed by their high levels of novelty preference, with even 2-week-old infants showing significant novelty preference for pattern stimuli. We hypothesized that we might have failed to detect a difference in novelty preference due to an excessively long familiarization which led to a ceiling effect. We also hypothesized that a difference in look duration during familiarization might appear with a shorter familiarization time. Therefore, in a second study, we reduced the familiarization time to 10 seconds at all ages (2, 3.5, 5, and 9 weeks). We again found no difference between groups in novelty preference, but we did find significantly longer look durations during familiarization in infants fed the low linolenic acid diet (11). In this case significant novelty preference only emerged at the later ages.

The finding that measures of look duration and novelty preference are dissociable can be used to argue against the idea that look duration reflects speed of processing. Given that exposure to the stimulus in the usual paired-comparison procedure is fixed in duration, if long lookers process more slowly, then they may process or encode the stimulus less completely in the fixed amount of time, and should therefore show less preference for the novel stimulus (which was not the case in our studies). On the other hand, if the difference between long and short lookers is in their general engagement with visual stimuli, or in their ability to disengage attention, then look durations in paired-comparison familiarization and choice phases should be correlated with each other (as reported by Colombo, this volume) and with look durations in habituation paradigms.

Both Colombo and McCall and Mash (this volume) consider a number of possible underlying mechanisms for the continuity between infant attention measures and later intellectual performance. All of these are also potential mechanisms underlying differences related to early fatty acid nutrition. One which is not discussed is sensory function. Since the most frequently reported effect of early n-3 fatty acid deprivation and LCPUFA supplementation has been an enhanced speed of acuity development, it is important to consider the possible relationship of acuity to the reported effects on look duration. The only evidence we can contribute is negative. We combined data from four different studies of 43 infant monkeys in which forced-choice preferential looking (FPL) and visual evoked potential (VEP) acuity measurements and visual paired comparison tests were each done longitudinally at four ages. We found no significant correlation between look duration during visual paired comparison tasks and visual acuity ($r = 0.002$ and $0.194$ for FPL and VEP acuity, respectively) (12). Similarly, Carlson *et al.* (13) found no correlation between look duration and acuity in her two studies of LCPUFA supplementation in preterm infants. Thus, it seems unlikely that the effects on look duration in these studies were a direct consequence of the differences in acuity.

On the other hand, we did find modest relationships between novelty preference and visual acuity, ($r = 0.533$ for FPL, $0.575$ for VEP). Therefore we would like to reinforce Mayer and Dobson's comment about the importance of considering the visual properties of the stimuli used in paired comparison tasks. Furthermore, the effect of early visual abnormalities on measures of novelty preference (including the standardized FTII) has not been examined.

Although the emphasis here is on visual attention measures, habituation and dishabituation to auditory stimuli have also been studied (14) and could provide another means to differentiate effects of LCPUFAs on attention from those on vision. Of course, at this point we simply do not know whether effects of LCPUFAs on the auditory system could also be present.

## Validity and Prediction

Habituation is a very basic form of nonassociative learning that can be demonstrated even in invertebrates. It seems surprising, therefore, that this fundamental process correlates as well as it does with capacities, such as reasoning and language, that we consider to be much more advanced and dependent on the human cerebral cortex. This relationship challenges some of our preconceptions about 'simple' versus 'complex' forms of learning. As noted by both McCall and Mash and by Colombo, it is not clear what the connection is between early habituation (and other visual attention measures) and later cognitive measures. One possibility is that differences in an early, basic process can have a ripple or cascade effect on any or all later developing capacities. For example, those infants who dwell on familiar stimuli (for whatever underlying reason) may be exposed to less variety of stimulation, and therefore learn less about the nature of the world around them, and those who seek new stimulation less avidly may evoke less varied input from their family members and caretakers. An alternative possibility noted by these authors is that infant measures of habituation are measuring something(s) more complex than simple habituation per se.

McCall and Mash and Colombo provide detailed reviews of the test–retest and predictive validity of visual attention measures. It is important to note that reliability values are somewhat better when a larger number of habituation or paired comparison items are presented (15). As noted by Colombo, the test–retest reliability for individual paired-comparison tasks (stimulus pairs) is very low, while reliability for multiple assessments using several tasks is much higher. Studies using only one or two habituation or paired comparison tasks tend to find low test-retest and/or predictive validity, whereas some of those using more test items find relatively better reliability, a point also made by Fagan and Singer (16). To take two contrasting examples, Laucht *et al.* (17) concluded that habituation measures were no more predictive than the Bayley mental development index (MDI) based on exposure to only one stimulus, whereas Rose *et al.* (18), using three habituation stimuli presented in three separate sessions, found correlations as high as −0.85 between look duration and standard IQ scores at 4–5 years of age. McCall and Carriger (15) note that 'a single memory [or habituation] task may be analogous to a single item on a paper and pencil test, and the reliability . . . for these infant measures is roughly similar to the correlation between single items on pencil and paper psychometric tests for older children and adults'. This point is important for evaluating the potential of these infant measures for prediction, as opposed to their previous success in doing so.

McCall and Mash review the correlation of IQ scores with real world outcomes such as school grades and job performance. What seems striking is that these predictions are not much better than the prediction from early infant attention measures to school age IQ. The validity of IQ scores is the result of a century of concentrated effort to use IQ tests to screen people for performance in school, jobs, and the military, and there has been tremendous practical impetus to refine and improve IQ testing for these purposes. Furthermore, these tests have been administered to tens or hundreds of millions of people, four or five orders of magnitude more than for any infant measure. Given that no other type of psychological or behavioral testing has had this amount of attention or resources devoted to it, it would be disappointing if standardized IQ tests were not at least as good as they are in fulfilling their assigned purpose. In this context, it seems surprising that any test administered in the first few months of life, and which, by comparison, has had a quite small amount of effort devoted to its development as a predictive instrument, should correlate with later IQ even half as well, despite the infant measures' far poorer test–retest reliability.

It is noted by McCall that IQ is related to low birth weight only when birth weight is very low. This is perhaps a good point to make more generally: that for most factors correlating with IQ, the relationship is nonlinear, so that the correlation is much stronger at the low end than at the high end of the spectrum (and in some cases disappears at middle and high values of either IQ or the related variable). This is true, for example, for correlations among IQ subtest scores and for the correlations of IQ with social class, job success, etc. (19). It is also related to the idea (McCall and Mash, this volume) that there may be thresholds for effects of nutrients or other environmental factors, and to the point that the predictive validity of standardized infant test such as the Bayley is higher for at-risk and frankly disordered infants. It seems clear that novelty preference scores (including the Fagan Test of Infant Intelligence (FTII)) also are significantly more predictive for high-risk infants (preterms, low socio-economic status or frankly abnormal) than for normal samples. Habituation measures appear to be an exception to the rule, predicting later IQ better for normal samples than for high-risk ones. McCall and Carriger (15) have shown that this is not a true relationship, and that even habituation measures predict slightly better for high-risk infants if the samples sizes of the various studies are factored out. However, it still appears that habituation and visual attention measures (such as look duration) may be somewhat more sensitive for differentiating infants within the normal range.

The predictive validity of infant attention measures, specifically, their ability to predict an individual's later standardized IQ scores, may be a valid question, but it is overemphasized in the context of strategies for evaluating possible effects of early fatty acid nutrition. Whether the existing data for effects of LCPUFAs can be taken to predict an effect on later IQ is not the only important question, and examining the data primarily from that perspective does not seem constructive, given the dearth of data and the insecurity of predictions from infant measures. It seems much more useful and productive to look for patterns and clues in these studies (as also advocated by Wainwright) to guide future research. In the limited data now available, a clear pattern seems to be emerging from five studies, two in monkeys (3,11) and three in human infants (5–7), all of which show effects of LCPUFA status on look duration (four in paired comparisons, one in a subset of infants in infant-controlled habituation). This is the only consistent effect of LCPUFAs found so far on infant behavior, and it seems at least sufficient to pose the hypothesis that look duration, as measured in these paradigms, is tapping some aspect of infant attentional behavior which is sensitive to this aspect of early nutrition. Clearly further studies, including the use of more standard look duration measures from habituation paradigms, are needed to test this hypothesis, and to explore possible underlying mechanisms. Whether the difference in look duration in these studies has any relation to later IQ is certainly unclear. However, (i) given the moderate predictive value of this measure, it may at least be worth investigating before we dismiss the question, and (ii) the difference may be important in understanding the effects of LCPUFA even if it has no relation to IQ. A more eloquent statement of this viewpoint comes from McCall and Carriger (15, p. 75): '... the importance of the habituation/recognition memory prediction phenomenon lies more in what it may reveal about the process of mental development than in the fact of the prediction per se or its size. Unfortunately, nearly all the empirical effort has been expended demonstrating the prediction, which is the first step; but very little research has been directed at discerning the nature of the processes or mechanisms responsible for the correlations . . ., which now should receive more emphasis.' We would add that the importance of any effects on infant attention in LCPUFA studies lies more in what they may tell us about the effects of early fatty acid nutrition than in their ability to predict later IQ.

## Individual versus Group Differences

Much of the literature on infant cognitive development, and the contributions of the developmental psychologists at this Workshop, are directed toward understanding individual differences. They put a primary emphasis on the ability of infant measures to predict later outcomes (almost entirely IQ) for a particular individual. This is a very stringent requirement of any measure, and some of the considerations in the study of individual differences are not entirely appropriate for the study of group differences. Colombo and Singer refer, in passing, to the contrast between individual and group reliability, between nomothetic and idiographic studies. However, given the emphasis on predictive validity for individuals, this issue should receive more attention. In the case of LCPUFA supplementation, we are studying a relatively subtle variable on a background of other very powerful variables. Perhaps a better model in this case would be studies of the effects of nutritional interventions on the incidence of disease, for example, effects of β-carotene supplements on rates of lung cancer. It would not be appropriate to attempt to predict the individuals who would contract cancer, or to dismiss a possible effect of the supplement because it is smaller than the known powerful effect of smoking. In a large population, even a small effect on the probability of a dichotomous outcome can be quite meaningful, and a small change in the mean value of a continuous variable can shift the population distribution so that a quite large effect is seen in the numbers of individuals falling above or below some criterion value at the tails of the distribution.

## Emphasis on IQ

In studies of infant malnutrition, drug and toxin exposure, traditional cognitive or IQ measures generally have received the preponderance of attention, despite the common finding that these are not the most sensitive indices of early nervous system insults. The field of LCPUFA supplementation has been repeating the path taken by studies of protein–calorie malnutrition over 20 years ago. In the early 1970s, a great deal of public, political, and scientific attention was directed to the issue of effects of early malnutrition on brain development and intelligence. The strong preconception, and dominant model, was that early malnutrition produced irreversible effects on brain development, and thereby led to long-lasting effects on intelligence (the brain damage–mental retardation model). As randomized studies in human infants were of course impossible, many studies compared malnourished and well-fed children and attempted, as with studies of breast- and bottle-feeding, to factor out the associated social and economic variables. In essence this proved to be impossible because these factors were inevitably almost completely confounded with early malnutrition. Animal studies, and some studies in children, generally found no effects on purely cognitive measures but, when they looked, found effects on other aspects of infant development including arousal, attention, responsiveness, reaction to novelty, and social interactions (20). As reviewed by Bornstein (this volume), some effects on cognition which were identified appeared to be secondary to effects on the infant's responsiveness or temperament and the resulting effects on infant–caretaker interactions. These results were taken by most observers to be uninteresting, because they failed to confirm the dominant model (and perhaps because the sensitive outcomes were ones that most people intuitively link with the infant's early home environment rather than with biochemical effects on the nervous system). Therefore they engendered little attention or research support. A somewhat similar path has been taken by studies of early cocaine exposure, as well described in the commentary by Bornstein.

Perhaps it is reasonable at this point to hypothesize the following broad generalization: a cluster of attributes that can be described under the term temperament (including arousal, attention, self-regulation, reactivity, responsiveness to novelty and to social interaction), is more sensitive to a variety of early perturbations of the nervous system than are traditional cognitive measures. At least they should receive equal emphasis in comprehensive evaluations of infant development. With regard to effects of LCPUFA, the possibility, based on work in n-3 fatty acid deficient rats, that LCPUFAs status affects dopamine systems (21) adds a neurochemical rationale for these measures as well (a hypothesis described in more detail in our commentary on Shaw and McEachern). Furthermore, much of the work with n-3 fatty acid deficient rats which purports to show differences in learning is interpretable within this framework (22). We would like to reinforce Bornstein's point that effects on temperament may provide mechanisms for indirect effects on cognition, but in any case they are important in their own right. If the presence or absence of LCPUFA in formulas has no direct effect on neural or cognitive development, but has only a concurrent influence on the reponsiveness of the infant to its mother and to other aspects of its environment, would we consider that effect trivial? Much of the child development literature suggests that such an effect could have far-reaching consequences.

## The Need for Multiple Assessments

IQ tests and general developmental tests, such as the Bayley Scales, are explicitly global assessments. They include a large number of different subtests which tap different functional domains. This is part of their strength, but it needs to be understood that this is not an advantage for all purposes. If an effect is present in one restricted domain, then the use of a global test may be very insensitive, because a substantial loss in that one area can be masked by normal scores in the others. Given our lack of understanding of the possible effects of LCPUFAs or other aspects of early nutrition, it seems appropriate that, when global assessments are used, the data from individual subtests be carefully inspected for selective effects. If any such effects appear to emerge, then follow-up studies can evaluate that functional domain with more specific tests.

One probable reason that infant attention measures do not predict later IQ very well is that they tap some of the multiple domains indexed by IQ tests, but not all of them. Therefore it might be more useful to hypothesize continuity within more restricted domains, and to compare a variety of specific infant measures with specific later tasks presumed to be mediated by the same processes or systems. This must be done with considerable caution, however, given our very limited knowledge of the true underlying processes, and the best correlations may not necessarily be found where we expect them; as, for example, with the relatively good association of novelty preference with later language skills (23,24).

## The Choice of Measures: Practical Realities versus Scientific Discovery

We strongly concur with Colombo's call for the use and/or creation of tasks designed to probe specific processes, particularly ones based on current knowledge of the differentiation of function within the nervous system and the vulnerabilities of different neural subsystems during development. We also strongly concur with McCall and Mash, Colombo, and Bornstein that only the use of multiple measures, including ones other than IQ, will have any chance of defining the effect of early nutrition in a meaningful way. Unfortunately, it is difficult to reconcile this viewpoint with the strict philosophy of clinical trials, which dictates the choice of a very small number of endpoints to avoid excessive numbers of statistical comparisons and therefore the likelihood of finding a difference by chance. It is also in direct conflict with the recommendations of the same authors to use only properly standardized measures with acceptable psychometric properties. These two recommendations are made without recognition of their incompatibility. If we are attempting to choose the most sensitive assessment tools for infants, what alternatives do we have? As noted by several authors, the Bayley and similar standardized infant tests have no predictive power in the first year, and generally are useful only for discriminating seriously impaired infants. Among infant attentional measures, only novelty preference from the FTII has been adequately standardized. But neither of these fills the need for a test that is both predictive of later outcomes and sensitive to the independent variable of interest. The handful of standardized measures provides an example of 'looking where the light is', although the lost keys are not under the light. On the other hand, differences in other measures (such as look duration) have been dismissed because the measures are not standardized and validated. An important distinction needs to be made here: it is appropriate to dismiss the conclusion that such a finding predicts a later difference in IQ, but not to dismiss the finding itself and attempts to follow it up appropriately. A comprehensive battery of standardized infant tests does not exist. We certainly do not have sufficient knowledge of cognitive development to choose one or two infant tests which will tell us whether a particular aspect of early nutrition affects cognitive development generally; and, indeed, many doubt that such a thing is possible. Only the use of a variety of exploratory tests, preferably in the same infants, so that their patterns and interrelationships can be examined, is likely to provide some light where the answers are. Appelbaum's call for exploratory data analysis and graphical display of data is also relevant here. These methods are similarly restricted by clinical trial methodology (the "Food and Drug Administration model"), but they are absolutely necessary to generate hypotheses and expedite scientific progress.

The difficult situation faced by this field is that the pace of clinical trials is determined by the pressure of real world events (commercial and regulatory concerns, consumer demand) rather than scientific considerations. This has led to a mismatch of means to ends, where clinical trial methodology is necessarily being used, but without a solid basis of exploratory research. One result is a premature conservatism in the choice of measures, sometimes to the point of impeding scientific progress. The most obvious example is the use of grating acuity as a primary assessment. In the early

1980s we found effects on grating acuity development in monkey infants fed diets low in n-3 fatty acids (25), and the first human infant studies therefore used this measure. But, more than 10 years later, virtually all the growing number of studies in this area are still using grating acuity, and only grating acuity, as their measure of visual development, not because we have a clear rationale for its special sensitivity to LCPUFA status, but because it has detected such effects and it is the one standardized, commercially available test of vision in infants which can be done without substantial expertise in vision science. Clearly there is a need for more basic exploratory studies, both of the effects of LCPUFAs and of infant development generally, before we can choose the most appropriate outcome measures for studies of early LCPUFA supplementation.

## References

1. Jacobson SW, Jacobson JL, O'Neill JM, Padgett RJ, Frankowski JJ, Bihun JT. Visual expectation and dimensions of infant information processing. *Child Dev* 1992; **63**: 711–24.
2. Reisbick S, Neuringer M, Graham M, Jacqmotte N, Karbo W, Sturman J. Visual recognition memory in infant rhesus monkeys: effects of dietary taurine. *Infant Behav Dev* 1995; **18**: 309–18.
3. Reisbick S, Neuringer M, Gohl E. Increased look duration in paired comparisons by rhesus monkey infants with n-3 fatty acid deficiency (n-3 FAD). *Soc Neurosci Abstr* 1994; **20**: 1696.
4. Reisbick S, Neuringer M, Gohl E, Wald R, Anderson GJ. Visual attention in infant monkeys: effects of dietary fatty acids and age. *Dev Psychol* 1997; **33**: 387–95.
5. Carlson SE, Werkman SH. A randomized trial of visual attention of preterm infants fed docosahexaenoic acid until two months. *Lipids* 1996; **31**: 85–90.
6. Werkman SH, Carlson SE. A randomized trial of visual attention of preterm infants fed docosahexaenoic acid until nine months. *Lipids* 1996; **31**: 91–7.
7. Forsyth JS, Willatts P. Do LCPUFA influence infant cognitive behavior? In: Bindels JG, Goedhart AC, Visser HKA, eds. *Recent Developments in Infant Nutrition*. Dordrecht: Kluwer, 1996: 225–34.
8. Jacobson SW, Jacobson JL, Sokol RJ, Martier SS, Ager JW. Prenatal alcohol exposure and infant information processing ability. *Child Dev* 1993; **64**: 1706–21.
9. Reisbick S, Neuringer M, Connor WE. Effects of n-3 fatty acid deficiency in nonhuman primates. In: Bindels JG, Goedhart AC, Visser HKA, eds. *Recent Developments in Infant Nutrition*. Dordrecht: Kluwer, 1996: 157–72.
10. Gunderson VM, Grant KS, Burbacher TM, Fagan JF, Mottet NK. The effect of low-level prenatal methylmercury exposure on visual recognition memory in infant crab-eating macaques. *Child Dev* 1986; **57**: 1076–83.
11. Reisbick S, Neuringer M. Visual attention is altered by n-3 fatty acid deficiency in monkey infants. In: *Proceedings of the AOCS Conference on PUFA*

in *Infant Nutrition: Consensus and Controversies*. Champaign, IL: American Oil Chemists' Society, 1996 (abstr).
12. Neuringer M, Reisbick S, Teemer C. Relationships between visual acuity and visual attention measures in rhesus monkey infants. *Invest Ophthalmol Vis Sci* 1996; **37**: S532 (abstr).
13. Carlson SE, Werkman SH, Peeples JM. Early visual acuity does not correlate with later evidence of visual processing in preterm infants although each is improved by the addition of docosahexaenoic acid to infant formula. In: *Proceedings of the Congress of the International Society for the Study of Fatty Acids and Lipids*. Champaign, IL: American Oil Chemists' Society, 1995 (abstr).
14. O'Connor MJ, Cohen S, Parmelee AH. Infant auditory discimination in preterm and full-term infants as a predictor of 5-year intelligence. *Dev Psychol* 1984; **20**: 159–65.
15. McCall RB, Carriger MS. A meta-analysis of infant habituation and recognition memory performance as predictors of later IQ. *Child Dev* 1993; **4**: 57–79.
16. Fagan JF, Singer LT. Infant recognition memory as a measure of intelligence. *Adv Infancy Res* 1993; **2**: 31–79.
17. Laucht M, Esser G, Schmidt MH. Contrasting infant predictors of later cognitive functioning. *J Child Psychol* 1994; **35**: 649–62.
18. Rose D, Slater A, Perry H. Prediction of childhood intelligence from habituation in early infancy. *Intelligence* 1986; **10**: 251–63.
19. Hunt E. The role of intelligence in modern society. *Am Sci* 1995; **83**: 356–68.
20. Galler J, Ramsey F. A follow-up study of the influence of early malnutrition on development: behavior at home and at school. *J Am Acad Child Adolesc Psychiatr* 1989; **28**: 254–61.
21. Delion S, Chalon S, Herault J, Guilloteau D, Besnard J-C, Durand G. Chronic dietary α-linolenic acid deficiency alters dopaminergic and serotoninergic neurotransmission in rats. *J Nutr* 1994; **124**: 2466–76.
22. Reisbick S, Neuringer M. Omega-3 fatty acid deficiency and behavior: a critical review and directions for future research. In: Yehuda S, Mostofsky DI, eds. *Handbook of Essential Fatty Acid Biology: Biochemistry, Physiology, and Behavioral Neurobiology*. Totowa, NJ: Humana, 1997: 397–426.
23. Thompson LA, Fagan JF, Fulker DW. Longitudinal prediction of specific cognitive abilities from infant novelty preference. *Child Dev* 1991; **62**: 530–8.
24. Bornstein MH, Sigman MD. Continuity in mental development from infancy. *Child Dev* 1986; **57**: 251–74.
25. Neuringer M, Connor WE, Van Petten C, Barstad L. Dietary omega-3 fatty acid deficiency and visual loss in infant rhesus monkeys. *J Clin Invest* 1984; **73**: 272–6.

# Index

**A**

acuity card procedure (ACP) 336, 486
  history and methodology 254–9
    forced-choice preferential looking 255–6
    preferential looking 254–5
  minimizing variability 278–80
    control and monitoring of testing conditions 278
    correcting for gestational age at birth 279–80
    excluding infants at risk of reduced acuity 279
    tester training and monitoring of testers' results 278–9, 289
  reliability 261–7
    methodological sources of variability 264–5
      interval between stimuli 264
      number of alternative stimulus locations 264
      spatial frequency of the initial stimulus 264–5
    test–retest reliability 265–7, 276–7
      between-tester reliability 266–7
      comparison of within- and between-tester reliability 267
      variability related to poorly controlled test distance and lighting 265
      within-tester reliability 265–6
    tester-related factors influencing 262–3
      tester differences 262–3
      tester's knowledge of grating spatial frequency 262
  random start card procedure 258–9
  results in studies of nutrition in infants 271–6
  stimulus configuration 256–7
  testing procedure 257–9
  validity 254, 259–61
    construct validity 260–1
    content validity 259–60
    criterion validity 261
age corrected scores 234
alcohol, maternal consumption
  effect on infant LCPUFA status 52
  effect on infant neurodevelopment 113, 380
  effect on visual recognition memory 64
  as potential confounder 219
amblyopia 152–3, 445, 468
ANOVA 421, 514
  repeated-measures 95, 419
ARA sources 129
arachidonic acid (AA) 3
  biosynthesis 34
  in breast milk 7

arachidonic acid (AA) (*continued*)
  declining levels in formula-fed infants 11–12
  dietary sources 53–4
  excess 149
  formation 5, 105
  functions 44
assessment instruments 211–14
  confounding variables 215–22, 235
  correlation in 231–2
  monitoring of testers' results 278–9, 289
  sample attrition 224, 243
  stipends and incentives 224–5
  timing of 214–15
  training of examiners 223–6
  *see also under names*
attention deficit hyperactivity disorder (ADHD) 48, 53, 149
average effects model 483

## B

background variables 503
  *see also* confounding variables
Bayley Scales of Infant Development 211–12, 215, 226, 245, 302, 331, 422
  gestational length and 233
  Mental Development Index (MDI) 65–6, 92, 212, 218, 409, 476, 520
  Psychomotor developmental index 14, 15, 67–8, 92
Beck Depression Inventory 219
behavioral research
  bias 389
  case-control studies 390
  concept of validity 388–9, 417–19
  construct validity 389, 394–9
    concurrent and predictive validity 397, 423–4
    convergent and divergent validity 394–5
    criterion validity 397
    floor or ceiling effects 396
    maternal factors as mediating variables 397–9
    mediating variables 395–6
    sensitivity of measurement 396–7
  control variables 390
  cross-sectional studies 390
  external validity 389, 399–402
    generalizability of animal models 399–402
  internal validity 388–9, 389–94
    quasi-experimental design 390–1
    true experimental design 389–90
  mediating variables 390
  power of an experiment 392–3
  prospective longitudinal studies 390
  reliability, concept of 393–4
    alternate-form 394
    interobserver 394
    test–retest 394
  statistical conclusion validity 391–3
    type I error 391–2
    type II error 392
bias in research 389
binocular acuity maturation 267–8
birth weight, adult diseases as outcome of 178
brain damage–mental retardation model 523
brain development
  approaches to studying effect of early diet on 446–8
  mechanisms of action for LCPUFAs in 453, 454
  timing of feeding and 454
brain growth spurt 400
Brazelton Neonatal Behavioral Assessment Scales 219, 495
breast feeding
  cognitive performance and 3, 8–11, 104–5
  as confounder 9
  duration of 480
  visual function and 8–11
breast milk
  composition 7, 44, 124, 479–80
  LCPUFAs in 3, 7–8
  transfer of LCPUFAs 52–3
Brief Symptom Inventory 219
bronchopulmonary dysplasia (BPD) 29, 69, 233–4
  as confounder 241, 247
  effect on visual attention 64

Brunet–Lezine psychomotor development scale 123, 409, 498
butterfly effect 433

## C

α-calcium calmodulin kinase II 466–7
canola oil, growth and 172
cataracts 445
causation of cognitive advantage 9, 437, 439
Center for Epidemiologic Studies Depression Scale 219
cocaethylene 219
cocaine
  effects of infant exposure 238, 494–6
  as potential confounder 219
cognition
  formula feeding and 105–6
  future directions 364–5
    design strategies 365
    measurement strategies 365
  inhibition 358–9
  measures of 359–61
    generalizability across paradigms 362–3
    multiple processes 360–1
    of measures 364
    single-process models 360
    standardization 363
  mechanisms 364
  memory processes 358
  positive response to novelty 359
  prediction from measures of infant attention and memory 363–4
  speed or rapidity of stimulus processing 357–8
  variation in reinforcement value of stimuli 359
cognitive status
  in childhood 494
  in infancy 488–93
    habituation 488–90, 491–3
    recovery 490–3
color vision tests 288
comparator theory 344
concurrent and predictive validity 397, 423–4
confounding variables 215–21, 235, 390, 480–2, 499
  vs moderating vs mediating variables 221–2
  in preterm populations 220–1
  socio-economic problem of 8–9, 10–11
construct validity 389, 394–9
convergent and divergent validity 394–5
contrast sensitivity tests 288
corn oil 121, 151
  visual function and 110, 132
correspondence 482
criterion validity 397
critical periods 482–4
cross-over designs 396
*Crypthecodium cohni* as source of docosahexaenoic acid 130

## D

dark rearing and visual deprivation 444–6, 468
definition 497–8
dependent variables 484–6
depression, maternal
  effect on breast-feeding 249
  postnatal 218, 219, 249–50
developmental canalization 211
developmental quotient (DQ) 217
diabetes 48
  effect on LCPUFA status 52
differential dropout, problem of 237, 512–13
docosahexaenoic acid (DHA) 3
  accumulation from milk 53
  animal studies 6–7
  biosynthesis 34
  in breast milk 7, 465
  contribution to intelligence 296
  declining levels in formula-fed infants 11–12
  dietary sources 53–4
  excess 149
  formation 5, 105
  functions 45
  recommended intake 44
  retinal damage and 33, 87
  visual function and 33, 85, 86, 486
dopamine hypothesis 458–9, 460–1
dual-process theory 344–5

## E

ecological systems based research 500–3
eicosapentaenoic acid 127, 128–9
electroretinogram (ERG), effect of diet 403, 478
essential fatty acid deficiency 45–6
essentiality, conditional, concept of 36, 42, 69, 83
examiners
  drift 224, 243
  as potential confounder 219
  training of 223–6
experimental design 498–9
external validity 389, 399–402

## F

Fagan Test of Infant Intelligence (FTII) 14, 29, 31, 61–3, 89–91, 93, 147, 213–15, 245–7, 338, 476, 521
  familiarization 329, 330–1
  gestational length and 233
familiarization phase 518–19
fat deposition in infancy 189–90, 205
fatty acids, synthesis of 42–4
fish oil
  absorption 58
  as source of DHA 127–8
  as source of LCPFAs 53–4
fixation duration 491
forced-choice preferential looking (FPL) 242, 255–6, 403, 404, 519
  superiority of FPL over 280–1

## G

Gesell 302
gestational length, effect on test results 232
graphical display of data, lack fo 515
grating acuity 486
  binocular acuity maturation 267–8
  binocular norms 268–70
    variability across studies 270
    variability in standard deviation across studies 270
  differences in grating spatial frequency between successive acuity cards 257
  in preterm babies 60–1

*see also* acuity card procedure
growth
  factors influencing 201
  impairment in LCPUFA-supplemented preterm infants 20
  mechanism of effect of n-3 fatty acids on 187–90, 199
  observed effects of n-3 LCPUFAs on 179–86
  reasons for concern about 173–9

## H

habituation, visual, procedure 340–9, 372–3, 518
  consistency of individual differences 346–9
  predictive validity 349
  test–retest reliabilities 346–9
  idealized curve 341
  infant cognitive status and 488–90, 491–3
  measures 342–6, 521–2
    attainment of the criterion 343
    decrement 343
    habituation pattern 344–5
    look duration 344, 463–4
    recovery 345–6
  methodology 341–6
    fixed-trial procedures 341–2
    infant-control procedures 342
  prediction of IQ and 306, 333–4
Haith's paradigm 409
Hawthorne effects 500
head circumference
  abnormal, adverse effects of 177
  in preterm babies 92–3
head shape 203
height, maternal, and infant plasma AA concentration 52
Hollingshead Index 217
HOME Inventory 217–18, 316, 391, 477

## I

individual vs group differences 523
information processing 488, 490
insulin sensitivity, muscle membrane content of LCPUFAs and 188–9
intelligence

definition 296–7
general (g) 357, 494
measuring 297–9
types of 298, 498
see also IQ; IQ tests
intent to treat analysis 512–13
internal validity 388–9, 389–94
intervention studies 215, 499–500
intrauterine transfer of LCPUFAs 51–2
intraventrical hemorrhage (IVH) as confounder 247
intraventricular hemorrhage/periventricular hemorrhage (IVH/PVH) 72
iodine deficiency, maternal 5
IQ 211
  cognitive processing speed and 492
  evidence that LCPUFAs influence 312–22
    studies of infants fed human milk vs formula 312–18, 323, 332–3
      assessments of developmental progress and IQ 314
      factors in mother's choice to provide milk 314
      Great Britain study 313–14
      mother's choice to provide milk 314–16
      random assignment to human milk versus formula 316–18
    experimental studies of adding LCPUFAs to infant formulas 318–22
      effects on infant attentional measures 319–22, 324
      effects on standardized tests of infant development 318–19, 324
  factors correlating with 300–2
    environmental factors 300–1
    heritability 300–1
    neural efficiency 300
    nutrition 301–2
    perinatal factors 301
      habituation and prediction of 306, 333–4

  limitations as a measure 494
  limitations as predictor 521
  maternal 9, 28
    as confounder 217
  measurement and prediction of 295–338
  nutrition and 478, 497
  prediction from infant assessments 302–12, 322–3
  prediction of later IQ from infant attentional measures 306–12
    fixed trial procedure 306
    habituation to criterion procedure 306, 333–4
    infant control procedure 306
    inhibition or disengagement of attention 311–12
    memory skills 311
    predictive mechanism or process 308–11
    prediction phenomenon 307–11
    recognition memory context 307
  prediction from standardized infant tests 302–3
    correlations within infancy 303–4
    predictive correlations to childhood IQ 304–5
  stability and change in 299
IQ (intelligence quotient) tests 297–8, 381–2, 441
  validity of 299–300
iron deficiency, early 5

**K**
kindling model 440, 441, 447

**L**
lagged effects model 483
language development 502–3
LCPUFA
  excess intake 164
  formation of 42–5
  functions 45–50
  disease and 47–8
  growth and 48–9
  in human infant studies 475–9
    differences at autopsy 477–8
    differences in behavior 475–7

as source of lipid-derived mediators 49–50
*see also* LCPUFA supplementation
LCPUFA supplementation
  argument for 6–12
    AA and DHA levels in formula-fed infants 11–12
    animal studies 6–7, 28, 31, 32
    autopsy data on brain content 12, 17
    cognitive and visual function 8–11
  detrimental effects 32–3
  recommendations for 21–2, 37–9
  trials in formula-fed infants 12–13
    duration of follow-up 18
    inconsistent findings 17–18
    limitations 17–22
    preterm studies 13–15, 17, 31–2, 54–69
    safety 19–20, 30, 34–5
    sample size 18–19, 30, 32
    term studies 15–16, 17–18, 31–2
learning disabilities 235
linoleic acid (LA) 105, 479
  animal studies 6–7
  in breast milk 7
  deficiency 45
  as essential nutrient 5
  recommended intake 44
  role in growth 187
α-linolenic acid (LNA) 105, 479
  in breast milk 7, 8
  deficiency 45, 435–6
  as essential nutrient 5
  recommended intake 44
  suppression of growth by 199
  visual development and 8
  visual function and 110–11, 112
lipid-derived mediators, LCPUFAs as source of 49–50
long-chain polyunsaturated fatty acid *see* LCPUFA
long-term potentiation 466
look duration 344, 463–4, 517–18, 519, 522
LTP 440

**M**

malnutrition 480
  brain development and 523–4
  early
    effect on CNS 175–6
    effects on neurodevelopmental outcome 174–5, 199
  prenatal, compensation for 481–2
marijuana as potential confounder 219
McCarthy CDI 476
measurement
  problem of 513–14
  multiple, use of 525–7
mediating variables 221–2, 235, 390, 395–6
  maternal factors as 397–9
moderating variables 221–2, 235
*Mortierella alpina*, arachidonic acid production by 129–30
motivation 497
multiple ANOVA 95, 419
multiple assessments, need for 525
multiple process models of infant attention
  effects of LCPUFAs 517–22

**N**

n-3 fatty acid deficiency 46–7, 401
  behavior 406–9
    in animals 406–9, 416–17, 420
    in humans 409
  cross-species developmental comparisons 414–15, 416
  functional effects 402–9
  vision 403–6, 478
    visual acuity and 113–16
    visual function and 113
n-6 fatty acid deficiency 45–6
necrotizing enterocolitis (NEC) 19–20, 29
neonatal adrenoleukodystrophy 47
Neonatal Behavioral Assessment Scale 498
neural development, early, and impact on cognition
  behavioral (cognitive) component 436–7
  causation 437, 439
  chemical/nutritional component 434
  functional/behavioral change, effect of 439–40
  neural component 434–6

role of chemical/nutritional elements in neural and behavioral alterations 438–9
neuronal growth spurt 175–6
neuroplasticity models 440–3
novelty preference 119–20, 306, 308, 310, 359, 491, 519
Nursing Child Assessment Feeding Scale 218
nutrition, early
  life-time effect on health outcomes 4
  life-time effect on cognitive development 4–5

## O

obesity, LCPUFAs in prevention of 189–90, 203
ocular dominance 440–1
opiates, as potential confounder 219
optimal design 515

## P

paired–comparison paradigm 349–56
  consistency of individual differences 355–6
    predictive validity 356
    test–retest reliabilities 355–6
  measures 352–4
    look duration/frequency 352–3
    shifting 354
    visual preferences 353–4
  methodology 350–2
    choice trials 351
    familiarization choice trial interactions 351–2
  rationale 349–50
palmitic acid, synthesis by brain 106
parents, infancy research and communication with 225–6
Peabody Picture Vocabulary Test 217
polychlorinated biphenyls (PCBs) 491, 518
power analysis 514–15
prediction habituation 306, 333–4
preferential looking acuity 254–5
  in term babies 116–19
prematurity stereotyping 223
preterm babies

cognitive advantage in breast feeding 9–11
confounders in 220–1, 240
LCPUFA supplementation trials in 13–15, 41–2, 54–69
  biochemistry 58–60
  conditional essentiality 69, 83
  diseases 68–9, 71, 85, 93
  growth and motor development 67–8
  mental development 65–6, 80
  methods of analysis 95
  nutrient interaction 66–7
  retinal function 65
  safety 20, 71, 81, 86
  trial design 55–8, 78–9, 81–2
  visual acuity 60–1, 86
  visual attention 61–4, 86, 89–92, 93
need for LCPUFA supplementation 20–1, 83
next generation trials 69–71, 79–80, 82
sources of LCPUFAs in 50–4
  maternal-to-fetal transfer 51–2
  maternal-to-infant transfer in breast milk 52–3
  synthesis 50–1
supplementation, effect on growth 170
variability of RCPUFAs in 85, 88, 124
programming, concept of 4, 27
psychological distress, post-partum 218

## R

race/ethnicity 244
random start card procedure 258–9
randomized double-blind placebo controlled studies 396
randomized trials 498–9, 512
  follow-up and selective dropouts 237
  limitations of 248
  sample size 236–7
  staffing 236
Raven's Progressive Matrices 297
recognition acuity 487
Refsun's disease, infantile 47

reliability, concept of 393–4
  alternate-form 394
  interobserver 394
  test–retest 394
retinal function
  DHA and damage to 33, 87
  in preterm babies 65
retinopathy of prematurity 87, 488

**S**

safety 19–20, 30, 34–5, 503–4
  in preterm studies 71, 81, 86
safflower oil 151
  visual function and 109, 110
self-righting 211
sensitive periods 107, 108, 150, 483–4
Similac Special Care formula 13
single-cell oils 162–3
Siriono, diet in 503
sleeper effects 215
small for gestational age (SGA)
  head circumference 176–7
smoking, maternal
  effect on LCPUFA status 52
  effect on neurodevelopment 113, 380
Snellen letter chart 260
socio-economic status (SES) 28, 244, 477
Solomon–Lessac design 499
split-plot analyses 95–6
Stanford–Binet Scale 297, 422
statistical conclusion validity 391–3
stereoacuity tests 288
strabismus 152, 445
sunflower oil, visual function and 110
synapses
  control of receptor development 431–3
  number 429–31

**T**

taurine 517
Teller Acuity cards 13–14, 154, 336, 403–4
  use of 145, 146
temperament 497, 524
term babies
  LCPUFA supplementation trials in 15–16
  range of LCPIFAs added to infant formula 124–31
  studies with 111–23
  supplementation, effect on growth 171–2
test–retest reliability 265–7, 276–7, 346–9, 355–6, 394, 520
  between-tester reliability 266–7
  comparison of within- and between-tester reliability 267
  variability related to poorly controlled test distance and lighting 265
  within-tester reliability 265–6
tobacco, as potential confounder 219
training
  of examiners 223–6
  and monitoring of testers' results 278–9, 289
triglycerides 129–30

**V**

validity 520
  concept of 388–9, 417–19
  construct 389, 394–9
    concurrent and predictive validity 397, 423–4
    convergent and divergent validity 394–5
    criterion validity 397
    floor or ceiling effects 396
    maternal factors as mediating variables 397–9
    mediating variables 395–6
    sensitivity of measurement 396–7
  external 389, 399–402
    generalizability of animal models 399–402
  internal 388–9, 389–94
    quasi-experimental design 390–1
    true experimental design 389–90
  statistical conclusion 391–3
    type I error 391–2
    type II error 392
very low birth weight infants
  DA in 54–5
  growth outcome 200
  head growth in 92–3
  neurodevelopmental status 176

neurologic sequelae 220
randomization in studies and
 confounding 216
visual acuity
 effect of diet 478
 n-3 fatty acid deficiency and 47
 in preterm babies 60–1
 relationship with visual attention
  64–5
 reliance on 486–8
visual attention
 in preterm babies 61–4
 relationship with visual acuity 64–5
visual deprivation, effect on neural
 development 444, 458, 467–8,
 469–70
visual evoked potential acuity (VEP)
 403, 486, 519
 advantages of 281–2
 disadvantages 282
 duration, preferential looking acuity
  and 119

effect of diet 478
effect of n-3 fatty acid deficiency on
 478
in preterm babies 60–1, 111
superiority over FPL measures 280–1
in term babies 120–3
visual function 486–8
 effect of diet on 443–6
 periods in development 445
visual recognition memory (VRM)
 212–13
vitamin A in preterm formula 66–7

## W

Wechsler Intelligence Scale for
 Children (WISC–R IQ) 297, 331

## Z

Zellweger's syndrome 47, 48
zinc in preterm formula 66–7